T0141945

Studies in Computational Intelligence

Volume 650

Series editor

Janusz Kacprzyk, Polish Academy of Sciences, Warsaw, Poland
e-mail: kacprzyk@ibspan.waw.pl

About this Series

The series "Studies in Computational Intelligence" (SCI) publishes new developments and advances in the various areas of computational intelligence—quickly and with a high quality. The intent is to cover the theory, applications, and design methods of computational intelligence, as embedded in the fields of engineering, computer science, physics and life sciences, as well as the methodologies behind them. The series contains monographs, lecture notes and edited volumes in computational intelligence spanning the areas of neural networks, connectionist systems, genetic algorithms, evolutionary computation, artificial intelligence, cellular automata, self-organizing systems, soft computing, fuzzy systems, and hybrid intelligent systems. Of particular value to both the contributors and the readership are the short publication timeframe and the worldwide distribution, which enable both wide and rapid dissemination of research output.

More information about this series at http://www.springer.com/series/7092

Yaxin Bi · Supriya Kapoor
Rahul Bhatia
Editors

Intelligent Systems and Applications

Extended and Selected Results from the SAI Intelligent Systems Conference (IntelliSys) 2015

Editors
Yaxin Bi
Faculty of Computing and Engineering
University of Ulster at Jordanstown
Newtownabbey
UK

Rahul Bhatia
The Science and Information
(SAI) Organization
Bradford, West Yorkshire
UK

Supriya Kapoor
The Science and Information
(SAI) Organization
Bradford, West Yorkshire
UK

ISSN 1860-949X ISSN 1860-9503 (electronic)
Studies in Computational Intelligence
ISBN 978-3-319-81501-5 ISBN 978-3-319-33386-1 (eBook)
DOI 10.1007/978-3-319-33386-1

Printed on acid-free paper

This Springer imprint is published by Springer Nature
The registered company is Springer International Publishing AG Switzerland

Preface

This volume comprises the extended versions of the papers selected from those presented at the SAI Intelligent Systems Conference (IntelliSys) 2015. These extended papers cover a wider range of topics, including unsupervised text mining, anomaly and intrusion detection, self-reconfiguring robotics, application of fuzzy logic to development aid, design and optimization, context-aware reasoning, DNA sequence assembly and multilayer perceptron networks. They present new and innovative research and development carried out presently in fields of knowledge representation and reasoning, machine learning and particularly in intelligent systems in a more broad sense. We hope that readers find the volume interesting and valuable; it provides the state-of-the-art intelligent methods and techniques for solving real-world problems along with a vision of the future research.

The Intelligent Systems Conference 2015 is an emerging international conference held on 10–11 November 2015, in London, UK. Although it was the first time to organize the IntelliSys conference, it has attracted nearly 180 participants from more than 50 countries, thereby attributing an overwhelming success to the conference. IntelliSys2015 is part of the SAI Conferences and featured with keynotes, industrial showcases and interactions with social media. These features make the event to provide an inspiration venue in relation to intelligent systems and artificial intelligence and their applications to the real world, bringing together researchers and practitioners from the related community, and allowing them to exchange their new ideas, explore the cut-edge technologies in solving intriguing practical problems, discuss latest scientific results and foster international collaboration.

This is the first volume in the Intelligent Systems Conference series, which is published by Springer as a companion to the conference proceedings. It is dedicated to another starting point of the conference series, putting together more technical details of some important development into a single volume, thereby facilitating readers to quickly grasp compelling aspects of the results disseminated in the conference and persuade their research and development.

A great many people have contributed to making IntelliSys2015 a successful conference. We would like to thank the IntelliSys2015 programme committee, steering committee, the organization team, keynote speakers and conference's sponsors. In particular, we would like to express our gratitude to the authors who submitted papers to the conference and extended their papers for being included as the book chapters, and their efforts devoted to the preparation of papers, revision and extension are much appreciated.

Yaxin Bi
Supriya Kapoor
Rahul Bhatia

Contents

A Hybrid Intelligent Approach for Metal-Loss Defect Depth Prediction in Oil and Gas Pipelines

Abduljalil Mohamed, Mohamed Salah Hamdi and Sofiène Tahar

Abstract The defect assessment process in oil and gas pipelines consists of three stages: defect detection, defect dimension (i.e., defect depth and length) prediction, and defect severity level determination. In this paper, we propose an intelligent system approach for defect prediction in oil and gas pipelines. The proposed technique is based on the *magnetic flux leakage* (MFL) technology widely used in pipeline monitoring systems. In the first stage, the MFL signals are analyzed using the Wavelet transform technique to detect any metal-loss defect in the targeted pipeline. In case of defect existence, an adaptive neuro-fuzzy inference system is utilized to predict the defect depth. Depth-related features are first extracted from the MFL signals, and then used to train the neural network to tune the parameters of the membership functions of the fuzzy inference system. To further improve the accuracy of the defect depth, predicted by the proposed model, highly-discriminant features are then selected by using the weight-based support vector machine (SVM). Experimental work shows that the proposed technique yields promising results, compared with those achieved by some service providers.

1 Introduction

It has been reported in [28] that the primary cause of approximately 30 % of defects in metallic long-distant pipelines, carrying crude oil and natural gas, is corrosion. To protect the surrounding environment from catastrophic consequences due to malfunctioning pipelines, an effective and efficient intelligent pipeline monitoringsystem

A. Mohamed (✉) · M.S. Hamdi
Information Systems Department, Ahmed Bin Mohammed Military College, Doha, Qatar
e-mail: ajamoham@abmmc.edu.qa

M.S. Hamdi
e-mail: mshamdi@abmmc.edu.qa

S. Tahar
Electrical and Computer Engineering Department, Concordia University, Montreal, QC, Canada
e-mail: tahar@ece.concordia.ca

Y. Bi et al. (eds.), *Intelligent Systems and Applications*,
Studies in Computational Intelligence 650, DOI 10.1007/978-3-319-33386-1_1

is required. Non-destructive evaluation (NDE)-based monitoring tools, such as magnetic sensors, are often employed to scan the pipelines [10, 29]. They are widely used, on a regular basis, to measure any *magnetic flux leakage* (MFL) that might be caused by defects such as corrosion, cracks, dents, etc. It has been observed that defect characteristics (for example defect depth) can be predicted by examining the shape and amplitude of the obtained MFL signals [2, 11, 27]. Thus, the majority of techniques reported in the literature rely heavily on the MFL signals. Other techniques that use other information sources such as ultrasonic waves and closed circuit television (CCTV) are also reported.

A reliability assessment of pipelines is a sequential process and consists mainly of three stages: defect detection and classification [5, 17, 18, 33, 35, 38], defect dimension prediction [22–24, 35], and defect severity assessment [9, 14, 34]. At the first stage, a defect may be reported by an inline inspection tool, mechanical damage accident, or visual examination. Then, the defect is associated to one of the possible defect types such as gouges on welds, dents, cracking, corrosion, etc. In the next stage, the dimensions (i.e., depth and length) of the identified defect are predicted. In the final stage, the defect severity level is determined. It is, based on the outcome of this stage that an immediate response, such as pressure reductions or pipe repair, may be deemed urgent. The defect depth is an essential parameter in determining the severity level of the detected defect. In [5], Artificial Neural Networks (ANNs) are first trained to distinguish defected MFL signals from normal MFL signals. In case of defect existence, ANNs are applied to classify the defect MFL signals into three defect types in the weld joint: external corrosion, internal corrosion, and lack of penetration. The proposed technique achieves a detection rate of 94.2 % and a classification accuracy of 92.2 %. The authors in [33] use image processing techniques to extract representative features from images of cracked pipes. To account for variations of values in these obtained features, a neuro-fuzzy inference system is trained such that the parameters of membership functions are tuned accordingly. A Radial Basis Function Neural Network (RBFNN) is trained to identify corrosion types in pipelines in [38], where artificial corrosion types are deliberately introduced on the pipeline, and the MFL data are then collected. The proposed Immune RBFNN (IRBFNN) was able to correctly locate the corrosion and determine its size. In [17], statistical and physical features are extracted from MFL signals and fed into a multi-layer perceptron neural network to identify defected signals. The identified defected signal is presented to a wavelet basis function neural network to generate a three-dimensional signal profile displayed on a virtual reality environment. In [18] a Support Vector Machine (SVM) is used to reconstruct shape features of three different defect types and used for defect classification and sizing. In [35], the authors propose a wavelet neural network approach for defect detection and classification in oil pipelines using MFL signals. In [22–24], artificial neural networks and neuro-fuzzy systems are used to estimate defect depths. In [22], static, cascaded, and dynamic feedforward neural networks are used for defect depth estimation. In [23], a self-organizing neural network is used as a visualization technique to identify appropriate features that are fed into different network architectures. In [24], a weight-based support vector machine is used to select the most relevant features that are fed into a neuro-fuzzy inference system.

For reliability assessment of pipelines, a fuzzy-neural network system is proposed in [34]. Sensitive parameters, representing the actual conditions of defected pipelines, are extracted from MFL signals and used to train the hybrid intelligent system. The intelligent system is then used to estimate the probability of failure of the defected pipeline. The authors in [9] propose an artificial neural network approach for defect classification using some defect-related factors including metal-loss, age, operating pressure, pipeline diameter. In [14], to reduce the number of features used in the defect assessment stage, several techniques are utilized including support vector machine, regression, principal component analysis, and partial least squares.

The rest of the paper is organized as follows. In Sect. 2, we review the reliability assessment process in oil and gas pipelines. The proposed hybrid intelligent system for defect depth prediction is introduced in Sect. 3. In Sect. 4, the performance evaluation of the proposed approach is described. We conclude with final remarks in Sect. 5.

2 Reliability Assessment of Oil and Gas Pipelines

The main components of the reliability assessment process of pipelines are shown in Fig. 1. After detecting the presence of a pipeline defect, the pipeline reliability assessment system estimates the size of the defect (i.e. its length and depth), and based on that, it predicts the severity of the defect and takes a measureable action to rectify the situation. In the following we briefly describe each component.

2.1 Data Processing

Nondestructive evaluation (NDE) tools are often used to scan pipelines for potential defects. One of these NDE tools are magnetic sensors that are attached to autonomous devices sent periodically into the pipeline under inspection. The purpose of magnetic sensors is to measure MFL signals [20], every 3 mm along the pipeline length. A sample of a MFL signal is shown in Fig. 2. The amount of MFL data is huge, thus feature extraction methods are needed to reduce data dimensionality. However, using

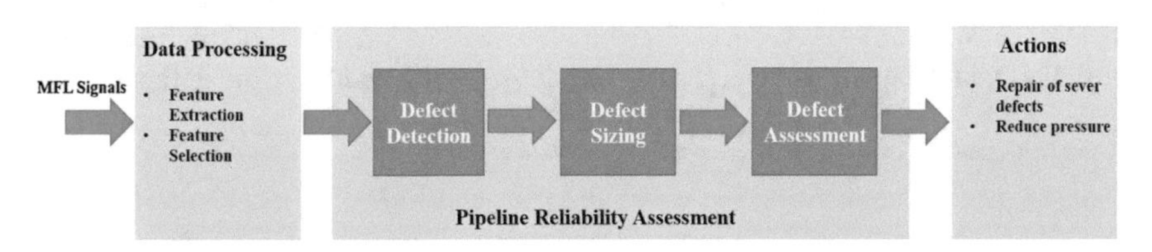

Fig. 1 Reliability assessment of pipelines

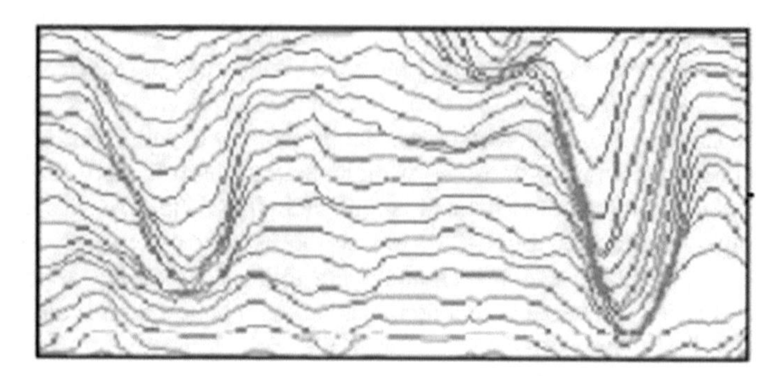

Fig. 2 A sample of a MFL signal

all the extracted features may not actually lead to a better reliability assessment performance. Thus, the most relevant features are selected and fed into the detection and sizing units.

2.2 *Defect Detection*

Once the processed MFL signals are received, techniques such as wavelets can be used to detect and locate defects on the pipe surface [11]. Wavelet techniques are powerful mathematical tools [8, 19, 21]. They were used in many applications such as data compression [4], data analysis and classification [36], and de-noising [26, 27, 32]. The wavelet technique can also be used to locate defects on the targeted pipeline as follows. The MFL signal contains three components and each of these components consists of a sum of curves, and these are a translated and dilated version of a reference pattern. Suppose the mother wavelet $\psi(x)$ refer to the reference pattern, and $\langle\psi_{j,k}(x)\rangle$ is the wavelet basis. Thus, the MFL signal $B(x)$ can be represented as:

$$B(x) = \sum_{j,k} c_{j,k}\psi_{j,k}(x) \tag{1}$$

When the wavelet transform of the MFL signal ($B(x)$) is computed with respect to the basis $\langle\psi_{j,k}(x)\rangle$, the set of non-zero coefficients $c_{j,k}$ indicate the locations of metal-loss defects on the surface of the pipeline. Moreover, the set of dilation factors of the reference pattern is determined, which yields the widths of the defects.

2.3 Defect Sizing

Defect sizing (in particular, determination of the defect depth) is an essential component of the reliability assessment system as, based on its outcome, the severity of the detected defect can be determined. However, the relationship between the measured MFL signals and the corresponding defect depth is not well-understood. Therefore, in the absence of feasible analytical models, hybrid intelligent tools such as Adaptive Neuro-Fuzzy Inference systems (ANFIS) become indispensable and can be used to learn this relationship [22–24, 35]. In this paper, after extracting and selecting meaningful features, an ANFIS model is trained using a hybrid learning algorithm that comprises least squares and the backpropagation gradient descent method [13].

2.4 Defect Assessment

Upon calculating the dimensions of the defect, its severity level can be determined by using the industry standard [1]. Basic equations for assessing defects can be used to construct defect acceptance curves as shown in Fig. 3 [7]. The first curve calculates the failure stress of defects in the pipeline at the maximum operating pressure (MAOP), and the other curve shows the sizes of defects that would fail the hydro pressure test [7].

2.5 Actions

The plot in Fig. 3 can be used to prioritize the repair of the defects. For example, two defects are predicted to fail at the MAOP. For these defects, an immediate repair is needed. Defects that have failure stresses lower than the hydro pressure curve are acceptable. Any defect between the assessing curves needs to be reassessed using advanced measuring tools.

3 A Hybrid Intelligent System for Defect Depth Estimation

The combination of neural networks and fuzzy inference systems has been successfully used in different application areas such as water quality management [31], queue management [25], energy [37], transportation [15]; and business and manufacturing [6, 12, 16, 30]. In this section, the applicability of ANFIS models in estimating metal-loss defect depth is demonstrated. The general architecture of the proposed approach is shown in Fig. 4.

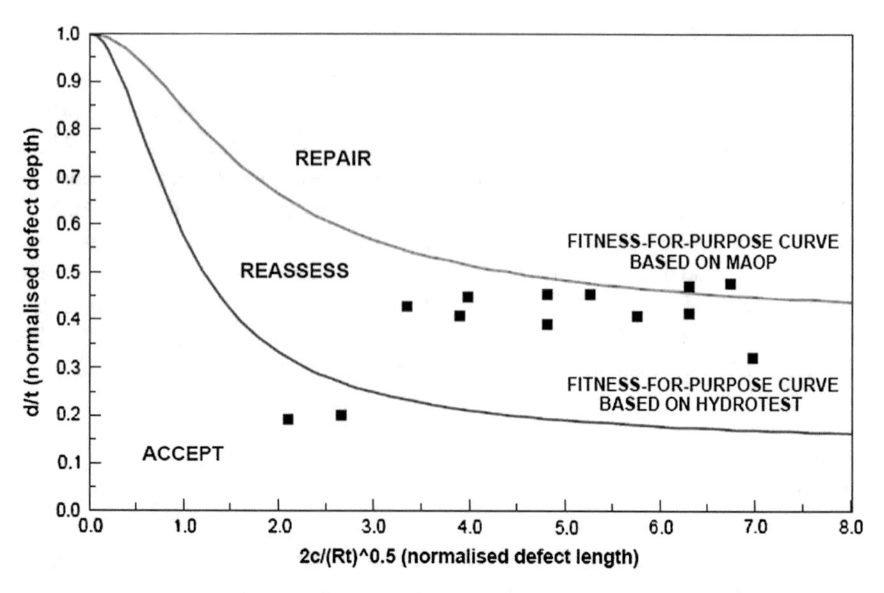

Fig. 3 Curves of defect assessment (Reproduced from [7])

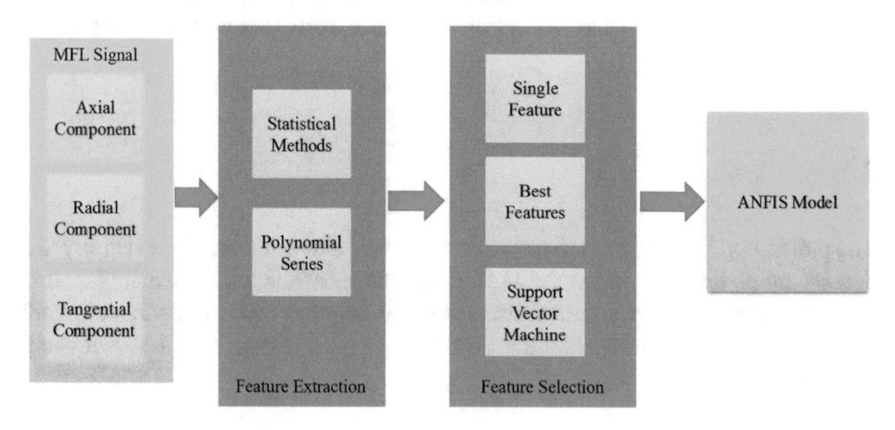

Fig. 4 The structure diagram of the proposed approach

The proposed approach consists of three stages, namely: feature extraction, feature selection, and the ANFIS model. The three components are described in the following subsections.

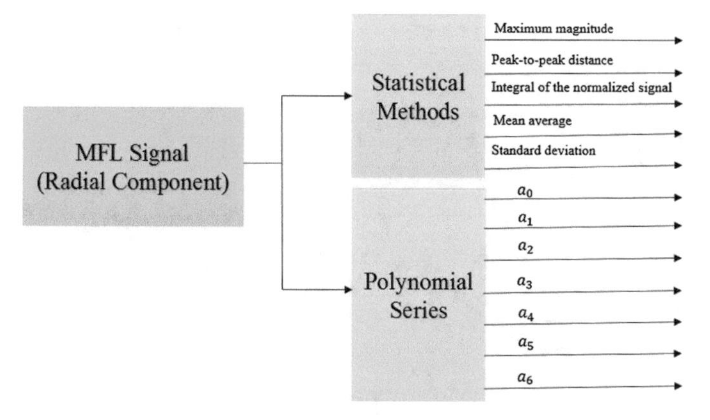

Fig. 5 Features extracted for the radial component of the MFL signal

3.1 *Feature Extraction*

The main purpose of feature extraction is to reduce the MFL data dimensionality. As shown in Fig. 4, the MFL signal is represented by the axial, radial and tangential components. For each component, statistical and polynomial series are applied as feature extraction methods. The features extracted from the radial component of the MFL signal are shown in Fig. 5. Statistical features are self-explanatory. Polynomial series of the form $a_n X^n + \cdots + a_1 X + a_0$ can approximate MFL signals. Polynomials of degrees 3, 6, and 6 have been found to provide the best approximation for axial, radial, and tangential components, respectively. Thus, the input features consist of the polynomial coefficients, $a_n, \ldots, a_0$ along with the five statistical features. Thus, in total we have 33 features, which will be referred to by F1, F2,…, F33.

3.2 *Feature Selection*

It is a known fact that different features might exhibit different discrimination capabilities. Most often, incorporating all obtained features in the training process may not lead to a high depth estimation accuracy. In fact, including some features may have a negative impact. Therefore, it is a common practice to identify the important features that are appropriate to the ANFSI model. Thus, the next step is to examine the suitability of each feature for the defect depth prediction task. The best features that yield the best depth estimation accuracy are then identified and used as an input feature pattern for the ANFIS model. Moreover, a support vector machine-based weight correlation method is used to assign weights for the obtained features. Features with the highest weights are selected to train a new ANFIS model. Different sets of features, having 16 to 29 features each, have been evaluated.

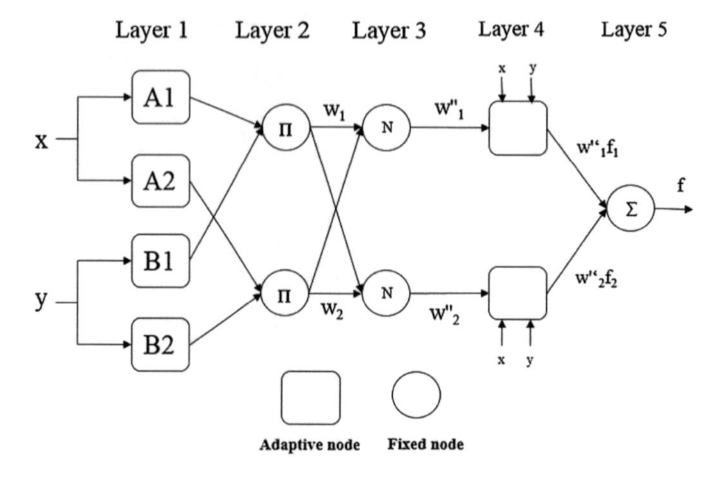

Fig. 6 The architecture of ANFIS

3.3 *The ANFIS Model*

ANIFIS, as introduced by [13], is a hybrid neuro-fuzzy system, where the fuzzy IF-THEN rules and membership function parameters can be learned from training data, instead of being obtained from an expert [3, 6, 12, 16, 25, 30, 31, 37]. Whether the domain knowledge is available or not, the adaptive property of some of its nodes allows the network to generate the fuzzy rules that approximate a desired set of input-output pairs. In the following, we briefly introduce the ANFIS architecture as proposed in [13]. The structure of the ANFIS model is basically a feedforward multi-layer network. The nodes in each layer are characterized by their specific function, and their outputs serve as inputs to the succeeding nodes. Only the parameters of the adaptive nodes (i.e., square nodes in Fig. 6) are adjustable during the training session. Parameters of the other nodes (i.e., circle nodes in Fig. 6) are fixed.

Suppose there are two inputs x, y, and one output f. Let us also assume that the fuzzy rule in the fuzzy inference system is depicted by one degree of Sugeno's function [13]. Thus, two fuzzy *if-then* rules will be contained in the rule base as follows:

Rule 1: if x is A_1 and y is B_1 then $f = p_1 x + q_1 y + r_1$.

Rule 2: if x is A_2 and y is B_2 then $f = p_2 x + q_2 y + r_2$.

where p_i, q_i, r_i are adaptable parameters.

The node functions in each layer are described in the sequel.

Layer 1: Each node in this layer is an adaptive node and is given as follows:

$$o_i^1 = \mu_{Ai}(x), \quad i = 1, 2 \tag{2}$$

$$o_i^1 = \mu_{Bi-2}(y), \quad i = 3, 4 \tag{3}$$

where x and y are inputs to the layer nodes, and A_i, B_{i-2} are linguistic variables. The maximum and minimum of the bell-shaped membership function are 1 and 0, respectively. The membership function has the following form:

$$\mu_{Ai}(x) = \frac{1}{1 + \{(\frac{x-c_i}{a_i})^2\}^{bi}} \tag{4}$$

where the set $\{a_i, b_i, c_i\}$ represents the premise parameters of the membership function. The bell-shaped function changes according to the change of values in these parameters.

Layer 2: Each node in this layer is a fixed node. Its output is the product of the two input signals as follows:

$$o_i^2 = w_i = \mu_{Ai}(x)\mu_{Bi}(y), \quad i = 1, 2 \tag{5}$$

where w_i refers to the firing strength of a rule.

Layer 3: Each node in this layer is a fixed node. Its function is to normalize the firing strength as follows:

$$o_i^3 = w_i'' = \frac{w_i}{w_1 + w_2}, \quad i = 1, 2 \tag{6}$$

Layer 4: Each node in this layer is adaptive and adjusted as follows:

$$o_i^4 = w_i'' f_i = w_i''(p_i x + q_i y + r_i), \quad i = 1, 2 \tag{7}$$

where w_i'' is the output of layer 3 and $\{p_i + q_i + r_i\}$ is the consequent parameter set.

Layer 5: Each node in this layer is fixed and computes its output as follows:

$$o_i^5 = \sum\nolimits_{i=1}^{2} w_i'' f_i = \frac{(\sum_{i=1}^{2} w_i f_i)}{w_1 + w_2} \tag{8}$$

The output of layer 5 sums the outputs of nodes in layer 4 to be the output of the whole network. If the parameters of the premise part are fixed, the output of the whole network will be the linear combination of the consequent parameters, i.e.,

$$f = \frac{w_1}{w_1 + w_2} f_1 + \frac{w_2}{w_1 + w_2} f_2 \tag{9}$$

The adopted training technique is hybrid, in which, the network node outputs go forward till layer 4, and the resulting parameters are identified by the least square method. The error signal, however, goes backward till layer 1, and the premise parameters are updated according to the descent gradient method. It has been shown in the literature that the hybrid-learning technique can obtain the optimal premise and consequent parameters in the learning process [13].

3.4 Learning Algorithm for ANFIS

To map the input/output data set, the adaptable parameters $\{a_i + b_i + c_i\}$ and $\{p_i + q_i + r_i\}$ in the ANFIS structure are adjusted in the learning process. When the premise parameters a_i, b_i and c_i of the membership function are fixed, the ANFIS yields the following output as shown in (9). Substituting (6) into (9), we obtain the following:

$$f = w_1'' f_1 + w_2'' f_2 \tag{10}$$

Substituting the fuzzy *if-then* rules in (10) yields:

$$f = w_1''(p_1 x + q_1 y + r_1) + w_2''(p_2 x + q_2 y + r_2) \tag{11}$$

or:

$$f = (w_1'' x) p_1 + (w_1'' y) q_1 + (w_1'') r_1) + (w_2'' x) p_2 + (w_2'' y) q_2 + (w_2'') r_2 \tag{12}$$

Equation (12) is a linear combination of the adjustable parameters. The optimal values of $\{a_i, b_i, c_i\}$ and $\{p_i, q_i, r_i\}$ can be obtained by using the least squared method. If the premise parameters are fixed, the hybrid learning algorithm can effectively search for the optimal ANFIS parameters.

4 Experimental Results

To evaluate the effectiveness of the proposed technique, in terms of defect depth estimation accuracy, extensive experimental work has been carried out. The obtained accuracies are evaluated based on different levels of error-tolerance including: ± 1, ± 5, ± 10, ± 15, ± 20, ± 25, ± 30, ± 35, and $\pm 40\%$. The impact of using selected features, while utilizing different number and type of membership functions for the adaptive nodes in the ANFIS model is studied. The results are reported in the following subsections.

4.1 Types of Membership Functions for the Adaptive Nodes

The shape of the selected membership function (MF) defines how each point in the universe of discourse of the corresponding feature is mapped into a membership value between 0 and 1 (the membership value indicates the degree to which the relevant input feature belongs to a certain metal-loss defect depth). The parameters that control the shape of the membership function are tuned in the adaptive nodes, during the training session. The number of parameters needed depends on the type of the membership function used in the adaptive node. Three types of membership functions namely Gaussian, Trapezoidal, and Triangle for the feature F1 are shown in Fig. 7.

It can be seen from Fig. 7 that the smoothness of transition from one fuzzy set to another varies, depending on the function type used in the adaptive node.

4.2 *Training, Testing, and Validation Data*

During the training session, four of the obtained features, namely F3, F6, F8, and F13, were not acceptable by the Matlab ANFIS model. Their sigma values were close to zero so they were discarded. Thus, only 29 features of the 33 above mentioned features were considered. The 1357 data samples, used for developing and testing the ANFIS model, were divided as follows: 70 % for training, 15 % for testing, and 15 % for validation (checking). The training data set consists of 949 rows and 30 columns. The rows represent the training samples, and the first 29 columns represent the extracted features of each sample and the last column represents the target (defect depth). The format of the testing and validation data sets is similar to that of the training data set, however, each consists of 204 rows (samples). We have used 100 epochs to train the ANFIS model.

A hybrid learning approach was adopted, in which the membership function parameters of the single-out Sugeno type fuzzy inference system were identified. The hybrid learning approach converges much faster than the original backpropagation method. In the forward pass, the node outputs go forward until layer 4 and the consequent parameters are identified with the least square method. In the backward pass, the error rates propagate backwards and the premise parameters are updated by gradient decent.

4.3 *Evaluation of the Feature Prediction Power*

In this section, the performance of each feature, in terms of its defect depth prediction accuracy, is examined. Due to limited space, only the features that yield the best prediction accuracy are reported. As shown in Table 1, seven features present the best estimation accuracy among all features, namely F1, F4, F9, F10, F11, F14, and F32. As expected, for the lower levels of error tolerance (particularly: ± 10, ± 15, and ± 20 %), none of the features yielded an acceptable prediction accuracy.

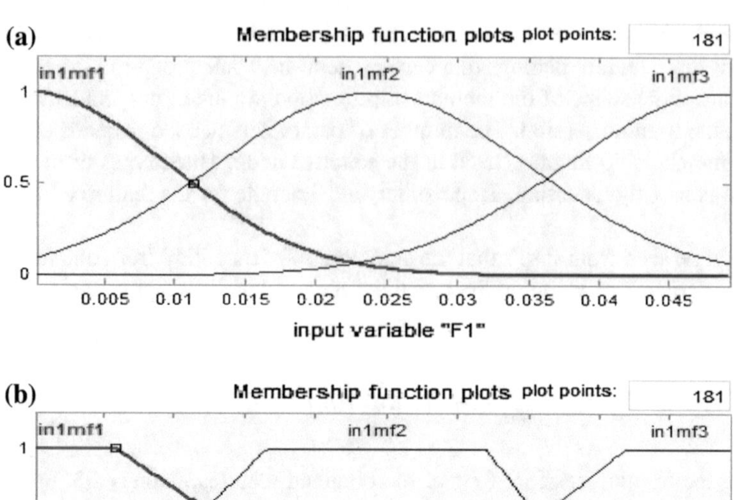

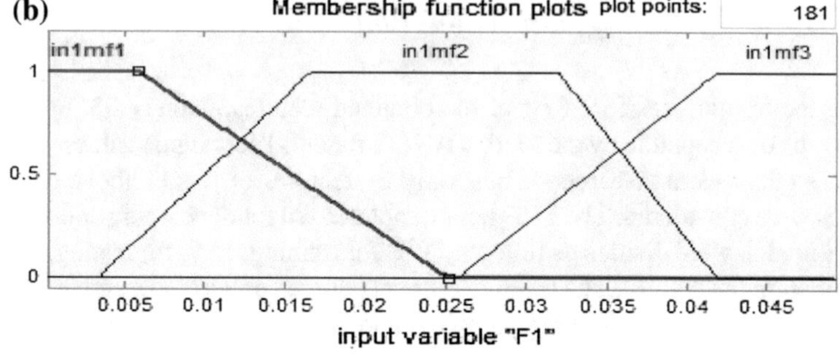

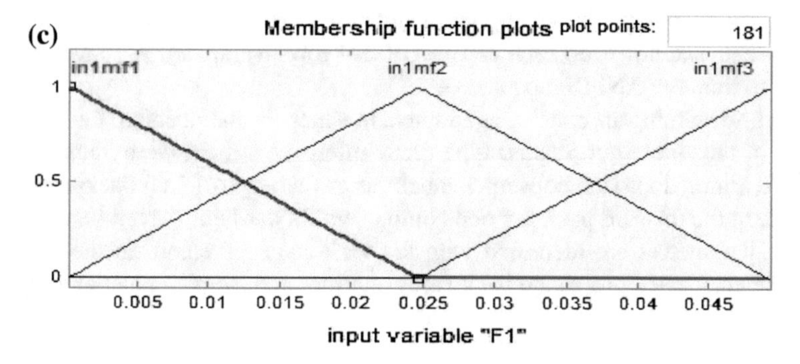

Fig. 7 Membership functions of the feature F1 **a** Gaussian, **b** trapezoidal, and **c** triangle

4.4 Using the Best Features

To improve the defect depth prediction accuracy of the ANFIS model, the features that yield the best prediction accuracy are used to train the ANFIS model. For 100 epochs, the training performance error of the ANFIS model is equal to 0.1482. The defect depth prediction accuracy of this model is clearly improved as shown in Table 2. For error-tolerance levels at ±10, ±15, and ±20 %, the model gives prediction accuracy at 62, 74, and 87 %, respectively.

Table 1 Defect depth prediction accuracy using the features F1, F4, F9, F10, F11, F14, and F32

Error-tolerance (%)	Input parameters						
	F1	F4	F9	F10	F11	F14	F32
±1	0.0196	0.0147	0.0294	0.0490	0.0196	0.0147	0.0392
±5	0.1176	0.1275	0.1618	0.1176	0.1373	0.1618	0.1275
±10	0.2892	0.2990	0.3627	0.2304	0.3235	0.2941	0.2941
±15	0.4510	0.4755	0.5637	0.4118	0.4902	0.5000	0.4608
±20	0.6275	0.6471	0.7549	0.5686	0.6667	0.6618	0.5882
±25	0.8137	0.8039	0.9265	0.6912	0.8137	0.8578	0.7010
±30	0.8775	0.8971	0.9461	0.7990	0.9020	0.9167	0.8480
±35	0.9265	0.9314	0.9510	0.9020	0.9412	0.9363	0.8922
±40	0.9461	0.9510	0.9559	0.9314	0.9559	0.9461	0.9265

Table 2 Defect depth prediction accuracy of the ANFIS model using the best features

Error-tolerance (%)	Input features (F1 F4 F9 F10 F11 F14 F32)
±1	0.0637
±5	0.3529
±10	0.6127
±15	0.7402
±20	0.8725
±25	0.9314
±30	0.9510
±35	0.9755
±40	0.9853

4.5 *Support Vector Machine-Based Feature Selection*

Another method for evaluating the prediction power of the obtained features is by assigning a weight for each feature, based on the support vector machine weight-correlation technique. Features with the highest weights are selected to train the ANFIS model. After examining different sets of selected features, the first 22 features have yielded the best prediction accuracy as demonstrated in Table 3.

As shown in Table 3, for error-tolerance levels at ±10, ±15, and ±20 %, the model gives a prediction accuracy at 62, 80, and 87 %, respectively. It is slightly improved over the best features, reported in Table 2. For the error-tolerance level at ±10 %, the prediction accuracy improved 1 %, and for ±15 %, improved 6 %. However, for the ±20 % error-tolerance level, it remained the same as that of the best features. Only, using all the 29 features can give comparable results (shown in the last column of Table 3).

Table 3 Defect depth prediction accuracy of the ANFIS model using 22 features and all features

Error-tolerance (%)	22 Input features	All features
±1	0.0637	0.0588
±5	0.3529	0.3627
±10	0.6225	0.6324
±15	0.8039	0.7990
±20	0.8775	0.8529
±25	0.9118	0.8971
±30	0.9559	0.9314
±35	0.9706	0.9510
±40	0.9804	0.9608

4.6 Using Different Membership Function Types

In this section, we examine the impact of using three different membership function types on the model performance. In the model adaptive nodes, to each of the seven best features (i.e., F1, F4, F9, F10, F11, F14, and F32), we assigned Gaussian, trapezoidal, and triangle membership functions. The shapes of these functions are shown in Fig. 7. The number of membership functions is set at 2 or 3, for each type. The outputs of the ANFIS model for the training, testing and validation data (for Gaussian membership function) are shown in Fig. 8. The x-axis represents the indices of the training data elements, whereas the y-axis represents the defect depths of the corresponding data elements. The 949 training data samples, representing the true defect depths, are plotted in blue (Fig. 8a). While the model out puts, representing the same defect depths estimated by the model, are plotted in red (Fig. 8a). Clearly, the model outputs (estimated defect depths) are satisfactory as they lie on or close to the true defect depths. Figure 8b, c show the model outputs (plotted in red) against the 204-element testing data (plotted in blue) and 204-element validation (checking) data (plotted in blue), respectively. The defect depth prediction accuracy, obtained by the ANFIS model for the three types of membership functions, is shown in Table 4, where the best accuracies are highlighted in red.

The model shows improvement for the error-tolerance levels ±1, ±5, and ±10 % as it yield 11, 48, and 71 %, respectively, compared to 6, 35, and 62 % produced by the ANFIS model using the 22 features. For the rest of the error-tolerance levels, it yields accuracies comparable to those produced by the ANFIS model using the 22 features. The best training error is calculated at 0.106, obtained by the ANFIS model using three triangle membership functions.

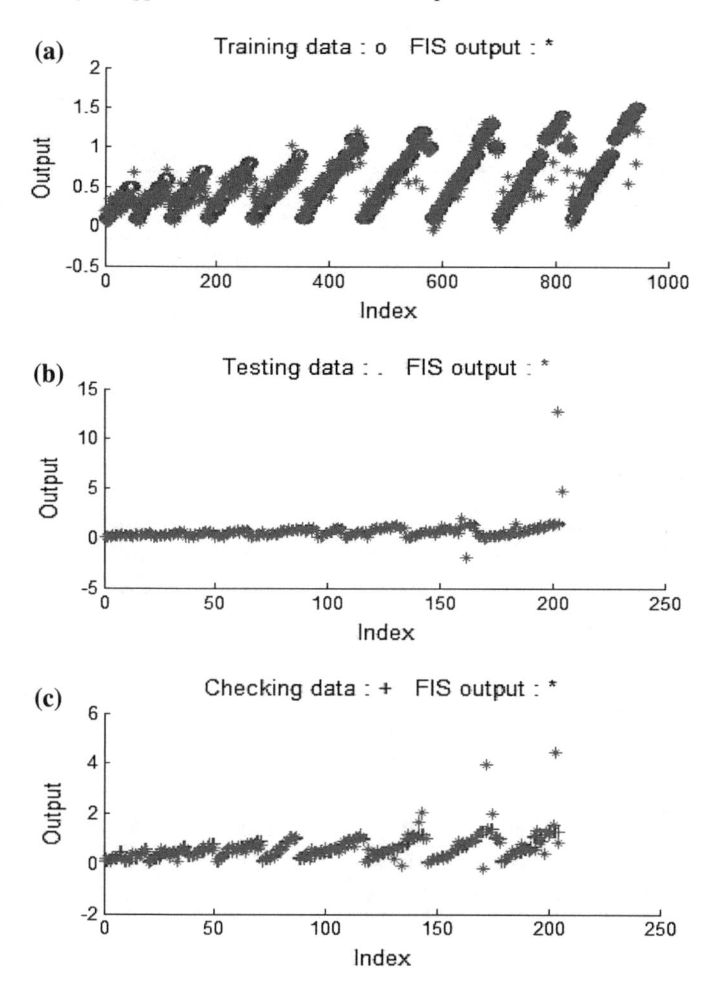

Fig. 8 The outputs of the ANFIS model (*red*) using the Gaussian membership function at the adaptive node: training data (**a**), testing data (**b**), and validation data (**c**), shown in *blue*

5 Conclusions

In order to assess the reliability of a functioning oil and gas pipeline, the depth of a detected defect should be first determined. Based on which, appropriate maintenance actions are carried out. In this work, a hybrid intelligent approach for metal-loss defect depth prediction in oil and gas pipelines is proposed. The proposed approach utilizes MFL data acquired by magnetic sensors scanning the metallic pipelines. To reduce the MFL data, feature extraction methods are applied. The extracted features are fed into an adaptive neuro-fuzzy inference system (ANFIS), individually, to examine their depth prediction capabilities. The most effective features are then

Table 4 Defect depth prediction accuracy of the ANFIS model using three different numbers and types of membership functions

No of functions	Function type	Learning algorithm	Training error	Depth prediction accuracy		
				±1 %	±5 %	±10 %
2	Gaussian	Hybrid	0.122	0.0588	0.3971	0.6324
2	Triangle	Hybrid	0.127	0.0539	0.3480	0.6176
2	Trapezoidal	Hybrid	0.131	0.0735	0.3775	0.6422
3	Gaussian	Hybrid	0.108	0.1029	0.4412	0.6814
3	Triangle	Hybrid	0.106	0.1176	0.4853	0.7108
3	Trapezoidal	Hybrid	0.114	0.0588	0.3382	0.6324
No of functions	Function type	Learning algorithm	Training error	Depth prediction accuracy		
				±15 %	±20 %	±25 %
2	Gaussian	Hybrid	0.122	0.7892	0.8873	0.9167
2	Triangle	Hybrid	0.127	0.7794	0.8578	0.8971
2	Trapezoidal	Hybrid	0.131	0.7647	0.8627	0.9069
3	Gaussian	Hybrid	0.108	0.7941	0.8480	0.8676
3	Triangle	Hybrid	0.106	0.7843	0.8235	0.8627
3	Trapezoidal	Hybrid	0.114	0.7794	0.8382	0.8676
No of functions	Function type	Learning algorithm	Training error	Depth prediction accuracy		
				±30 %	±35 %	±40 %
2	Gaussian	Hybrid	0.122	0.9461	0.9657	0.9657
2	Triangle	Hybrid	0.127	0.9314	0.9461	0.9608
2	Trapezoidal	Hybrid	0.131	0.9363	0.9510	0.9510
3	Gaussian	Hybrid	0.108	0.9020	0.9216	0.9363
3	Triangle	Hybrid	0.106	0.8775	0.8922	0.9020
3	Trapezoidal	Hybrid	0.114	0.8873	0.9167	0.9216

identified and used to retrain the ANFIS model. Utilizing different types and numbers of membership functions, the proposed ANFIS model is tested for different levels of error-tolerance. At the levels of ±10, ±15, and ±20 %, the best defect depth prediction obtained by the new approach are 71, 79, and 88 %, using triangle, triangle, and Gaussian membership functions, respectively. In future work, sophisticated feature extraction methods will be investigated to enhance the model performance. Moreover, the uncertainty properties, inherently pertaining to the magnetic sensors, will also be addressed.

Acknowledgments This work was made possible by NPRP grant # [5-813-1-134] from Qatar Research Fund (a member of Qatar Foundation). The findings achieved herein are solely the responsibility of the authors.

References

1. American Society of Mechanical Engineers: ASME B31G Manual for Determining Remaining Strength of Corroded Pipelines (1991)
2. Amineh, R., et al.: A space mapping methodology for defect characterization from magnetic flux leakage measurement. IEEE Trans. Magn. **44**(8), 2058–2065 (2008)
3. Azamathulla, H., Ab Ghani, A., Yen Fei, S.: ANFIS-based approach for predicting sediment transport in clean sewer. Appl. Soft Comput. **12**(3), 1227–1230 (2012)
4. Bradley, J., Bradley, C., Brislawn, C., Hopper, T.: FBI wavelet/scalar quantization standard for gray-scale fingerprint image compression. SPIE Proc., Visual Inf. Process. II **1961**, 293–304 (1993)
5. Carvalho, A., et al.: MFL signals and artificial neural network applied to detection and classification of pipe weld defects. NDT Int. **39**(8), 661–667 (2006)
6. Chen, M.: A hybrid ANFIS model for business failure prediction utilizing particle swarm optimization and subtractive clustering. Inf. Sci. **220**, 180–195 (2013)
7. Cosham, A., Kirkwood, M.: Best practice in pipeline defect assessment. Paper IPC00-0205, International Pipeline Conference (2000)
8. Daubechies, I.: Ten Lectures on Wavelets. SIAM: Society for Industrial and Applied Mathematics, Philadelphia (1992)
9. El-Abbasy, M., et al.: Artificial neural network models for predicting condition of offshore oil and gas pipelines. Autom. Constr. **45**, 50–65 (2014)
10. Gloria, N., et al.: Development of a magnetic sensor for detection and sizing of internal pipeline corrosion defects. NDT E Int. **42**(8), 669–677 (2009)
11. Hwang, K., et al.: Characterization of gas pipeline inspection signals using wavelet basis function neural networks. NDT E Int. **33**(8), 531–545 (2000)
12. Iphar, M.: ANN and ANFIS performance prediction models for hydraulic impact hammers. Tunn. Undergr. Space Technol. **27**(1), 23–29 (2012)
13. Jang, J.: ANFIS: adaptive-network-based fuzzy inference system. IEEE Trans. Syst., MAN, Cybern. **23**, 665–685 (1993)
14. Khodayari-Rostamabad, A., et al.: Machine learning techniques for the analysis of magnetic flux leakage images in pipeline inspection. IEEE Trans. Magn. **45**(8), 3073–3084 (2009)
15. Khodayari, A., et al.: ANFIS based modeling and prediction car following behavior in real traffic flow based on instantaneous reaction delay. In: IEEE Intelligent Transportation Systems Conference, pp. 599–604 (2012)
16. Kulaksiz, A.: ANFIS-based estimation of PV module equivalent parameters: application to a stand-alone PV system with MPPT controller. Turk. J. Electr. Eng. Comput. Sci. **21**(2), 2127–2140 (2013)
17. Lee, J., et al.: Hierarchical rule based classification of MFL signals obtained from natural gas pipeline inspection. In: International Joint Conference on Neural Networks, vol. 5 (2000)
18. Lijian, Y., et al.: Oil-gas pipeline magnetic flux leakage testing defect reconstruction based on support vector machine. In: Second International Conference on Intelligent Computation Technology and Automation, vol. 2 (2009)
19. Mallat, S.: A Wavelet Tour of Signal Processing. Academic, New York (2008)
20. Mandal, K., Atherton, D.: A study of magnetic flux-leakage signals. J. Phys. D: Appl. Phys. **31**(22), 3211 (1998)
21. Misiti, M., Misiti, Y., Oppenheim, G., Poggi, J.: Wavelets and Their Applications. Wiley-ISTE, New York (2007)
22. Mohamed, A., Hamdi, M.S., Tahar, S.: A machine learning approach for big data in oil and gas pipelines. In: 3rd International Conference on Future Internet of Things and Cloud (2015)
23. Mohamed, A., Hamdi, M.S., Tahar, S.: Self-organizing map-based feature visualization and selection for defect depth estimation in oil and gas pipelines. In: 19th International Conference on Information Visualization (2015)

24. Mohamed, A., Hamdi, M.S., Tahar, S.: An adaptive neuro-fuzzy inference system-based approach for oil and gas pipeline defect depth estimation. In: SAI Intelligent Systems Conference (2015)
25. Mucsi, K., Ata, K., Mojtaba, A.: An adaptive neuro-fuzzy inference system for estimating the number of vehicles for queue management at signalized intersections. Transp. Res. Part C: Emerg. Technol. **19**(6), 1033–1047 (2011)
26. Muhammad, A., Udpa, S.: Advanced signal processing of magnetic flux leakage data obtained from seamless gas pipeline. NDT E Int. **35**(7), 449–457 (2002)
27. Mukhopadhyay, S., Srivastave, G.: Characterization of metal-loss defects from magnetic signals with discrete wavelet transform. NDT Int. **33**(1), 57–65 (2000)
28. Nicholson, P.: Combined CIPS and DCVG survey for more accurate ECDS data. J. World Pipelines **7**, 1–7 (2007)
29. Park, G., Park, E.: Improvement of the sensor system in magnetic flux leakage-type noddestructive testing. IEEE Trans. Magn. **38**(2), 1277–1280 (2002)
30. Petković, D., et al.: Adaptive neuro-fuzzy estimation of conductive silicone rubber mechanical properties. Expert Syst. Appl. **39**(10), 9477–9482 (2012)
31. Roohollah, N., Safavi, S., Shahrokni, S.: A reduced-order adaptive neuro-fuzzy inference system model as a software sensor for rapid estimation of five-day biochemical oxygen demand. J. Hydrol. **495**, 175–185 (2013)
32. Shou-peng, S., Pei-wen, Q.: Wavelet based noise suppression technique and its application to ultrasonic flaw detection. Ultrasonics **44**(2), 188–193 (2006)
33. Sinha, S., Karray, F.: Classification of underground pipe scanned images using feature extraction and neuro-fuzzy algorithm. IEEE Trans. Neural Netw. **13**(2), 393–401 (2002)
34. Sinha, S., Pandey, M.: Probabilistic neural network for reliability assessment of oil and gas pipelines. Comput.-Aided Civil Infrastruct. Eng. **17**(5), 320–329 (2002)
35. Tikaria, M., Nema, S.: Wavelet neural network based intelligent system for oil pipeline defect characterization. In: 3rd International Conference on Emerging Trends in Engineering and Technology (ICETET) (2010)
36. Unser, M., Aldroubi, A.: A review of wavelets in biomedical applications. Proc. IEEE **84**(4), 626–638 (1996)
37. Zadeh, A., et al.: An emotional learning-neuro-fuzzy inference approach for optimum training and forecasting of gas consumption estimation models with cognitive data. Technol. Forecast. Soc. Change **91**, 47–63 (2015)
38. Zhongli, M., Liu, H.: Pipeline defect detection and sizing based on MFL data using immune RBF neural networks. In: IEEE Congress on Evolutionary Computation pp. 3399–3403 (2007)

Predicting Financial Time Series Data Using Hybrid Model

Bashar Al-hnaity and Maysam Abbod

Abstract Prediction of financial time series is described as one of the most challenging tasks of time series prediction, due to its characteristics and their dynamic nature. Support vector regression (SVR), Support vector machine (SVM) and back propagation neural network (BPNN) are the most popular data mining techniques in prediction financial time series. In this paper a hybrid combination model is introduced to combine the three models and to be most beneficial of them all. Quantization factor is used in this paper for the first time to improve the single SVM and SVR prediction output. And also genetic algorithm (GA) used to determine the weights of the proposed model. FTSE100, S&P 500 and Nikkei 225 daily index closing prices are used to evaluate the proposed model performance. The proposed hybrid model numerical results shows the outperform result over all other single model, traditional simple average combiner and the traditional time series model Autoregressive (AR).

1 Introduction

In modern financial time series prediction, predicting stock prices has been regarded as one of the most challenging applications. Thus, various numbers of models have been depicted to support the investors with more precise predictions. However, stock prices are influenced by different numbers of factors and the nonlinear relationships between factors existed in different periods such that predicting the value of stock prices or trends is considered as an extremely difficult task for the investors. Therefore, many methods and approaches have been introduced and employed in the prediction of stock prices and trends since the start of the stock market.

Additionally, researchers proposed numerous conventional numerical prediction models. However, traditional statistical models such ARCH, GARCH, ARMA,

B. Al-hnaity (✉) · M. Abbod
Electronic and Computer Engineering Department, Brunel University, Uxbridge, London UB8 3PH, UK
e-mail: bashar.Al-hnaity@brunel.ac.uk

M. Abbod
e-mail: maysam.abbod@brunel.ac.uk

Y. Bi et al. (eds.), *Intelligent Systems and Applications*,
Studies in Computational Intelligence 650, DOI 10.1007/978-3-319-33386-1_2

ARIMA and AR have failed to capture the complexity and behavior of the stock price. George Box stated in his work that "essentially, all models are wrong, but some are useful" [1]. Therefore, researchers have introduced more advanced nonlinear techniques including Support Vector Machine (SVM), Support Vector Regression (SVR), Neural Network (NN). Due to inherent characteristics such as non-linear, non-stationary, noisy, high degree of uncertainty and hidden relationships of the financial time series, single artificial intelligence and other conventional techniques have failed to capture its non-stationary property and accurately describe its moving tendency. Therefore, researches and market participants have paid a tremendous attention in order to tackle such problem. Thus, various models architecture and new algorithm have been introduced and developed in the literature to alleviate the influence of noise.

A various number of models and theories have been implemented by researchers in order to improve the prediction performance. Different techniques has been combined with single machine learning algorithms, for example Zhang and Wu [2] have integrated a Back-Propagation neural network (BPNN) with an improved Bacterial Chemotaxis Optimization (IBCO). Anther method was proposed combining data preprocessing methods, genetic algorithms and Levenberg-Marquardt (LM) algorithm in learning BPNN by [3]. In their studies data transformation and selection of input variables were used under data preprocessing in order to improve the model's prediction performance. Moreover, the obtained result has proved that the model is capable of dealing with data fluctuations as well as yielding a better prediction performance. And also, a hybrid artificial intelligent model was proposed by [4] combining genetic algorithms with a feed forward neural network to predict stock exchange index. The latest research indicates that in short term prediction models based on artificial intelligence techniques outperform traditional statistical based models [5]. However, and also as a result of the dramatic move in stock indices in response to many complex factors, there is plenty of room for enhancing intelligent prediction models. Therefore, researchers have introduced and tried to combine and optimize, different algorithms and take the initiative to build new hybrids models in order to increase prediction speed and accuracy.

There are many example of these attempts such as; Armano et al. [6] who optimized GA with ANN to predict stock indices; SVM were also combined with PSO in order to carry out the prediction of stock index by Shen and Zhang [7]. Kazem's [8] prediction model, chaotic mapping, firefly algorithm and support vector regression (SVR) were proposed to predict stock market prices. In Kazem's study SVR-CFA model was for the first time introduced and the results were compared with the SVR-GA (Genetic Algorithm), SVR-CGA (Chaotic Genetic Algorithm), SVR-FA (Firefly Algorithm) and ANN, and the ANFIS model, and the new adopted model SVR-CFA outperformed the other compared models. The seasonal Support Vector Regression (SSVR) model was developed by [9] to predict seasonal time series data. Moreover, in SSVR models in order to determine the model parameters, hybrid genetic algorithm and tabu search (GA, TS) algorithms were implemented. And also on the same data sets, Sessional Auto regressive Integrated Moving Average (SARIMA) and SVR were used for prediction, however the empirical results based on prediction accuracy indicated that the proposed model SSVR outperformed both models (SVR and SARIMA). A novel hybrid model to predict future evolution of a various stock indices was developed by integrating a genetic algorithm based on optimal time scale feature extractions with support vector machines. Neural network, pure SVMs and traditional GARCH models were used as a benchmark and prediction performances were compared and thus the proposed hybrid model's prediction performance was the best. Root mean squared error (RMSE) is one of the main utilized prediction models for performance measurement, however the reduction in this standard statistical measurement was significantly high. A novel hybrid ensemble model to predict FTSE 100 next day closing price was proposed by Al-hnaity and Abbod [10]. Genetic Algorithm (GA) was used to determent the weight of the hybrid combination model, and thus, the empirical results showed that the proposed hybrid model utilizing GA has outperformed the rest utilized approaches in their study.

In this chapter, the FTSE 100, S&P 500 and Nikkei225 daily closing prices will be predicted. Different data mining techniques will be utilized to capture the non-liner characteristics in a stock prices time series. The proposed approach is to combine the support vector machine SVM, support vector regression SVR and BPNN as the weight of these models are determined by the GA. The evaluation of the proposed model performance will be done by using it on the FTSE 100, S&P 500 and Nikkei225 daily closing price as the illustrative examples. The below examples show that the proposed model outperform all single models and with regard to the possible changes in the data structure, proposed model is shown to be more robust. The reminder of this chapter is organized as follow: in Sect. 2 the component models SVM, SVR and BPNN are briefly introduced. Section 3 demonstrates the prediction procedures and the hybrid methodology. The experimental result exhibits in Sect. 4. Finally conclusions and future work are in Sect. 5.

2 Methodology

2.1 *Benchmark Prediction Model*

In this chapter a traditional prediction model Simple Auto-regressive model (AR) is used, in order to benchmark the performance efficiency of the utilized models. Moreover, a simple average in this thesis is used as a benchmark combination method.

2.1.1 Simple Auto-Regressive Model

The autoregressive model (AR) in this study is used as a benchmark model to evaluate the prediction power between the utilized models based on the relative improvements on root mean square error. Equation 1 illustrates the AR model used.

$$y_t = a_1 y\,(t-1) + a_2 y(t-2) + \cdots, a_t y\,(t-p) \tag{1}$$

Eqution 1 $y\,(t)$ is the predicted stock price based on the past close daily price, $y\,(t) - y\,(t-n)$ and the coefficients of AR model are $a_1 - a_n$. 5 lagged daily price is the order of which is used in the AR model was varied and found to give better prediction result. The model coefficients were determined by using the implemented regress function in the MATLAB.

2.1.2 Prediction Combination Techniques

Combining different prediction techniques has been investigated widely in the literature. In the short range prediction combining the various techniques is more useful according to [2, 11]. Timmermann [12] stated in his study that using simple average may work as well as more sophisticated approaches. In this chapter simple average is used as a benchmark combination model. Equation 2 illustrates the calculation of combination prediction method at time t [13].

$$f_{SM}^{t} = \left(f_{M1}^{t} + f_{M2}^{t} + f_{M3}^{t}\right) \div 3 \tag{2}$$

2.2 Artificial Neural Network

In recent years predicting financial time series utilizing Artificial Neural Network (ANN) applications has increased dramatically. The idea of ANN can be seen before reaching the output, where the filtration of the inputs through one or more hidden layers each of which consists of hidden units, or nodes is considered as the main idea. Thus, final output is related to the intermediate output [14].

The ability of learning from data through adaptive changing structure based on external or internal information that flows through the network during the learning phase and generates output variables based on learning is one of the most important advantages of ANN. Furthermore, the non-linear nature of ANN is also a valuable quality. ANNs are classified as non-linear data modeling tool, thus, one of the main purposes of utilizing such a model is to find the patterns in data or modeling complex relationships between inputs and outputs. Hence, an explicit model-based approach fails, but ANNs adapt irregularities and unusual of features in a time series of interest.

The application of ANNs has been popularly utilized in financial time series prediction modeling. A comprehensive review of ANNs and their application in a various finance domains is given in Sect. 1. However, as with many other techniques, ANNs have some disadvantages such as not allowing much of understanding of the data which might be because they are not explicit models. Therefore, providing a black box for the prediction process is considered as a disadvantage. The danger of over-fitting the in-sample training data is also a major ANN methods drawback [15]. In terms of the goodness of fit the performance of in-sample data sets is good, and is what ANNs are trained on. However, in out-of-sample data it is conditional on not breaking the structure in the data sets. In accordance with Balestrassi et al. [16] excessive training time and a large number of parameters that must be experimentally selected in order to generate good prediction are considered as other drawbacks facing ANN applications. In this chapter the ANN BPNN is used to predict financial data, Sect. 2.2.1 demonstrates this model.

2.2.1 Back Propagation Neural Network

In modeling time series with non-linear structures, the most commonly used structure is three layers feed-forward back propagation [17]. The weights are determined in the back propagation process by building connections among the nodes based on data training, producing a least-mean-square error measure of the actual or desired and the estimated values from the output of the neural network. The initial values are assigned for the connection weights. In order to update the weights, the error between the predicted and actual output values is back propagated via the network. Minimizing of the error the desired and predicted output attempts occurs after the procedure of supervised learning [18]. The architecture of this network contains a hidden layer of neurons with non-linear transfer function and an output layer of neurons with non-linear transfer function and an output layer of neurons with linear transfer functions. Figure 1 illustrates the architecture of a back propagation network, where x_j $(j = 1, 2, \ldots, n)$ represent the input variables; z_i $(j = 1, 2, \ldots, m)$ represent the outputs of neurons in the hidden layer; and y_t $(t = 1, 2, \ldots, l)$ represent the outputs of the neural network [19].

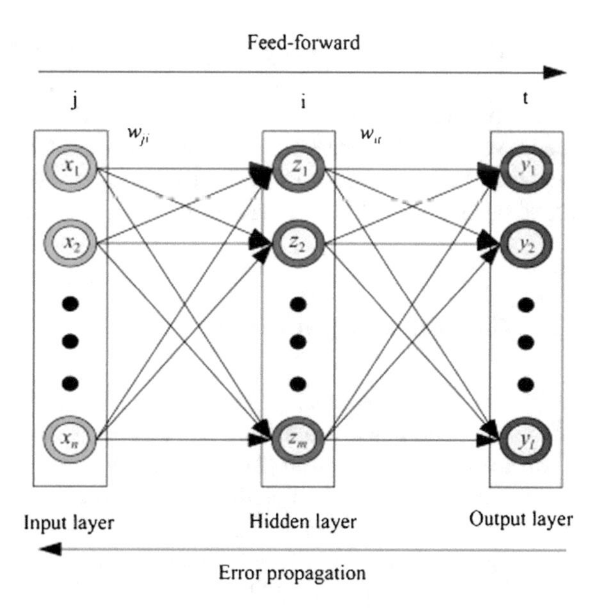

Fig. 1 Architecture of feed forward back propagation neural network [19]

In theory a neural network has the ability to simulate any kind of data pattern given a sufficient training. Training the neural network will determine the perfect weight to achieve the correct outputs. The following steps illustrate the training process of updating the weights values [20]. The first stage is hidden layers: the bellow Equation explains how the outputs of all neurons in the hidden layer are calculated:

$$net_i = \sum_{j=0}^{n} w_{ji} x_j vi = 1, 2, \ldots m \tag{3}$$

$$z_i = f_H (net_i)\, i = 1, 2, \ldots m \tag{4}$$

where net_i is the activation value of the ith node, z_i is the output of the hidden layer, and f_H is called the activation Equation of a node; in this chapter a sigmoid function is utilized. Equation 5 explains the utilized sigmoid activation equation.

$$f_H (x) = \frac{1}{1 + exp(-x)} \tag{5}$$

Second stage the output: The outputs of all neurons in the output layer are given as Eq. 6 illustrated bellow:

$$y_t = f_t \left(\sum_{i=0}^{m} w_{it} z_i \right) t = 1, 2, \ldots l \tag{6}$$

The activation Equation is f_t $(t = 1, 2, \ldots, l)$, which is usually a linear equation. The weights are assigned with random values initially, and are modified by the delta rule according to the learning samples traditionally. The topology in this study is determine by a set of trial and error search conducted to choose the best number of neurons experiments. Different range of 20–5 neurons in hidden layer two layer feed forward back propagation network were tried. The model stopped training after reaching the pre-determined number of epochs. The ideal topology was selected based on the lowest Mean Square Error.

2.3 *Support Vector Machine*

The SVM theory was developed by Vladimir Vapnik in 1995. It is considered as one of the most important breakthroughs in machine learning field and can be applied in classification and regression [21]. In modeling the SVM, the main goal is to select the optimal hyperplane in high dimensional space ensuring that the upper bound of generalization error is minimal. SVM can only directly deal with linear samples but mapping the original space into a higher dimensional space can make the analysis of a non-linear sample possible [22, 23]. For example if the data point (x_i, y_i) was given randomly and independently generated from an unknown function, the approximate function formed by SVM is as follows:

$g\left(x\right) = w\varnothing\left(x\right) + b.\varnothing\left(x\right)$ is the feature and non-linear mapped from the input space x. w and b are both coefficients and can be estimated by minimizing the regularized risk equation.

$$R\left(C\right) = C\frac{1}{N}\sum_{i=1}^{N} L\left(d_i, y_i\right)\frac{1}{2} + ||w||^2 \tag{7}$$

$$L\left(d, y\right) = \begin{cases} |d - y| - \varepsilon & |d - y| \geq \varepsilon \\ 0 & \text{other,} \end{cases} \tag{8}$$

C and ε in Eqs. 7 and 8 are prescribed parameters. C is called the regularization constant while ε is referred to as the regularization constant. $L\left(d, y\right)$ Is the intensive loss function and the term $C\frac{1}{N}\sum_{i=1}^{N} L\left(d_i, y_i\right)$ is the empirical error while the $\frac{1}{2} + ||w||^2$ indicates the flatness of the function. The trade-off between the empirical risk and flatness of the model is measured by C. Since introducing positive slack variables ζ and ζ^* Eq. 8 transformed to the following:

$$R\left(w, \zeta, \zeta^*\right) = \frac{1}{2}ww^T + C \times \left(\sum_{i=1}^{N}\left(\zeta, \zeta^*\right)\right) \tag{9}$$

Subject to:

$$w\phi\left(x_i\right) + b_i - d_i \leq \varepsilon + \zeta_{i*} \tag{10}$$

$$d_i - w\phi\left(x_i\right) - b_i \leq \varepsilon + \zeta_i \tag{11}$$

$$\zeta_i, \zeta_{i*} \geq 0 \tag{12}$$

The decision equation (kernel function) comes up finally after the Lagrange multipliers are introduced and optimality constraints exploited. The Eq. 13 is the formed of kernel equation:

$$f\left(x\right) = \sum_{i}^{l}\left(\alpha_i - \alpha_i{}'\right) K\left(x_i, x_j\right) + b \tag{13}$$

where $\alpha_i{}'$ are Lagrange multipliers. The satisfy equalities $\alpha_i \times \alpha_i{}' = 0, \alpha_i \geq 0, \alpha_i{}' \geq 0$. The kernel value is the same with the inner product of two vectors x_i and x_j in the feature space $\phi\left(x_i\right)$ and $\phi\left(x_j\right)$. The most popular kernel function is Radial Basis Function (RBF) it is form in Eq. 14.

$$K\left(x_i, x_j\right) = exp\left(-\gamma||x_i - x_j||^2\right) \tag{14}$$

Theoretical background, geometric interpretation, unique solution and mathematical tractability are the main advantages which has made SVM attract researchers and investors interest and be applied to many applications in different fields such as prediction of financial time series.

2.4 Support Vector Regression

As explained in Sect. 2.3 the idea of SVM is to constructs a hyperplane or set of hyperplanes in a high or infinite dimensional space, which can be used for classification. In the regression problem the same margin concept used in SVM is used. The goal of solving a regression problem is to construct a hyperplane that is as close to as many of the data points as possible. Choosing a hyperplane with small norm is considered as a main objective, while simultaneously minimizing the sum of the distances from the data points to the hyperplane [24].

In case the of solving regression problems using SVM, SVM became known as the support vector regression (SVR) where the aim is to find a function f with parameters w and b by minimizing the following regression risk:

$$R(f) = \frac{1}{2}(x, w) + C \sum_{i=1}^{N} l(f(x_i), y_i) \tag{15}$$

C in Eq. 15 is a trade-off term, the margin in SVM is the first term which is used in measuring VC-dimension [20].

$$f(x, w, b) = (w, \phi(x)) + b, \tag{16}$$

In the Eq. 16 $\phi(x) : x \rightarrow \Omega$ is kernel function, mapping x into in the high dimensional space, SVR and as proposed by [24]. The $-\varepsilon$ insensitive loss function is used as follows:

$$l(y, f(x)) = \begin{cases} 0, & \text{if} |y - f(x) < \varepsilon \\ |y - f(x)| - \varepsilon, & \text{Otherwise} \end{cases} \tag{17}$$

Equation 18 constrained minimisation problem is equivalent to previous minimisation Eq. 15.

Min

$$y\left(w, b, \zeta^* = \frac{1}{2}(w, w) + C \sum_{i=1}^{N} (\zeta_i + \zeta_{i^*})\right) \tag{18}$$

Subject to:

$$y_i - ((w, \phi(x_i) + b)) \leq \varepsilon + \zeta_i, \tag{19}$$

$$((w, \phi(x_i)) + b) - y_i \leq \varepsilon + \zeta_{i^*}, \tag{20}$$

$$\zeta_i^* \geq 0 \tag{21}$$

In sample (x_i, y_i) the ζ_i and ζ_{i*} measure the up error and down error. Maximizing the dual function or in other words constructing the dual problem of this optimization problem (primal problem) by large method is a standard method to solve the above minimization problem. There are four common kernel functions; among these, this chapter will be utilizing the radial basis function (RBF). In accordance with [25, 26] and [27] the RBF kernel function is the most widely applied in SVR. Equation 22 is defining the kernel RBF, where the width of the RBF is denoted by σ. Furthermore, Cherkassky and Ma [25] suggested that the value of σ must be between 0.1 and 0.5 in order for SVR model to achieve the best performance. In this chapter σ value is determined as 0.1.

$$K(x_i, x_i) = exp\left(\frac{-||x_i - x_j||^2}{2\sigma^2}\right) \tag{22}$$

3 Prediction Procedures

Financial time series data are characteristically non-linear, non-stationary, and noisy, with high degree of uncertainty and hidden relationships. These characteristics are a reason for information unavailability, affecting the behavior of financial markets between past and future captured prices. According to Li and Ma [28] traditional linear methods and the majority of sophisticated non-linear machine learning models have failed to capture the complexity and the non-linearity that exists in financial time series, particularly during uncertainty periods such the credit crisis in 2008 [29]. Financial time series characteristics imply that the statistical distributions of time series can change over time. The cause of these changes may be economic fluctuations, or political and environmental events [30, 31]. As a result, it has been verified in the literature that no single method or model works well to capture financial time series characteristics properly and accurately, which leads to different and inaccurate financial time series prediction results [32, 33]. In order to address these issues, a new two step hybrid model is proposed to capture the non-linear characteristics in a stock price time series. The proposed approach is to combine the support vector machine SVM, support vector regression SVR and BPNN as the weight of these models are determined by the GA.

3.1 Hybrid Combination Model GA-WA

Combining different prediction techniques have been investigated widely in the literature. In the short range prediction combining the various techniques is more useful according to [2, 11]. In accordance to Timmermann's study, using the simple average may work as well as more sophisticated approaches. However, using one model can produce a more accurate prediction than any other methods. Therefore, simple averages would not be sufficient in such cases [12]. Compared with different prediction models, the hybrid prediction method is based on a certain linear combination. The assumption for the actual value in period t by model i is f_{it} $(i = 1, 2, \ldots m)$, the corresponding prediction error will be $e_{it} = y_t - f_{it}$. And also the weight vector will be $W = [w_1, w_2, \ldots w_m]^T$ .. Then in the hybrid model the predicted value is computed as follows [34, 35]:

$$\hat{y}_t = \sum_{i=1}^{m} w_i f_{it} \ (t = 1, 2, \ldots n) \tag{23}$$

$$\sum_{i=1}^{m} w_i = 1 \tag{24}$$

Equation 23 can be expressed in another from, such:

$$\begin{aligned} &\hat{y} = FW \\ &\text{where} \quad \hat{y} = \left[\hat{y_1}, \hat{y_2}, \ldots \hat{y_n}\right]^T, F = [f_{it}]_{n \times m} \end{aligned} \tag{25}$$

The error for the prediction model can be formed as Eq. 26 illustrates.

$$\begin{aligned} e_t = y_t - \hat{y}_t = \textstyle\sum_{i=1}^{m} w_i y_t - \sum_{i=1}^{m} w_i f_{it} = \\ \textstyle\sum_{i=1}^{m} w_1 w_i \left(y_t - f_{it}\right) = \sum_{i=1}^{m} w_i e_{it} \end{aligned} \tag{26}$$

This research proposes a hybrid model of combining SVR, BPNN and SVM.

$$\hat{Y}_{combined_t} = \frac{w_1 \hat{Y}_{SVR} + w_2 \hat{Y}_{BPNN} + w_3 \hat{Y}_{SVM}}{(w_1 + w_2 + w_3)} \tag{27}$$

The prediction values in period t are $\hat{Y}_{combined_t}$, $\hat{Y}_{SVR}$, $\hat{Y}_{BPNN}$ and $\hat{Y}_{SVM}$ for the hybrid, SVM, BPNN and SVR models, where the assigned weights are w_1, w_2, w_3 respectively, with $\sum_{i=1}^{3} w_i = 1.0 \leq w_i \leq 1$.

The most important step in developing a hybrid prediction model is to determine the perfect weight for each individual model. Setting $w_1 = w_2 = w_3 = 1/3$ in Function 27 is the simplest combination method for the three prediction models. Nevertheless, in many cases equal weights cannot achieve the best prediction result. Therefore, this chapter adopts a hybrid approach utilizing GA as an optimizer to determine the optimal weight for each prediction model. Figure 2 illustrates the architecture of the GA-WA hybrid model.

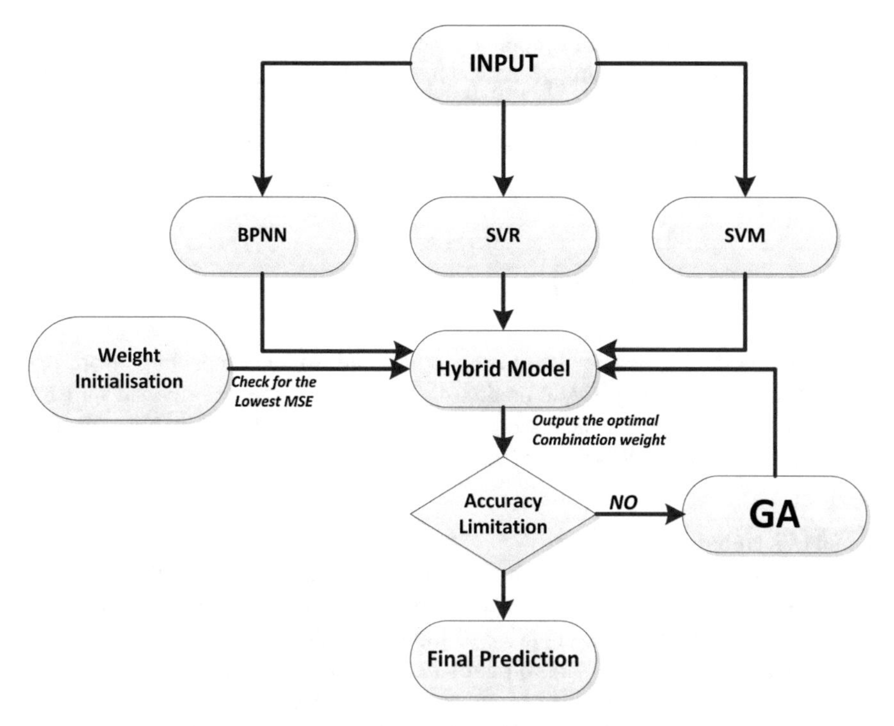

Fig. 2 The flow chart of the GA-WA hybrid model

3.1.1 Genetic Algorithm

Genetic Algorithm GA is a well known tool in computational method modeled on Darwinian selection mechanism. GA principles were proposed by Holland [36], and developed by Goldbreg [37] and Koza [38]. Thus, the main purpose of using such algorithm, is to solve optimization problems such determining the optimal weights for the proposed hybrid model in this research. In comparison to other conventional optimization methods, GA have many differences, which contributed in making GA more efficient in searching for the optimal solution. The following points exhibited those differences [39].

- Computing the strings in GA algorithms is done by encoding and decoding discrete points than using the original parameters values. Thus, GAs tackling problems associated to discontinuity or non-differentiability function, where, traditional calculus methods have failed to work. Therefore, due to adaptation of binary strings, such characteristic allows GAs to fit computing logic operations better.
- The prior information is not important, and thus there is no need for such information as the primary population is randomly generated. GA using a fitness function in order to evaluate the suggested solution.

- Initialization, selection and reproduction whether, crossover and mutation are what GA depends on in searching process, which involve random factors. As a results, searching process in GAs for every single execution will be stand alone even the ones who are under the identical parameter setting, which perhaps may effect the results.

3.2 Data Pre-processing

Selecting and pre-processing the data are crucial steps in any modeling effort, particularly for generalizing a new predictive model. Data sets are divided into two sets: training and testing, which is explained in the below Subsection. Thus, the pre-processing steps for financial time series data are as follows:

- Data Cleaning:
  In real world, data tend to be incomplete and inconsistent. Therefore, the purpose of data cleaning (data cleansing) is to fill in missing values and correct inconsistencies in the data. The daily closing price of the FTSE 100, S&P 500 and Nikkie 225 are used as an input, and due to public holidays in the stock market, there were missing values in the data set. Thus, treating the missing data can be done by different methods such as; ignore the tuple, fill in the missing value manually, use a global constant to fill in the missing value, use the attribute mean to fill in the missing value, use the attribute mean for all samples belonging to the same class as the given tuple and use the most probable value to fill in the missing values. This Chapter adopted the ignore the tuple method. In other words, in order to treat the missing information in the data set the tuples were excluded from the data set [40].
- Attribute creation:
  In the presented data in section must be able to be sufficiently utilized in prediction proposed models. Therefore, and according to Fuller [41], a real valued time series is considered as a set of random variables indexed in time or it can be a set of observations ordered in time such as performance degradation data. In accordance with traditional setting of time series regression, the broad concept of the target value is represented as an unknown function for the input vector x_t of the p-lagged last value of y itself, as Eq. 28 illustrated, where p is the number of back time steps which in this research is 5. Thus the input array and the output array are respectively as follows in Eqs. 29 and 30, where t is the number of training data.

$$y_t = f(x_t) = f\left(y_{t-1}, y_{t-2}, \ldots, y_{t-p}\right) \tag{28}$$

$$X = \begin{bmatrix} y_1 & y_2 & \cdots & y_p \\ y_2 & y_3 & \cdots & y_{p+1} \\ y_{t-p} & y_{t-p+1} & \cdots & y_{t-1} \end{bmatrix} = \begin{vmatrix} x_{p+1} \\ x_{p+2} \\ \vdots \\ x_t \end{vmatrix} \tag{29}$$

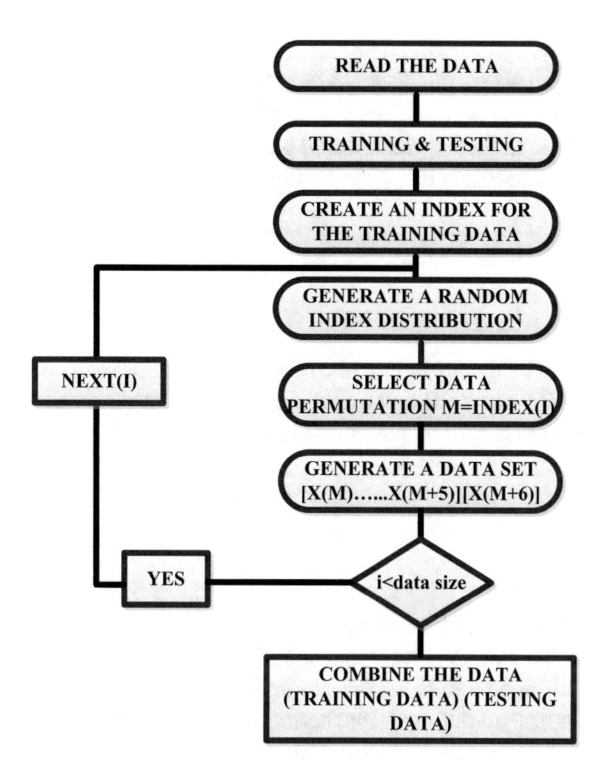

Fig. 3 The data preprocessing frame

$$Y = \begin{vmatrix} y_{p+1} \\ y_{p+2} \\ \vdots \\ y_t \end{vmatrix} \tag{30}$$

Moreover, these steps are for the training in-sample data set and testing out-of-sample data set. However, this research adopted a new approach in forming the training in-sample data set and testing out-of-sample data set in order to achieve the best prediction results. The following forms illustrate the steps of this forming: for training data: $Tr = \left[x_t \ldots\ldots x_{t+5}\right]\left[x_{t+6}\right]$, where m is a random number permutation $1 < m < p$, p is the data size. For testing data: $Ts = \left[x_t \ldots x_{t+5}\right]\left[x_{t+6}\right]$, where t is $1 : p$, P is the testing data set size. Figure 3 illustrate the architecture of the data prepossessing.

3.3 Quantization Factor

This chapter adopts a new approach to enhance the prediction output of SVM and SVR models. The quantization factor is for the first time introduced to SVM and

SVR. As explained above in the methodology in section SVM and SVR the model input is (x_i, y_i). After adding the optimal factors which were determined by trial and error, the optimal factor was chosen from a range of factors between 10 and 100. The optimal factor was selected based on the lowest mean square error. The below steps illustrate the change in the input after introducing the quantization factor:

$$Xprim = X_i \div Factor \tag{31}$$

$$Yprim = Y_i \div Factor \tag{32}$$

The model inputs become $(Xprim_i, Yprim_i)$. After applying the above model SVM and SVR and to get the final prediction result the chosen factor is multiplied by the output of each model as illustrated in Eqs. 33 and 34:

$$Xprim_{pred} = Xprim \times Factor \tag{33}$$

$$Yprim_{pred} = Yprim \times Factor \tag{34}$$

This method has been proposed to enhance the performance and prediction ability of SVM and SVR techniques. To the best of our knowledge this might be the first time this approach is introduced and utilized in financial time series predictions (stocks indies predictions). Furthermore, the test was carried out on the training data set of the three stock indies (FTSE100, S&P 500 and Nikkei225). first, for FTSE100, the best $Qfactor$ in SVR model is 100. Second, for S&P 500 the best $Qfactor$ is 100. Third, for Nikkei225 the best $Qfactor$ is 100. Moreover, in SVM prediction model the best $Qfactor$ is 100 for FTSE100 data set, 10 is the best $Qfactor$ for S&P 500 and 10 is the best $Qfactor$ for the Nikkei225 data set.

3.4 Prediction Evaluation

3.4.1 Standard Statistical Measure

The accuracy of prediction is referred to as "goodness of fit" which in other words means how well the prediction models fit a set of observations. Moreover, prediction accuracy can also be regarded as an optimists term for prediction errors as stated in [42]. Hence, prediction error represents the difference between the predicted value and the actual value. Therefore, the prediction error can be defined as: if x_n is the actual observations at time n and $\hat{x_n}$ is the predicted value for the same period, then the prediction error takes the following form:

$$E_n = x_n - \tilde{x_n} \tag{35}$$

Generally, in Eq. 35 $\hat{x_n}$ is one step ahead prediction so E_n is the one step ahead prediction error. The accuracy measurement can be classified under five categories according to [43]: scale-dependent measures, measures based on percentage errors, measures based on relative errors, relative measures and scaled error. However, in this research the most popular and commonly available statistical accuracy measures are used as Table 1 demonstrates:

MSE, RMSE and MAE are scale-dependent measures which are commonly utilized to compare different methods on the same data set. In addition to standard statistical measures, standard deviation is used to observe variations. Moreover, a well known cross correlation method of computing the degree of relationship existing between the variables is also used in this study [44]. Table 2 illustrates the formulas utilized to measure prediction accuracy.

These statistical analyses after they are computed will provide the required information regarding the prediction accuracy and will strengthen the conclusions. However, after considering the pros and cons of the above-mentioned statistical accuracy measures, MSE and RMSE are chosen for performance comparison on the same data set, and for comparison across data sets MAE is utilized in this research [45].

Table 1 Statistical accuracy measures

Name	Category
Mean Square Error (MSE)	Scale-dependent measures
Root Mean Square Error (RMSE)	Scale-dependent measures
Mean Absolute Error (MAE)	Scale-dependent measures
Cross correlation coefficient (R)	Method of computing the degree of relationship between variable
Standard Deviation (SD)	An statistical test used to observe variations

Table 2 Error metrics equation to measure prediction accuracy

Abbreviations	Formulas
MSE	$= \frac{1}{N}\sum_{i-1}^{N}\left(x_i - \hat{x}_i\right)^2$
RMSE	$= \sqrt{\frac{1}{N}\sum_{i=1}^{N}\left(x_i - \hat{x}_i\right)^2}$
MAE	$= \frac{1}{N}\sum_{i=t}^{N}\frac{\lvert x_i - \hat{x}_i\rvert}{\lvert x_i\rvert} =$
R	$= \frac{\sum_{i=1}^{N}(x_i - \bar{x_i})\left(\hat{x_i} - \hat{\bar{x_i}}\right)}{\sqrt{\sum_{i=1}^{N}(x_i - \bar{x_i})^2\sum_{i=1}^{N}\left(\hat{x_i} - \hat{\bar{x_i}}\right)^2}}$
SD	$= \sqrt{\frac{\sum\left(x_i - \hat{x}_i\right)^2}{N-1}}$

4 Experimentation Design and Results

4.1 Data Sets

This chapter proposes a hybrid model to predict the next day index closing price. In order to test the proposed model, The first data set is the daily closing price of the FTSE 100 index. It is used to demonstrate the predictability of the single approach model. The first 7788 observations (03/01/1984–29/10/2013) are used as the in-sample data set (training set). The last 250 observations (30/10/2013–30/10/2014) are used as the out-sample set (testing set). The second data set is the daily closing price of the S&P 500 index which is also used to demonstrate and validate the predictability of the proposed single approach method. The first 16230 observations (03/01/1950–02/07/2014) are used as the in-sample data set (training set). The last 250 observations (03/07/2014–30/06/2015) are used as the out-sample set (testing set). The third data set is the daily closing price of the Nikkei 225 index and it is also utilized to demonstrated the predictability of the proposed single approached methods. The first 7508 observations (04/01/1984–04/07/2014) are used as the in-sample data set (training set). The last 250 observations (07/07/2014–30/06/2015) are used as the out-sample set (testing set). Thus, in order to discover a robust model it is highly suggested to have a long training duration and large sample [46]. Figure 4 illustrates the data division into in-sample and out-of-sample sets.

4.2 BPNN, SVM and SVR Parameters Determination

This chapter proposes a hybrid method combining three different models SVM, SVR and BPNN to predict FTSE 100, S&P 500 and Nikkei 225 stocks index closing prices. First step to build the proposed method is by performing the single classical models on the data sets individually. Building a BPNN prediction model by determining the architecture as mentioned in the methodology section in BPNN part. BPNN

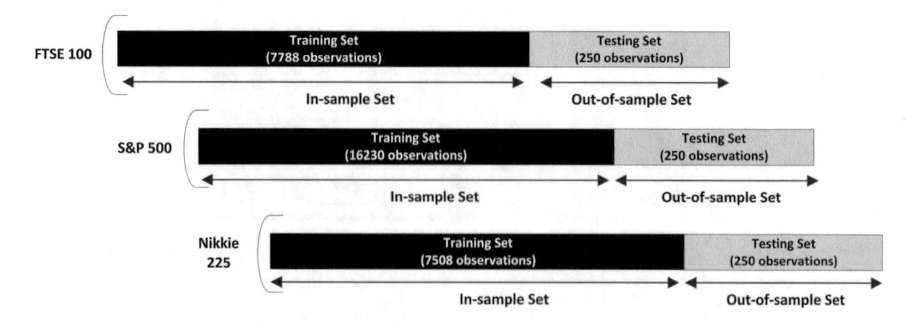

Fig. 4 In-sample and out-of-sample data sets

topology has been set by trial and error techniques, two hidden layer feed forward back propagation network with only two hidden layer and a range of 20-5 is the range of neurons. The best model topology for FTSE100 data set is, [20-5] (20, 5 nodes for the two hidden layers) with learning rate of 0.3 and momentum constant of 2, the best S&P 500 model topology is [10-5] (10, 5 nodes for the two hidden layers) with learning rate of 0.3 and momentum constant of 2 and for Nikkei225 it is [10-5] (10, 5 nodes for the two hidden layers) with learning rate of 0.7 and momentum constant of 10.

To build the SVR and SVM model, the method which was addressed in the methodology section and in the framework prediction model section to determine the parameters of both model and the factor is applied. In the literature there are no general rules for choosing those parameters. Thus, this research adopted the most common approach to search for the best C and γ values, which is the grid search approach [2]. The grid search approach was proposed by Lin et al. [47] using a validation process in order to achieve the ability to produce good generalizations to decide parameters. The criteria of choosing the optimal C and γ parameters is by trying pairs of C and $\gamma$and the best combination of these parameters which can generate the minimum mean square error MSE is chosen to set up the prediction model.

On the grid and after identifying a better region, on this region a more specific grid search can be conducted [47]. Finally after determining these lowest cross validation prediction error parameters C and γ, these parameters are chosen to be used in the creation of the prediction models. Figure 5 illustrates the process of grid search algorithm when building the SVM, SVR prediction model.

A set of exponentially growing sequences of C and γ are used and the best parameter combination results are: for the FTSE100 prediction next day closing price is ($C = 100$, $\gamma = 0.0001$), which gives the minimum mean square error MSE in the training data set and is therefore the best to set up the prediction model. For the S&P 500 next day closing price prediction, the best combination parameters are ($C = 100$, $\gamma = 0.0001$), which are the best combination parameters which give the minimum MSE in training data set. Furthermore, the best combination results to set up SVR prediction next day closing price model for Nikkei225 parameters are ($C = 100$, $\gamma = 0.0003$) which give the minimum MSE in the training data set. Moreover, the parameters combination to set up a prediction model using SVM are: Firstly, for the FTSE100 data set the best combination of parameters are $C = 100$, $\gamma = 0.0003$. Secondly, for S&P 500 the best combination of parameters are $C = 100$, $\gamma = 0.0003$. Thirdly, for Nikkei225 the best combination of parameters are $C = 90$, $\gamma = 0.0002$.

4.3 Prediction Results

To further explain the presented artificial intelligence techniques (SVM, SVR and BPNN) and its capability in understanding the pattern in the historical financial time series data, SVM, SVR and BPNN are applied to demonstrate their predictability of real world financial time series. In addition, this chapter presents a new hybrid

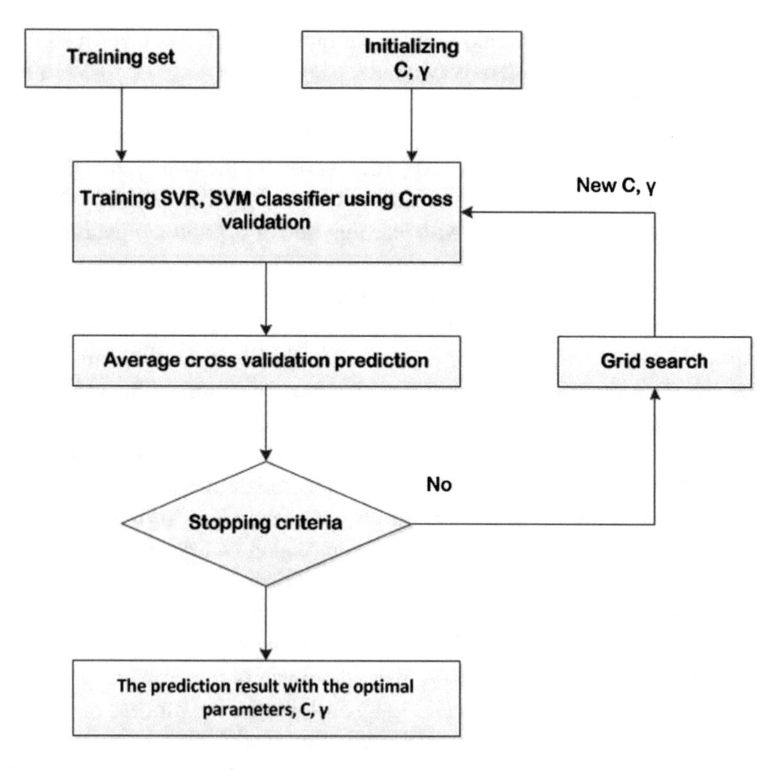

Fig. 5 Grid search algorithm for parameters selection

model based on GA-WA method which is constructed to predict FTSE 100, S&P 500 and Nikkei 225 next day closing prices. To demonstrate the validity of the proposed methods, this section compare the obtained results of single approaches models with the results of the hybrid proposed model.

The prediction results of FTSE 100, Nikkei 225 and S&P 500 testing data sets are summarized in Table 3. It can be observed from Table 3 that GA-WA hybrid model outperformed the rest of proposed methods by achieving the smallest MSE, RMSE and MAE values for FTSE 100 and S&P 500 testing data sets. Furthermore, the results of GA-WA hybrid prediction model for Nikkei 225 testing data set, is the best results in comparison with the rest of SVM, SVR, BPNN, SA and AR models. The cross correlation coefficient R of each data sets are exhibited in Table 3. It can be observed that the results of R for all methods are drawing near to 1, which implies that the predicted values and the actual are not deviating too much. Moreover, each method were run twenty times and the stander deviation was calculated. It can be observed that the results of SD for all models are relatively small, which implies hat the models are not running randomly.

Figure 6 depicts the actual testing data set of Nikkei 225 daily closing price and predicted value from the AR, SVM, BPNN, SVR, SA and GA-WA. As it can be observed from the Fig. 6 all the utilized method have generated a good predicting

Table 3 The prediction result of FTSE 100, S&P 500 and Nikkei 225 using SVR, SVM, BPNN, AR, GA-WA and SA

Index name	Models	MSE	RMSE	MAE	R	SD
FTSE100	SA	2410.96	49.10	20.45	0.99	—
	AR	2429.91	49.29	31.93	0.99	—
	SVR	2438.87	49.38	20.71	0.99	1.61
	SVM	6532.26	80.82	20.63	0.49	1.86
	BPNN	2210.75	47.02	20.41	0.99	1.74
	GA-WA	2066.65	45.46	20.15	0.99	1.64
S&P 500	SA	2566.93	50.66	45.69	0.95	—
	AR	240.56	15.51	11.68	0.96	—
	SVR	244.99	15.65	11.83	0.97	1.21
	SVM	23516.58	153.35	142.20	0.36	2.09
	BPNN	602.97	22.85	18.12	0.94	3.45
	GA-WA	219.04	14.80	10.57	0.96	1.19
Nikkei225	SA	36889.34	192.06	139.98	0.99	—
	AR	34806.38	186.56	134.88	0.99	—
	SVR	34720.67	186.33	135.36	0.99	2.92
	SVM	48662.44	220.59	168.62	0.93	2.48
	BPNN	35547.91	188.50	137.028	0.99	4.00
	GA-WA	34362.89	185.37	133.49	0.99	1.791

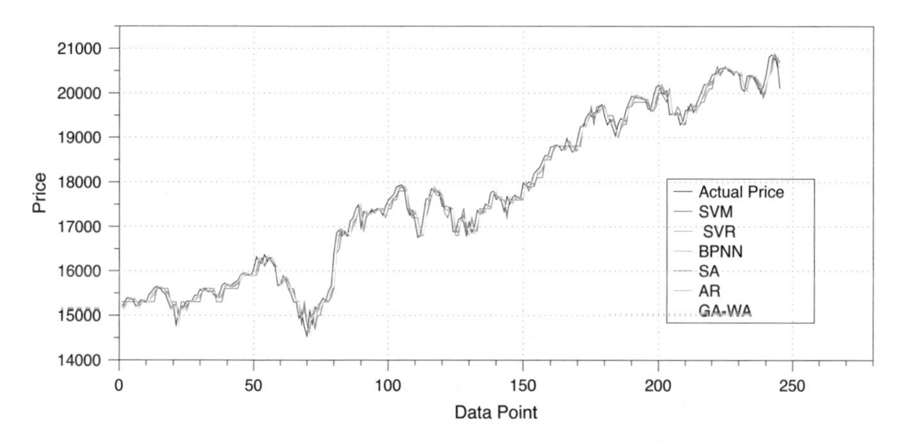

Fig. 6 The actual Nikkei 225 closing price index and its predicted value from AR, SA, SVM, SVR, BPNN and GA-WA models

results. The predicted values are very close to the actual values and to one another as well. Figure 7 presents the actual FTSE 100 closing price of the testing data set and predicted value from the AR, SVR, BPNN, SVM, SA and GA-WA. It also can be observed that the predicted obtained value from all the utilized models are very close the actual and to one another.

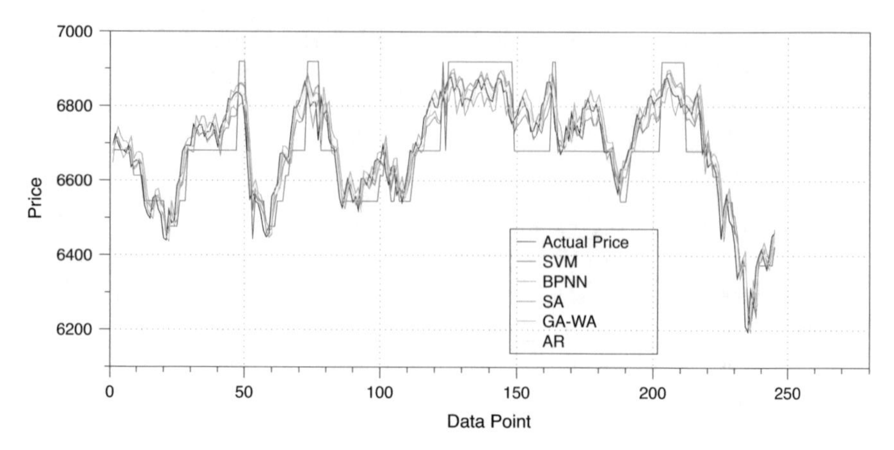

Fig. 7 The actual FTSE 100 closing price index and its predicted value from AR, SA, SVM, SVR, BPNN and GA-WA models

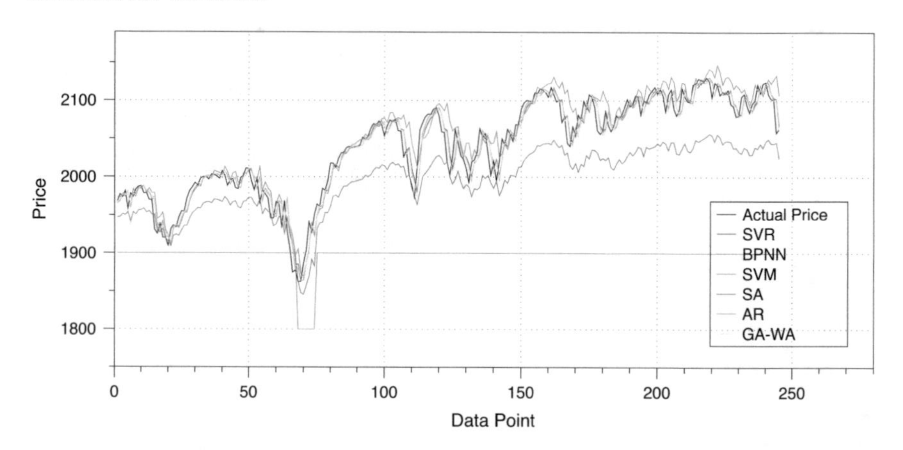

Fig. 8 The actual S&P 500 closing price index and its predicted value from AR, SA, SVM, SVR, BPNN and GA-WA models

The predicted values of AR, SVR, SVM, BNN, SA and GA-WA for the S&P 500 closing price testing data set are illustrated in Fig. 8. It can be observed from the Fig. 8 that AR, SVR, SVM, SA and GA-WA predicted values are close to the actual values than BPNN model.

5 Conclusions

This paper examines the predictability of stock index time series whereas the dynamics of such a data in real situations is complex and unknown. As a result using single classical model will not produce the accurate prediction result. In this paper a hybrid combination model was introduced combining BPNN, SVM and SVR. FTSE 100 daily closing price was used to evaluate the proposed model performance. The proposed model was compared with different models and as the Table 3 shows that the hybrid model outperforms all other models. From the result that were achieved in this paper, the proposed hybrid model outperform the other three single models and the benchmark models, thus future work should revolve around different hybrid combination models paradigm. Different singles models can be added and the possibility of combining them. Moreover, to test the model robustness further extension of this study can be done by testing different data sets.

References

1. Box, G.E.P., Draper, Norman R.: Empirical Model-building and Response Surfaces. Wiley, New York (1987)
2. Yudong, Z., Lenan W.: Stock market prediction of S&P 500 via combination of improved bco approach and bp neural network. Expert Syst. Appl. **36**(5), pp. 805–818 (2009)
3. Asadi, S., Hadavandi, E., Mehmanpazir, F., Nakhostin, M.M.: Hybridization of evolutionary Levenberg–Marquardt neural networks and data pre-processing for stock market prediction. Knowl. Based Syst. **35**, 245–258 (2012)
4. Hadavandi, E., Ghanbari, A., Abbasian-Naghneh, S.: Developing an evolutionary neural network model for stock index forecasting. In: Advanced Intelligent Computing Theories and Applications, pp. 407–415. Springer, Heidelberg (2010)
5. Hill, T., O'Connor, M., Remus, W.: Neural network models for time series forecasts. Manag. Sci. **42**(7), 1082–1092 (1996)
6. Armano, G., Marchesi, M., Murru, A.: A Hybrid Genetic-neural Architecture for Stock Indexes Forecasting. Inf. Sci. **170**(1), 3–33 (2005)
7. Shen, W., Zhang, Y., Ma, X.: Stock return forecast with LS-SVM and particle swarm optimization. In: International Conference on Business Intelligence and Financial Engineering, 2009. BIFE'09, pp. 143–147. IEEE, New York (2009)
8. Kazem, A., Sharifi, E., Hussain, F.K., Saberi, M., Hussain, O.K.: Support vector regression with chaos-based firefly algorithm for stock market price forecasting. Appl. Soft Comput. **13**(2), 947–958 (2013)
9. Pai, P.-F., Lin, K.-P., Lin, C.-S., Chang, P.-T.: Time series forecasting by a seasonal support vector regression model. Expert Syst. Appl. **37**(6), 4261–4265 (2010)
10. Al-Hnaity, B., Abbod, M.: A novel hybrid ensemble model to predict FTSE100 index by combining neural network and EEMD. In: 2015 European Control Conference (ECC), pp. 3021–3028. IEEE, New York (2015)
11. Armstrong, J.S.: Combining forecasts: the end of the beginning or the beginning of the end? int. j. forecast. **5**(4), 585–588 (1989)
12. Timmermann, A.: Forecast combinations. Handbook of Economic Forecasting, vol. 1, pp. 135–196 (2006)
13. Xiang, C., Fu, W.M.: Predicting the stock market using multiple models. In: 9th International Conference on Control, Automation, Robotics and Vision, 2006. ICARCV'06, pp. 1–6. IEEE, New York (2006)
14. Gooijer, D., Jan, G., Hyndman, R.J.: 25 years of time series forecasting. Int. J. Forecast. **22**(3), 443–473 (2006)
15. LeBaron, B., Weigend, A.S.: A bootstrap evaluation of the effect of data splitting on financial time series. In: IEEE Transactions on Neural Networks, vol. 9(1), pp. 213–220. IEEE, New York (1998)

16. Balestrassi, P.P., Popova, E., de Paiva, A.P., Lima, J.W.M.: Design of experiments on neural network's training for nonlinear time series forecasting. Neurocomputing **72**(4), 1160–1178 (2009)
17. Diaz-Robles, L.A., Ortega, J.C., Fu, J.S., Reed, G.D., Chow, J.C., Watson, J.G., Moncada-Herrera, J.A.: A hybrid ARIMA and artificial neural networks model to forecast particulate matter in urban areas: the case of Temuco, Chile. Atmos. Environ. **42**(35) 8331–8340 (2008)
18. Kubat, Miroslav: Neural Networks: A Comprehensive Foundation by Simon Haykin Macmillan. Cambridge University Press, Cambridge (1999)
19. Wang, J-J., Wang, J-Z., Zhang, Z-G., Guo, S.-P.: Stock index forecasting based on a hybrid model. Omega **40**(6), 758–766 (2012)
20. Faruk, D.Ö.: A hybrid neural network and ARIMA model for water quality time series prediction. Eng. Appl. Artif. Intell. **23**(4):586–594 (2010)
21. Cortes, C., Vapnik, V.: Support-vector Networks. In: Machine Learning, vol. 20(3), pp. 273–297. Springer, Heidelberg (1995)
22. Pontil, M., Verri, A.: Properties of support vector machines. In: Neural Computation, vol. 10(4), pp. 955–974. MIT Press, Cambridge (1998)
23. Osuna, E., Freund, R., Girosi, F.: Support Vector Machines: Training and Applications (1997)
24. Xia, Y., Liu, Y., Chen, Z.: Support Vector Regression for prediction of stock trend. In: 2013 6th International Conference on Information Management, Innovation Management and Industrial Engineering (ICIII), vol. 2, pp. 123–126. IEEE, New York (2013)
25. Cherkassky, V., Ma, Y.: Practical selection of SVM parameters and noise estimation for SVM regression. Neural Netw. **17**(1), 113–126 (2004)
26. Tay, F.E.H., Cao, L.: Application of support vector machines in financial time series forecasting. Omega, **29**(4), 309–317 (2001)
27. Cao, L.-J., Tay, F.E.H.: Support vector machine with adaptive parameters in financial time series forecasting. IEEE Trans. Neural Netw. **14**(6), 1506–1518 (2003)
28. Li, Y., Ma, W.: Applications of artificial neural networks in financial economics: a survey. In: 2010 International Symposium on Computational Intelligence and Design (ISCID), vol. 1, pp. 211–214. IEEE, New York (2010)
29. Theofilatos, K., Karathanasopoulos, A., Middleton, P., Georgopoulos, E., Likothanassis, S.: Modeling and trading FTSE100 index using a novel sliding window approach which combines adaptive differential evolution and support vector regression. In: Artificial Intelligence Applications and Innovations, pp. 486–496. Springer, Heidelberg (2013)
30. Hall, J.W.: Adaptive selection of US stocks with neural nets. In: Trading on the Edge: Neural, Genetic, and Fuzzy Systems for Chaotic Financial Markets, pp. 45–65. Wiley, New York (1994)
31. Abu-Mostafa, Y.S., Atiya, A.F.: Introduction to financial forecasting. In: Applied Intelligence, vol. 6(3), pp. 205–213. Springer, Heidelberg (1996)
32. Zhang, G.P.: Time series forecasting using a hybrid ARIMA and neural network model. Neurocomputing **50**, 159–175 (2003)
33. Chatfield, C.: What is the bestmethod of forecasting. J. Appl. Stat. **15**(1), 19–38 (1988)
34. Gupta, S., Wilton, P.C.: Combination of forecasts: an extension. Manag. Sci. **33**(3), 356–372 (1987)
35. Christodoulos, C., Michalakelis, C., Varoutas, D.: Forecasting with limited data: combining ARIMA and diffusion models. Technol. Forecast. Soc. Change **77**(4), 558–565 (2010)
36. John, H.H.: Adaptation in Natural and Artificial Systems: An Introductory Analysis with Applications to Biology, Control, and Artificial Intelligence. MIT press, Cambridge (1992)
37. Goldberg, D.E.: Genetic Algorithm in Search, Optimization and Machine Learning, vol. 1(98), p. 9. Addison Wesley Publishing Company, Reading (1989)
38. Koza, J.R.: Genetic Programming: On the Programming of Computers by Means of Natural Selection, vol. 1. MIT press, Cambridge (1992)
39. Wu, B., Chang, C.-L.: Using genetic algorithms to parameters (d, r) estimation for threshold autoregressive models. Comput. Stat. Data Anal. **38**(3), 315–330 (2002)
40. Han, J., Kamber, M., Pei, J.: Data Mining: Concepts and Techniques: Concepts and Techniques. Elsevier (2011)

41. Fuller, W.A.: Introduction to Statistical Time Series, vol. 428. Wiley, New York (2009)
42. Armstrong, J.S.: Principles of Forecasting: A Handbook for Researchers and Practitioners. Science and Business Media, vol. 30. Springer, Heidelberg (2001)
43. Hyndman, R.J. Koehler, A.B.: Another Look at Measures of Forecast Accuracy. Int. J. Forecast. **22**(4), 679–688 (2006)
44. Wang, J., Wang, J.: Forecasting stock market indexes using principle component analysis and stochastic time effective neural network. Neurocomputing **156**, 68–78 (2015)
45. Cao, Q., Leggio, K.B., Schniederjans, M.J.: A comparison between Fama and French's model and artificial neural networks in predicting the Chinese stock market. Comput. Oper. Res. **32**(10), 2499–2512 (2005)
46. Zhang, D., Zhou, L.: Discovering golden nuggets: data mining in financial application. In: IEEE Transactions on Systems, Man, and Cybernetics, Applications and Reviews, Part C, vol. 34(4), pp. 513–522. IEEE, New York (2004)
47. Hsu, C-W., Chang, C-C., Lin, C-J. et al.: A Practical Guide to Support Vector Classification (2003)

Diagnosis System for Predicting Bladder Cancer Recurrence Using Association Rules and Decision Trees

Amel Borgi, Safa Ounallah, Nejla Stambouli, Sataa Selami and Amel Ben Ammar Elgaaied

Abstract In this work we present two methods based on Association Rules (ARs) for the prediction of bladder cancer recurrence. Our objective is to provide a system which is on one hand comprehensible and on the other hand with a high sensitivity. Since data are not equitably distributed among the classes and since errors costs are asymmetric, we propose to handle separately the cases of *recurrence* and those of *no-recurrence*. ARs are generated from each training set using an associative classification approach. The rules' uncertainty is represented by a confidence degree. Several symptoms of low intensity can be complementary and mutually reinforcing. This phenomenon is taken into account thanks to aggregate functions which strengthen the confidence degrees of the fired rules. The first proposed classification method uses these ARs to predict the bladder cancer recurrence. The second one combines ARs and decision tree: the original base of ARs is enriched by the rules generated from a decision tree. Experimental results are very satisfactory, at least with the AR's method. The sensibility rates are improved in comparison with some other approaches. In addition, interesting extracted knowledge was provided to oncologists.

1 Introduction

In many areas, it is necessary to make critical decisions in a sometimes difficult context. In this paper, we are interested in a sensitive domain: aid to diagnosis in the medical field. It is more precisely about the diagnosis of recurrence bladder cancer.

A. Borgi (✉) · S. Ounallah
Institut Supérieur d'Informatique, Université de Tunis El Manar, LIPAH, Tunis, Tunisia
e-mail: Amel.Borgi@insat.rnu.tn

S. Ounallah
e-mail: safaounallah6@gmail.com

N. Stambouli · S. Selami · A.B.A. Elgaaied
Faculté des Sciences de Tunis, Laboratoire de Génétique Immunologie et Pathologies Humaines, Université de Tunis El Manar, Tunis, Tunisia
e-mail: nejlastam@gmail.com

Y. Bi et al. (eds.), *Intelligent Systems and Applications*,
Studies in Computational Intelligence 650, DOI 10.1007/978-3-319-33386-1_3

Bladder cancer is one of the deadliest diseases. Worldwide, it is the fourth cancer in terms of incidence and the third in terms of prevalence [1], 50–70 % of superficial tumors recur [2]. Because of the importance of the recurrence frequency, this cancer is considered as one of the most expensive cancers, for patient, in terms of treatment and clinical monitoring. The work presented in this article aims to assist oncologists in predicting the bladder cancer recurrence. This important task could contribute to save lives and to better understand the bladder cancer recurrence. It should be noted that errors costs are asymmetric. Indeed correctly predict the *recurrence* cases is more important than correctly predict the *no-recurrence* cases.

Clinical decision support systems have been proposed with varying degrees of success [3, 4]. The underlying methodologies include data mining approaches such as Bayesian approach and Neural networks, as well as knowledge-based approaches. This work concerns a supervised learning task. The goal of supervised learning is to build classification models that assign the correct class to previously unseen and unlabeled objects. Supervised learning regarding a given domain is based on a set of examples. Each example is a case already solved or classified. It is associated with a pair (description, class), where the description is a set of pairs (attribute, value) which is the available knowledge (in our case the description of a patient). The class of the example is the decision associated with the given description (for us, the decision is the *recurrence*, or *no-recurrence*, of the bladder cancer). Such a set of examples is called a training set [5, 26, 27].

In this work, our aim is to provide to the oncologists a diagnosis system that is comprehensible and able to correctly predict the *recurrence* instances (high sensitivity[1]). So the classification model must be understandable and easy to interpret by the doctor. This is the reason why we are interested in rules-based approaches. We consider two classification methods by generation of rules: decision trees and Association Rules.

Production rules *IF premise THEN conclusion* are a mode of knowledge representation widely used in decision systems because they ensure the transparency and the easy explanation of the classifier. Indeed, the rules enhance the interpretability of the results by providing more insight into the classifier structure and decision-making process [6–8]. The construction of production rules using the knowledge and the know-how of an expert is a very difficult task. The complexity and cost of such a knowledge acquisition have led to an important development of learning methods used for an automatic knowledge extraction, and in particular for rules extraction [7, 9].

Initially, we were interested in rules obtained from decision trees, but the experimental results were not satisfactory [10, 11, 27]. This is mainly due to two problems: the diversity of the characteristics in patients with bladder cancer that does not help to specify relationships between these characteristics and the imbalance between the

[1]Sensitivity: the proportion of positive class instances that are correctly classified (called True Positive rate or TP rate).

classes. Indeed the data we have are unevenly distributed between the two classes (cases of cancer *recurrence* and cases of *no-recurrence*). Nevertheless, this approach is presented here as it will be compared with the other proposed approaches. Moreover, the rules obtained from the decision trees will be used to enrich the knowledge base on the bladder cancer.

We were then interested in the Association Rules (ARs) [12, 20]. The use of such rules to treat supervised classification problem is known as associative classification [13, 14]. In the medical field, a classification method using association rules has been applied for the detection of ischemic beats in 2006 [15]. The main advantage of the proposed method is its high accuracy and the ability to provide interpretation for decision making. Recently in 2012, Dwarakanath and Ramesh proposed to use class association rules in order to select characteristics for diagnosis assistance [16].

In this work, we propose a diagnosis system based on associative classification. The proposed system uses and combines well known approaches that are adapted to our particular application domain. ARs are simple and easy to interpret, at least when their size and number remain reasonable. To solve the problem of imbalance between the classes, we propose to split the data according to their class value. Thus, three databases are considered in the learning phase: two sub-bases containing respectively *recurrence* instances and *no-recurrence* instances, and a third base with mixed instances, *recurrence* and *no-recurrence* (complete base). Furthermore, the generated rules are accompanied by a confidence degree. It allows to take into account uncertainty of some relationships between patients characteristics and the fact that several symptoms of low intensity can be complementary and mutually reinforcing. During the classification phase which consists in attributing a class to a new example, the degrees are aggregated according to the origin of the rules.

We also propose, another approach based on a combination of the association rules and the rules obtained from decision trees. The original base of ARs obtained with the previous approach is enriched by the rules generated from a decision tree. The other difference of treatment of this approach lies in the classification phase, that is in the way a new observation (a new patient) is classified as a *recurrence* or a *no-recurrence* case.

The article is organized as follows. In Sect. 2 we introduce the bladder cancer data that we have studied. Then, in Sect. 3 we remind the principal concepts related to association rules, and present classification methods using these rules. In Sect. 4, we present our first approach based on association rules for predicting bladder cancer recurrence. Then, in Sect. 5, we briefly describe the basic concepts of decision trees and their use for predicting bladder cancer recurrence. In Sect. 6, we introduce our second approach that consists in combining both association rules and decision trees. Before concluding the study, we provide in Sect. 7 the experimental results and their interpretation.

2 Bladder Cancer Recurrence: Datasets

In this section we present the context of our study. We particularly describe the data and explain the attributes. In Tunisia, bladder cancer is at the second place of all man cancers and the 7th place of women's cancer [17]. Between January 1995 and May 2004, 896 patients were hospitalized in the urology department of the Rabta[2] for bladder invasive tumors that require radical treatment. Among these 896 patients, 645 presented a superficial urothelial carcinoma [18]. It is the type of cancerous tumors studied in this work. A retrospective study was performed on 543 patients who underwent RTU[3] of superficial bladder tumor of stage Ta and T1, followed by BCG[4] therapy. The tumor stage defines the depth of the tumor in the muscular wall of the bladder. The monitoring of the patients is based on endoscopic control at 3 months, 6 months then annually. In case of superficial recurrence histologically confirmed a second cycle of BCG therapy is installed.

The studied data set concerns the 543 patients who have been the subject of a retrospective study. They are Tunisian women and men, divided into two classes: *recurrence* and *no-recurrence*. 24 attributes describe these patients: 22 symbolic attributes and 2 continuous attributes. Table 1 shows the considered attributes and their meanings. These attributes concern in particular the age and the sex of the patient, his or her smoking habits, the symptoms onset before the first consultation, the obstruction of the bladder at the time of the cystectomy,[5] the antecedents of tuberculosis and the perforation of the bladder at the time of the operation. Among the attributes specific to the tumor, let us quote: the seat, the aspect, the base, the stage, the rank, the number of the tumors and the size of the largest tumor. In Table 2, we provide the type and values of each attribute. The distribution of the patients into the classes (attribute *Recidive*) is unbalanced, the *no-recurrence* class contains 357 patients, and the *recurrence* class 186 patients.

The available data are not perfect, that is to say, they are neither completely nor perfectly described. The data may contain errors of description or labeling. Furthermore they are inhomogeneous, resulting from several sources gathered in different contexts. In addition, they have not been collected in order to be analyzed by a computer. So in order to be able to exploit these data, a pretreatment step was essential. It consisted in cleaning and editing the data so that they can be exploited by learning approaches of the platform WEKA[6] [19].

---

[2]Rabta: University Hospital in Tunis, Tunisia.

[3]RTU: Trans-Urethral Resection of bladder cancer, it is an operation which consists in removing one or more tumors on the level of the bladder.

[4]BCG (Bacillus Calmette-Guerin): the BCG therapy is a non-specific treatment of localized, non-invasive cancer of the bladder. BCG solution contains alive but little virulent bacteria which have the effect of stimulating the immune system to destroy cancer cells in the bladder.

[5]Cystectomy: examination of the interior of the bladder using an optical system.

[6]http://www.cs.waikato.ac.nz/ml/index.html.

Table 1 Attributes of bladder cancer dataset

Attributes	Meanings
Arrettabac	The smoking habits of the patient
Intage	The age of the patient
Sexe	The sex of the patient
Expprof	Professional exposure
DelaiSymtome	The symptoms onset before the first consultation (months)
NbreTV	The number of bladder tumors
Siege	The seat of the tumor
Obstruct	The obstruction of the bladder at the time of the cystectomy
Desobs	The desobstruction of the bladder at the time of the cystectomy
Aspect	The aspect of the bladder tumor
Base	The base of the bladder tumor
Perfvesi	The perforation of the bladder at the time of the operation
Stade	The stage of the bladder tumor
Grade	The rank of the bladder tumor
Grade3	The rank of tumor is 3
T1g3	The stage of tumor is pT1 and the grade is G3
Cis	The presence or absence of superficial cancer CIS (Carcinome In Situ)
Tabac	Type or amount of tobacco used by the patient
Fibrinol	The tumor contains fibers
Atcdtbc	The antecedents of tuberculosis
Suivi	Monitoring patient from the first day of diagnosis (months)
Nbretum	The number of the bladder tumors
Tailletm	The size of the largest tumor
Recidive	Recurrence of the bladder cancer (after treatments)

3 Association Rules for Classification

Since the presentation of Association Rules (ARs) mining by Agrawal and Strikant in their paper [12], this area remained one of the most active research areas in machine learning and knowledge discovery. ARs mining is a valuable tool that has been used widely in various industries like supermarkets, mail ordering, telemarketing, insurance fraud, and many other applications where finding regularities are targeted. In [12], the authors have presented one of the first algorithms for ARs mining: the Apriori algorithm. The task of ARs mining over a market basket has been described in [12]. Formally, let D be a database of sales, and let $I = i_1, i_2, \ldots, i_m$ be a set of binary literals called items. A transaction T in D contains a set of items called itemset, such that $T \subset I$. Generally, the number of items in an itemset is called length of an itemset. Each itemset is associated with a statistical threshold named support. The support of an AR $X \rightarrow Y$ is the ratio of the tuples which contain the itemsets X and Y to the total number of tuples in the database. The confidence of an AR $X \rightarrow Y$ is

Table 2 Types and values of the attributes of bladder cancer dataset

Attributes	Values
Arrettabac	{non, oui, non-tabagique}
Intage	$\{< 50, > 70, [50 - 70]\}$
Sexe	{homme, femme}
Expprof	{non, oui}
DelaiSymptome	$\{]1 - 2], <= 1, > 3,]2 - 3]\}$
NbreTV	{multiple, unique}
Siege	{meat+col, autres-localisations}
Obstruct	{non, oui}
Desobs	{non, resection-prostate, UI+RE/AP, RE/AP+LB/LV, UI}
Aspect	{coalescente, fine, solide}
Base	{pediculee, large}
Perfvesi	{non, oui}
Stade	{pTa, pT1}
Grade	{G1, G2, G3}
Grade3	{non, oui}
T1g3	{non, oui}
Cis	{non, oui}
Tabac	$\{<30PA$, NEFFA, non-tabagique, $>30PA\}$
Fibrinol	{non, oui}
Atcdtbc	{non, oui}
Suivi	{3, 4, 5, 6, 9, 10, 12, 13, 14, 15, 16, 17, 18, 24, 27, 28, 30, 31, 35, ..., 168}
Nbretum	{1, 2, 3, 4, 5, 6, 7, 8, 9, 10, 11, 12, 15, 20}
Tailletm	{0.5, 0.8, 2, 1, 2.5, 1.5, 7, 10, 4, 5, 3, 9, 6, 8, 15, 12}
Recidive	{non, oui}

the ratio of the tuples which contain the itemsets X and Y to the tuples which contain the itemset X. Given the transactional database D, the AR problem is to find all rules that have a support and confidence greater than the thresholds specified by the user: they are denoted by minsupp and minconf respectively [12, 20]. The problem of generating all ARs from a transactional database can be decomposed into two subproblems [12]:

1. Frequent itemset generation: the objective is to find all the itemsets with support greater than the minsupp. These itemsets are called frequent itemsets.
2. Rule generation: the objective is to extract all the high confidence rules from the frequent itemsets generated in step 1.

Let us consider for example the database shown in Table 3, and let minsupp and minconf be 0.7 and 1.0, respectively. The frequent itemsets in Table 3 are {bread}, {milk}, {juice}, {bread, milk} and {bread, juice}. The Association Rules that pass minconf among these frequent itemsets are $milk \rightarrow bread$ and $juice \rightarrow bread$. For

Table 3 Transaction database [20]

Transaction Id	Item
I1	Bread, milk, juice
I2	Bread, juice, milk
I3	Milk, ice, bread, juice
I4	Bread, eggs, milk
I5	Ice, basket, bread, juice

instance, the support count for {bread, milk} is 4. The total number of transactions is 5, so the support is $4/5 = 0.8$. The confidence of the AR *milk* $\rightarrow$ *bread* is obtained by dividing the support count for {bread, milk} by the support count for {milk}, since there are 4 transactions that contain milk, the confidence for this rule is $4/4 = 1$.

In this article, we focus on ARs to handle a supervised classification problem: it is about the associative classification. Associative classification was introduced in [13]. It uses a special subset of ARs, those whose conclusion is restricted to one of the class modalities. These rules are called class association rules (CARs). They form the classification model and are used to predict the class of a new example.

Among the association classification algorithms, we can quote the most used algorithms namely CBA [13] and CMAR [14]. The CBA[7] algorithm operates in three main steps. First, it discretizes attributes and second, it uses the Apriori approach of Agrawal and Srikant to discover frequent itemsets and generate the Association Rules. Finally a subset of the produced rules is selected according to their accuracy. To predict the class of a new example, the first rule checked by the example is chosen for prediction. CMAR[8] generates and evaluates rules in a similar way as CBA. Their major difference lies in the pruning process and the classification of new examples. CMAR uses a tree structure and a Chi2 test for pruning redundant rules [14]. In this work, we relied on the CBA method, based on Apriori algorithm. In fact, we are only interested in the rule generation phase, common to both CBA and CMAR approaches. For the recognition phase, we propose to take into account the rules uncertainty and to aggregate their confidence degrees (see Sect. 4).

4 A First Proposed Approach: Class Association Rules for Predicting Bladder Cancer Recurrence

In this work, our goal is to help the oncologists to predict the bladder cancer recurrence for hospitalized patients. The classification model must be understandable and easy to interpret by the doctor. This is the reason why we are interested in classification methods by generation of rules. In this section we present an approach based on association rules. Indeed, associative classification has the advantage of being simple to implement and of producing an easily understandable model. The class associa-

[7]CBA: Classification Based on Association.

[8]CMAR: Classification based on Multiple Association Rules.

tions rules help the expert to interpret or explain certain phenomena associated with bladder, such as for example, the co-occurrence of bladder cancer characteristics for the early detection of recurrence. Our objective is also to provide a reliable diagnosis aid tool that is to say with a high sensitivity.

The distribution of the considered data between the two classes is unbalanced. Associative classification may provide an answer to this because it can be used regardless of the class to predict, unlike supervised classification methods such as decision trees, where knowing the class is essential. Indeed we propose to generate ARs on different sub-databases in an independent way. For that, we consider separately the data whose class is *recurrence*, those whose class is *no-recurrence* and finally the whole data (mixed data) i.e. both classes (*recurrence* and *no-recurrence*). This section presents the proposed approach which has two phases: the first is the "generation of classification rules", the second is the "classification" which consists in predicting the class of an example of unknown class using the rules previously generated.

4.1 Generation of Classification Rules

The learning phase consists in generating association rules from the training set (the observations or examples) using CBA method [13]. Thus, a set of rules which express the association between the characteristics of the bladder cancer is generated. As we have already stated, the used training set is unbalanced, the number of *recurrence* examples (186) is much lower than the examples of *no-recurrence* (357). In this case, the generation of rules which conclusion is *recurrence* is almost impossible even with low levels of support and confidence. No rule having the class *recurrence* as a conclusion was obtained during experimental tests realized with various values of support and confidence. Furthermore, errors costs are asymmetric. To remedy this problem, we propose to split the training set into two subsets: a training set containing the 357 *no-recurrence* instances denoted as TS_{no} and another training set for the *recurrence* instances denoted as TS_{yes} and containing 186 observations. The ARs are then extracted separately from each one of these two sub-bases and also from a third base: the complete base denoted as TS_{mixed} (containing all the 543 examples) (see rules generation step in Fig. 1).

After the ARs generation, we obtain a set of classification rules which represents the domain knowledge available to the system. The rules are regrouped according to their origin, i.e. the used training set (see Fig. 1). The set of ARs generated from the TS_{yes} sub-base is denoted as BR_{yes}, their conclusion is *recurrence*. The set of ARs generated from the TS_{no} sub-base is denoted as BR_{no} their conclusion is *no-recurrence*. Finally, the rules generated from the complete base TS_{mixed} constitute the base of rules denoted as BR_{mixed} and their conclusion may be *recurrence* or *no-recurrence*. However, it should be noted that with our training data, the rules generated from all the learning examples TS_{mixed} are always rules whose conclusion is *no-recurrence*.

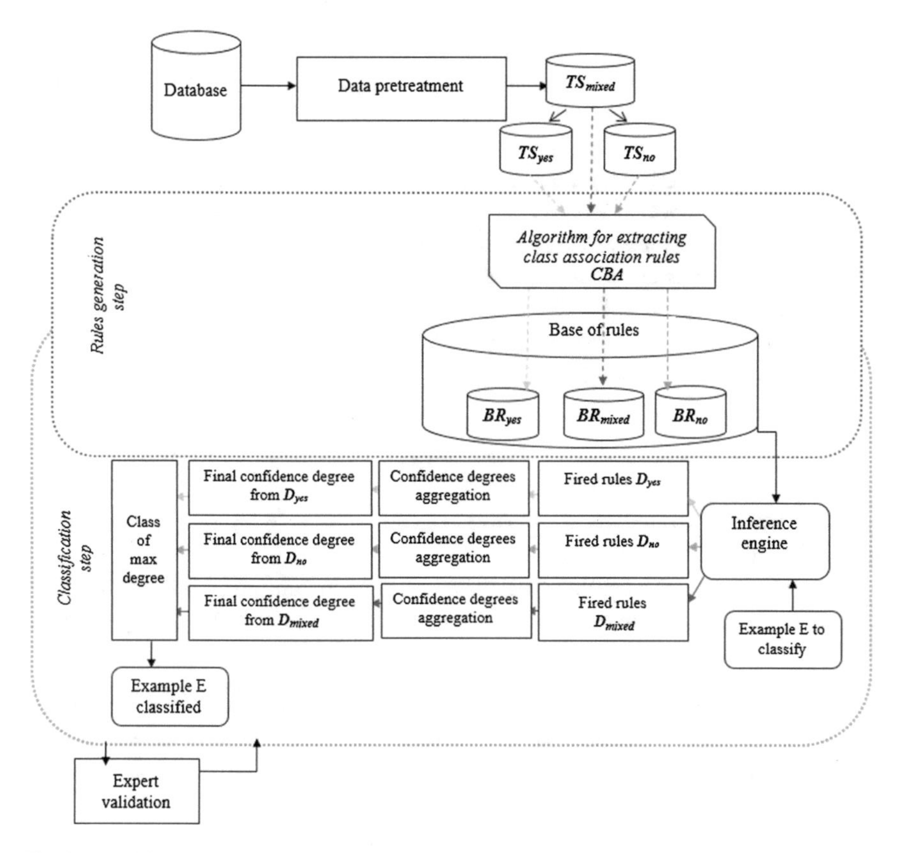

Fig. 1 Step of our classification approach based on Association Rules

4.2 Classification Phase

The classification phase consists in determining the class of a new example using the base of rules generated during the previous phase. An inference engine is used for this purpose. This engine gets as input the rule bases previously generated and a vector (of 23 attributes) describing the new example E to classify. The goal of the inference engine is then to associate a class with this vector, i.e. to predict whether a patient represented by this vector E corresponds to a case of *recurrence* or *no-recurrence*.

In the classification phase, an example can satisfy at the same time the conditions of several rules. The fired rules can conclude to different classes. To decide which class we attribute to the new example, we propose to associate to the rules, during their generation, a confidence degree. Then in the classification phase, these degrees will be aggregated and will allow deciding to which class belongs the new example. These confidence degrees allow taking into account uncertainty of some relationships between patients characteristics. Moreover, their aggregation reflects the fact that several low intensity symptoms can be complementary. Two questions arise: What weight should we associate with the rules? Which aggregation function should we use?

Uncertain knowledge representation and reasoning under uncertainty have been studied in many works [21–23]. In our work, the rules are those generated using CBA algorithm. Two measures are used in this procedure: the support and the confidence. We propose to consider as a confidence degree of a rule $A \rightarrow B$, its confidence measure value denoted as β. This one expresses well the confidence that we have in the conclusion when the premise is verified. Indeed, confidence is the probability of the conclusion knowing the premise:

$$\beta = \frac{P(A \cap B)}{P(A)} \tag{1}$$

Confidence is particularly simple to use as it is estimated for each rule during the generation phase by CBA algorithm.

The next issue is the confidence degrees aggregation during classification phase. During the inference, the rules fired by a new example to classify are regrouped in three subsets according to the origin of their generation:

- The set of fired rules built from TS_{yes} denoted as D_{yes} (their class is *recurrence*).
- The set of fired rules built from TS_{no} denoted as D_{no} (their class is *no-recurrence*).
- And the set of fired rules built from the complete base denoted as D_{mixed} (whose conclusion was *no-recurrence* in all experimental tests).

We have then to combine the degrees of the rules of D_{yes}, those of D_{no} and those of D_{mixed}. Thus, we obtain three final confidence degrees. The class assigned to the new instance is then that of maximum final degree of confidence. The combination of the degrees into a unique value is done via an aggregation function. Many aggregation functions were studied [24, 25]. In this work, we retain the aggregation functions classically used in the case of uncertain rules: triangular conorms (T-conorms) [25].

Specifically, we have to consider the specificities of the considered domain and our objectives. Considering the variety of medical symptoms of bladder cancer, it is possible, for a new example (a new patient) to classify, to trigger a large number of heterogeneous rules[9] each having a low degree of confidence. These rules may represent several symptoms of low intensity but that are complementary and mutually reinforcing. To take into account this phenomenon, it is necessary to use an aggregation method which reinforces the degrees. For this, we propose to use probabilistic T-conorm or T-conorm of Hamacher, both defined below [25]. Both assess the final confidence degree by taking into account all the degrees of the fired rules and reinforcing them.

[9]Heterogeneous rules: rules that do not have the same attributes in their premises.

Probabilistic T-conorm:

$$TCproba(a, b) = a + b - a \times b \quad (2)$$

T-conorm of Hamacher:

$$TC_{Hamacher}(a, b) = \frac{(a + b - 2 \times a \times b)}{(1 - a \times b)} \quad (3)$$

Besides, let us note that the association rules generated from the two sub-training bases TS_{yes} and TS_{no} (containing examples belonging to the same class) have confidence degrees equal to 1: in the conclusion part, the class attribute covers all the examples of the considered sub-base (the sub-base containing only the *recurrence* examples or the one containing only the *no-recurrence* examples). In the case of the base TS_{mixed} that contains all the examples (of both classes), the situation is different, and the confidence degree is not necessarily equal to 1. In order to balance these rules with those generated from the two pure sub-bases (TS_{yes} and TS_{no}), we use a weakening function: the idea is to penalize the rules derived from the two pure sub-bases before carrying degrees aggregation. We propose to use the following weakening function for the rules degrees obtained from the two pure sub-bases TS_{yes} and TS_{no}.

$$Weak(a) = a \times 0.8 \quad (4)$$

This choice is explained by the experimental results. Indeed, the classification rules generated from TS_{mixed} have confidence degrees of the order of 0.8.

In Fig. 1, we recapitulate the steps of this first proposed approach based on AR.

5 Using Decision Trees for Predicting Bladder Cancer Recurrence

In this section, we describe how we use the well known decision trees to predict bladder cancer recurrence. Decision tree is a very popular machine learning method due to its simplicity and transparency. It is a tool easily understandable and easily explainable [5, 26, 27]. For example, in medicine, in financial engineering or in pattern recognition, using decision trees on a given database enables the construction of a rule base to help making a decision on the given domains.

A decision tree is an arrangement of tests that provides an appropriate classification at every step in an analysis [26]. More specifically, decision trees classify instances by sorting them down the tree from the root node to some leaf node, which provides the classification of the instance. Each node in the tree specifies a test of some attribute of the instance, and each branch descending from that node corresponds to one of the possible values for this attribute. One of the most well known decision tree algorithms is ID3 and its successor C4.5 [11, 27]. These decision tree induction algorithms grow the tree, starting from a single parent node which contains a set of data, by selecting an attribute from among a set of candidate attributes at each node. ID3 uses a simple notion of information content while C4.5 attempts to overcome the bias of uneven sampling by normalizing across attributes.

5.1 Generation of Rules

In our work, we use the induction algorithm C4.5 for building a decision tree from the bladder cancer dataset. It is to note that the whole training set is considered, with both cases (classes): *recurrence* and *no-recurrence*.

The final output of the learning phase is a decision tree classifier. Figure 2 presents an example of a decision tree built from the bladder cancer dataset. Each path from the root of a decision tree to one of its leaves can be transformed into a rule simply by conjoining the tests along the path to form the antecedent part, and taking the leaf's class prediction as the class value. For example, one of the paths in Fig. 2 can be transformed into the following rule:

IF *Nbretum* less or equal 4.8 and *t1g3* is non **THEN** the bladder cancer is *no-recurrence*.

After using The C4.5 algorithm, we retranscribe the obtained decision tree into production rules. There are three reasons for such a transformation. First, production rules are a widely used and well-understood vehicle for representing knowledge in knowledge-based systems [5, 7, 28, 29]. Secondly, a decision tree can be difficult for a human expert to understand and modify, whereas the extreme modularity of production rules makes them relatively transparent. Finally, and most importantly, this transformation can improve classification performance by eliminating tests in the decision tree attributable to peculiarities of the training set, and by making it possible to combine different methods for the same task [28]. In the next part we will show how to combine the decision trees and the association rules for predicting the bladder cancer.

We remind that each Association Rule (AR) extracted from the bladder cancer dataset has a confidence degree, and that these degrees are aggregated during the classification phase and allow deciding to which class (*recurrence* or *no-recurrence*) belongs the new example (new patient). As ARs and rules obtained from decision trees will be combined, we also need to represent the uncertainty of these last, and to associate to each rule derived from the decision tree a confidence degree. So we associate with each rule generated using C4.5 algorithm a confidence degree denoted β. This confidence degree reflects the confidence given to the conclusion, if the premises are checked. This confidence estimates the rate of correctly classified instances using the relevant rules, a test set of instances is used for that purpose. β is calculated as follows:

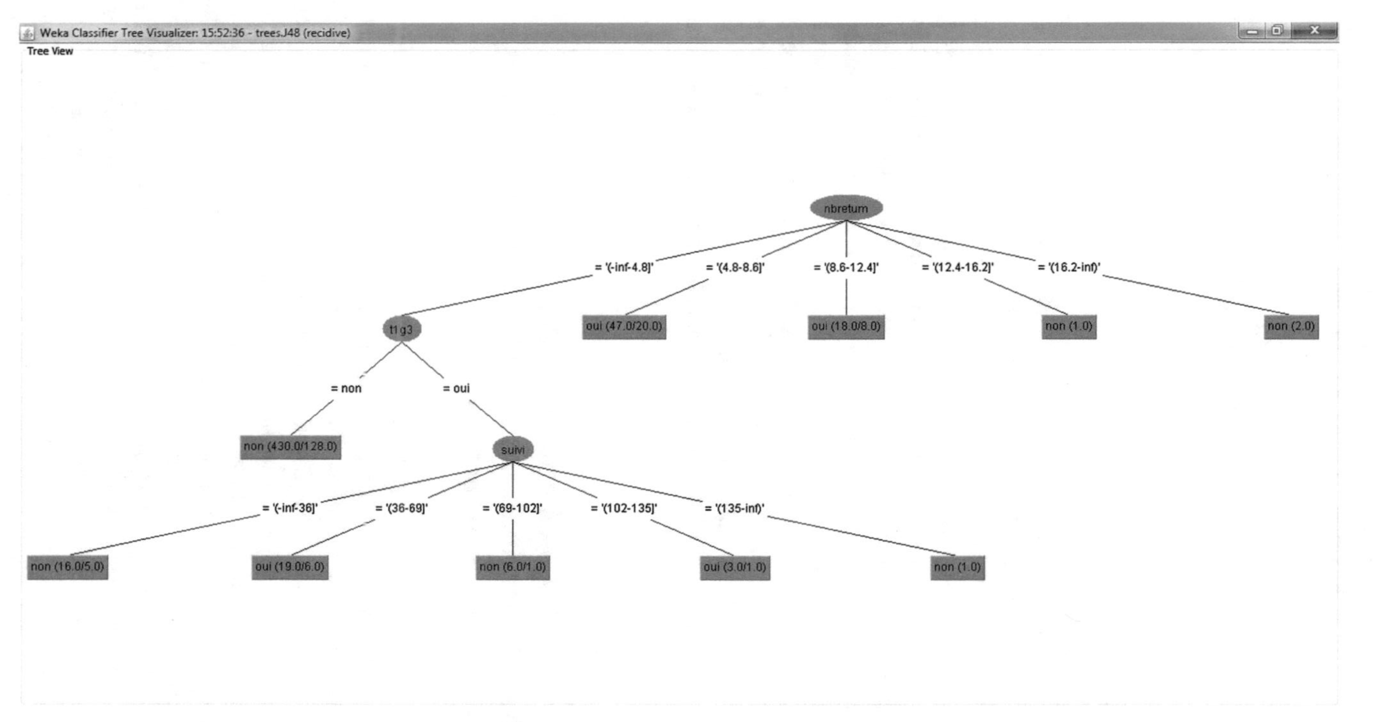

Fig. 2 Example of a decision tree built from the bladder cancer dataset

- NCCI: Number of Correctly Classified Instances.
- NICI: Number of Incorrectly Classified Instances.

$$\beta = \frac{NCCI}{NCCI + NICI} \quad (5)$$

5.2 *Classification Phase*

The classification phase consists in using the base of rules obtained by the decision tree during the previous phase in order to classify new observations (new patients), in other words to determine if a new patient presents a recurrence or not. A classical inference engine is used for that purpose. It gets as input the rule bases previously generated and a vector (of 23 attributes) describing the new example E to classify. As output, it provides a class associated with this vector, i.e. it predicts whether a patient represented by this vector E corresponds to a case of *recurrence* or *no-recurrence*.

In the case of the rules obtained from a decision tree, a new example fires only one rule, contrary to the approach based on the ARs. So here we do not need to use the confidence degrees neither to aggregate them. The class associated to the new observation is simply the one of the fired rule.

In Fig. 3, we recapitulate the steps of the bladder cancer classification based on decision trees.

6 A Second Proposed Approach: Combination of Association Rules and Decision Trees for Predicting Bladder Cancer Recurrence

In order to take the advantage of both methods presented above, we propose in this section to combine decision trees and association rules. In the literature, several approaches treat the combination of decision trees and association rules (ARs) for classification. For example in [29], the authors propose an innovative approach to forecast cross selling opportunities in telecommunications industry based on combination of decision trees and association rules. Thus, this approach involves two data mining tools: the decision tree, to discover the characteristics of WAP (Wireless Application Protocol) users and the ARs to identify new services related to WAP. The results showed that this approach can improve the accuracy rate of forecasting and help telecommunications sellers to make effective cross-sell policy. Recently in 2014, in [30] a new hybrid algorithm titled EDPI (Epidemic Disease Prediction and Identification) was proposed. It is a combination of decision tree and AR used to predict an epidemic disease in some areas. On the theoretical level the authors created a non-linear data mining model using decision tree algorithm. The dependency

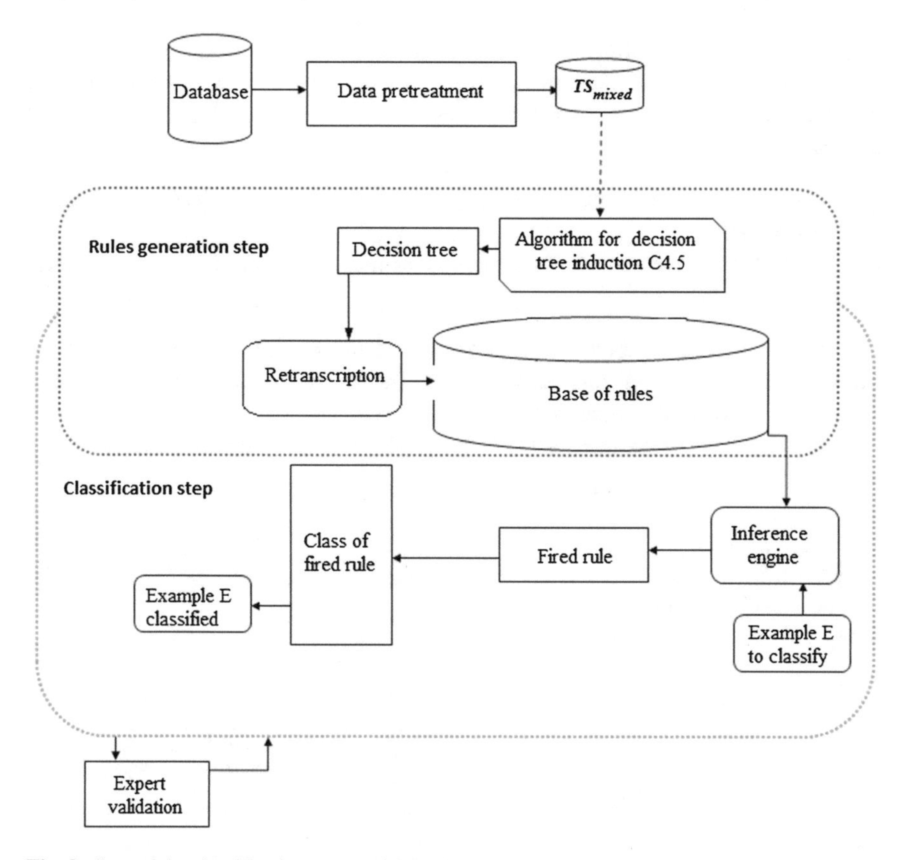

Fig. 3 Step of the classification approach based on decision tree

between various attributes was found using the Apriori algorithm. Also in a medical domain, in 2010 the authors of [31] have proposed a hybrid approach of AR mining and decision tree algorithm to classify brain cancer images. More precisely, the association rules were used to extract the frequent items from the brain cancer images, and the decision tree were used to classify the medical images for diagnosis into three types: normal, benign and malignant. The experimental results showed that the proposed classifier allowed improving the sensitivity and the specificity compared with the well known classifier C4.5 and the associative classifier.

In our work, the combined methods (decision tree and ARs) use the same data set contrary to [29, 31]. Thus, during the generation phase, we use both approaches: classification based on ARs and classification based on decision tree. Thus, we obtain a base of rules generated from decision tree and a base of ARs extracted thanks to CBA algorithm. These two rule bases are merged into a single one.

During the classification phase, an inference engine uses this whole base to classify a new example. Moreover, it manages the uncertainty of the rules. The rules fired by a new example E to classify are splitted into two subsets according to the class of their conclusion:

- The set of fired rules which conclusion is *recurrence*, denoted as ND_{yes} (their class is *recurrence*).
- The set of fired rules which conclusion is *no-recurrence*, denoted as ND_{no} (their class is *no-recurrence*).

We have then to combine the degrees of the rules of ND_{yes} and those of ND_{no}. Thus, we obtain two final confidence degrees. The class assigned to the new instance is then that of maximum final degree of confidence.

In Fig. 4, we recapitulate the steps of our proposed approach based on association rules and decision tree.

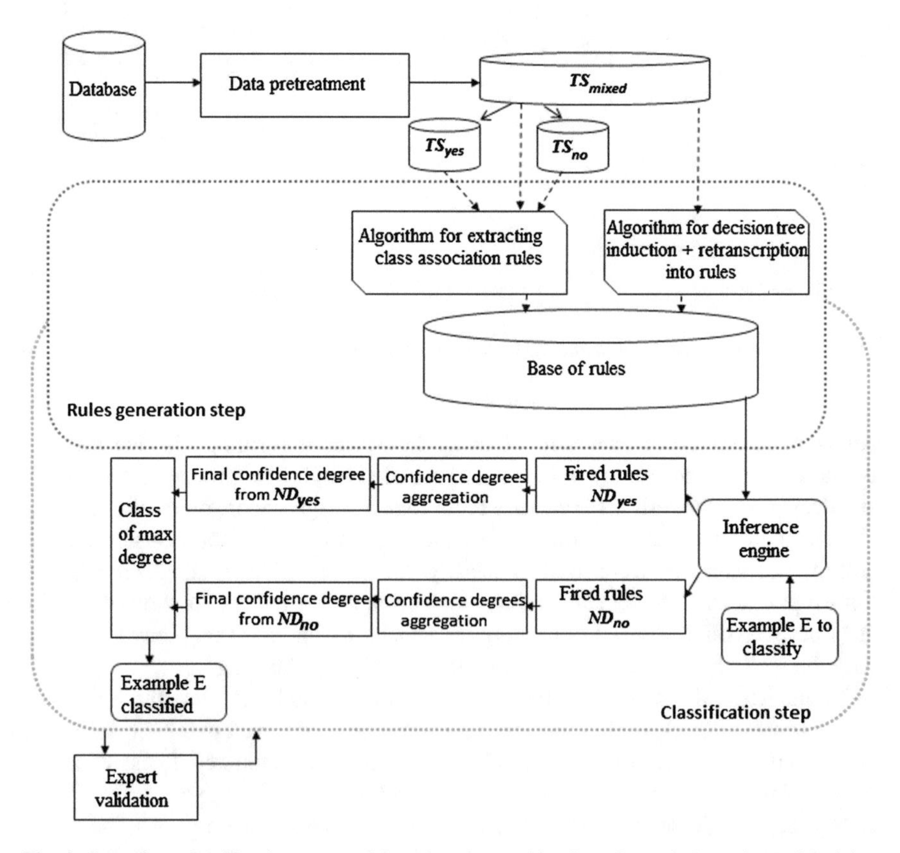

Fig. 4 Step of our classification approach based on the combination of association rules and decision tree

7 Experimental Results

We have evaluated the three presented classification methods with the bladder cancer data:

- The well-known decision tree approach (Fig. 3),
- The proposed associative classification approach (Fig. 1),
- And the proposed approach based on the combination of Association Rules and decision tree (Fig. 4).

We used for that the percentage split method, 70 % of the data for training sets and 30 % for testing sets [19]. The performance evaluation of a classifier is usually done using good classification rates. This criterion is insufficient when the classes are not balanced, as in our problem, or in some areas, such as medical diagnosis. In our application, it is very important to evaluate the rate of false negatives (FN rate), i.e. the proportion of patients that are classified as *no-recurrence* cases by the system while they correspond to *recurrence* cases: it is clear that the consequences of such a decision can be serious.

The ratios traditionally used in information retrieval systems are, for each class, TP[10] (sensitivity), TN[11] (specificity), FP[12] and FN.[13] In our study, we use the following measures:

- Good classification rates: to evaluate the performance of the learning system.
- TP rate: to evaluate the performance for the class *recurrence*.
- TN rate: to evaluate the performance for the class *no-recurrence*.

Let us remind that our objective is to minimize the false negative rate (FN), it means maximizing the sensitivity, i.e. TP rate ($FN = 1 - TP$).

First, we applied a certain number of conventional supervised methods on our data, including decision trees [11], Jrip [32], SMO [33], Simple Naïve Bayes [34], IBK and Part [35]. We used for that the platform WEKA [19]. As we said in Sect. 5, the used decision tree algorithm is C4.5 (its implementation in the platform WEKA is called J48). The main experimental results are given in Table 4. The rate of good classifications is followed by the number of rules (between brackets) obtained when the considered method provides a rule base. The analysis of these results shows that the specificity (TN rate) is high for all the approaches, which means that the corresponding system is efficient for the recognition of *no-recurrence* class (i.e. when a patient does not present recurrence, the model detects this case successfully). However, the sensitivity (TP rate) is low, which means that the system makes bad recognition of *recurrence* cases (i.e. certain patients with a tumor that recurs are detected by the model as cases with *no-recurrence*), this can obviously generates

---

[10]TP: True Positive rate.

[11]TN: True Negative rate.

[12]FP: False Positive rate.

[13]FN: False Negative rate.

Table 4 Comparison of classification results

Classification methods	Good classification rates	TN rate (%)	TP rate (%)
J48 (decision tree with pruning)	63.13 % (44)	91.2	16
J48 (decision tree without pruning)	53.91 % (344)	66.2	33.3
Naive Bayes Simple	61.29 %	80.9	28.4
JRip	62.67 % (2)	92.6	12.3
SMO	64.51 %	89	23.5
IBK (avec K = 1)	61.29 %	76.5	35.8
Part	53.46 % (51)	66.9	30.9
Association rules' approach	51.07 % (150)	38.53	68.83
Combination of AR and decision tree	55.25 % (194)	70.03	26.88

a major risk to the health of the patient. For example the approach based on the decision trees (J48) with pruning gives a good classification rates of 63.13 %, a low TP rate equal to 16 % and a high TN rate of 91.2 %.

The experimental results obtained with both methods we proposed are given in the last two lines of Table 4. They were realized with probabilistic T-conorm (T-conorm of Hamacher gives substantially identical results) and with the following support and confidence thresholds values:

- minsupp = 0.8, minconf = 0.8, for TS_{yes}
- minsupp = 0.9, minconf = 1, for TS_{no}
- minsupp = 0.1, minconf = 0.8, for TS_{mixed}

These values were set to generate 50 Association Rules (ARs) from each training set during the learning phase (extraction of ARs). So the total number of the extracted ARs is 150.

These experimental results obtained with the approach based on ARs show that the proposed model has significantly improved, compared to all other tested approaches, the ability to detect *recurrence* cases with approximately 69 % of success. This improvement varies from 48 to 82 %. This sensitivity improvement (TP rate) comes along with a specificity decrease (TN rate). As for the good classification rates, they decrease compared to four other approaches and are of the same order of magnitude as two others (J48 without pruning and Part).

Sensitivity is the most important criterion in our study. Moreover, one of our goals was to have a satisfying recognition performance with a low rate of false negative (FN) that is to say, a high rate of true positives (TP). The decrease in specificity, i.e. *no-recurrence* classified as *recurrence*, is not a problem as far as it has no effect on the health of patients suffering from bladder cancer.

Let us analyze now the experimental results obtained with the method based on the combination of ARs and decision tree. In this case, the base of rules contains

194 rules (150 ARs and 44 rules obtained from the decision tree). The obtained classification rate is 55.25 %, it is rather better than the Association Rules' approach (51.07 %) but much less than the decision trees one (63.13 %). We note a high TN rate of 70.03 %, and a TP rate (sensitivity) of 26.88 %. So we note an improvement of the TP rate compared with J48 (decision tree) but we remark an important loss of sensitivity compared with the Association Rules' approach. We can say that the combination of AR and decision trees acts as the conventional supervised methods that we tested, the system is efficient for the recognition of *no-recurrence* class (the specificity is high), but the system makes bad recognition of *recurrence* cases (the sensitivity is low).

As our objective is to maximize sensitivity, it appears that the best approach to predict the bladder cancer recurrence is the one based on Association Rules (Fig. 1). Regarding the number of rules, this last method generates 150 class association rules. All these rules have been used for the classification of a new observation (classification phase). However from the viewpoint of knowledge extraction, this number of rules is relatively high for a quick and thorough analysis by experts. Nevertheless, only the rules whose conclusion is recurrence really interest the experts, and in our case their number is 50. They were analyzed by experts, in this particular case a doctor and a biologist.

We report here and synthesize their main conclusions. Improvement of screening, of prevention and of patients monitoring, pass first of all by the increase of knowledge on the diseases, including their epidemiological and etiological aspects and the cohort study. However, the extraction of value-added information may require new approaches. Thus the use of association rules to extract relations between attributes on the considered database allowed to confirm some attributes associations already described in the literature and their incrimination in bladder cancer recidivism [36, 37]. In addition, the class association rules have led to new assumptions that are potentially very consistent and defensible. For example, it emerges from the extracted association rules that patients having the attribute *Antecedent of tuberculosis* would be more likely to have cancer recurrence, these patients represent 12 % of the database. The hypothesis advanced by the experts is the following: these patients have a genetic predisposition that makes them resistant to BCG treatments (they have an antecedent of tuberculosis despite BCG vaccination) and will be more susceptible to the recurrence (despite BCG therapy). Let us also quote the example of the attributes *Monitoring time*, *Operative complication*, *Tumor base* and *Tumor aspect* which are correlated and strongly associated with *recurrence*.

8 Conclusion

In this paper, we propose two diagnosis systems for predicting bladder cancer recurrence. Both are based on a supervised classification approach using Association Rules (ARs), but the first one only uses the ARs and the second one combines ARs with decision tree: the original base of ARs is enriched by the rules generated from a decision tree.

In both cases, to extract the ARs, we treat separately the positive examples (*recurrence*) and the negative examples (*no-recurrence*) and then during the classification phase we aggregate the confidence degrees of the fired rules in order to make the final decision for predicting the class. One of our goals was to provide a system with high sensitivity (i.e. the number of undetected *recurrence* cases is low). The experimental results of the approach based on the combination of ARs and decision tree are similar to the conventional supervised methods that we tested. Even if it is good in terms of TN rate (70.03 %), the sensitivity remains low (26.88 %). However, the results obtained with the Association Rules' approach (associative classification) are very satisfactory. The obtained TP rate (sensitivity) is nearly 69 %. It significantly improves the rates obtained with several other approaches. Furthermore, the results were validated by experts. Some rules are relevant and helpful to better understand the relationship between certain attributes, in case of bladder cancer recurrence. Some rules have confirmed known results and others have revealed interesting association between certain attributes ignored so far. Several perspectives of this work, both theoretical and practical, are possible. For instance, it would be interesting to use other associative classification approaches as well as other rule based classification methods. It would also be interesting to study a more detailed prediction regarding the time of bladder cancer recurrence (cases of tumor recurrence after 3 months, 9 months 1 year or 5 years).

References

1. Greenlee, R.T., Hill-Harmon, M.B., Murray, T., Thun, M.: Cancer statistics, 2001. CA Cancer J. Clin. **51**, 15–36 (2001)
2. Millan-Rodriguez, F., Chechile-Tonolio, G., Salvador-Bayarri, J., Palou, J., Vicente Rodriguez, J.: Multivariate analysis of the prognostic factors of primary superficial bladder cancer (2000). J. Urology. **163**, 73–78
3. Maumus, S., Napoli, A., Szathmary, L., Visvikis-Siest, S.: Fouille de données biomédicales complexes : extraction de règles et de profils génétiques dans le cadre de l'étude du syndrome métabolique. In: Perrire, G., Gunoche, A., Gourgeon, C. (eds.) Journées Ouvertes Biologie Informatique Mathématiques JOBIM 2005, pp. 169–173. Lyon, France (2005)
4. Ming-Yih, L., Chi-Shih, Y.: Entropy-based feature extraction and decision tree introduction for breast cancer diagnostic with standardized thermography images. Comput. Methods Programs Biomed. **100**, 269–282 (2010)
5. Bouchon-Meunier, B., Marsala, C.: Fuzzy decision trees and databases. In: Andreasen, T., Christiansen, H., Legind-Larsen, L. (eds.) Flexible Query Answering Systems, pp. 277–288. Kluwer Academic, Berlin (1997)
6. Bersini,H., Bontempi, G., Birattari, M.: Is readability compatible with accuracy? From neuro-fuzzy to lazy learning. In: Brauer, W. (eds.) Fuzzy-Neuro Systems'98. Computational Intelligence: Proceedings of the 5th International Workshop on Fuzzy-Neuro Systems, pp. 10–25. IOS Press (1998)
7. Duch, W., Setiono, R., Zurada, J.M.: Computational intelligence methods for rule-based data understanding. IEEE (2004). doi:10.1109/JPROC.2004.826605

8. Soua, B., Borgi, A., Tagina, M.: An ensemble method for fuzzy rule-based classification systems. Knowl. Inf. Syst. **36**, 385–410 (2013)
9. Haton, J.-P., Bouzid, N., Charpillet, F., Haton, M., Lasri, H., Marquis, P., Mondot, T., Napoli, A.: Le raisonnement en intelligence artificielle. InterEditions, Paris (1991)
10. Ounallah, S.: Fouille de données pour l'aide au diagnostic médical. Master Report, Institut Supérieur d'Informatique, Université de Tunis El Manar (December 2014)
11. Quinlan, J.R.: C4.5 : Programs for Machine Learning. Morgan Kaufmann, San Francisco (1993)
12. Agrawal, R., Strikant, R.: Fast algorithmes for mining association rules. In: Proceedings of the 20th International Conference on Very Large Data Bases VLDB '94, pp. 487–499. Morgan Kaufmann (1994)
13. Bing, L., Wynne, H., Yiming, M.: Integrating classification and association rule mining. In: Proceeding of the 4th International Conference on Knowledge Discovery and Data Mining KDD'98, pp. 80–86. AAAI Press (1998)
14. Wenmin, L., Jiawei, H., Jian, P.: CMAR: accurate and efficient classification based on multiple class-association rules. In: Proceedings of the 2001 IEEE International Conference on Data Mining ICDM '01, pp. 369–376. IEEE Computer Society (2001)
15. Exarchos, T.P., Papaloukas, C., Fotiadis, D.I., Michalis, L.K.: An association rule mining-based methodology for automated detection of ischemic EGC beats. IEEE Trans. Biomed. Eng. **53**, 1531–1540 (2006)
16. Dwarakanath, B., Ramesh, K.: Classe association rules based feature selection for diagnosis of Dravet syndrome. Int. J. Sci. Res. **3**, 1670–1673 (2014)
17. Sellami, A., Boudawara, T.S., Hsairi, M., Jlidi, R., Achour, N.: Registre du cancer du sud Tunisien : incidence des cancers dans le gouvernorat de Sfax 2000-2002. In: Data and Statistics. World Health Organisation (2007) Available via DIALOG. http://www.emro.who.int/. Accessed 03 Jul 2014
18. Pasin, E., Josephson, D.Y., Mitra, A.P., Cote, R.J., Stein, J.P.: Superficial bladder cancer: an update on etiology, molecular development, classification, and natural history. Urology **10**, 31–43 (2008)
19. Remco, R.B., Eibe, F., Mark, H., Richard, K., Peter, R., Alex, S., David, S.: WEKA Manual for Version 3-6-10. University of Waikato, New Zealand (2013)
20. Mustafa Nofal, A.A.D., Bani-Ahmad, S.: Classification based on association-rule mining techniques: a general survey and empirical comparative evaluation. Ubiquitous Comput. Commun. J. **5**, 9–17 (2010)
21. Motro, A.: Imprecision and uncertainty in databade systems. In: Bosc, P., Kacprzyk, J. (eds.) Fuzziness in Database Managment Systems, pp. 3–22. Physica-Verlag HD, Heidelberg (1995)
22. Bonnal, P., Gourc, D., Locaste, G.: Where do we stand with fuzzy project scheduling? J. Constr. Eng. Manag. **130**, 114–123 (2004)
23. Lin-li, Z., Xing-kui, H., Guo-hong, D.: Research on uncertain knowledge representation and processing in the expert system. In: 2011 Fourth International Symposium on Knowledge Acquisition and Modeling (KAM), pp. 277–281. IEEE (2011)
24. Dubois, D., Prade, H.: A review of fuzzy set aggregation connectives. Inf. Sci. **36**, 65–121 (1985)
25. Weber, S.: A general concept of fuzzy connectives, negations and implications based on t-norms and t-conorms. Fuzzy Sets Syst. **11**, 103–113 (1983)
26. Mitchell, T.M.: Machine Learning. McGraw-Hill, New York (1997)
27. Quinlan, J.R.: Induction of Decision Trees. Mach. Learn. **1**, 81–106 (1986)
28. Quinlan, J.R.: Generating production rules from decision trees. In: Proceedings of the 10th International Joint Conference on Artificial Intelligence IJCAI'87, pp. 304–307. Morgan Kaufmann (1987)
29. Xue-cheng, Y., Jun, W., Xiao-hang, Z., Ting, J.L.: Using decision tree and association rules to predict cross selling opportunities. In: Proceedings of the Seventh International Conference on Machine Learning and Cybernetics, pp. 1807–1811. IEEE (2008)
30. Kameswara R, N.K., Saradhi V, G.P.: A hybrid Algorithm for Epidemic Disease Prediction with Multi Dimensional Data. Int. J. Adv. Res. Comput. Sci. Softw. Eng. **4**, 1033–1037 (2014)

31. Rajendran, P., Madheswaran, M.: A hybrid algorithm for epidemic disease prediction with multi dimensional data. CoRR **2**, 127–136 (2010)
32. Cohen, W.: Fast effective rule induction. In: Proceedings of the 12th International Conference on Machine Learning ICML'95, pp. 115–123. Morgan Kaufmann (1995)
33. Platt, J.C.: Fast training of support vector machines using sequential minimal optimization. In: Schlkopf, B., Burges, C., Smola, A. (eds.) Advances in Kernel Methods: Support Vector Learning, pp. 185–208. MIT Press, Cambridge (1998)
34. John, G.H., Langley, P.: Estimating continuous distributions in Bayesian classifiers. In: Proceedings of the Eleventh Conference on Uncertainty in Artificial Intelligence UAI'95, pp. 338–345. Morgan Kaufmann (1995)
35. Eibe, F., Ian, H.W.: Generating accurate rule sets without global optimization. In: Proceedings of the Fifteeth International Conference on Machine Learning ICML'98, pp. 144–151. Morgan Kaufmann (1998)
36. Brennan, P., Bogillot, O., Cordier, S., Greiser, E., Schill, W., Vineis, P., Gonzalo, L.A., Tzonou, A., Jenny, C.C., Ulrich, B.A., Jckel, K.H., Donato, F., Consol, S., Wahrendorf, J., Hours, M., T'Mannetje, A., Kogevinas, M., Boffetta, P.: Cigarette smoking and bladder cancer in men: a pooled analysis of 11 case-control studies. Int. J. Cancer **86**, 289–294 (2000)
37. Ben-Abdelkrim, S., Rammeh, S., Trabelsi, A., Ben-Yacoub-Abid, L., Ben-Sorba, N., Jadane, L., Mokni, M.: Reproductibilité des classifications OMS 1973 et OMS 2004 des tumeurs urothéliales papillaires de la vessie. Can. Urol. Assoc. J. (2012). doi:10.5489/cuaj.10078

An Unsupervised Text-Mining Approach and a Hybrid Methodology to Improve Early Warnings in Construction Project Management

Mohammed Alsubaey, Ahmad Asadi and Charalampos Makatsoris

Abstract Extracting critical sections from project management documents is a challenging process and an active area of research. Project management documents contain certain early warnings that, if modelled properly, may inform the project planners and managers in advance of any impending risks via early warnings. Extraction of such indicators from documents is termed as text mining, which is an active area of research. In the context of construction project management, extraction of semantically crucial information from documents is a challenging task that can in turn be used to provide decision support by optimising the entire project lifecycle. This research presents a two-step modelling and clustering methodology. It exploits the capability of a Naïve Bayes classifier to extract early warnings from management text data. In the first step, a database corpus is prepared via a qualitative analysis of expertly fed questionnaire responses. In the latter stage a Naïve Bayes classifier is proposed which evaluates real-world construction management documents to identify potential risks based on certain keyword usages. The classifier outcome was compared against labelled test documents and gave an accuracy of 68.02 %, which is better than the majority of text mining algorithms reported in the literature.

Keywords Naïve Bayes classifier · K-nearest neighbours · Project management documents · Knowledge management (KM) · Knowledge discovery in databases (KDD)

The original version of this chapter was revised. An erratum to this chapter can be found at DOI 10.1007/978-3-319-33386-1_22.

M. Alsubaey · A. Asadi
Department of Mechanical, Aerospace and Civil Engineering,
Brunel University, London, UK
e-mail: alsubaey.mohammed@gmail.com

A. Asadi
e-mail: AhmadAsadi@outlook.com

C. Makatsoris (✉)
Sustainable Manufacturing Systems Centre, School of Aerospace, Transportation and Manufacturing, Cranfield University, Bedfordshire MK43 0AL, UK
e-mail: h.makatsoris@cranfield.ac.uk

Y. Bi et al. (eds.), *Intelligent Systems and Applications*,
Studies in Computational Intelligence 650, DOI 10.1007/978-3-319-33386-1_4

1 Introduction

Exploratory text mining in unstructured project management documents is a process used to extract novel knowledge [45]. In the project management domain, data and text mining techniques have long been used to extract useful such novel information in order to advance the state-of-knowledge of the underlying project's management lifecycle. Information in critical project management documents are frequently been analysed and modelled to extract early warnings in the domains of fraud prevention [60], email/web analytics [15], financial analysis [32], and socio-political impact analyses [58].

Extending unstructured data into useful information has routinely been modelled via various statistical machine-learning techniques. Techniques such as Support Vector Machines, Artificial Neural Networks, Fuzzy Inference and Markov Chains are used to train text-mining models via labelled datasets [7, 11, 28, 31, 36, 42, 57]. These techniques are used to predict and classify embedded information about the document or the relevant project based upon previously trained and labelled datasets. Yet, statistical machine learning techniques often fail to model and extract information from morphological structure of the text. For instance, it is not always possible to gain a word-level understanding of unstructured data as sentence-level structuring provides a deeper understanding of the message encoded in the document. However, at the same time, discrete word-pairs do contain crucial information particularly when semi-structured documents such as memos and meeting Minutes are concerned. For instance, a past-date of three weeks ago specified as a word-column pair "Delivery—23/10/2015" indicates a delay of three weeks if the current date is 15/11/2015. However, memos can still have statements such as "the procurement is still not done due to the required parts not being delivered" which cannot only be understood on word-level and a temporal modelling must be considered.

Based on the above mentioned limitation, this research presents a hybrid text classification methodology with a focus on text mining from construction project management documents. In corporate sector organisations, text mining is generally used to improve decision support on organisation activities and improved understanding of processes. The phenomenon is used to improve the understanding of Knowledge Management (KM) systems in order to help professional improve their skills and competencies [55]. Particularly, any project management documents similar to Minutes tend to contain information that possess critical information about how well the project is going-on. However, the documents tend to contain associated data-pairs and text-islands holding information that may be exploited to identify certain early warnings with a degree of confidence.

This paper proposes a novel, hybrid early-warning-sign classification methodology to predict project failure likelihood. The aim of this research is to improve the extraction of project early warnings via semi-structured data-pairs and information islands. In order to achieve this, the technique first presents a qualitative approach to label project management document data for model training. The methodology then extends to a supervised Naïve-Bayes classifier to predict early warnings over labelled data extracted from the qualitative research baseline. The paper is organised as follows: Sect. 2 presents a critical overview of the existing state-of-the-art, Sect. 3 presents the research methodology including the models and training mechanisms, Sect. 4 analyses the research findings and their comparative analysis with the existing work. Section 5 finally concludes with the results and future directions.

2 Research Overview

2.1 Text Mining and Project Failure Information Extraction from Management Documents

A large fraction of companies tend to miss their planned goals due to issues arising from budget and schedule overruns [38, 53]. A large number of projects fail completely due to the inability of their management to realise hidden problems and risks within the projects' trajectory [6]. The problems mainly identified in such failures primarily include poor project scopes, abrupt scope changes, poor budgeting, inefficient planning and/or scheduling and poor selection of management teams [50]. According to [26], inadequate project support, improper risk management, poor stakeholder consideration and inappropriately selected contractors and engineering firms also play a major role in project failures. With the automation and electronic-control of project management with the advent of information and communication technologies (ICT), knowledge extraction for risks and impending project failures have taken a completely different aspect commonly regarding as information retrieval, text and data mining. The domain uses advanced text and information processing algorithms to gain novel information about the project and is commonly known as data and/or text mining [20].

2.2 Data Mining Techniques in Project Management

Delay has significant effect on completion cost and time of construction projects. Delays can be minimized when their causes are identified. Knowing the cause of any particular delay in a construction project would help avoiding the same. Although various studies have been undertaken to identify the factors affecting the causes of delays, since the problems are rather contextual, the studies need to focus on specific geographical area, country, or region. For instance, a major criticism of the Middle East construction industry is due to the growing rate of delays in projects delivery.

A study in London [3] presented new project delay prediction model with the use of data mining.

The researchers identified the causes of construction delays as seen by clients, consultants, and contractors, and then combined the top delay factors with secondary data collected from an on-going project and built a model using Weka software to predict delays for similar projects. Their research has explored different models, various classification methods and attained interesting results on predictability of project delays by measuring the performance of the models using real data. The results for two classification algorithms showed an average test accuracy of 76 % when classifying delays from a combination of real-project monitored data and one of the top delay factors, for 34 months duration.

The best algorithm for the data used from construction project is J48 classifier with accuracy of 79.41 %. Naïve Bayes classifier has lowest accuracy and higher error rate compared to J48. Despite the small differences in readings obtained for the two classifiers, the results suggested that among the machine-learning algorithm tested, J48 classifier has the potential to significantly improve the conventional classification methods for use in construction field.

Data mining offers promising ways to expose hidden patterns within a large amount of data, which can be used to predict future behaviours. The results achieved for project delay predictions would help managers to categorize risks associated with execution stages allowing them to identify delays before they happen and their main causes rather than minimize the consequences as well as measure the progress to success. Establishing predictions and thus assumptions engage managers to chase specific information and more importantly define their value, which in turn help to determine the areas of concerns in which the managers should be willing to improve.

2.3 Information Systems and Knowledge Management

Project management documents tend to contain both structured and unstructured data, which are tightly correlated and embedded in project management documents. As the computer-based knowledge management is becoming a norm, proper organisation and control of such documents is becoming a challenge. Moreover, in the present day age of information systems modelling, extracting useful information from such documents is an active area of research [41].

In the context of construction project management, data and knowledge generally derive from a number of interrelated elements that include human resources, cost, materials, planning and design. A combination of these factors makes it difficult to assess and analyse any risk factors within project plans. Kim et al. [34] proposed a technique commonly regarded as the Knowledge Discovery in Databases (KDD) to select a set of these processes that have the most impact on the performance of the on-going process. The methodology used a Back Propagation Neural Network algorithm to train against a set of databases and relevant classification labels in order to train a model to identify the cause of project delays. The main technique was compared with different machine learning algorithms including Bayesian network, Correlation matrix and Factor analysis and reported the Back Propagation training approach to have the best identification accuracy.

Table 1 A summary of various AI approaches used in customer-oriented project management activities based on work done by [43]

Project managements aspect mapping	Sub-areas focussed	Algorithms and/or models employed
Early warning identification	Response against early warnings	Ansoff's management model [24]
	Accident prevention and safety management	Fault tree analysis [18], nearest neighbour [54], fiber Bragg grating system [17], Pareto analysis [37]
	Information collection, sharing and communication	Hybrid data fusion model, [16], decision tree analysis [14]
	Decision support	Reference [23], Rough set and SVM [46]
Resource utilisation	Multi-skilled staff utilisation	Process integration [5], knowledge discovery processes [19], 4D modelling [56]
Supply chain control	Multi-project management	Automated resource allocation [47]
Best practice modelling	IT integration, contracting, quality management, problem solving and cost reduction.	Project cost growth via 3DCAD [14], quality management system modelling [29], PPP Best Practice [21], contractor to stake-holder problem solving [25], contractor performance analyser [48]

An in-depth analysis of research done in the domain of project and human factors management was done by [43]. The work presented an analysis of various machine-learning approaches that could be used to model and identify project management activities as summarised in Table 1.

2.4 Application of Machine Learning in Data Mining

Earlier information and data modelling techniques in project management context relied predominantly on human experts who performed manual classification of documents. However, with the massive increase of management-related documents, this has become in infeasible and prone to mistakes. Caldas et al. [9] integrated unstructured information into a machine-learning approach to develop a document classification system. The work utilised a construction project database as a case study and integrated topics extracted from specifications, meeting minutes, information requests, change orders, contracts and field reports to implement the underlying classification system. Extending on this work, extensive research has been done focussing on various sub-areas of automated information extraction in the project management industry. Since this work, machine-learning algorithms have increasingly been investigated and employed in information retrieval and data mining. Work in project management data mining has focussed on the automated classification of various linguistic documents where work done by [9] has presented a SVM algorithm to classify documents into various categories such as schedule, demolition, and conveyance categories.

Artificial neural networks (ANN), support vector machines (SVM), Naïve Bayes and k-nearest neighbour classification approaches have been applied to address a number of data mining problems. Tianyi and Tuzhilin [52] investigated various clustering approaches to information segmentation. Moreover, [13] investigated SVM whereas fuzzy logic was employed to model uncertainties via rule-based systems in information modelling by [39]. Ko and Cheng [35] used a fuzzy logic, neural networks and genetic algorithms based hybrid machine learning approach to continuously assess project performance and dynamically select factors that influence project success. Data mining systems have also been extensively used in customer service support [27], concept relation extraction [1], data mining in course management [51], document integration via model-based information systems [10] and text mining for topic identification and clustering in patent searching algorithms [54].

Based on the existing research state discussed above and a review work by [33], the overall text-mining research at present focuses on the following areas:

- Improvement and exploration of feature selection methods.
- Reduction of training and testing time of classifiers.
- Filtering specific keywords to detect early warnings.
- Use of semantics and ontologies for the classification of documents.
- Extracting meaningful information such as factors and trends in business, marketing and financial trends in project management documents.

Despite this work, extracting terms, textual landmarks and hidden warnings from text is still an active area of research facing numerous challenges.

2.5 Extracting Early Warning Signals from Project Management Texts

It is generally easier to identify clearly defined statements in project statements such as "project failure" or "ceased cash flow" compared to hidden/indirect signals such as missing milestone dates or staff absence. For instance, work done by [4] gives an example of the petroleum crisis of 1970s where the large oil firms were caught by surprise despite the advance forecast reports present on the managers' desks that were largely overlooked. The work thus forms a baseline of data-mining research in project management where, timely analysis of early warnings contained within planning and design documents is critical. Reference [44] contributed major research in the area of early warning where his work reported and analysed a wide range of problems within project management that could pre-emptively indicate unfolding issues. The problems or issues investigated in his research included finance, project control, communication, performance, scheduling and delays. The work focussed on Igor Ansoff's theory by analysing dependencies between early warnings, causes of problems, problems and responses. According to the research such information hidden in lines of contracts can be assigned a numerical weight based upon their frequency of occurrence. Similarly, [46] utilised SVM as a structural risk minimisation methodology and Rough Set theory to reduce noise to achieve 87.5 % accuracy in the improvement of a binary early warning identification system.

The basis formed by research presented above and in Table 1. broadly identifies a number of research areas focussing on data mining techniques to gather information from project documents based on factors such as:

- Project performance in the presence of negligence.
- Delay causing factors.
- Training, staff skills and its impact on deliverables.
- Responsibility under set project plans.
- Staff resource backup in case of emergencies.
- Impact of supply chain disturbance.
- Inventory prediction and control.

However, in data mining, the majority of research either focuses on assessing the state of the project or advice on certain "adjustments" which can be made to minimise the identified early warnings.

2.6 *Naïve Bayes Classification in Machine Learning and Data Mining*

It is generally the responsibility of project managers to assess project plans for potential deviations such as delays in milestone achievements, impacts of deliveries of project deliverables and thus meeting deadlines. A lot of time is thus spent searching for valuable information from archived hard copies such as meeting Minutes, memos, and other documents containing unstructured data. According to [41], decisions made on information extracted in this way is prone to human error and may lead to judgmental errors due to the inherent inability of humans to analyse swathes of management document data. Bayesian analysis has routinely been used to extract information and factors from documents based upon trained models. However, the majority of work focuses on human factors or expert knowledge integration to enable classification. Raghuram et al. [49]'s work focussed on retraining the network based on new information where the initial model is constructed via expert knowledge. However, in classification based on retraining, the issue of incomplete data has always been a challenge. Gunasinghe and Alahakoon [22] presented a fuzzy ARTMAP and back-propagation-based approach to fill lack of input information. However, Bayesian identification generally identifies discrete keywords based on their probabilistic likelihood to belong to certain classes. This poses a challenging in cases where information is not only hidden in keywords but in complete sentences. For instance, for a sentence found in a meeting Minutes document given below:

Table 2 Examples of Early warning scoring and classification system for identification minutes items

Items from minutes of meeting	Classification	EWSs score
There is a good level of delay in delivery of material	Negative	Low "materials" score
There is a good level of coordination in the delivery of material	Positive	High "scope of material delivery" score

Sentence: There is a good level of delay in delivery of material

If the sentence given above is ranked based on positive, negative and neutral class associations, then words such as good or delivery can be taken as positive. However, if the sentence is taken as a whole, it indicates an early warning into possible delays in the supply chain. This characteristic of text documents presents a time-based nature of words. For instance, sentences with the majority of words being common can still belong to different categories as follows (Table 2):

Table 2 presents a well-known data/text-mining problem where only words cannot be used to completely understand the hidden information within a document. Time-based analysis has commonly been used in this domain where machine-learning classifiers are used to learn from complete statements in documents, which are labelled by relevant field experts. For instance, Markov chains are commonly used to identify "islands of information" in free text [11].

The approach provides a good method to extract useful information from cluttered documents. Hidden Markov models (HMM) have been used in text/data processing for event tracking, text categorisation, and topic discovery and document syntactic features [8, 12, 30, 40, 59]. The latter is commonly used to categorise document into various categories based upon the information present within (for example: email, web-pages, and research articles). Despite providing a capability to extract hidden indications from unstructured data, the majority of these methods still cannot establish unstructured-to-numeric data association.

3 Methodology: Text Parsing via Naive Bayes Text Analysis

This research proposes a novel text analysis methodology based on an expert-labelled text corpus to label and train project Minutes documents. To simplify the latter stages, a brief summary of this methodology is given below:

- **Step 1:** Conduct questionnaire-driven qualitative analysis to extract terms/ keywords indicating early warnings and their association with certain early warning classes

- **Step 2:** Use the corpus obtained in Step 1 to label meeting Minutes documents
- **Step 3:** Based on the ground-truth documents obtained in Step 2, train a Naïve Bayes classifier
- **Step 4:** Divide the labelled data into training and testing sections and evaluate the classifier for various "early warning" classes obtained in Step 1
- **Step 5:** Assess the algorithm outcome based on True Positives (TP) and Negatives TN

3.1 Naïve Bayes Model for Extraction of Early-Warnings

In construction project management, a large number of documents are prepared during projects' lifecycle to track, plan and control the progress of ongoing work. The text data presented in these document contain a cause-and-effect relationship between certain keyword combinations that can be learned and then used to train a model. The model can then mine similar information from unclassified/unlabelled construction documents and identify early warnings hidden within these documents. For instance,

1. The text "unavailability of the required human resources" can indirectly predict the "lack of required human skills" early warning for a project.
2. This early warning can be used to pre-emptively provide additional staff at the early stages of a project to avoid a project delay

Employee absence	Resource shortage	True	False
F	F	0.02	0.98
T	F	0.56	0.44
F	T	0.39	0.61
T	T	0.99	0.01

The technique draws its concept from the baseline Bayesian belief network that identifies various word-to-word relationships. As shown in Fig. 1, for a simple two-event case, the outcome class "Delay in milestone achievement" in a project may be due to two reasons being either the employee being absent or resource being short. Each state will have a different probability to have resulted in the outcome class. The situation can be modelled as a Bayesian network with two probable values of truth attributing to the delay happening and false being not happening. This joint probability function for this case can therefore be defined as:

$$P(\gamma, \delta, \kappa) = P(\gamma|\delta, \kappa) \times P(\delta|\kappa) \times P(\kappa) \tag{1}$$

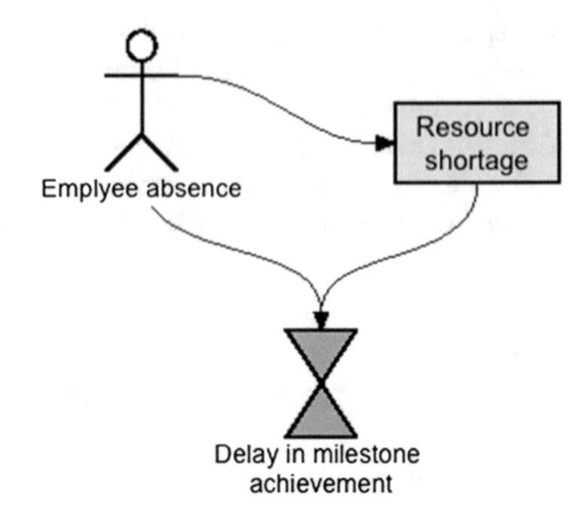

Fig. 1 Bayesian belief network for a simple early warning identification case

where the γ, δ *and* κ stand for delay, employee absence and resource shortage. This model is capable of answering a question:

"What is the probability of delaying a project when as a result of resource shortage?"

Based on the conditional probability formulation, the anatomy of testing a simple statement from a project document for impending project failure can be done as follows:

$$P(A|X) = \frac{P(X|A)P(A)}{P(X|A)P(A) + P(X|not\ A)P(not\ A)} \tag{2}$$

Equation (2) can further be defined as:

$P(X|A)$: Chance of employee absence X in case of a project delayed by 1+ month A (In this case 66 %)

$P(A)$: Chance of employee absence A (In this case 35 %)

$P(not\ A)$: Chance of not having even one employee absent $A\sim$ (In this case 65 %)

$P(X|not\ A)$: Chance of project delay given that you did not have any employee absent (7.3 % in this case) s

$$P(A|X) = \frac{0.66 \times 0.35}{(0.66 \times 0.35) + (0.073 \times 0.65)} = \frac{0.231}{0.7055} = 82.95\,\% \tag{3}$$

Therefore, based on the delays given in row 2 Table 3, the chance of delaying project by 1+ month $P(A|X)$ given employee absence X to be 66 % calculates out to be 82.95 % according to the Naïve Bayes rules. Using this concept, the proposed methodology trains a Naïve Bayes model on the basis of an expertly-fed word corpus against a set of early warning classifications as described in the latter sections.

Table 3 Measures of various (supposed) statistical values depicting defining occurrences of project delays due to various project lifecycle issues

	Project delay of 1+ month (A) (35 %)	No project delay (Not A) (65 %)
Employee absence X: manpower below the stated requirement (%)	66	7.3
Resource shortage κ: urgent materials not in inventory (%)	10	92.6

Table 4 Elaboration of the sample size used for ground-truth training data preparation from user questionnaires

Sample item/answer	Reductions in the quality of the constructed facility
Professional association	Project manager
Experience in field	Up to 20 years

3.1.1 Data Corpus Preparation and Modelling

The model training part of this work forms its basis from information gathered via expert input. An initial text extraction mechanism based on the following three mechanisms (Table 4):

- A generic Naïve Bayes classifier based on document parsing from management documents [2]
- A generic Naïve Bayes classifier based on expert-defined ground-truth keywords/class combination
- An extended Naïve Bayes classifier based on labelled Minutes document database

A detailed data parsing procedure is given below:

3.2 *Step 1: Data Cleansing for Word Association Extraction*

In standard word documents, each word can be associated to a range of meanings depending upon its context. For instance, particular words such as "low", "shortage", "less" can generally be attributed to a situation indicating a negative meaning from a sentence. However, the documents often contain word associations or tuples, which provide better context when used in pairs or rows. Once such example of an item with a description and expected due data is shown in Fig. 2. Where a due-date earlier than the current date for a subject stating word pairs such as "outstanding issues" indicate an impending delay in project goal achievement.

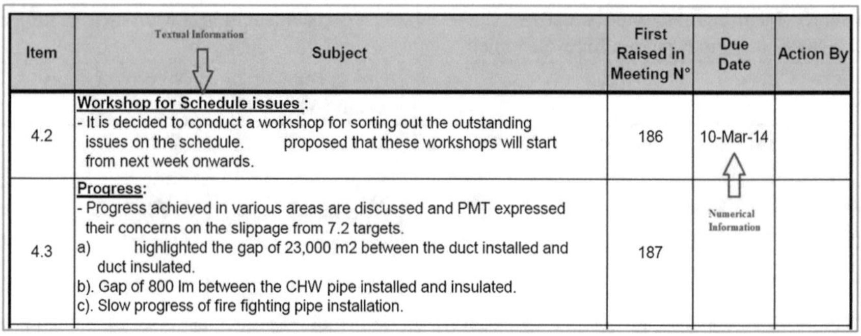

Item	Textual Information ⇩ Subject	First Raised in Meeting N°	Due Date	Action By
4.2	Workshop for Schedule issues : - It is decided to conduct a workshop for sorting out the outstanding issues on the schedule. proposed that these workshops will start from next week onwards.	186	10-Mar-14 ⇧ Numerical Information	
4.3	Progress: - Progress achieved in various areas are discussed and PMT expressed their concerns on the slippage from 7.2 targets. a) highlighted the gap of 23,000 m2 between the duct installed and duct insulated. b). Gap of 800 lm between the CHW pipe installed and insulated. c). Slow progress of fire fighting pipe installation.	187		

Fig. 2 Unstructured data—Difference of textual and numerical information and their complex association in project management documents

Construction project management documents are unique in a sense that the language used is generally different from standard management documents. For instance, the documents use technical terms such as "letter of credit", "bidding contract", "list of requirements" and/or "change orders" which are frequently accompanied with meaningful numerical values such as dates, and quantities which provide additional information on delivery schedules, inventory statistics or other measurement parameters. All these values can be used in both negative and positive capacities and these parameters play a crucial role in such an assessment. For instance, a high number in a "change order" may mean an indication of lack of stable project requirements. This effectively changes the scope to text and numerical pairs to be modelled together as shown in Fig. 2. Where unstructured data from column two must be used in conjunction with numerical information from the Due Date column to provide a better assessment capability.

3.2.1 Term/Word Extraction from Management Documents

The most important step in the model training is to convert unstructured data to structured information. In Fig. 2, row 2, column 3 contains free-form text under the "Subject" category that must be converted into meaningful keywords. In Natural Language Processing (NLP), term extraction tools such as 'StanfordNLP', MALLET, 'OpenNLP' or "Microsoft Sql Server Integration Services (SSIS)" are used for term lookup. However, the majority of these tools are limited to single words and do not extract wider, sentence-level contexts. In order to train a viable model to identify early warnings from similar management texts, an expert-driven word corpus was created to train the Naïve Bayes model for early warning identification. A hierarchical PDF-parser was developed to extract nouns, verbs and adjectives from sentences extracted from Minutes meeting documents while eliminating unnecessary articles such as "the", "a" and "then". The terms thus extracted from Fig. 2 are shown in with the occurrence of each depicted as a Score (Column 4). The Item Id

Table 5 Sample extraction of early warning terms from a minutes document shown in Fig. 2

Item Id	Terms	Type	Score	Date
4.2	Outstanding	Adjective	1	10-Mar-14
4.2	Issues	Noun	2	10-Mar-14
4.2	Schedule	Noun	2	10-Mar-14
4.3	Progress	Verb	3	10-Mar-14
4.3	Expressed	Verb	1	-
4.3	Concerns	Noun	1	-
4.3	Gap	Verb	1	-
4.3	Fire fighting	Verb	1	-
4.3	Pipe	Noun	1	-
4.3	Installation/ Installed	Verb	2	-
4.3	Slow	Adjective	1	-
4.3	Insulated	Adjective	1	-

is the serial-numbered items discussed in meeting Minutes. Terms are extracted as nouns, adjectives and verbs from parsed sentences. For instance, for the sentence "slow progress of firefighting pipe installation" will generate three verbs uniquely expressing actions or occurrences that are "progress", "firefighting" and "installation", with adjective "slow" used to further indicate the negative nature of the case made for the associated noun "pipe" (Table 5).

Hence, in Fig. 2, is transformed into structured data as shown in Table 5.

3.3 *Analyse Word Association in Management Documents*

The overall meaning of a sentence structure will broadly depend upon the entire neighbourhood of the sentence. For instance, "required skills" will indicate a shortage of an aspect necessary for a project. On the other hand, "there is no shortage" actually demonstrates the availability rather than unavailability. Therefore, the proposed technique draws on a word-pair model where it first looks into a basic word classification and then creates a sentence-level word map to improve the overall context (Fig. 3).

As shown in Fig. 3, a frequency model can be derived to demonstrate sentence-level word-pair class association. The association can be taken as a bag-of-words type of model where each bag would representation a certain class as shown in Table 6.

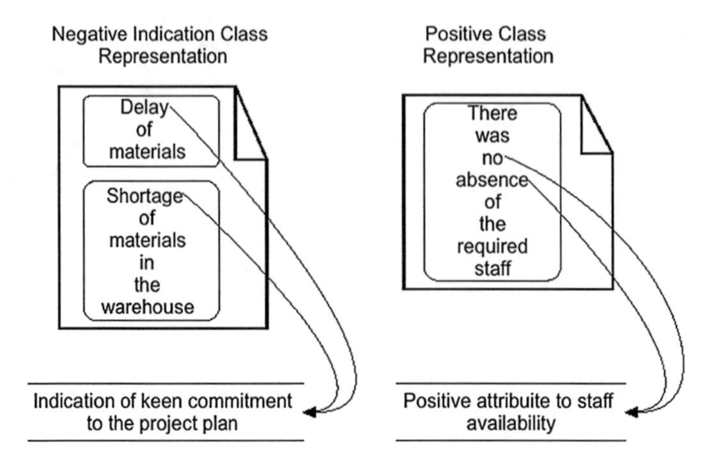

Fig. 3 A sentence-level word map representing two different classes

3.4 Modelling Data Preparation for Naïve Bayes Training

In order to prepare a Naïve Bayes early warning identifier, a text tagging system for PDF documents was necessary to establish usage relationships between certain words and terminologies used in meeting Minutes, drafts, memos and other documents. The initial term-to-early-warning relationship was developed based on a set of 20 questions. This survey was posted to individuals belonging to individuals with experience ranging from 5 to 20 years in construction project management domain. A total of 13 individuals responded to these questions with 15 questions shown in Table 6 with the resultant potential impacts cited in Table 7. The sample responses from random respondents, their extracted terms for model training, scores and the relevant class (risk/warning) are shown in the last column. Respondents were allowed to state any class to the questions which were then manually filtered for minor deviations in the meaning and generated a total of 12 unique early warnings as follows:

- Keen commitment to project scope (PS)
- Key management support (MS)
- Manpower resource (MR)
- Team required skills (RS)
- Commitment to project plan (PP)
- Team's required knowledge (RK)
- Keen commitment to project milestone (PM)
- Stable project responsibility (PR)
- Key management support (MS)
- Making purchases (MP)
- Materials on site (MO)
- Stable project requirement (SS)

Table 6 List of questions used in the interview to identify text trends indicating early warnings in project management documents

Selected questions	Sample answer from one of the respondents	Extracted terms for Naïve Bayes training Term (score)	Class identified by the expert (class code)
How do you think specific staff delay affects different aspects of a project? Please name various types of delays?	… Delay in finishing the tasks assigned, delay in response and not attending meetings …	VB: Delay (3) VB: Finishing (1) VB: Assigned (1) VB: Response (1) AD: Not Attending (1) CN: Tasks (1) CN: Meetings (1)	Keen commitment to project scope (PS)
How supply chain problems can be identified in management documents?	… Untracked long lead item …	VB: Untracked (1) AD: Long (1) CN: Lead (1) CN: Item (1) VB: Idle (1)	Key management support (MS)
Explain the impact of lack of scope of personnel representing the managements to have to the overall efficiency of the project?	… Idle workers whereas the shortage of useful staff generally lead to tasks not completed on time …	CN: Workers (1) VB: Shortage (1) AD: Useful (1) CN: Staff (1) CN: Tasks (1) VB: Not completed on time (1)	Manpower resource (MR)
Please specify keywords in meeting minutes that indicate to the key-project lack of following up of standards and procedures?	… Drawing not meeting the spicfication …	CN: Drawing (1) VB: Meeting (1) CN: Specification (1)	Team required skills (RS)
Please specify specific terms used in the meeting Minutes that indicate direct impact due to pending issues on the overall technical performance of the project?	… Failure, restrictions, unavailability, poor performance …	VB: Failure (1) VB: Restrictions (1) VB: Unavailability (1) AD: Poor (1) VB: Performance (1)	Stable project requirements (SS)
What impact do you think increase employee absence has on the overall project performance? Also, please name a set of factors in project meeting Minutes that indicate problems due to absenteeism	… Quality manager not come or absent, no acting manager …	CN: Quality manager (1) VB: Absent VB: No acting manager (1)	Keen commitment to project scope (PS)

(continued)

Table 6 (continued)

Selected questions	Sample answer from one of the respondents	Extracted terms for Naïve Bayes training Term (score)	Class identified by the expert (class code)
Please indicate the factors that contribute to red-tape issues that uselessly delay the project?	… Delay in document issuance, lack of properly documented guidelines …	VB: Delay (1)	Team's required knowledge (RK)
		VB: Documented (1)	
		CN: Issuance (1)	
		VB: Lack (1)	
		AD: Properly (1)	
		CN: Document (1)	
		CN: Guidelines (1)	
		VB: Lack (1)	
What keywords indicate uncertainty in decision making for certain project activities?	… Lack of the impetus needed to get the project underway …	CN: Impetus (1)	Keen commitment to project milestone (PM)
		VB: Needed (1)	
		CN: Project (1)	
		VB: Underway (1)	
What modes of communication and activities indicate serious problems in a project?	… Generally reminder letters indicate a more serious situation due to their documented nature …	VB: Unavailability (1)	Stable project responsibility (PR)
		VB: Lack (1)	
		VB: Serious (1)	
		CN: Nature (1)	
Please exemplify how shortages and progress delays are indicated in project meeting Minutes?	… Unavailability of materials, lack of inventory and task completion delays …	CN: Unavailability (1)	Key management support (MS)
		CN: Lack (1)	
		VB: Delays (1)	
What practices in project management indicate lack of prioritization in various parties	… Excessive over-drafts and finance management …	AD: Excessive (1)	Making purchases (MP)
		CN: Overdraft (1)	
		CN: Finance (1)	
		CN: Management (1)	
Please state how repeated requests of similar items/quantities/ actions be identified from the project documents	… Requests not addressed, delays in delivery, completion or management of certain tasks …	VB: Delays (1)	Materials on site (MO)
		VB: Completion (1)	
		VB: Management (1)	
Please state what impact the lack of resources such as not hiring an experienced engineers have and how can this impact be identified?	… Any suddenly increased requirement will likely delay the project in such a case …	ADJ: Suddenly	Manpower resource (MR)
		VB: Increased (1)	
		CN: Requirement (1)	
		ADJ: Likely (1)	
		VB: Delay (1)	
		CN: Project (1)	

(continued)

Table 6 (continued)

Selected questions	Sample answer from one of the respondents	Extracted terms for Naïve Bayes training Term (score)	Class identified by the expert (class code)
Please state how shortages in certain items can be identified in project documents and how can this be compensated	"Item not supplied, item unavailable in inventory"	CN: Item (1)	Materials on site (MO)
		ADJ: Not Supplied (1)	
		ADJ: Unavailable (1)	
		CN: Inventory (1)	

Table 7 Table indicating associations between certain questions drawn from Table 6

Questions	Keywords	Impact
Key management support	Frozen action	Delay management and engineering decision have a more serious effect on the planning and design aspect of the project
Keen commitment to project milestone	Inconsistent supply chain hold-ups	There may be hold-ups causing surpluses or shortages of materials due to unmanaged supply chain issues
Keen commitment to project scopes	Inflexibility in how to control their projects	Sponsor with unclear expectations and schedules
Stable project requirements	Contractor asks about things that are explained in the bid documents	Discrepancy in the project requirements
Team required knowledge/ skills	Repetitive disapproval or rejected	delay in document issuance and lack of properly documented guidelines
Materials on site	Poor inventory control	L/C and down payment excessive overtime
Manpower resource	Mr X has taken leave of absence due to no reason	Negative impact on the delivery of the project so a replacement resource is required

4 Research Outcome, Experiments and Analysis

As discussed earlier, the approach initially addressed a discretely trained methodology based on word-level frequencies trained via a Naïve Bayes classifier. In order to train the supervised Naïve Bayes training model, the data was initially taken directly from questionnaire responses, which was then used to label Minutes' text. The questions asked experts and engineers from the field of construction project management to indicate specific words used in management documents and the level of risk and warning they attributed towards a particular situation. For instance, delay in clearance of funds for any purchases may directly attribute to any impending project failures due to lack of project responsibility. Similarly, delays in the completion of project deliverables may be indicated by keywords such as “staff unavailability”, “key skills shortage” or “lack of personnel availability”. The association so such groups can potentially provide better training of a model compared to a manually labelled project meeting minutes. 70 % of this textual information was then used to train the model whereas the remaining 30 % was used to assess the outcome of the classifier.

4.1 Data Consolidation and Analysis

For testing a total of 46 PDF documents were labelled based on word-labels from questionnaire responses to extract indicators based on resulting delay types. For example, a Minutes document containing discussions on staff shortage leading to delays due to lack of manpower was used to extract critical sentences containing relevant information. A total of 23 sentences stating keyword pairs such as “employee unavailability”, “manpower shortage”, “frozen action” and “unexpected absence” were fed to the classifier and the outcome were reported as shown in Table 8.

On application of the classification model, the sentences that were correctly known to identify the early warnings were regarded as accurate hits whereas any other matches were termed inaccurate. The best classification accuracy was found to be for the early warning of MP (Making Purchases) class with an accurate iden-

Table 8 A confusion matrix elaborating on the TN and TN accuracies of the 12 of early warning classes

Class indices	PS	MS	MR	RS	PP	RK	PM	PR	MS	MP	MO	SS	Total	Accuracy (%)
PS	27	5	0	0	6	0	0	3	0	0	0	0	41	65.85
MS	11	46	1	0	3	0	0	4	9	0	0	0	74	62.16
MR	0	0	75	22	0	12	0	0	0	0	0	6	115	65.22
RS	0	0	4	91	0	5	0	0	0	0	0	22	122	74.59
PP	14	2	0	0	79	0	7	0	13	0	0	0	115	68.70
RK	0	0	23	5	0	56	0	1	0	4	1	3	93	60.22
PM	7	1	7	0	2	1	45	0	6	0	0	0	69	65.22
PR	17	8	3	0	4	0	0	77	14	0	0	0	123	62.60
MS	11	7	1	0	0	0	0	23	117	0	1	0	160	73.13
MP	0	0	0	0	0	0	0	2	0	89	21	0	112	79.46
MO	0	0	0	0	0	0	0	0	0	28	73	0	101	72.28
SS	0	0	0	17	0	0	0	0	0	0	0	35	52	67.31

Table 9 Comparative analysis of existing text mining algorithms

ML	SVM (%)	ROCCHIO (%)	IBM Miner (%)	Naïve Bayes (%)	K-Nearest Neighbours (%)	Modified Naïve Bayes (%)
Score	91.12	64.71	60.95	58.82	49.11	68.02

tification of 79.46 % words. The category only reported 18.75 % text terms to be miss-classified as Materials on site (MO) and 1.78 % incorrectly identified in the PR (Project Responsibility) category. The lowest accuracy match was that of Teams' Required Knowledge class where 24.73 % of the terms were misclassified as (lack of) Manpower Resource (MR). The overall accuracy in prediction came out to be 68.06 %.

The accuracy, when compared to other techniques, presents a promising step forward with regards to matching various early warnings. In Table 9 the proposed technique was only outperformed by SVM that at presents stands at 91.12 %.

5 Conclusion

The technique addresses an information mining approach from project management documents to extract early warnings. The underlying approach targeted two core problems in construction data mining namely the development of a term-extractor and early-warning mapping corpus and the development of a machine learning early warning identification classifier.

The approach is novel in a sense that it improves on discrete word-level early warning extraction based on a supervised, expert-labelled project management document database. The approach was evaluated against 12 unique aspects that indicate early warnings including lack of management support, requirement delays, skills and material shortage and planning weaknesses. The underlying principles of extracting associated data-pairs were modelled via keywords extracted through qualitative information from an expertly guided questionnaire. The proposed methodology outperformed the majority of these methodologies including standalone Nearest-neighbour and K-means approaches by improving respective accuracies of 58.82 and 49.11 to 68.06 %. This approach is only outperformed by SVM, which currently delivers an accuracy of 91.12 % that is primarily limited to a two-class problem whereas the proposed approach addresses a multi-class solution.

The methodology can further be extended to identify sentence-level warnings by training sentence chains in a time-series state-transition fashion via a Markovian approach. This approach is likely to improve the accuracy of certain document sections depending upon how well the database is trained against a labelled dataset. The future extension of this research is to use the labelled qualitative information to train class-specific Genetic algorithm. The approach is expected to improve the existing accuracy as it would consider a sentence-level meaning instead of word-pairs and discrete (single) words as well as state solution.

References

1. Al Qady, M., Kandil, A.: Concept relation extraction from construction documents using natural language processing. J. Constr. Eng. Manag.-ASCE **136**(3), 294–302 (2010)
2. Alsubaey, M., Asadi, A., Makatsoris, H.: A Naive Bayes approach for EWS detection by text mining of unstructured data: A construction project case. In: SAI Intelligent Systems Conference (IntelliSys), 2015, pp. 164–168 (2015). Accessed 10–11 Nov 2015
3. Asadi, Ahmad, Alsubaey, Mohammed, Makatsoris, Charalampos: A machine learning approach for predicting delays inconstruction logistics. Int. J. Adv. Logist. **4**, 115–130 (2015)
4. Ansoff, H.I.: Managing strategic surprise by response to weak signals. Calif. Manag. Rev. **18**(2), 21–33 (1975)
5. Arashpour, M., Wakefield, R., Blismas, N., Minas, J.: Optimization of process integration and multi-skilled resource utilization in off-site construction. Autom. Constr. **50**, 72–80 (2015)
6. Baghdadi, A., Kishk, M.: Saudi arabian aviation construction projects: identification of risks and their consequences. Procedia Eng. **123**, 32–40 (2015)
7. Bavan, A.S.: An artificial neural network that recognizes an ordered set of words in text mining task. In: 2009 International Conference on the Current Trends in Information Technology (CTIT), pp. 15–16 (2009). Accessed 1–5 Dec 2009
8. Cai Lan, Z., Shasha, L.: Research of information extraction algorithm based on Hidden Markov model. In: 2010 2nd International Conference on Information Science and Engineering (ICISE), pp. 4–6 (2014). Accessed 1–4 Dec 2010
9. Caldas, C.H., Soibelman, L., Han, J.W.: Automated classification of construction project documents. J. Comput. Civ. Eng. **16**(4), 234–243 (2002)
10. Caldas, C.H., Soibelman, L., Gasser, L.: Methodology for the integration of project documents in model-based information systems. J. Comput. Civ. Eng. **19**(1), 25–33 (2005)
11. Cerulo, L., Ceccarelli, M., Di Penta, M., Canfora, G.: A hidden Markov model to detect coded information islands in free text. In: 2013 IEEE 13th International Working Conference on Source Code Analysis and Manipulation (SCAM), pp. 22–23. Accessed 157–166 Sept 2013
12. Chen, D., Liu, Z.: A new text categorization method based on HMM and SVM. In: 2010 2nd International Conference on Computer Engineering and Technology (ICCET), pp. V7-383–V7-386 (2010). Accessed 16-18 April 2010
13. Cheung, K.W., Kwok, J.T., Law, M.H., Tsui, K.C.: Mining customer product rating for personalized marketing. Decis. Support Syst. **35**(2), 231–243 (2003)
14. Chi, S., Suk, S.-J., Kang, Y., Mulva, S.P.: Development of a data mining-based analysis framework for multi-attribute construction project information. Adv. Eng. Inform. **26**(3), 574–581 (2012)
15. Dey, L., Bharadwaja, H.S., Meera, G., Shroff, G.: Email Analytics for Activity Management and Insight Discovery. In: 2013 IEEE/WIC/ACM International Joint Conferences on Web Intelligence (WI) and Intelligent Agent Technologies (IAT), pp. 557–564 (2013). Accessed 17–20 Nov 2013
16. Ding, L.Y., Zhou, C.: Development of web-based system for safety risk early warning in urban metro construction. Autom. Constr. **34**, 45–55 (2013)
17. Ding, L.Y., Zhou, C., Deng, Q.X., Luo, H.B., Ye, X.W., Ni, Y.Q., Guo, P.: Real-time safety early warning system for cross passage construction in Yangtze Riverbed Metro Tunnel based on the internet of things. Autom. Constr. **36**, 25–37 (2013)
18. Dokas, I.M., Karras, D.A., Panagiotakopoulos, D.C.: Fault tree analysis and fuzzy expert systems: early warning and emergency response of landfill operations. Environ. Model. Softw. **24**(1), 8–25 (2009)

19. Erohin, O., Kuhlang, P., Schallow, J., Deuse, J.: Intelligent utilisation of digital databases for assembly time determination in early phases of product emergence. Procedia CIRP **3**, 424–429 (2012)
20. Feilmayr, C.: Text mining-supported information extraction: an extended methodology for developing information extraction systems. In: 2011 22nd International Workshop on Database and Expert Systems Applications (DEXA), pp. 217–221 (2011). Accessed Aug 29–Sept 2 2011
21. Gordon, C., Mulley, C., Stevens, N., Daniels, R.: How optimal was the Sydney Metro contract? Comparison with international best practice. Res. Trans. Econ. **39**(1), 239–246 (2013)
22. Gunasinghe, U., Alahakoon, D.: A biologically inspired neural clustering model for capturing patterns from incomplete data. In: 2010 5th International Conference on Information and Automation for Sustainability (ICIAFs), pp. 126–131 (2010). Accessed 17–19 Dec 2010
23. Haji-Kazemi, S., Andersen, B., Krane, H.P.: Identification of early warning signs in front-end stage of projects, an aid to effective decision making. Procedia Soc. Behav. Sci. **74**, 212–222 (2013)
24. Haji-Kazemi, S., Andersen, B., Klakegg, O.J.: Barriers against effective responses to early warning signs in projects. Int. J. Proj. Manag. **33**(5), 1068–1083 (2015)
25. Handford, M., Matous, P.: Problem-solving discourse on an international construction site: patterns and practices. Engl. Specif. Purp. **38**, 85–98 (2015)
26. Harding, J.: Avoiding project failures. Chem. Eng. **119**(13), 51–54 (2012)
27. Hui, S.C., Jha, G.: Data mining for customer service support. Inf. Manag. **38**(1), 1–13 (2000)
28. Indhuja, K., Reghu, R.P.C.: Fuzzy logic based sentiment analysis of product review documents. In: 2014 First International Conference on Computational Systems and Communications (ICCSC), pp. 17–18 (2014). Accessed 18–22 Dec 2014
29. Ingason, H.T.: Best project management practices in the implementation of an ISO 9001 quality management system. Procedia Soc. Behav. Sci. **194**, 192–200 (2015)
30. Jing, J.: Modeling syntactic structures of topics with a nested HMM-LDA. In: Ninth IEEE International Conference on Data Mining, 2009. ICDM '09, pp. 824–829 (2009). Accessed 6–9 Dec 2009
31. Karaali, O., Corrigan, G., Massey, N., Miller, C., Schnurr, O., Mackie, A.: A high quality text-to-speech system composed of multiple neural networks. In: Proceedings of the 1998 IEEE International Conference on Acoustics, Speech and Signal Processing, 1998, vol. 2, pp. 1237–1240. Accessed 12–15 May 1998
32. Keqin, W., Qikun, W., Hongyu, M., Mubing, Z., Kuan, J., Xuepei, Z., Lin, Y., Ting, W., Huaiqing, W.: Intelligent text mining based financial risk early warning system. In: 2015 2nd International Conference on Information Science and Control Engineering (ICISCE), pp. 279–281 (2015). Accessed 24–26 April 2015
33. Khan, A., Baharudin, B., Lee, L.H., Khan, K.: A review of machine learning algorithms for text-documents classification. J. Adv. Inf. Technol. **1**(1), 4–20 (2010)
34. Kim, H., Soibelman, L., Grobler, F.: Factor selection for delay analysis using knowledge discovery in databases. Autom. Constr. **17**(5), 550–560 (2008)
35. Ko, C.-H., Cheng, M.-Y.: Dynamic prediction of project success using artificial intelligence. J. Constr. Eng. Manag.-ASCE **133**(4), 316–324 (2007)
36. Larouk, O., Batache, M.: An evaluation the uncertainty of textual data with logic and statistics. In: Third International Symposium on Uncertainty Modeling and Analysis, 1995, and Annual Conference of the North American Fuzzy Information Processing Society. Proceedings of ISUMA - NAFIPS '95., pp. 739–744 (1995). Accessed 17–19 Sep 1995
37. Lee, H.W., Oh, H., Kim, Y., Choi, K.: Quantitative analysis of warnings in building information modeling (BIM). Autom. Constr. **51**, 23–31 (2015)
38. Legodi, I., Barry, M.L.: The current challenges and status of risk management in enterprise data warehouse projects in South Africa. In: 2010 Proceedings of PICMET '10: Technology Management for Global Economic Growth (PICMET), pp. 1–5 (2010). Accessed 18–22 July 2010
39. Leung, R.W.K., Lau, H.C.W., Kwong, C.K.: On a responsive replenishment system: a fuzzy logic approach. Expert Syst. **20**(1), 20–32 (2003)

40. Liu, J., Wang, Q., Liu, Y., Li, Y.: A short text topic discovery method for social network. In: 2014 33rd Chinese Control Conference (CCC), pp. 512–516. Accessed 28–30 July 2014
41. Mao, W., Zhu, Y., Ahmad, I.: Applying metadata models to unstructured content of construction documents: a view-based approach. Autom. Constr. **16**(2), 242–252 (2007)
42. Merkl, D., Schweighofer, E.: En route to data mining in legal text corpora: clustering, neural computation, and international treaties. In: Proceedings., Eighth International Workshop on Database and Expert Systems Applications, 1997, pp. 465–470 (1997). Accessed 1–2 Sept 1997
43. Ngai, E.W.T., Xiu, L., Chau, D.C.K.: Application of data mining techniques in customer relationship management: a literature review and classification. Expert Syst. Appl. **36**(2), 2592–2602 (2009)
44. Nikander, I.O., Eloranta, E.: Project management by early warnings. Int. J. Proj. Manag. **19**(7), 385–399 (2001)
45. Otsuka, N., Matsushita, M.: Constructing knowledge using exploratory text mining. In: 2014 Joint 7th International Conference on and Advanced Intelligent Systems (ISIS), 15th International Symposium on Soft Computing and Intelligent Systems (SCIS), pp. 1392–1397 (2014). Accessed 3–6 Dec 2014
46. Pang, X.-L., Feng, Y.-Q.: (2006) An improved economic early warning based on rough set and support vector machine. In: Proceedings of 2006 International Conference on Machine Learning and Cybernetics, vols. 1–7, pp. 2444–2449
47. Ponsteen, A., Kusters, R.J.: Classification of human- and automated resource allocation approaches in multi-project management. Procedia Soc. Behav. Sci. **194**, 165–173 (2015)
48. Proverbs, D.G., Holt, G.D.: Reducing construction costs: European best practice supply chain implications. Eur. J. Purch. Supply Manag. **6**(3–4), 149–158 (2000)
49. Raghuram, S., Yuni, X., Palakal, M., Jones, J., Pecenka, D., Tinsley, E., Bandos, J., Geesaman, J.: Bridging text mining and bayesian networks. In: International Conference on Network-Based Information Systems, 2009. NBIS '09, pp. 298–303. Accessed 19–21 Aug 2009
50. Ratsiepe, K.B., Yazdanifard, R.: Poor risk management as one of the major reasons causing failure of project management. In: 2011 International Conference on Management and Service Science (MASS), pp. 1–5 (2011). Accessed 12–14 Aug 2011
51. Romero, C., Ventura, S., Garcia, E.: Data mining in course management systems: Moodle case study and tutorial. Comput. Educ. **51**(1), 368–384 (2008)
52. Tianyi, J., Tuzhilin, A.: Segmenting customers from population to individuals: does 1-to-1 keep your customers forever? IEEE Trans. Knowl. Data Eng. **18**(10), 1297–1311 (2006)
53. Tsai, W.H., Lin, S.J., Lin, W.R., Liu, J.Y.: The relationship between planning and control risk and ERP project success. In: IEEE International Conference on Industrial Engineering and Engineering Management, 2009. IEEM 2009, pp. 1835–1839 (2009). Accessed 8–11 Dec 2009
54. Tseng, Y.-H., Lin, C.-J., Lin, Y.-I.: Text mining techniques for patent analysis. Inf. Process. Manag. **43**(5), 1216–1247 (2007)
55. Verma, V.K., Ranjan, M., Mishra, P.: Text mining and information professionals: role, issues and challenges. In: 2015 4th International Symposium on Emerging Trends and Technologies in Libraries and Information Services (ETTLIS), pp. 133–137 (2015). Accessed 6–8 Jan 2015
56. Wang, H.J., Zhang, J.P., Chau, K.W., Anson, M.: 4D dynamic management for construction planning and resource utilization. Autom. Constr. **13**(5), 575–589 (2004)
57. Wang, Z., Chu, L.: The algorithm of text classification based on rough set and support vector machine. In: 2010 2nd International Conference on Future Computer and Communication (ICFCC), pp. V1-365–V1-368 (2010). Accessed 21–24 May 2010
58. Wex, F., Widder, N., Liebmann, M., Neumann, D.: Early warning of impending oil crises using the predictive power of online news stories. In: 2013 46th Hawaii International Conference on System Sciences (HICSS), pp. 1512–1521 (2013). Accessed 7–10 Jan 2013
59. Yamron, J.P., Carp, I., Gillick, L., Lowe, S., van Mulbregt, P.: A hidden Markov model approach to text segmentation and event tracking. In: Proceedings of the 1998 IEEE International Conference on Acoustics, Speech and Signal Processing, 1998, vol. 1, pp. 333–336 (1998). Accessed 12–15 May 1998

60. Yongpeng, Y., Manoharan, M., Barber, K.S.: Modelling and analysis of identity threat behaviors through text mining of identity theft stories. In: 2014 IEEE Joint Intelligence and Security Informatics Conference (JISIC), pp. 184–191 (2014). Accessed 24–26 Sept 2014

Quantitative Assessment of Anomaly Detection Algorithms in Annotated Datasets from the Maritime Domain

Mathias Anneken, Yvonne Fischer and Jürgen Beyerer

Abstract The early detection of anomalies is an important part of a support system to aid human operators in surveillance tasks. Normally, such an operator is confronted with the overwhelming task to identify important events in a huge amount of incoming data. In order to strengthen their situation awareness, the human decision maker needs an support system, to focus on the most important events. Therefore, the detection of anomalies especially in the maritime domain is investigated in this work. An anomaly is a deviation from the normal behavior shown by the majority of actors in the investigated environment. Thus, algorithms to detect these deviations are analyzed and compared with each other by using different metrics. The two algorithms used in the evaluation are the Kernel Density Estimation and the Gaussian Mixture Model. Compared to other works in this domain, the dataset used in the evaluation is annotated and non-simulative.

1 Introduction

In order to be able to make the best decision, an operator in surveillance tasks needs an overview about all incoming data. Therefore, only if operators are able to understand and interpret the whole data correctly, they will be able to make the best possible next move. As stated by Fischer and Beyerer [7], the main problem is not the acquisition of the data, but the large amount of data. In order to aid human decision makers, a support system is needed. This system needs to provide the aid to identify important

M. Anneken (✉) · Y. Fischer · J. Beyerer
Fraunhofer Institute of Optronics, System Technologies and Image Exploitation (IOSB), Karlsruhe, Germany
e-mail: mathias.anneken@iosb.fraunhofer.de

Y. Fischer
e-mail: yvonne.fischer@iosb.fraunhofer.de

J. Beyerer
Vision and Fusion Laboratory, Karlsruhe Institute of Technology (KIT), Karlsruhe, Germany
e-mail: juergen.beyerer@iosb.fraunhofer.de

Y. Bi et al. (eds.), *Intelligent Systems and Applications*,
Studies in Computational Intelligence 650, DOI 10.1007/978-3-319-33386-1_5

events. Without such a system, the operator needs to analyze the whole data by hand. This is tedious work and might result in a reduced concentration of operators. Hence, the chance is high to overlook crucial events.

The crucial events are most often deviations from the normal behavior. Therefore, this events can be classified as anomalies. If a support system is able to provide an operator reliable with information about anomalies, this will increase the situation awareness of the operator. Thus, algorithms to identify anomalies are evaluated in this work. The idea is to reduce the workload of an operator by providing information about the important events. While this reduction helps an operator to concentrate on the important tasks, it is crucial for the acceptance, that the algorithms will identify the anomalies as good as possible. In order to be able to assess the performance of an algorithm in real life, the algorithm has to be evaluated with real data. Else, the support system might not be able to detect some anomalies, which are not covered in a simulated dataset.

Here, the detection of anomalies is investigated in the maritime domain. In this special case, the normal traffic drives on sea lanes. Anomalies can, e.g., be seen as the deviation from these lanes or a different speed compared to the normal traffic. Thus, the algorithms in the evaluation must be able to assess spatio-temporal data in form of trajectories.

2 Related Work

Chandola et al. [6] give a wide overview about different tasks and algorithms used to detect anomalies. The applications range from sensor network and cyber security to fraud detection and image processing. Morris and Trivedi [13] give a survey especially for anomaly detection using vision-based algorithms. For each application appropriate algorithms are introduced. The underlying concepts for the algorithms vary, depending on the use case, e.g., classification-, clustering- or nearest neighbor-based algorithms are used. Each of these algorithms has its advantages and disadvantages.

Especially in the maritime domain, the detection of anomalies is an important field of research. Several different approaches were introduced to identify abnormal behavior of vessels and to incorporate expert knowledge to correctly assess specific situations.

Laxhammar et al. [11] compare the Kernel Density Estimation (KDE) and the Gaussian Mixture Model (GMM). The models are trained with a real life dataset comprising the *position*, *heading* and *speed* of each vessel. For the evaluation, artificial anomalies are simulated. The models' performances to resemble the normal behavior is evaluated by comparing the log-likelihood with the 1st percentile as well as the median log-likelihood. For the anomaly detection performance, the needed number of observations for detecting an anomaly is compared. Anneken et al. [2] evaluate the same algorithms by using an annotated dataset.

Brax and Niklasson [5] introduce a state-based anomaly detection algorithm. The different discrete states are *heading*, *speed*, *position* and *relative position to the next vessel*. Different roles (here called agents) are developed and incorporate the states. These roles comprise, e.g., smuggler and raid agents. The probability for each role is calculated using the prior defined states. For the evaluation, different scenarios resembling specific situations are generated to obtain an accurate ground truth. The algorithms are only tested using this simulated ground truth.

Andersson and Johansson [1] use an algorithm based on a Hidden Markov Model (HMM) to detect abnormal behavior. They train the HMM with simulated normal behavior of ships in a certain area. For the evaluation, the data is divided into discrete states. The states resemble the change of specific values, i.e., *distance to other objects*, *vessel size*, *identification number*, *speed* and *heading*. Afterwards, the model is evaluated by using a simulated pirate attack.

Laxhammer and Falkman [10] introduce the sequential conformal anomaly detection. The underlying conformal prediction framework is, e.g., explained further by Schafer and Vovk [16]. The algorithm provides a reliable lower boundary for the probability that a prediction is correct. This threshold is given as a parameter for the algorithm and directly influences the false positive rate. The similarity between two trajectories is calculated by using the Hausdorff distance. The model is trained with real life data, but the anomalies for the evaluation are simulated.

De Vries and van Someren [17] use *piecewise linear segmentation* methods to partition trajectories of maritime vessels. The resulting trajectories are grouped in clusters. Afterwards, the anomaly detection is performed by using kernel methods. Additionally, expert domain knowledge like geographical information about harbors, shipping lanes and anchoring areas is incorporated. The algorithms are validated with a dataset from the Netherlands' coast near Rotterdam.

Guillarme and Lerouvreur [9] introduce an algorithm consisting of three main steps. They first partition the training trajectories into stops and moves segments using the *Clustering-Based Stops and Moves of Trajectories* algorithm. Afterwards, a similarity measure and a clustering algorithm based on the density of the data is used to cluster the resulting sub-trajectories. The clusters discovered by the algorithm need to be assessed by hand. With this results, motion patterns and junctions for the trajectories are defined. For the evaluation, satellite AIS data is used. No information about the performance of the algorithm compared to other algorithms is given.

Fischer et al. [8] present an approach based on dynamic Bayesian networks. Different situations and their relationship with each other are modeled in a situational dependency network. With this network, the existence probability for each defined situation, e.g., *a suspicious incoming smuggling vessel*, can be estimated. This estimated probability is used to detect unusual behavior. The algorithm is tested with simulated data.

Anneken et al. [3] reduce the complexity of trajectories by using b-splines estimation. The control points of the b-splines are used as the feature vector and the normal model is trained by using different machine learning algorithms. For the evaluation an annotated dataset is used. The results of the different algorithms are compared with the results of a KDE and a GMM as previously shown in [2, 11].

The majority of the previous work in the maritime domain uses simulated data to evaluate their proposed algorithms. E.g., the evaluations in [1, 8] rely entirely on simulated data, and in [10, 11] the anomalies are created artificially. In this work, the same annotated dataset as in [2, 3] is used. Additionally to the previously used areas, an additional area is introduced and the algorithms are compared by using a different set of metrics.

3 Dataset

The dataset for the evaluation was recorded by using the automatic identification system (AIS). The AIS provides different kinds of data like *navigation-status*, *estimated time of arrival*, ..., *destination*. For the analysis only a subset of the whole available data is used, namely *position*, *speed*, *heading*, *maritime mobile service identity (MMSI)*, *timestamp* and *vessel-type*. The whole dataset comprises more than 2.4 million unique measurements recorded during a time span of seven days.

A depiction of the dataset in form of a heat-map is shown in Fig. 1. The map encodes the traffic density with colors ranging from green for low density to red for high density. Thus sea lanes and harbors are easily recognizable for their higher traffic density. Geographically, the recorded area comprises the western parts of the Baltic Sea, the Kattegat and parts of the Skagerrak. Temporally, it spans a whole week starting from 16th May 2011. Altogether, 3,702 different vessels (unique MMSIs) grouped into 30 different vessel types were detected. In the first step, clearly wrong measurements as well as measurements generated by offshore structures (e.g., lights) are removed. Afterwards, 3,550 unique vessels remain. For further processing the data points created by each ship are grouped by their corresponding MMSI and connected to tracks. If the time between two measurements with the same MMSI is too large (here, larger than 30 min), the track of the ship is split. Therefore, each vessel can generate more than one track and altogether 25,918 different tracks are detected.

For further investigations, only cargo vessels and tankers are used. In the prepared dataset, there are 1,087 cargo vessels and 386 tankers. The two types have similar movement behavior; therefore, they are treated as one. This means, there will always be one compound model for both types instead of a single one for each.

Due to the huge amount of data and the necessary effort to annotate the whole dataset, only three subareas (called Fehmarn, Kattegat and Baltic) are annotated which reflect specific criteria. A description of the Kattegat and the Fehmarn area can be found in [2].

Figure 2 depicts the normal behavior within the Baltic area. This area consists of two main sea lines. In the east both lines are starting in the north west of the Danish island Bornholm. One line goes in the direction of Copenhagen, the other in the direction of Fehmarn and Lolland. Furthermore, the sea lanes of vessels calling the port at Trelleborg can be seen in the north. As the traffic density of the other possible lanes (e.g. the traffic to Ystad) is low, all other traffic is defined as abnormal behavior

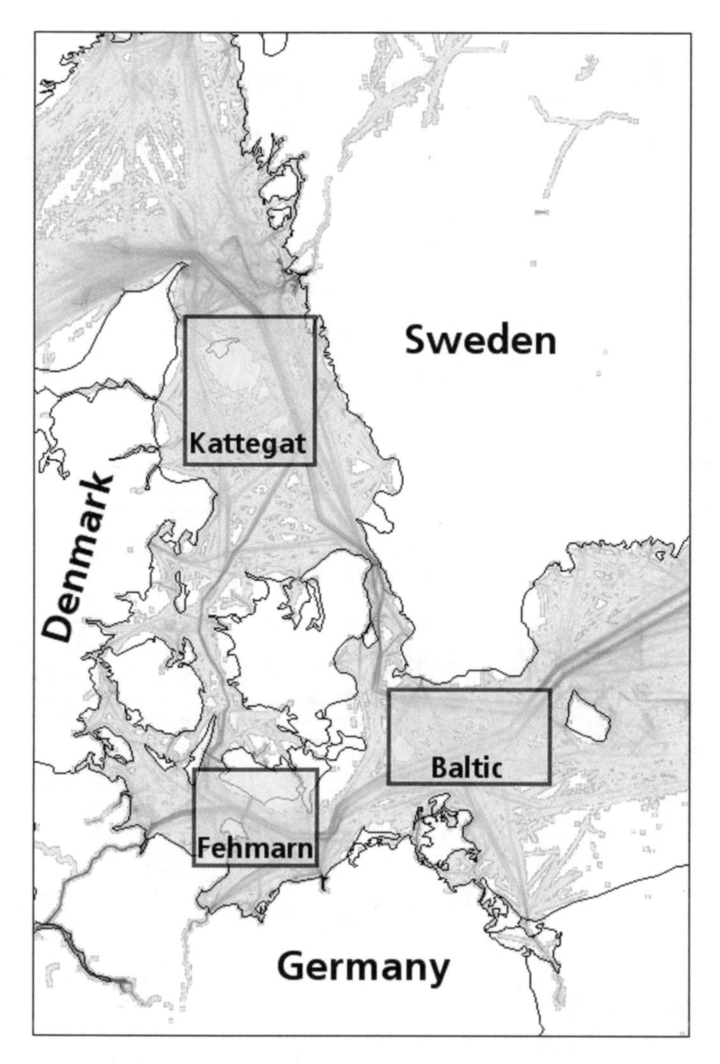

Fig. 1 Heat-map of the vessel traffic in the dataset. The traffic density is encoded by the color, whereas the gradient from *red* to *green* represents the gradient from high density to low density. The *marked areas* (namely Fehmarn, Kattegat and Baltic) are further analyzed

(compare Fig. 3). The resulting area has 698 unique tankers and cargo vessels which generate a total of 26,808 data points.

During the annotation of the tanker and cargo vessels in the designated areas, moored vessels moored in a harbor are removed from the dataset, for the behavior in harbors is out of scope for this work. For the Fehmarn area, 14.5 % of the data points by cargo vessels and 8.7 % of the data points by tankers are marked as unusual

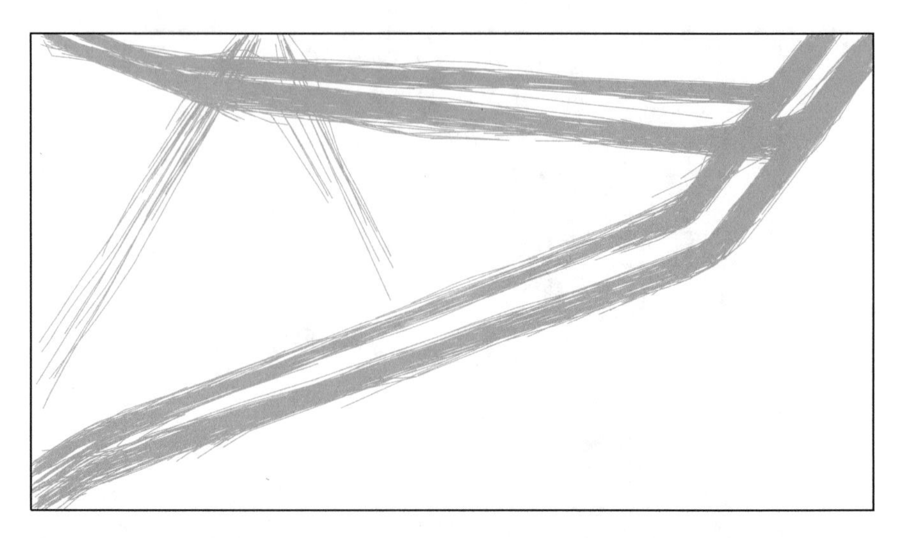

Fig. 2 Normal trajectories (*grey lines*) in the annotated and evaluated area consisting of a part of the Baltic Sea. The *white* background represents water

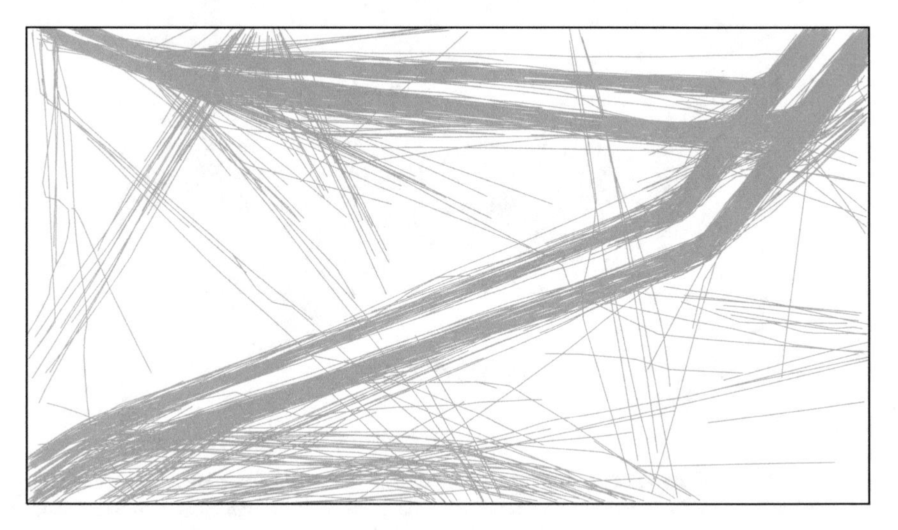

Fig. 3 All trajectories (*grey lines*) in the annotated and evaluated area consisting of a part of the Baltic Sea. The *white* background represents water

ones. In the Kattegat area, 5.4 % of the data points by tankers and 6.6 % of the data points by cargo vessels were annotated as abnormal. For the Baltic area, 15.2 % of data points generated by cargo vessels and 14.4 % of the ones by tankers are marked as anomalies.

4 Test Set-Up

In this section, the two evaluated algorithms, the metrics and the general process of detecting algorithms are introduced. The Gaussian Mixture Model (GMM) and the Kernel Density Estimation (KDE) are chosen as algorithms for the evaluation. The algorithms themselves and the possible parameters are described. As feature vector, the position in latitude p_{lat} and longitude p_{lon} as well as the speed vector split into its latitude v_{lat} and longitude v_{lon} components are used, resulting in

$$x_i = \{p_{\text{lat}}, p_{\text{lon}}, v_{\text{lat}}, v_{\text{lon}}\}$$

for each data point i. A new model has to be trained for each area. Further, each area can be divided by a grid and for each cell in the grid a distinct model has to be trained. The optimal grid-size as well as the parameters of the models have to be estimated. For these purposes, the Python package *scikit-learn* [15] is used.

4.1 Algorithms

4.1.1 Gaussian Mixture Model

A GMM consists of n superimposed multivariate normal distributions called components. Each distribution i has its own mean vector μ_i and covariance matrix Σ_i. Together, they form the parameter set $\theta_i = \{\mu_i, \Sigma_i\}$ for each component i. The dimension of μ_i and Σ_i depend on the number of observed features k. The probability density function is then given by

$$f(x) = \sum_{i=1}^{n} f_g(x, \theta_i)$$

with the density function for each component given by

$$f_g(x, \theta_i) = \frac{1}{(2\pi)^{\frac{k}{2}} \sqrt{|\Sigma_i|}} \exp\left(-\frac{1}{2}(x - \mu_i)^T \Sigma_i^{-1}(x - \mu_i)\right).$$

In order to estimate the parameter sets θ_i, the expectation-maximisation (EM) algorithm is used. Prior to this estimation, the number of components n must be available. More details on the GMM and the EM algorithm is given, e.g., by Barber [4].

4.1.2 Kernel Density Estimator

The KDE or Parzen-window density estimation estimates the probability density function (PDF) of a dataset with n data points. Each of these data points is assigned a kernel function $K(x)$ with the bandwidth h. For each kernel function the same bandwidth is chosen. As kernel function, any valid PDF may be chosen. By taking the sum of all kernels evaluated at the point x, the PDF is estimated resulting in

$$f(x) = \frac{1}{n}\sum_{i=1}^{n}\frac{1}{h^k}K\left(\frac{x - x_i}{h}\right).$$

The Gaussian kernel

$$K(x) = \frac{1}{(2\pi)^{\frac{k}{2}}\sqrt{|\Sigma|}}\exp\left(-\frac{1}{2}x^T\Sigma^{-1}x\right)$$

with the covariance matrix set to the identity matrix $\Sigma = I_k$ is used as kernel function. The bandwidth has a huge impact on the resulting PDF. If it is chosen too small, the resulting estimation will overfit the problem; if the chosen bandwidth is too large, underfitting will occur. Further information on the KDE is available, e.g., by Murphy [14].

4.2 Anomaly Detection

For both algorithms, the detection of abnormal behavior is defined as depicted in Fig. 4. First, a normal model is estimated by using only data with normal behavior. Then, the minimum log-likelihood l_{min} for each model for the training data is calculated. The log-likelihood is the natural logarithm of the likelihood function which is defined as the conditional probability that an outcome is generated by a specific model. By using only normal data for the training, it can be expected that abnormal data will generate a lower log-likelihood.

For each new data point x, the log-likelihood l_{x} for the estimated model is calculated. If $l_{\text{min}} > l_{\text{x}}$ holds true, the data point is considered an anomaly. Thus l_{min} is the boundary between abnormal and normal behavior.

4.3 Metrics

For the evaluation, precision, recall, f1-score, accuracy, false positive rate (FPR), receiver operating characteristic (ROC) and area under ROC are used as metrics to

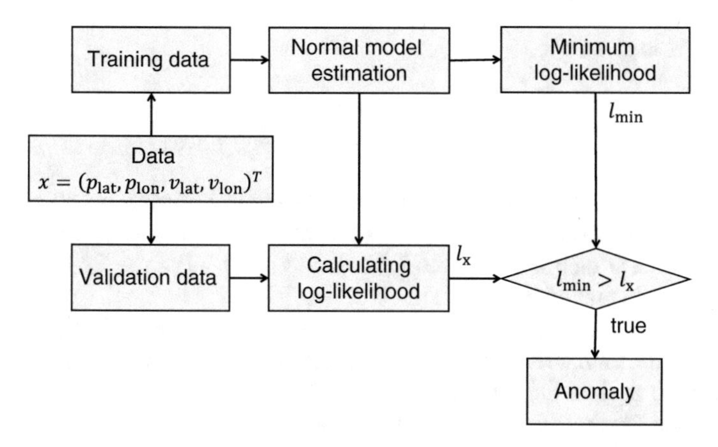

Fig. 4 Flow chart of the anomaly detection algorithm

compare and assess the algorithms. Here, only a brief introduction to the metrics is given. Further explanation are, e.g., given by Manning et al. [12].

The precision is defined as

$$\text{precision} = \frac{\text{true positives}}{\text{true positives} + \text{false positives}},$$

whereas the recall is defined as

$$\text{recall} = \frac{\text{true positives}}{\text{true positives} + \text{false negatives}}.$$

The f1-score is the harmonic mean of the precision and recall. It is defined as

$$\text{f1-score} = 2 \times \frac{\text{precision} \times \text{recall}}{\text{precision} + \text{recall}}.$$

The recall describes the fraction of the positives which are actually classified as positive (true positives). Thus, a small recall means, that there are lots of false negative classifications. The precision describes the fraction of all positively classified results which are actually positives. Hence, a small precision equals a great number of false positive classifications. The accuracy is defined as

$$\text{accuracy} = \frac{\text{true positives} + \text{true negatives}}{\text{true positives} + \text{true negatives} + \text{false negatives} + \text{false negatives}}.$$

It describes the amount of correctly identified object compared to all available objects.

The FPR is given by

$$\text{FPR} = \frac{\text{false positives}}{\text{false positives} + \text{true negatives}}.$$

It describes the probability, that the classifier falsely identifies an object as positive. These metrics are commonly used in classification tasks. For an optimal classifier, the value of each metric except the FPR should be "1". For the FPR, the optimal classifier should yield the value "0".

The ROC curve is a graphical tool to describe the performance of a classifier while varying the discrimination threshold between to classes. It is a plot of the recall (true positive rate) against the false positive rate. If the ROC curve of a classifier is a diagonal line starting in the origin of the coordinate system, it will be equal to a random guess. This is the worst result for a classifier. For further comparison, the area under ROC (AUROC) is used as a metric. The worst value for the AUROC is "0.5", the best is "1". A value of "0.5" would equal a random guessing strategy and an optimal classifier would be described by an AUROC of "1".

5 Empirical Evaluation

Before the results of both algorithms are compared with each other, the optimal parameters for both algorithms have to be determined. Therefore, a k-fold cross-validation as, e.g., described by Witten and Frank [18] is conducted for different parameter combination. The parameters to estimate are the bandwidth for the KDE and the number of components as well as the optimal grid-size for the GMM.

For the cross-validation, each of the validation fold consists of the same ratio of normal and abnormal data in order to ensure that the folds are comparable to each other. The training fold has no anomalies at all. All in all, the available data is divided into k folds. In each step of the cross-validation, the model is trained with $k - 1$ folds and validated by using the remaining fold. The results for the optimal parameters using a 3-fold cross-validation are shown in Table 1. For each step in the cross-validation, the precision, recall, and f1-score are calculated. Finally, the cumulated means of these scores are determined and the parameter set with the highest f1-score is chosen as the best.

In Fig. 5a, b the precision, recall and f1-score for different bandwidths using the data in the Baltic area are shown. The difference between those two figures is the underlying grid. The model for Fig. 5a has no grid, while the model for Fig. 5b has a 3×3 grid. Comparing those figures, the one without grid performs better. Thus the parameter as shown in Table 1 are used in the evaluation. For the KDE, the grid-size is not an important parameter to optimize. Due to the main principle behind the KDE, only data points from the training set which are close to the evaluated data point will have an influence on the resulting probability density function. If the distance between data points from the training set and the evaluated data point is large, the

Table 1 Optimal Parameter

Area	KDE	GMM	
	Bandwidth	# Components	Grid Size
Fehmarn	0.06	75	5×5
Kattegat	0.09	50	3×3
Baltic	0.085	70	1×5

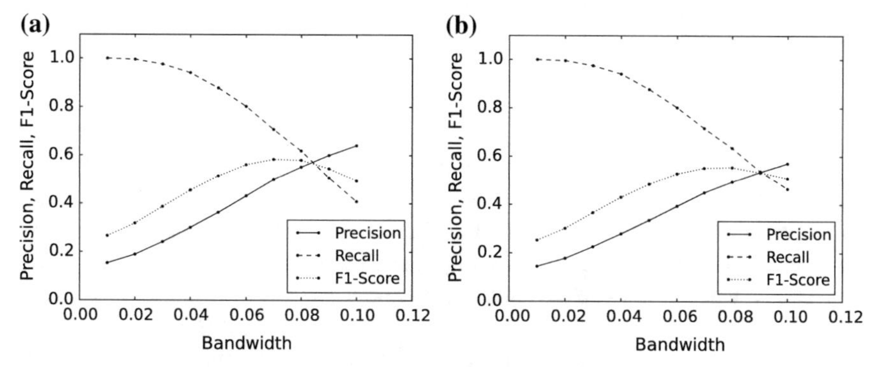

Fig. 5 Precision, recall, and f1-score for different bandwidths and different grids in the Baltic area

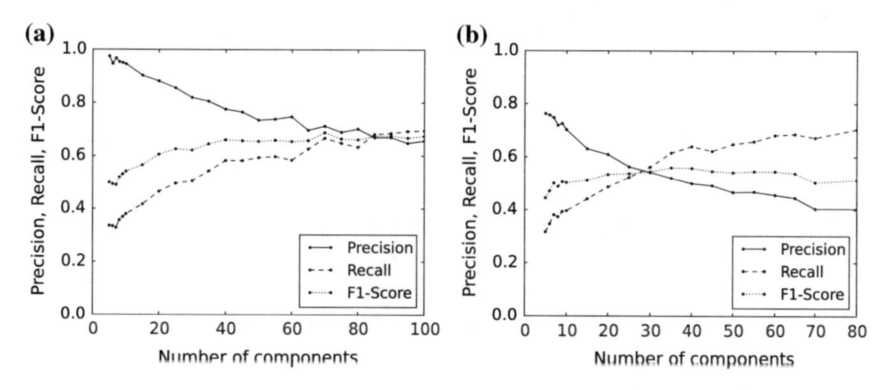

Fig. 6 Precision, recall, and f1-score for different numbers of components and different grids in the Baltic area

resulting value of the kernel function will tend to zero. Thus, omitting some points will only decrease the calculation time, which is not important for this work.

The results for varying the number of components in the Baltic area is shown in Fig. 6a, b. The first has a 1×5 grid, the second a 5×5 grid. These figures are only exemplary to show the different behavior of the algorithm with different parameter configurations. The best performance is achieved with the 1×5 grid. The number of components is chosen like shown in Table 1.

Table 2 Evaluation results

	Fehmarn		Kattegat		Baltic	
	KDE	GMM	KDE	GMM	KDE	GMM
Precision	0.5128	0.5607	0.3844	0.4675	0.5690	0.7013
Recall	0.6405	0.5428	0.4040	0.5250	0.5655	0.6227
F1-Score	0.5696	0.5515	0.3940	0.4946	0.5673	0.6597
Accuracy	0.8705	0.9144	0.9224	0.9376	0.8707	0.9037
FPR	0.0940	0.0456	0.0431	0.0369	0.0755	0.0467
AUROC	0.8719	0.8165	0.8514	0.7771	0.8656	0.8981

As the main task is the detection of anomalies, a data point which is detected and annotated as anomaly is a true positive. The results for the different areas using the different metrics are given in Table 2. Furthermore, the ROC for each area and algorithm combination is depicted in Fig. 7. For each fold, the ROC is drawn, with the AUROC value stated in the legend. Furthermore, the mean of the fold is depicted.

Comparing the scores of the algorithms as depicted in Table 2, it is clear, that neither of the algorithms delivers a good performance for the detection of anomalies. The overall performance of the GMM measured by the f1-score is always higher as the one of the KDE. In the Fehmarn area, both algorithms have nearly an equal score, while in the Kattegat area the GMM's f1-score is 25.5 % higher and in the Baltic area it is 16.3 % higher.

The accuracy of the GMM in all areas is higher than the KDE's. In the Fehrmarn area, it is 5 %, in the Kattegat area 1.6 %, and in the Baltic area 3.8 % higher. The FPR of the GMM is lower in all areas than the one of the KDE. Thus, the GMM's performance is always better for these metrics in the annotated areas.

Comparing the AUROC of the two algorithms, the KDE performs better in the Fehmarn area and the Kattegat area, while the GMM has a better result in the Baltic area. These results can also be derived by comparing the shape of the ROC curves in Fig. 7. The ROC curve of the GMM is less steep than the one of the KDE in the Fehmarn area and the Kattegat area, resulting in the lower AUROC value of the GMM in these particular areas.

Figures 8, 10, and 12 show the results for the GMM algorithm for the different folds in the Baltic area. The equivalent figures for the KDE algorithm are Figs. 9, 11, and 13. Similar figures for the other areas can be found in [2]. Both algorithms will perform quite well, if the data point which is to be analyzed is far away from the normal sea lanes as it can be seen in the southern part of each figure. Most of the points in the southern part are correctly identified as either anomalies or normal data points. A problem for both algorithms are the more sparsely and less dense trajectories to and from Trelleborg. These normal trajectories are found as abnormal behavior as seen, e.g., in Figs. 8 and 9.

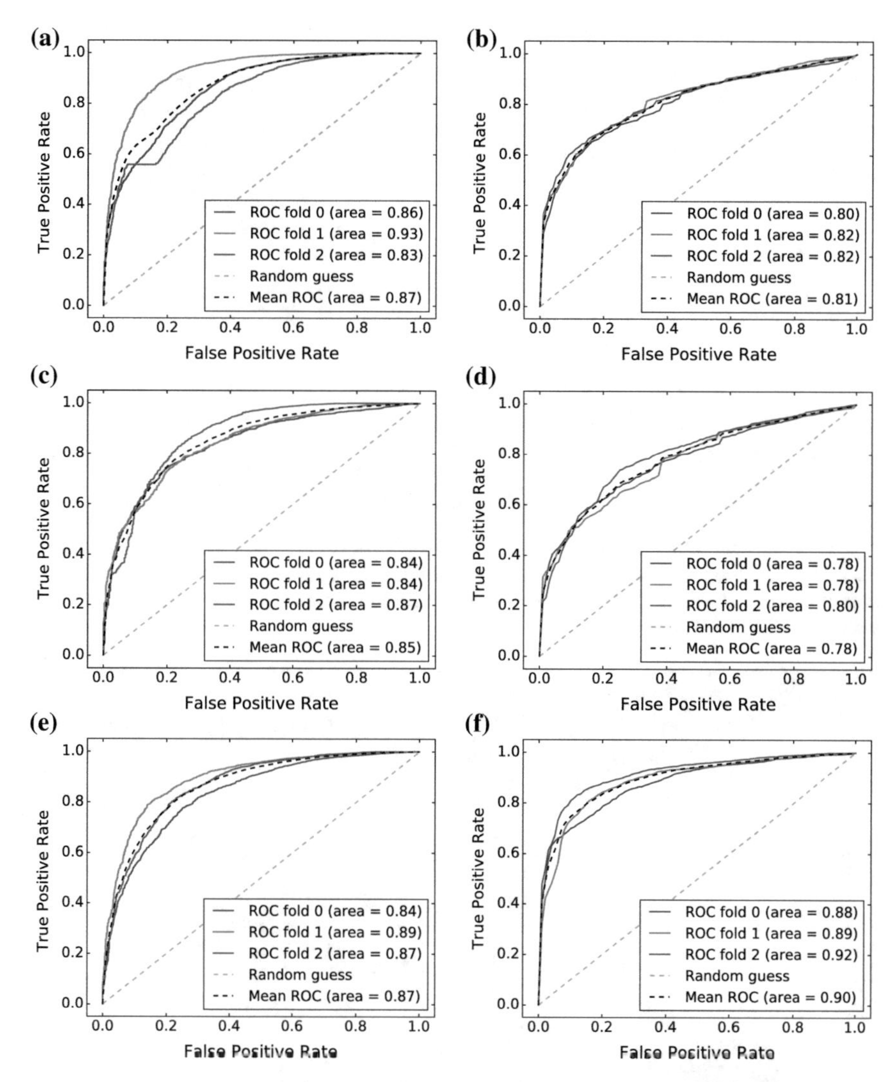

Fig. 7 The ROC for the KDE and GMM in the different areas. In each figure, the ROC curve for each fold as well as the mean ROC is shown. In the legend, the area under each curve is stated. **a** KDE—Fehmarn. **b** GMM—Fehmarn. **c** KDE—Kattegat. **d** GMM—Kattegat. **e** KDE—Baltic. **f** GMM—Baltic

If a trajectory is quite near to the normal behavior, and if the speed and heading of the vessel is similar to the normal model, the trajectory will not be identified as anomaly. In Figs. 12 and 13, this problem can be seen in the eastern region, where several vessels are entering the sea lanes at no distinct point.

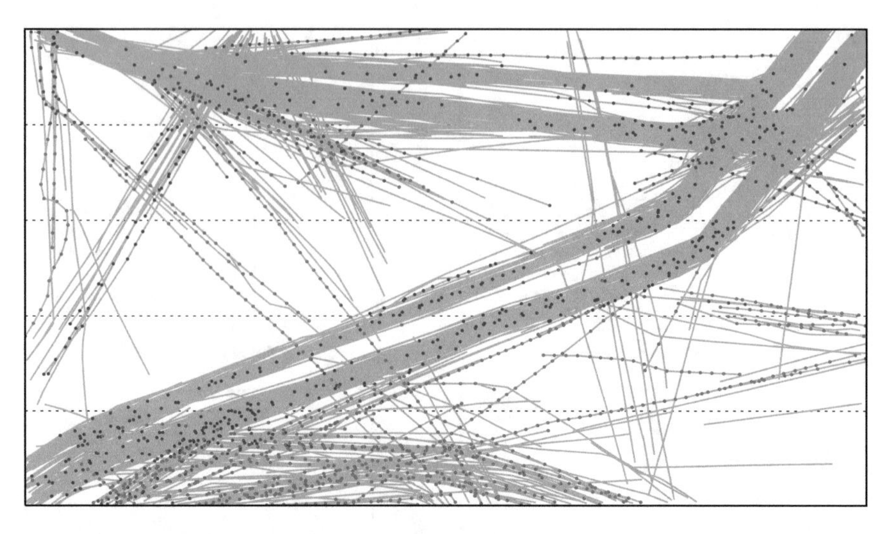

Fig. 8 GMM results for the Baltic area for one fold. The *grey lines* correspond to all trajectories, the *grey dotted lines* represent the used grid, the *dots* represent some evaluated data points. *Green dots* represent correctly found anomalies, *red dots* missed anomalies and *blue dots* normal points which are falsely declared as anomaly

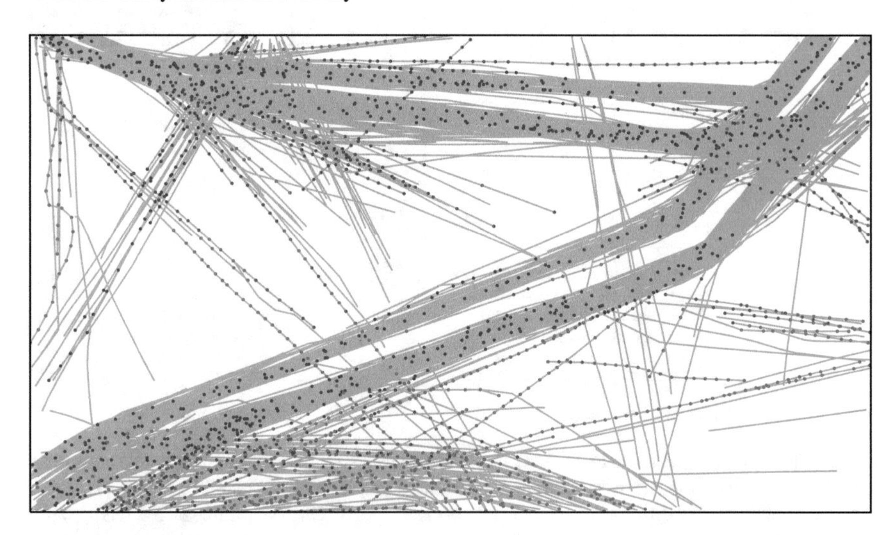

Fig. 9 KDE results for the Baltic area for one fold. The *grey lines* correspond to all trajectories, the *dots* represent some evaluated data points. *Green dots* represent correctly found anomalies, *red dots* missed anomalies and *blue dots* normal points which are falsely declared as anomaly

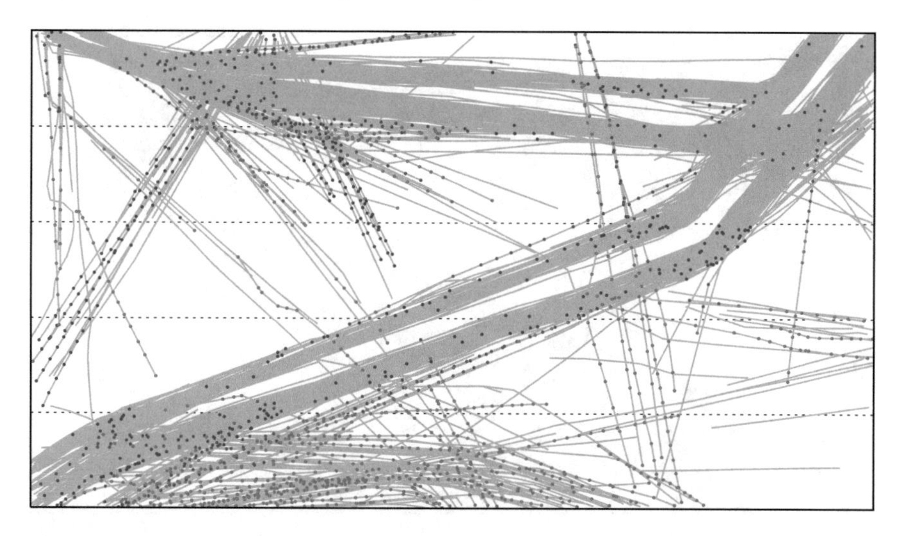

Fig. 10 GMM results for the Baltic area for one fold. The *grey lines* correspond to all trajectories, the *grey dotted lines* represent the used grid, the *dots* represent some evaluated data points. *Green dots* represent correctly found anomalies, *red dots* missed anomalies and *blue dots* normal points which are falsely declared as anomaly

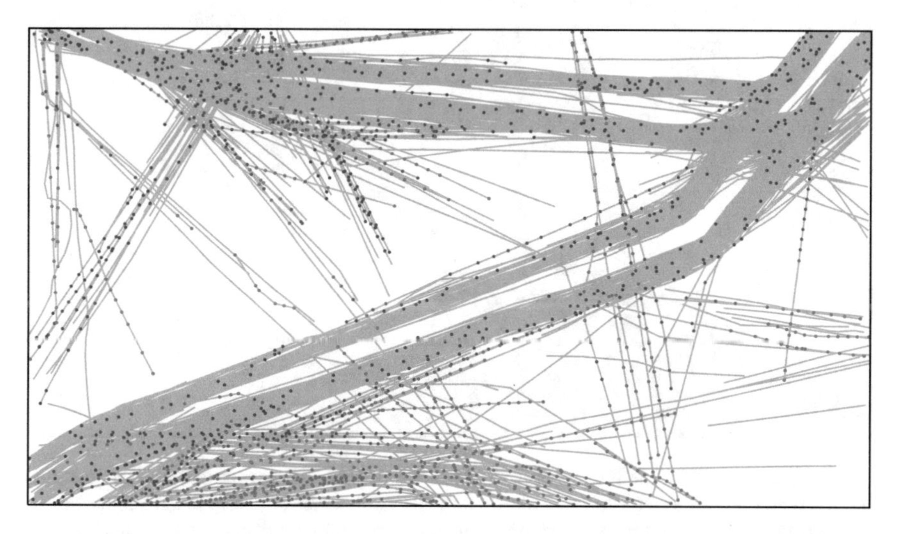

Fig. 11 KDE results for the Baltic area for one fold. The *grey lines* correspond to all trajectories, the *dots* represent some evaluated data points. *Green dots* represent correctly found anomalies, *red dots* missed anomalies and *blue dots* normal points which are falsely declared as anomaly

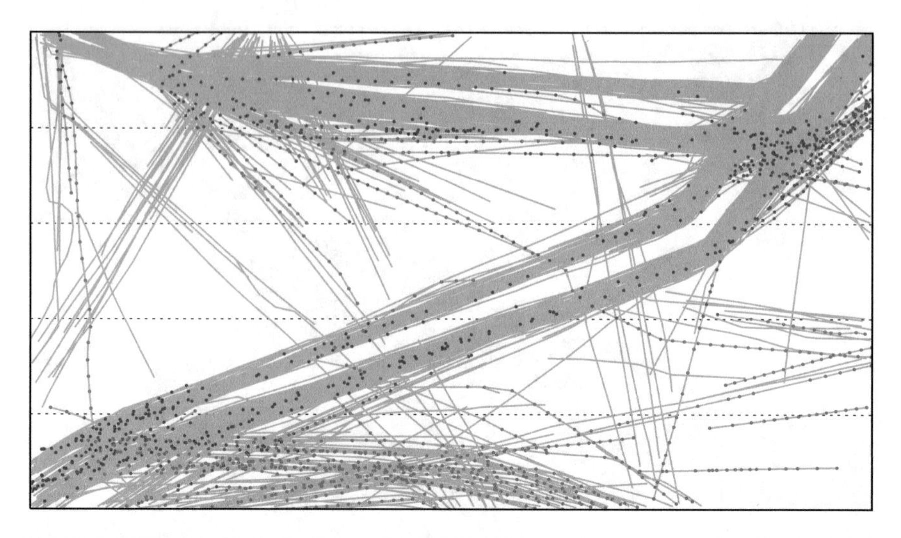

Fig. 12 GMM results for the Baltic area for one fold. The *grey lines* correspond to all trajectories, the *grey dotted lines* represent the used grid, the *dots* represent some evaluated data points. *Green dots* represent correctly found anomalies, *red dots* missed anomalies and *blue dots* normal points which are falsely declared as anomaly

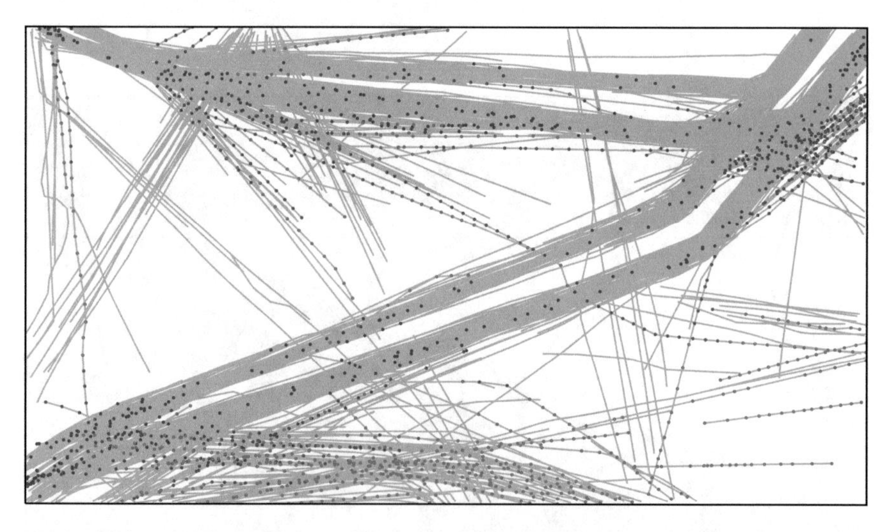

Fig. 13 KDE results for the Baltic area for one fold. The *grey lines* correspond to all trajectories, the *dots* represent some evaluated data points. *Green dots* represent correctly found anomalies, *red dots* missed anomalies and *blue dots* normal points which are falsely declared as anomaly

Another problem can occur, if there is not enough data in one of the grid-cells to build a normal model. In this case, there will be no valid model for the decision about abnormal behavior. Therefore, a different strategy has to be chosen; e.g., all data points in the cell are anomalies, or there are no anomalies in the cell. Here, all points in those cells are marked as anomaly. This problem arises especially in the Fehmarn area as shown in [2]. Furthermore, several falsely classified anomalies occur in all areas with both algorithms (blue dots).

6 Conclusion

The two algorithms generate a large amount of false positives and false negatives. Therefore, the results are not as good as it would be expected for a support system. Both algorithm estimate the underlying PDF of the sea traffic. For the GMM, the PDF is expected to consists of superimposed multivariate normal distributions, while the real PDF is unknown and might be of a different kind. This might be a reason for the result. A KDE is able to estimate an arbitrary PDF, if enough training data is available. If not enough data is available, the resulting PDF might differ significantly from the true PDF.

Both algorithms only evaluate a single point of the trajectory at a time. Therefore, the whole trajectory is never considered. Thus, Trajectories as shown in Fig. 14 will probably not be recognized correctly. Each data point of the orange trajectory for itself might be detected as normal behavior resulting in a normal label for the whole trajectory. An anomaly like this can only be recognized by evaluating the whole trajectory.

By using a grid to divide the area, the following problems might occur: In the border region between two cells, the grid can perform worse than using no grid at all. The EM algorithm fits the components of the GMM to the underlying data of each cell separately. The data in each cell abruptly ends at the grid border. Hence, it

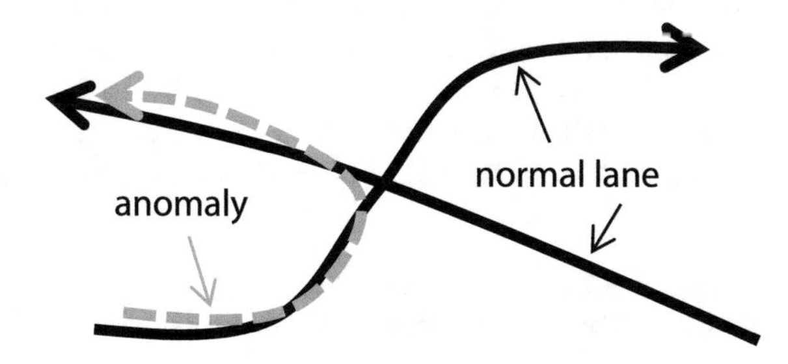

Fig. 14 Problem with point only evaluation. Two trajectories (each is only valid in the direction of the arrow) are crossing. An abnormal behavior is depicted as an *orange dashed line*. It starts on one trajectory and changes to the other during the crossing. Depending on the context, this can be considered as anomaly

is likely to be less dense at the border compared to the center of a cell. This might likely result in the placement of the components of a GMM in the center instead of the border, even though globally observed the data might have the same density at the center and at the border.

Furthermore, if the grid was chosen unfavorably, the resulting model will not be able to learn a sea lane properly, because the grid might divide a sea lane or cut out parts of a sea lane. A similar problem might occur, if there are not enough data points in a cell for the estimation of a normal model. For this case, different strategies are possible, e.g., every point in these areas is marked as anomaly. This might result in cutting sea lanes with normal behavior and detecting this normal behavior as anomaly.

7 Future Work

Even though, the optimal parameters are estimated, the same parameters are used for all grid-cells. Therefore, an improvement could be achieved by estimating the model parameters for each cell separately, respectively to use an adaptive approach for the bandwidth estimation for the KDE. Thus, the difference in density and complexity of each local area would be taken into account.

Currently, only quite simple algorithms for the anomaly detection are examined. The performance of these algorithms was suboptimal. Therefore, the next step is to compare more sophisticated algorithms. These algorithms should consider past points of a track while evaluating a new point. By incorporating the additional information provided using whole or partial trajectories, the results should improve compared to the density estimation using a GMM or a KDE.

Another open point is the annotation of the whole dataset and not only some artificial subsets. Currently, only a small subset of the whole dataset is annotated and inspected during the evaluation. By using the whole dataset, a better overview of the observed area can be used to get a better understanding of the normal behavior, and thus to improve the models. Due to the greater amount of data, the methods of annotating the data must be reconsidered and improved. e.g., to achieve better model results the tracks could be annotated by domain experts or by using another strategy to ensure a consistent and reliable annotation. Also, using sea-maps to gain a better understanding of the sea lanes, shoals etc. will help to improve the annotated ground truth.

Acknowledgments The underlying projects to this article are funded by the WTD 81 of the German Federal Ministry of Defense. The authors are responsible for the content of this article.

References

1. Andersson, M., Johansson, R.: Multiple sensor fusion for effective abnormal behaviour detection in counter-piracy operations. In: Proceedings of the International Waterside Security Conference (WSS), pp. 1–7 (2010)
2. Anneken, M., Fischer, Y., Beyerer, J.: Evaluation and comparison of anomaly detection algorithms in annotated datasets from the maritime domain. In: Proceedings of the SAI Intelligent Systems Conference (2015)
3. Anneken, M., Fischer, Y., Beyerer, J.: Anomaly detection using b-spline control points as feature space in annotated trajectory data from the maritime domain. In: Proceedings of the 8nd International Conference on Agents and Artificial Intelligence (2016). (accepted paper)
4. Barber, D.: Bayesian Reasoning and Machine Learning. Cambridge University Press, cambridge (2014)
5. Brax, C., Niklasson, L.: Enhanced situational awareness in the maritime domain: an agent-based approach for situation management. In: Proc. SPIE, vol. 7352, pp. 735,203–735,203–10 (2009)
6. Chandola, V., Banerjee, A., Kumar, V.: Anomaly detection: a survey. ACM Comput. Surv. **41**(3), 15:1–15:58 (2009)
7. Fischer, Y., Beyerer, J.: Ontologies for probabilistic situation assessment in the maritime domain. In: Proceedings of the IEEE International Multi-Disciplinary Conference on Cognitive Methods in Situation Awareness and Decision Support (CogSIMA), pp. 102–105 (2013)
8. Fischer, Y., Reiswich, A., Beyerer, J.: Modeling and recognizing situations of interest in surveillance applications. In: Proceedings of the IEEE International Inter-Disciplinary Conference on Cognitive Methods in Situation Awareness and Decision Support (CogSIMA), pp. 209–215 (2014)
9. Guillarme, N.L., Lerouvreur, X.: Unsupervised extraction of knowledge from s-ais data for maritime situational awareness. In: Proceedings of the 16th International Conference on Information Fusion (FUSION), pp. 2025–2032 (2013)
10. Laxhammar, R., Falkman, G.: Sequential conformal anomaly detection in trajectories based on hausdorff distance. In: Proceedings of the 14th International Conference on Information Fusion (FUSION), pp. 1–8 (2011)
11. Laxhammar, R., Falkman, G., Sviestins, E.: Anomaly detection in sea traffic - a comparison of the gaussian mixture model and the kernel density. In: Proceedings of the 12th International Conference Information Fusion (FUSION) (2009)
12. Manning, C.D., Raghavan, P., Schütze, H.: Introduction to Information Retrieval. Cambridge University Press, Cambridge (2008)
13. Morris, B., Trivedi, M.: A survey of vision-based trajectory learning and analysis for surveillance. IEEE Trans. Circuits Syst. Video Technol. **18**(8), 1114–1127 (2008)
14. Murphy, K.P.: Machine learning: a probabilistic perspective. MIT Press, Cambridge (2012)
15. Pedregosa, F., Varoquaux, G., Gramfort, A., Michel, V., Thirion, B., Grisel, O., Blondel, M., Prettenhofer, P., Weiss, R., Dubourg, V., Vanderplas, J., Passos, A., Cournapeau, D., Brucher, M., Perrot, M., Duchesnay, E.: Scikit-learn: machine learning in Python. J. Mach. Learn. Res. **12**, 2825–2830 (2011)
16. Shafer, G., Vovk, V.: A tutorial on conformal prediction. J. Mach. Learn. Res. **9**, 371–421 (2008)
17. de Vries, G.K.D., van Someren, M.: Machine learning for vessel trajectories using compression, alignments and domain knowledge. Expert Syst. Appl. **39**(18), 13,426 – 13,439 (2012)
18. Witten, I.H., Frank, E.: Data Minig: Practical Machine Learning Tools and Techniques with Java Implementations, 2nd edn. Morgan Kaufmann Publishers, Burlington (2005)

Using Fuzzy PROMETHEE to Select Countries for Developmental Aid

Eric Afful-Dadzie, Stephen Nabareseh, Zuzana Komínková Oplatková and Peter Klimek

Abstract Wealthy nations continue to demonstrate their unwavering support to improving conditions and the general well-being of poor countries in spite of the recent economic crises. However, as developmental aid relatively shrinks, both Aid donors and recipient countries have shown keen interest in methodologies used in evaluating developmental assistance programs. Evaluation of aid programs is seen as a complex task mainly because of the several non-aid factors that tend to affect overall outcomes. Adding to the complexity are the subjective sets of criteria used in Aid evaluations programs. This paper proposes a two stage framework of fuzzy TOPSIS and sensitivity analysis to demonstrate how aid-recipient countries can be evaluated to deepen transparency, fairness, value for money and sustainability of such aid programs. Using the Organisation for Economic Co-operation and Development (OECD) set of subjective criteria for evaluating aid programs; a numerical example pre-defined by linguistic terms parameterized by triangular fuzzy numbers is provided to evaluate aid programs. Fuzzy PROMETHEE is used in the first stage to evaluate and rank aid-recipients followed by a comparative analysis with Fuzzy VIKOR and Fuzzy TOPSIS to ascertain an accurateness of the method used. A sensitivity analysis is further added that anticipates possible influences from lobbyists and examines the effect of that bias in expert ratings on the evaluation process. The result shows a framework that can be employed in evaluating aid effectiveness of recipient-countries.

E. Afful-Dadzie · Z.K. Oplatková
Faculty of Applied Informatics, Tomas Bata University in Zlin,
Zlin, Czech Republic
e-mail: afful@fai.utb.cz

Z.K. Oplatková
e-mail: kominkova@fai.utb.cz

S. Nabareseh (✉) · P. Klimek
Faculty of Management and Economics, Tomas Bata University in Zlin,
Zlin, Czech Republic
e-mail: nabareseh@fame.utb.cz

P. Klimek
e-mail: klimek@fame.utb.cz

Y. Bi et al. (eds.), *Intelligent Systems and Applications*,
Studies in Computational Intelligence 650, DOI 10.1007/978-3-319-33386-1_6

Keywords Developmental aid programs · Fuzzy set theory · Organization for Economic Cooperation and Development (OECD) · Fuzzy PROMETHEE · Fuzzy VIKOR · Fuzzy TOPAIA · Fuzzy MCDM · Evaluation · Sensitivity analysis

1 Introduction

Despite the incessant call on donor countries for a budget reduction, most organisations still release billions in aid of developing and poor countries. Some of these organisations include the Organization for Economic Cooperation and Development (OECD), UK's Department for International Development (DFID), the United States Agency for International Development (USAID), Canada's Country Indicators for foreign Policy (CIFP), the African Development Bank (AfDB) and the World Bank. A net amount of 132 billion dollars was spent jointly in 2012 and rose to 134 billion in 2013 by wealthy nations to reduce poverty and improve developmental conditions of poor nations [1]. The fall in developmental aid in the past years triggered stringent measures to ensure that donors and recipients alike are thoroughly evaluated to ensure the overall sustainability of developmental aid programs [2–4]. This phenomenon, adding to the lack of consensus among researchers about the impact of aid on economic growth [5–8] and the several instances of changes in aid allocation criteria [9, 10] explain the need for robust methodologies to evaluate developmental aid.

Most proposed criteria for appraising development aid are more often subjective apparently because of other non-aid inputs that have the propensity to influence total outcome. Economic recession, food and energy prices, interest rates, trade credits and among others are some non-aid factors capable of affecting the general outcome of an aid evaluation program [2]. The OECD currently adopts five subjective criteria in evaluating their developmental aid programs. These criteria are *relevance*, *effectiveness*, *efficiency*, *impact* and *sustainability* used to evaluate developmental aid programs for a particular country. The selection of beneficiaries for aid is sometimes challenging when fuzzy criteria is used. To address this challenge, this paper recommends a fuzzy PROMETHEE framework for evaluating countries involved in developmental aid program performance ranking. The method helps to track the progress of countries whiles ensuring that future aid allocations are based on performance of previous aid programs. The rest of the paper is presented as follows: Modeling uncertainty with fuzzy set theory is briefly explained followed by the definition, review of relevant literature and steps in fuzzy PROMETHEE method. Finally, a numerical example of how fuzzy PROMETHEE could help to evaluate and rank participating countries in developmental assistance programs is presented.

2 Modelling Uncertainty with Fuzzy Sets

Zadeh [11] introduced the fuzzy set theory to tackle issues of uncertainty, imprecision and vagueness in information that are not statistical in nature. The fuzzy sets concept is hinged on a relative graded membership and has been applied extensively in subjective modeling mostly in multi-criteria decision making (MCDM) environments. In fuzzy MCDM, the subjective criteria are represented by linguistic variables which are further expressed with linguistic terms [12]. The following presents the definitions with basic operations of the fuzzy set theory.

2.1 *Fuzzy Set*

Let X be a nonempty set, the universe of discourse $X = \{x_1, x_2, \ldots, x_n\}$. A fuzzy set A of X is a set of ordered pairs $\{(x_1, f_A(x_1)), (x_2, f_A(x_2)), \ldots, (x_n, f_A(x_n))\}$, characterized by a membership function $f_A(x)$ that maps each element x in X to a real number in the interval [0, 1]. The function value $f_A(x)$ stands for the membership degree x in A. This paper uses the Triangular Fuzzy Number (TFN) defined below for evaluation.

2.2 *Triangular Fuzzy Number*

In triangular fuzzy number (TFN), the membership function is expressed as a triplet (f, g, h). The membership function $f_A(x)$ of the triangular fuzzy number is illustrated in Fig. 1 and defined as:

$$f_A(x) = \begin{cases} 0 & x < f \\ \frac{x-f}{g-f}, & f \le x \le g \\ \frac{h-x}{h-g}, & g \le x \le h \end{cases}$$

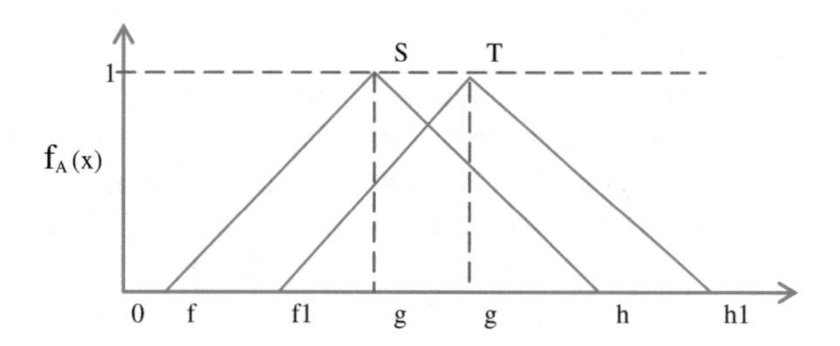

Fig. 1 Two triangular fuzzy numbers

The value of x at g gives the maximal value of $f_A(x)$, that is $f_A(x) = 1$. The value x at f represents the minimal grade of $f_A(x)$, i.e. $f_A(x) = 0$. The constants f and h stand for the lower and upper bounds of the available area data respectively. According to [13], fuzzy models using TFNs are effective for solving decision-making problems with subjective and vague available information. The TFNs are used in very practical applications because of the computational efficiency and its simplicity.

2.3 Basic Fuzzy Sets Operations

Supposing $S = (f, g, h)$ and $T = (f_1, g_1, h_1)$ are two TFNs as shown in Fig. 1, then the basic operations on these two TFNs are as follows:

$$S \oplus T = (f, g, h) + (f_1, g_1, h_1) = (f + f_1, g + g_1, h + h_1) \quad (1)$$

$$S - T = (f, g, h) - (f_1, g_1, h_1) = (f - h_1, g - g_1, h - f_1) \quad (2)$$

$$S \times T = (f, g, h) \times (f_1, g_1, h_1) = (ff_1, gg_1, hh_1) \quad (3)$$

$$S \div T = (f, g, h) \div (f_1, g_1, h_1) = \left(\frac{f}{h_1}, \frac{g}{g_1}, \frac{h}{f_1}\right) \quad (4)$$

Let $S = (f, g, h)$ and $S = (f_1, g_1, h_1)$ be two TFNs depicted in Fig. 1. The distance between them is computed using the vertex method in Eq. 6:

$$d(S, T) = \sqrt{1/3\left[(f - f_1)^2 + (g - g_1)^2 + (h - h_1)^2\right]} \quad (5)$$

3 Fuzzy PROMETHEE

The Preference Ranking Organization Method for Enrichment (PROMETHEE) is an extensively accepted outranking methods in Multiple Criteria Decision Making (MCDM). Brans and Vincke [14] proposed the method which performs a pairwise comparison of pairs of alternatives and grades them between a [0, 1] interval [15] using a preference function. The PROMETHEE method is preferred in ranking and selecting alternatives due to its robustness in comparing the performances of alternatives and considers it in the composite ranking. Just as in other MCDM methods, there is a fuzzy extension of the PROMETHEE method when dealing with uncertain and subjective data. Fuzzy PROMETHEE has been applied in varied areas such as health care management [16], information systems outsourcing [15], logistics [17], customer reviews [18], landslide susceptibility mapping [19] among many others. Fuzzy PROMETHEE has equally seen improvements in a number of variants, that is versions (PROMETHEE I, II, III, IV, V, VI), and extensions as seen in [20–23].

This paper uses a combination of PROMETHEE I and II. PROMETHEE I deals with a partial ranking of alternatives [14, 21, 22]. The sum of indices, $\pi\ (m, l)$, initially determines the preference of alternative m over the other alternatives considered. This is referred to as the 'leaving flow' $\phi^+(m)$, and implies the relative good performance of m over the other alternatives. The alternative with the highest 'leaving flow' is pronounced the best in the evaluation. Likewise, the sum of indices, $\pi\ (l, m)$ is calculated to indicate the preferences of all other alternatives measured against m. This is also denoted as the 'entering flow' $\phi^-(m)$, and implies the dependency of alternative m in relation to the rest of the alternatives. PROMETHEE II however introduces a net flow $\phi(m)$ which denotes the difference between the leaving and the entering flows and helps to realize a full ranking. The alternative with the highest net flow is therefore declared the best alternative.

Below is a step by step outline of definitions and formulae of the fuzzy PROMETHEE methodology culled from [14, 15, 21, 22]. The methodology is adopted in the numerical example in selecting countries for developmental aid programs.

Step 1a: Determination of linguistic Variables (criteria), linguistic terms, alternatives and decision makers

The first step is to determine the linguistic variables and its associated linguistic terms, the alternatives and number of decision makers needed in the decision making process. This set of information is what is used to construct the decision matrix. The linguistic terms are translated into fuzzy numbers and used to rate the linguistic variables. The linguistic terms are qualitative words which reflect the subjective view of an expert or decision maker about the criteria per each alternative under consideration [12, 24, 25]. This linguistic terms with their TFNs for this paper are shown in Tables 1 and 2 respectively that captures on a scale of 0–1 the importance criteria and the alternatives.

Step 1b: Determination of Importance Criteria Weights

Decision makers determine the importance or weight of each criterion using the linguistic terms in Table 2. In Eq. 6 below, w_j denotes the weight of the jth criterion C_j based on the linguistic preference assigned by a decision maker. It is noted that each weight $\tilde{w}_j^k = (w_{j1}^k, w_{j2}^k, w_{j3}^k)$ is expressed as a TFN.

$$\tilde{W} = \left[\tilde{w}_1, \tilde{w}_2, \ldots, \tilde{w}_n\right], \quad j = 1, 2, \ldots, n \tag{6}$$

Step 2a: Construction of the fuzzy decision matrix

In a situation where m alternatives and n criteria are offered to k decision-makers, $(D_1, D_2, \ldots, D_k)$ to choose the best alternative, a fuzzy MCDM problem as seen in Eq. 7 can be stated in the form of a matrix.

Table 1 Linguistic scale for the importance of criterion

Linguistic terms	Triangular fuzzy number
Very Low (VL)	(0.0, 0.1, 0.3)
Low (L)	(0.1, 0.3, 0.5)
Medium (M)	(0.3, 0.5, 0.7)
High (H)	(0.5, 0.7, 0.9)
Very High (VH)	(0.7, 0.9, 1.0)

Source Authors'

Table 2 Linguistic terms for alternative ratings

Linguistic terms	Triangular fuzzy number
Very Low (VL)	(0.0, 0.0, 2.5)
Low (L)	(0.0, 2.5, 5.0)
High (H)	(2.5, 5.0, 7.5)
Very High (VH)	(5.0, 7.5, 10.0)
Extremely High (EH)	(7.5, 10.0, 10.0)

Source Authors'

$$\tilde{D} = \begin{matrix} \\ A_1 \\ A_2 \\ \vdots \\ A_m \end{matrix} \begin{matrix} C_1 & C_2 & & C_n \\ \left[\begin{matrix} \tilde{x}_{11} \\ \tilde{x}_{21} \\ \vdots \\ \tilde{x}_{m1} \end{matrix}\right. & \begin{matrix} \tilde{x}_{12} \\ \tilde{x}_{22} \\ \vdots \\ \tilde{x}_{m2} \end{matrix} & \begin{matrix} \cdots \\ \cdots \\ \ddots \\ \cdots \end{matrix} & \left.\begin{matrix} \tilde{x}_{1n} \\ \tilde{x}_{2n} \\ \vdots \\ \tilde{x}_{mn} \end{matrix}\right] \end{matrix}, \quad i = 1, 2, \ldots, m; j = 1, 2, \ldots, n \tag{7}$$

where $\tilde{x}_{ij}$ is the rating of alternative A_i with respect to criterion, both expressed in TFNs. This implies that the rating of a decision maker k is $\tilde{x}_{ij}^k = \left(r_{ij}^k, u_{ij}^k, v_{ij}^k\right)$.

Step 2b: Aggregation of decisions

This stage aggregates the fuzzy weights of the criteria and the alternative ratings. This is done respectively by using the interval valued technique as illustrated in Eqs. 8 and 9 below.

$$\tilde{w}_j = \frac{1}{n}\left[\tilde{w}_j^1 + \tilde{w}_j^2 + \cdots, +\tilde{w}_j^n\right] \tag{8}$$

$$\tilde{x}_{ij} = \frac{1}{n}\left[\tilde{x}_{ij}^1 + \tilde{x}_{ij}^2 + \cdots, +\tilde{x}_{ij}^n\right] \tag{9}$$

Step 3: Normalization of the decision matrix

This step normalizes the aggregated fuzzy decision matrix gotten from step 2b above. The normalized fuzzy decision matrix is defined as in Eq. 10 and computed using Eq. 11 below. The result of the normalized matrix is still a TFN.

$$\tilde{S} = \left[\tilde{s}_{ij}\right]_{m\times n}, \ i = 1, 2, \ldots, m; j = 1, 2, \ldots, n \tag{10}$$

$$\tilde{s}_{ij} = \left(\frac{\tilde{r}_{ij}}{v_j^+}, \frac{\tilde{u}_{ij}}{v_j^+}, \frac{\tilde{v}_{ij}}{v_j^+}\right) v_j^+ = \max_i v_{ij} \tag{11}$$

Step 4: Construction of the fuzzy preference function

The fuzzy preference function $\tilde{P}_j(m, n)$ is calculated to describe the decision-makers' preference among the pairs of alternatives in this step. The usual-criterion, quasi-criterion, criterion with linear preference, level-criterion, criterion with Linear Preference and indifference area, and the Gaussian-criteria are six different types of preference functions that range between [0, 1] as presented by [14]. The usual-criterion function (Type I) is employed in the paper and defined in Eq. 12 below.

$$\tilde{P}_j(m, n) = \begin{cases} 0, \ \tilde{s}_{mj} \leq \tilde{s}_{nj} \\ 1 \ \ \tilde{s}_{mj} > \tilde{s}_{nj} \end{cases} \quad j = 1, 2, \ldots, k \tag{12}$$

Step 5: Computation of weighted aggregated preference function

The weighted aggregated preference function is computed using Eq. 13 below.

$$\tilde{\pi}(m, l) = \sum_{j=1}^{k} \tilde{P}_j(m, n)\tilde{w}_j \tag{13}$$

where $\tilde{w}_j$ signifies the importance weight of the criteria

Step 6: Computation of the leaving, entering and net flows

In this step, each alternative is related to $(n - 1)$ alternatives that results in either a positive or negative flow [16, 23]. The approach calculates the leaving, entering and net flows using Eqs. 14, 15 and 16 below respectively.

$$\text{Leaving flow: } \tilde{\phi}^+(m) = \frac{1}{n-1}\sum_{m\neq l} \tilde{\pi}(m, l), \quad \forall m, l \in A, \tag{14}$$

$$\text{Entering flow:} \quad \tilde{\phi}^-(m) = \frac{1}{n-1}\sum_{m\neq l} \tilde{\pi}(l, m), \quad \forall m, l \in A, \tag{15}$$

where n is the number of alternatives.

Step 7: Establishing ranking

This step uses PROMETHEE II for a full ranking using the net flow as shown in Eq. 16.

$$\text{Net flow: } \tilde{\phi}(m) = \tilde{\phi}^+(m) - \tilde{\phi}^-(m), \forall m \in A. \tag{16}$$

4 Application

The fuzzy PROMETHEE method is applied in this numerical example to evaluate countries applying for developmental aid programmes. The example adopts the OECD criteria currently used in evaluating developmental aid programs. The said set of criteria are *relevance*, *effectiveness*, *efficiency*, *impact* and *sustainability* presented in the framework in Fig. 2 below. We use the five criteria by the OECD and eight arbitrary alternatives (Countries) within the lower middle income group of the World Bank. Alternatives used in this numerical example are Nepal, Myanmar, Afghanistan, Somalia Benin, Chad, Ethiopia and Haiti in no particular order.

The first step of the fuzzy PROMETHEE method illustrated above is the determination of the linguistic variables, linguistic terms, the alternatives and the decision makers. Tables 1 and 2 below present the linguistic terms for the importance criteria and the alternative ratings respectively and their respective TFNs. Figure 2 further outlines the criteria and alternatives adopted in this paper.

Using the TFNs in Table 1 and the criteria ratings by decision makers in Table 3, the importance weights are computed by using Eq. 6. The aggregation is done with Eq. 8. Criteria 4 (C4) is rated as the most important criterion in evaluating and selection countries for developmental aid as seen in Table 3 below. This is followed by C3, C5, C2 and C1 in that order.

The second step in the fuzzy PROMETHEE method is constructing a decision matrix using Eq. 7 and aggregated with Eq. 9. The 5-set linguistic terms in Table 2 (*Very Low-VL; Low-L; High-H; Very High-VH; Extremely High-EH*) were used by the decision makers in assessing countries for developmental aid. Table 4 presents the decision matrix of the three decision makers used in this example.

The third step normalized the aggregated decision matrix using Eq. 11. Computation of the preference function to describe the decision-makers' aggregated preference between pairs of alternatives is presented in the fourth step. As indicated above, the paper adopted the 'usual criterion' presented in Eq. 12 for the computation. The pairwise preferences are presented in Table 5 below.

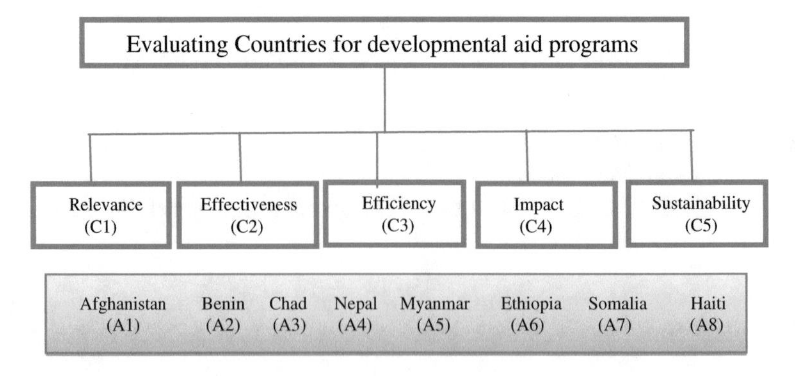

Fig. 2 Conceptual framework for selecting beneficiaries. *Source* Authors'

Table 3 Importance weight criterion

	D1	D2	D3	Importance weight
C1	L	L	VL	(0, 0.233, 0.5)
C2	M	L	M	(0.1, 0.433, 0.07)
C3	H	H	M	(0.1, 0.633, 0.9)
C4	VH	VH	H	(0.5, 0.833, 1)
C5	L	H	M	(0.1, 0.500, 0.9)

Source Authors'

Table 4 Alternative ratings by decision makers

Decision maker 1

	A1	A2	A3	A4	A5	A6	A7	A8
C1	VL	L	H	VH	H	VL	EH	H
C2	H	H	VH	H	VL	L	EH	H
C3	L	VL	L	H	H	H	VH	L
C4	H	VH	L	EH	VL	L	VH	L
C5	VL	VH	L	H	L	VL	VH	H

Decision maker 2

	A1	A2	A3	A4	A5	A6	A7	A8
C1	VH	H	VH	L	VL	H	EH	VH
C2	VL	EH	H	VH	L	H	VH	H
C3	EH	H	L	VL	H	VH	VH	H
C4	VL	H	L	H	H	VH	VH	L
C5	EH	H	L	VH	EH	H	EH	EH

Decision maker 3

	A1	A2	A3	A4	A5	A6	A7	A8
C1	EH	H	H	VL	L	L	VH	EH
C2	H	VH	H	L	L	H	VH	L
C3	H	EH	VL	L	VH	L	VH	H
C4	H	VL	H	H	EH	H	VH	VL
C5	VL	L	VH	EH	L	L	VH	H

Source Authors'

The weighted aggregated preference function is then calculated in step five using Eq. 13 and presented below in Table 6.

Step six computes the leaving, entering and net flows Eqs. 14–16 respectively. Table 7 below presents the computed leaving, entering, net flows and the alternative ranking. Figure 3 also presents the partial preorder outranking and the preorder outranking giving alternative 7 (A7) as the best alternative.

The partial preorder outranking is denoted by Fig. 3a as a partial ranking of alternatives while Fig. 3b is the full ranking that signifies the best alternative with chronological succeeding alternatives.

Table 5 Pairwise preference functions for alternatives

	C1	C2	C3	C4	C5
P(A1, A2)	1	0	0	0	0
P(A1, A3)	0	0	1	0	0
P(A1, A4)	0	0	1	0	0
P(A1, A5)	1	1	0	0	0
P(A1, A6)	1	0	0	0	1
P(A1, A7)	0	0	0	0	0
P(A1, A8)	0	0	1	1	0
P(A2, A1)	0	1	0	1	0
P(A2, A3)	0	0	1	1	0
P(A2, A4)	0	0	1	0	0
P(A2, A5)	0	1	0	0	0
P(A2, A6)	0	1	0	0	1
P(A2, A7)	0	0	0	0	0
P(A2, A8)	0	1	1	1	0
P(A3, A1)	1	1	1	1	1
P(A3, A2)	0	1	1	1	1
P(A3, A4)	0	0	0	0	0
P(A3, A5)	1	1	0	0	0
P(A3, A6)	1	1	0	0	1
P(A3, A7)	0	0	0	0	0
P(A3, A8)	0	1	0	1	0
P(A4, A1)	0	1	0	1	0
P(A4, A2)	1	0	0	0	0
P(A4, A3)	0	0	1	1	0
P(A4, A5)	1	1	0	0	0
P(A4, A6)	1	1	0	0	0
P(A4, A7)	0	0	0	0	0
P(A4, A8)	0	1	0	1	0
P(A5, A1)	0	0	0	1	0
P(A5, A2)	0	0	0	0	0
P(A5, A3)	0	0	1	1	0
P(A5, A4)	0	0	1	0	0
P(A5, A6)	0	0	0	0	1
P(A5, A7)	0	0	0	0	0
P(A5, A8)	0	0	1	1	0
P(A6, A1)	0	0	0	1	0
P(A6, A2)	0	0	0	0	0
P(A6, A3)	0	0	1	1	0
P(A6, A4)	0	0	1	0	0
P(A6, A5)	0	1	0	0	0

(continued)

Table 5 (continued)

	C1	C2	C3	C4	C5
P(A6, A7)	0	0	0	0	0
P(A6, A8)	0	0	1	1	0
P(A7, A1)	0	1	0	1	0
P(A7, A2)	1	0	0	0	0
P(A7, A3)	0	0	1	1	0
P(A7, A4)	0	0	1	0	0
P(A7, A5)	1	1	0	0	1
P(A7, A6)	1	1	0	0	1
P(A7, A8)	0	1	1	1	0
P(A8, A1)	0	0	0	0	0
P(A8, A2)	1	0	0	0	0
P(A8, A3)	0	0	1	0	0
P(A8, A4)	0	0	0	0	0
P(A8, A5)	0	1	0	0	0
P(A8, A6)	1	0	0	0	1
P(A8, A7)	0	0	0	0	0

Source Authors'

Table 6 Weighted aggregated preference function

	A1	A2	A3	A4	A5	A6	A7	A8
A1		0.733	1.633	1.633	1.970	2.230	0	3.970
A2	3.567		3.967	1.633	1.230	2.730	0	5.200
A3	7.433	6.7		0	1.970	3.470	0	3.570
A4	3.567	0.733	3.967		1.970	1.970	0	3.570
A5	2.333	0	3.967	1.633		1.500	0	3.970
A6	2.333	0	3.967	1.633	1.230		0	3.970
A7	3.567	0.733	3.967	1.633	3.470	3.470		5.2000
A8	0	0.733	1.633	0	1.230	2.230	0	

Source Authors'

5 Comparative Analysis

5.1 Fuzzy TOPSIS

Since Yoon and Hwang [25] introduced the Technique for Order Preference by Similarity to Ideal Solution (TOPSIS) method, it has become one of the industry standards widely applied in the area of Multiple Criteria Decision Making (MCDM). To determine the best alternative measured against sets of criteria, the TOPSIS method does this by introducing concurrently the shortest distance from the Fuzzy Positive Ideal

Table 7 The leaving/entering, net flows and alternatives ranking

	Leaving flow $(\phi^{+}(m))$	Entering flow $(\phi^{-}(m))$	Net Flow	Ranking
A1	1.7381	3.257143	−1.51905	7
A2	2.61905	1.37619	1.24286	2
A3	3.30476	3.30000	0.00476	5
A4	2.25238	1.166667	1.08571	3
A5	1.91429	1.866667	0.04762	4
A6	1.87619	2.514286	−0.63810	6
A7	3.14762	0	3.14762	1
A8	0.83333	4.204762	−3.37143	8

Source Authors'

Solution (FPIS) and the farthest distance from the Fuzzy Negative Ideal Solution (FNIS). The FNIS works by maximizing the cost criteria and minimizing the benefit criteria, whiles the FPIS seeks to maximize the benefit criteria whiles minimizing the cost criteria (Afful-Dadzie et al. [12]). The alternatives are evaluated and subsequently selected by ranking their relative closeness combining two distance measures. The numerical example used above is applied in TOPSIS to compare the ranking of the methods.

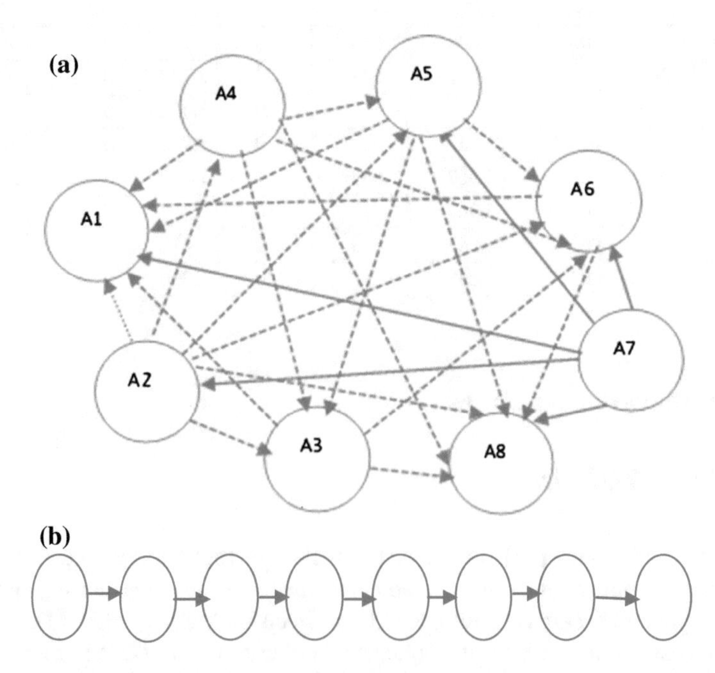

Fig. 3 (a) Partial preorder outranking; (b) Full preorder outranking

Table 8 Weights of each criteria

	TFN			BNP
C1	0	0.233	0.5	0.244
C2	0.1	0.433	0.7	0.411
C3	0.1	0.633	0.9	0.544
C4	0.5	0.833	1	0.778
C5	0.1	0.500	0.9	0.500

Source Authors'

The numerical example uses the same criteria, number of decision makers and alternatives as used in PROMETHEE. However, the TOPSIS method and the procedure in coming out with the ranking of the alternatives is quite different.

The judgements on the importance weights are made by the three decision makers, aggregated using the graded mean integration method is used to aggregate each criterion while the Center of Area (COA) technique is applied in calculating the BNP. The criteria weights are presented in Table 8 below.

The alternative ratings produced by the three decision makers in Table 4 are aggregated before normalization. The linguistic terms in Table 2 are applied to the alternative ratings and aggregated using the graded mean integration. The results of the aggregated alternative ratings is shown in Table 9.

The next step is the normalization of the aggregated decision matrix using Eq. 17 below. The results of the normalized decision matrix is shown in Table 10.

$$\tilde{r}_{ij} = \left(\frac{\tilde{f}_{ij}}{h_j^+}, \frac{\tilde{g}_{ij}}{h_j^+}, \frac{\tilde{h}_{ij}}{h_j^+} \right) h_j^+ = \max_i h_{ij} \tag{17}$$

The fuzzy normalized matrix is then weighted using the BNP values generated in Table 11. The weighted normalized fuzzy decision matrix is presented in Table 11.

The fuzzy positive and fuzzy negative ideal solutions are determined. The relative closeness coefficient is calculated based on the fuzzy positive and fuzzy negative ideal solutions. Table 12 presents the distance measurement including the associated ranks of all the alternatives used.

Based on the results in Table 12, alternative A7 is ranked top followed by A4 and A2 respectively.

5.2 *Fuzzy VIKOR*

The 'VIseKriterijumska Optimizacija I Kompromisno Resenje' technique in MCDM consist of a multi-criteria optimization and compromise solution. The technique was extended in the fuzzy environment so as to address subjectivity and impreciseness in data. In this technique, a compromise ranking is established using weight intervals,

Table 9 Fuzzy aggregated decision matrix

	A1			A2			A3			A4			A5			A6			A7			A8		
C1	0.0	5.8	10.0	0.0	4.2	7.5	2.5	5.8	10.0	0.0	3.3	10.0	0.0	2.5	7.5	0.0	2.5	7.5	5.0	9.2	10.0	2.5	7.5	10.0
C2	0.0	3.3	7.5	2.5	7.5	10.0	2.5	5.8	10.0	0.0	5.0	10.0	0.0	1.7	5.0	0.0	4.2	7.5	5.0	8.3	10.0	0.0	4.2	7.5
C3	0.0	5.8	10.0	0.0	5.0	10.0	0.0	1.7	5.0	0.0	2.5	7.5	2.5	5.8	10.0	0.0	5.0	10.0	5.0	7.5	10.0	0.0	4.2	7.5
C4	0.0	3.3	7.5	0.0	4.2	10.0	0.0	3.3	7.5	2.5	6.7	10.0	0.0	5.0	10.0	0.0	5.0	10.0	5.0	7.5	10.0	0.0	1.7	5.0
C5	0.0	3.3	10.0	0.0	5.0	10.0	0.0	4.2	10.0	2.5	7.5	10.0	0.0	5.0	10.0	0.0	2.5	7.5	5.0	8.3	10.0	2.5	6.7	10.0

Source Authors'

Table 10 Normalized fuzzy decision matrix

	A1			A2			A3			A4			A5			A6			A7			A8		
C1	0.00	0.58	1.00	0.00	0.42	0.75	0.25	0.58	1.00	0.00	0.33	1.00	0.00	0.25	0.75	0.00	0.25	0.75	0.50	0.92	1.00	0.25	0.75	1.00
C2	0.00	0.33	0.75	0.25	0.75	1.00	0.25	0.58	1.00	0.00	0.50	1.00	0.00	0.17	0.50	0.00	0.42	0.75	0.50	0.83	1.00	0.00	0.42	0.75
C3	0.00	0.58	1.00	0.00	0.50	1.00	0.00	0.17	0.50	0.00	0.25	0.75	0.25	0.58	1.00	0.00	0.50	1.00	0.50	0.75	1.00	0.00	0.42	0.75
C4	0.00	0.33	0.75	0.00	0.42	1.00	0.00	0.33	0.75	0.25	0.67	1.00	0.00	0.50	1.00	0.00	0.50	1.00	0.50	0.75	1.00	0.00	0.17	0.50
C5	0.00	0.33	1.00	0.00	0.50	1.00	0.00	0.42	1.00	0.25	0.75	1.00	0.00	0.50	1.00	0.00	0.25	0.75	0.50	0.83	1.00	0.25	0.67	1.00

Source Authors'

Table 11 Weighted normalized fuzzy decision matrix

	A1			A2			A3			A4			A5			A6			A7			A8		
C1	0.00	0.14	0.50	0.00	0.10	0.38	0.00	0.14	0.50	0.00	0.08	0.50	0.00	0.06	0.38	0.00	0.06	0.38	0.00	0.21	0.50	0.00	0.18	0.50
C2	0.00	0.14	0.53	0.03	0.33	0.70	0.03	0.25	0.70	0.00	0.22	0.70	0.00	0.07	0.35	0.00	0.18	0.53	0.05	0.36	0.70	0.00	0.18	0.53
C3	0.00	0.37	0.90	0.00	0.32	0.90	0.00	0.11	0.45	0.00	0.16	0.68	0.03	0.37	0.90	0.00	0.32	0.90	0.05	0.48	0.90	0.00	0.26	0.68
C4	0.00	0.28	0.75	0.00	0.35	1.00	0.00	0.28	0.75	0.13	0.56	1.00	0.00	0.42	1.00	0.00	0.42	1.00	0.25	0.63	1.00	0.00	0.14	0.50
C5	0.00	0.17	0.90	0.00	0.25	0.90	0.00	0.21	0.90	0.03	0.38	0.90	0.00	0.25	0.90	0.00	0.13	0.68	0.05	0.42	0.90	0.03	0.33	0.90

Source Authors'

Table 12 The distance measurement

	d_i^+	d_i^-	CC_i	Rank
A1	3.79	2.17	0.3637	4
A2	3.68	2.37	0.3919	3
A3	3.85	1.99	0.3410	7
A4	3.61	2.34	0.3939	2
A5	3.81	2.15	0.3610	5
A6	3.83	2.11	0.3557	6
A7	3.23	2.63	0.4484	1
A8	3.84	1.90	0.3310	8

Source Authors'

Table 13 Fuzzy best value f_j^+ and fuzzy worst value f_j^-

Fuzzy best value			Fuzzy worst value		
5.00	9.17	10.00	0.00	2.50	7.50
5.00	8.33	10.00	0.00	1.67	5.00
5.00	7.50	10.00	0.00	2.50	5.00
5.00	7.50	10.00	0.00	1.67	5.00
5.00	8.33	10.00	0.00	2.50	7.50

Source Authors'

the fuzzy best and fuzzy worst solutions. The fuzzy VIKOR method deals with ranking of alternatives with multi-conflicting or non-commensurable criteria [26].

Adapting the same criteria, number of decision makers, alternatives and fuzzy decisions, a numerical example is presented below for comparison with the fuzzy PROMETHEE and fuzzy TOPSIS results above.

Based on the aggregated fuzzy decisions in Table 9, the fuzzy best and fuzzy worst values are calculated using Eqs. 18 and 19. The results are stated in Table 13.

$$\tilde{f}_j^+ = \max_i \tilde{x}_{ij}, f_j^- = \min_i \tilde{x}_{ij}, \quad \text{for } j \in B \tag{18}$$

$$\tilde{f}_j^+ = \min_i \tilde{x}_{ij}, f_j^- = \max_i \tilde{x}_{ij}, \quad \text{for } j \in C \tag{19}$$

The fuzzy decisions are then normalized and weighted using Eqs. 20 and 21. The results are detailed in Tables 14(a, b) and 15(a, b) below.

$$\tilde{d}_{ij} = (\tilde{f}_j^* - \tilde{x}_{ij})/(c_j^* - a_j^\circ) \quad \text{for } j \in B \tag{20}$$

$$\tilde{d}_{ij} = (\tilde{x}_{ij} - \tilde{f}_j^*)/(c_j^\circ - a_j^*) \quad \text{for } j \in C \tag{21}$$

where B is the benefit criteria and C, the cost criteria.

Table 14 Normalized fuzzy difference

(a)

	A1- fj*-xij			A2- fj*-xij			A3- fj*-xij			A4- fj*-xij		
C1	–2.50	6.67	10.00	–2.50	5.00	10.00	–5.00	3.33	10.00	–5.00	5.83	10.00
C2	0.00	6.67	10.00	–5.00	0.83	7.50	–2.50	5.00	10.00	–5.00	3.33	10.00
C3	–5.00	1.67	7.50	–5.00	2.50	10.00	–5.00	1.67	10.00	–2.50	5.00	10.00
C4	–5.00	2.50	10.00	–5.00	3.33	10.00	–2.50	4.17	10.00	–5.00	0.83	7.50
C5	–5.00	3.33	10.00	–5.00	3.33	10.00	–5.00	5.00	10.00	–5.00	0.83	7.50

(b)

	A5- fj*-xij			A6- fj*-xij			A7- fj*-xij			A8- fj*-xij		
C1	–2.50	6.67	10.00	–5.00	3.33	7.50	–5.00	0.00	5.00	–5.00	1.67	7.50
C2	–2.50	4.17	10.00	–5.00	2.50	7.50	–5.00	0.00	5.00	–2.50	4.17	10.00
C3	–5.00	2.50	10.00	0.00	5.83	10.00	–5.00	0.00	5.00	–2.50	3.33	10.00
C4	–5.00	2.50	10.00	–2.50	4.17	10.00	–5.00	0.00	5.00	0.00	5.83	10.00
C5	–2.50	5.83	10.00	–5.00	4.17	10.00	–5.00	0.00	5.00	–5.00	1.67	7.50

Source Authors'

Table 15 Weighted normalized fuzzy difference

(a)

	Weighted A1			Weighted A2			Weighted A3			Weighted A4		
C1	0.0	1.6	5.0	0.0	1.2	5.0	0.0	0.8	5.0	0.0	1.4	5.0
C2	0.0	2.9	7.0	–0.5	0.4	5.3	–0.3	2.2	7.0	–0.5	1.4	7.0
C3	–0.5	1.1	6.8	–0.5	1.6	9.0	0.5	1.1	9.0	–0.3	3.2	9.0
C4	–2.5	2.1	10.0	–2.5	2.8	10.0	–1.3	3.5	10.0	–2.5	0.7	7.5
C5	–0.5	1.7	9.0	–0.5	1.7	9.0	–0.5	2.5	9.0	–0.5	0.4	6.8

(b)

	Weighted A5			Weighted A6			Weighted A7			Weighted A8		
C1	0.0	1.6	5.0	0.0	0.8	3.8	0.0	0.0	2.5	0.0	0.4	3.8
C2	–0.3	1.8	7.0	–0.5	1.1	5.3	–0.5	0.0	3.5	–0.3	1.8	7.0
C3	–0.5	1.6	9.0	0.0	3.7	9.0	–0.5	0.0	4.5	–0.3	2.1	9.0
C4	–2.5	2.1	10.0	–1.3	3.5	10.0	–2.5	0.0	5.0	0.0	4.9	10.0
C5	–0.3	2.9	9.0	–0.5	2.1	9.0	–0.5	0.0	4.5	–0.5	0.8	6.8

Source Authors'

Computing the separation measures of $\tilde{S}_i$ and $\tilde{R}_i$ are calculated from the fuzzy best and fuzzy worst values. The results of the measures is presented in Table 16(a, b) below.

The next step is to compute the value of $\tilde{Q}_i$ with Eq. 22

$$\tilde{Q}_i = v(\tilde{S}_i - \tilde{S}^*)/(S^{oc} - S^{*a}) \oplus (1 - v)(\tilde{R}_i - \tilde{R}^*)/(R^{oc} - R^{*a}) \tag{22}$$

Table 16 Separation measures of $\tilde{S}_i$ and $\tilde{R}_i$

(a)

	dA1			dA2			dA3			dA4		
C1	0.00	0.16	0.50	0.00	0.12	0.50	0.00	0.08	0.50	0.00	0.14	0.50
C2	0.00	0.29	0.70	–0.05	0.04	0.53	–0.03	0.22	0.70	–0.05	0.14	0.70
C3	–0.05	0.11	0.68	–0.05	0.16	0.90	–0.05	0.11	0.90	–0.03	0.32	0.90
C4	–0.25	0.21	1.00	–0.25	0.28	1.00	–0.13	0.35	1.00	–0.25	0.07	0.75
C5	–0.05	0.17	0.90	–0.05	0.17	0.90	–0.05	0.25	0.90	–0.05	0.04	0.68
$\tilde{S}_j$	–0.35	0.93	3.78	–0.40	0.76	3.83	–0.25	1.00	4.00	–0.38	0.71	3.53
$\tilde{R}_j$	0.00	0.29	1.00	0.00	0.28	1.00	0.00	0.35	1.00	0.00	0.32	0.90

(b)

	dA5			dA6			dA7			dA8		
C1	0.00	0.16	0.50	0.00	0.08	0.38	0.00	0.00	0.25	0.00	0.04	0.38
C2	–0.03	0.18	0.70	–0.05	0.11	0.53	–0.05	0.00	0.35	–0.03	0.18	0.70
C3	–0.05	0.16	0.90	0.00	0.37	0.90	–0.05	0.00	0.45	–0.03	0.21	0.90
C4	–0.25	0.21	1.00	–0.13	0.35	1.00	–0.25	0.00	0.50	0.00	0.49	1.00
C5	–0.03	0.29	0.90	–0.05	0.21	0.90	–0.05	0.00	0.45	–0.05	0.08	0.68
$\tilde{S}_j$	–0.35	0.99	4.00	–0.23	1.11	3.70	–0.40	0.00	2.00	–0.10	1.00	3.65
$\tilde{R}_j$	0.00	0.29	1.00	0.00	0.37	1.00	0.00	0.00	0.50	0.00	0.49	1.00

Source Authors'

Table 17 The value of q

	Q		
A1	–0.52	0.25	0.97
A2	–0.52	0.22	0.98
A3	–0.51	0.29	1
A4	–0.52	0.24	0.9
A5	–0.52	0.26	1
A6	–0.5	0.31	0.97
A7	–0.52	0	0.52
A8	–0.49	0.36	0.96

Source Authors'

where $\tilde{S}^* = MIN_i\tilde{S}_i$, $S^{oc} = MAX_iS_i^c$, $\tilde{R}^* = MIN_i\tilde{R}_iR^{oc} = MAX_iR_i^c$ and $v(v = n + 1/2n)$. The computed values of $\tilde{Q}_i$ are presented in Table 17.

The values of $\tilde{Q}_i\tilde{S}_i$ and $\tilde{R}_i$ are then defuzified. The fuzzy numbers are converted into crisp values using the Center of Area method. The values are then ranked with the smaller value of $\tilde{Q}_i$ being the best ranked alternative as seen in Table 18.

The comparative ranks of the three techniques; fuzzy PROMETHEE, fuzzy TOPSIS and fuzzy VIKOR are presented in Table 19.

Table 18 The defuzified values and the respective ranks

	Q	*S*	*R*	*Q*	*S*	*R*
A1	0.24	1.45	0.43	4	4	3
A2	0.23	1.39	0.43	3	3	2
A3	0.26	1.58	1.58	7	8	8
A4	0.2	1.29	1.29	2	2	6
A5	0.25	1.55	1.55	5	7	7
A6	0.26	1.53	0.46	6	6	4
A7	0	0.53	0.17	1	1	1
A8	0.28	1.52	0.5	8	5	5

Source Authors'

Table 19 Comparative ranks

	PROMETHEE	TOPSIS	VIKOR
A1	7	4	4
A2	2	3	3
A3	5	7	7
A4	3	2	2
A5	4	5	5
A6	6	6	6
A7	1	1	1
A8	8	8	8

Source Authors'

From Table 19, the fuzzy TOPSIS and fuzzy VIKOR methods give us the same ranking. This is so because the two methods are distance based measures with similar methodology. Fuzzy PROMETHEE has different ranks from A1 to A5.

6 Sensitivity Analysis

This section of the paper analyses the cross effect of influenced decisions on the alternatives ratings by the three decision makers. The sensitivity analysis seeks to find out the impact on the ranking if a decision maker is perceived to have been influenced. The alternative ratings by the decision makers in Table 4, the ratings of the first three alternatives (A7, A4 and A2) in Table 12 and the top three criteria (C4, C3, and C5) in Table 3 are used in the sensitivity analysis for the three scenarios and nine cases as seen in Table 20. The alternative ratings of VH and EH for these alternatives and criteria are altered to L.

In scenario 1 case 1, the alternative ratings for decision maker 1 for A7 are altered for the top three criteria (c4, c3, c5) from VH (5.0, 7.5, 10.0) and EH (7.5, 10.0, 10.0) to L (0.0, 2.5, 5.0). All other alternative ratings for the criteria remain the same

Table 20 Comparative ranks

Case	(Changes made in Alternatives, At), A = A1, A2, …, A8, t = 1, 2, 3, …, 8	
Scenario 1 (DM 1)	Case 1	A7 (c4, c3, c5) = (0, 2.5, 5), A1–A6, A8
	Case 2	A4 (c4, c3, c5) = (0, 2.5, 5), A1–A3, A5–A8
	Case 3	A2 (c4, c3, c5) = (0, 2.5, 5), A1, A3–A8
Scenario 2 (DM 2)	Case 4	A7 (c4, c3, c5) = (0, 2.5, 5), A1–A6, A8
	Case 5	A4 (c4, c3, c5) = (0, 2.5, 5), A1–A3, A5–A8
	Case 6	A2 (c4, c3, c5) = (0, 2.5, 5), A1, A3–A8
Scenario 3 (DM 2)	Case 7	A7 (c4, c3, c5) = (0, 2.5, 5), A1–A6, A8
	Case 8	A4 (c4, c3, c5) = (0, 2.5, 5), A1–A3, A5–A8
	Case 9	A2 (c4, c3, c5) = (0, 2.5, 5), A1, A3–A8

Source Authors'

for decision maker 1 and the other decision makers. In case 2, criteria (c4, c3, c5) for A7 of decision maker 1 are replaced from VH (5.0, 7.5, 10.0) and EH (7.5, 10.0, 10.0) to L (0.0, 2.5, 5.0) while other criteria remain the same for the other alternative ratings for all other decision makers. In case 3, alternative ratings of criteria (c4, c3, c5) for A2 for decision maker 1 are also replaced from VH (5.0, 7.5, 10.0) and EH (7.5, 10.0, 10.0) to L (0.0, 2.5, 5.0) while other criteria remain the same for the other alternative ratings for all other decision makers.

The same format is applied to scenarios 2 and 3 and for cases 4 to 9, for the criteria (c4, c3, c5) and alternatives A7, A4 and A2. Linguistics terms of VH (5.0, 7.5, 10.0) and EH (7.5, 10.0, 10.0) are replaced with L (0.0, 2.5, 5.0) on separate basis for decision makers 2 and 3 as shown in Table 10. Steps 5–9 are carried out on separate basis for each case under the scenarios and compared with the original ranking to ascertain whether an influenced decision can affect the ranking of the alternatives.

From Table 11 and Fig. 4, the effect of the alterations of the cases for the three scenarios on the rankings of the alternatives are shown with the original ranking. The symbol 'c1' stands for case 1, 'c2' for case 2 and so on. A change in the alternative ratings of the three top criteria (c4, c3, c5) for alternatives A7, A4 and A2 by a decision maker results in few changes in the alternative rankings of A4 and A2 as seen in Table 21 and Fig. 4 below. Alternative 7 remains the top ranked alternative throughout the three scenarios and nine cases as in the original ranking. Alternative 4 also remains the second ranked except for scenario 1(c2), scenario 2(c5) and scenario 3 (c8) where the ranking of the alternative was changed to third. Alternative 2 remained the third ranked except for scenario 1 (c2), scenario 2 (c5) and scenario 3 (c8) which encountered some changes. The rest of the alternatives remained unchanged as seen in Table 11 and Fig. 4 below. From the sensitivity analysis, the ranking of the alternatives in the proposed model will not be affected by an influence of an individual decision maker when the same criteria weights are applied.

Table 21 Comparative ranks

		SCENARIO 1 = DM1 = L			SCENARIO 2 = DM2 = L			SCEANRIO 3 = DM3 = L		
Alternatives	Original	c1	c2	c3	c4	c5	c6	c7	c8	c9
	Rank	Rank	Rank	Rank	Rank	Rank	Rank	Rank	Rank	Rank
(A1)	4	4	4	4	4	4	4	4	4	4
(A2)	3	3	2	3	3	2	3	3	2	3
(A3)	7	7	7	7	7	7	7	7	7	7
(A4)	2	2	3	2	2	3	2	2	3	2
(A5)	5	5	5	5	5	5	5	5	5	5
(A6)	6	6	6	6	6	6	6	6	6	6
(A7)	1	1	1	1	1	1	1	1	1	1
(A8)	8	8	8	8	8	8	8	8	8	8

Source Authors'

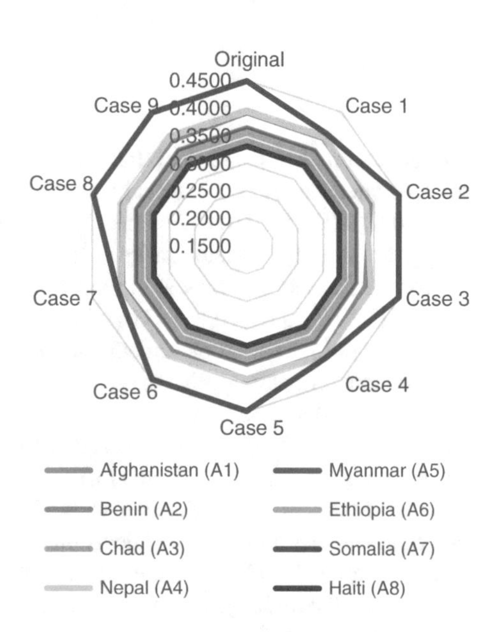

Fig. 4 Plot of result of sensitivity analysis on criteria. *Source* Authors'

7 Conclusion

The paper designed a four-stage MCDM framework of Fuzzy PROMETHEE, fuzzy VIKOR, Fuzzy TOPSIS and a novel sensitivity analysis to help evaluate developmental aid programs sponsored by aid-donors around the world. With a numerical example, the framework demonstrates how donor-recipient countries participating in developmental aid program can be evaluated to ascertain progress made and therefore which countries should deserve future funding based on previous performances. The study relied on a set of 5-criteria evaluation format used by OECD countries in evaluating developmental aid programs to model the proposed evaluation technique. With a custom-made rating scale, the framework relies on the experience and the expertise of country development officers and executives in evaluating the performance of participating countries against the set criteria. The strength of the proposed model is seen in how aid-recipient countries can be evaluated by ranking them on their performances thereby ensuring fairness, value for money and sustainability of aid programs. The use of the evaluation process where sensitivity analysis is employed strengthens the framework by ensuring that bias in expert ratings are easily detected to warrant resetting the process again.

Acknowledgments This work was supported by GACR P103/15/06700S, NPU I No. MSMT-7778/2014, CEBIA-Tech No. CZ.1.05/2.1.00/03.0089. It was also supported by Internal Grant Agency of TBU under the project Nos. IGA/FAI/2015/054, IGA/FaME/2016/019 and IGA/FaME/2015/023.

References

1. OECD International Development Statistics. OECD Press, France (2014)
2. Kharas, K.: Measuring Aid Effectiveness Effectively: A Quality of Official Development Assistance Index. Brookings Institution Press, Washington (2011). Available at: http://www.brookings.edu/research/opinions/2011/07/26-aid-effectiveness-kharas. Accessed 10 December 2014
3. Howes, S.: A framework for understanding aid effectiveness determinants, strategies and trade-offs. Asia Pac. Policy Stud. **1**(1), 58–72 (2014)
4. Dalgaard, C.J., Hansen, H.: Evaluating Aid Effectiveness in the Aggregate: A Critical Assessment of the Evidence. University Library of Munich, Germany (2010)
5. Deaton, A.: Instruments, randomization, and learning about development. J. Econ. Lit. 424–455 (2010)
6. Arndt, C., Jones, S., Tarp, F.: Aid, growth, and development: have we come full circle? J. Glob. Dev. **1**(2) (2010)
7. Doucouliagos, H., Paldam, M.: The aid effectiveness literature: the sad results of 40 years of research. J. Econ. Surv. **23**(3), 433–461 (2009)
8. Minoiu, C., Reddy, S.G.: Development aid and economic growth: a positive long-run relation. Q. Rev. Econ. Financ. **50**(1), 27–39 (2010)
9. Morrison, K.M.: As the World Bank turns: determinants of IDA lending in the cold war and after. Bus. Pol. **13**(2) (2011)
10. Claessens, S., Cassimon, D., Van Campenhout, B.: Evidence on changes in aid allocation criteria. World Bank Econ. Rev. **23**(2), 185–208 (2009)
11. Zadeh, A.: Fuzzy sets. Inf. Control **8**(3), 338–353 (1965)
12. Afful-Dadzie, E., Nabareseh, S., Afful-Dadzie, A., Oplatková, Z.K.: A fuzzy TOPSIS framework for selecting fragile states for support facility. Quality and Quantity, pp. 1–21 (2014). doi:10.1007/s11135-014-0062-3
13. Krohling, R.A., Campanharo, V.C.: Fuzzy TOPSIS for group decision making: a case study for accidents with oil spill in the sea. Expert Syst. Appl. **38**(4), 4190–4197 (2011)
14. Vincke, J.P., Brans, Ph: A preference ranking organization method. The PROMETHEE method for MCDM. Manag. Sci. **31**(6), 647–656 (1985)
15. Ying-Hsiu, C., Tien-Chin, W., Chao-Yen, W.: Strategic decisions using the fuzzy PROMETHEE for IS outsourcing. Expert Syst. Appl. **38**(10), 13216–13222 (2011)
16. Amaral, T.M., Ana, P.C.C.: Improving decision-making and management of hospital resources: An application of the PROMETHEE II method in an Emergency Department. Oper. Res. Health Care **3**(1), 1–6 (2014)

17. Elevli, B.: Logistics freight center locations decision by using Fuzzy-PROMETHEE. Transport **29**(4), 412–418 (2014)
18. Yi, P., Kou, G., Li, J.: A fuzzy promethee approach for mining customer reviews in chinese. Arab. J. Sci. Eng. **39**(6), 5245–5252 (2014)
19. Shadman, R.M., Rahimi, S., Beglou, M.J.: PROMETHEE II and fuzzy AHP: an enhanced GIS-based landslide susceptibility mapping. Nat. Hazards **73**(1), 77–95 (2014)
20. Xiaojuan, T., Liu, X., Wang, L.: An improved PROMETHEE II method based on Axiomatic Fuzzy Sets. Neural Comput. Appl. **25**(7–8), 1675–1683 (2014)
21. Ting-Yu, C.: A PROMETHEE-based outranking method for multiple criteria decision analysis with interval type-2 fuzzy sets. Soft Comput. **18**(5), 923–940 (2014)
22. Sonia, H., Halouani, N..: Hesitant-fuzzy-promethee method. In: 2013 5th International Conference on Modeling, Simulation and Applied Optimization (ICMSAO), pp. 1–6. IEEE, New York (2013)
23. Wei-xiang, L., Li, B.: An extension of the Promethee II method based on generalized fuzzy numbers. Expert Syst. Appl. **37**(7), 5314–5319 (2010)
24. Afful-Dadzie, A., Afful-Dadzie, E., Nabareseh, S., Oplatková, Z.K.: Tracking progress of African Peer Review Mechanism (APRM) using fuzzy comprehensive evaluation method. Kybernetes **43**(8), 1193–1208 (2014)
25. Yoon, K.P., Hwang, C.L.: Multiple Attribute Decision Making: An Introduction, vol. 104. Sage Publications, New York (1995)
26. Afful-Dadzie, E., Nabareseh, S., Oplatková, Z. K., Klímek, P.: Model for assessing quality of online health information: a fuzzy VIKOR based method. J. Multi-Criteria Decis. Anal. (2015)

Self-Reconfiguring Robotic Framework Using Fuzzy and Ontological Decision Making

Affan Shaukat, Guy Burroughes and Yang Gao

Abstract Advanced automation requires complex robotic systems that are susceptible to mechanical, software and sensory failures. While bespoke solutions exist to avoid such situations, there is a requirement to develop generic robotic framework that can allow autonomous recovery from anomalous conditions through hardware or software reconfiguration. This paper presents a novel robotic architecture that combines fuzzy reasoning with ontology-based deliberative decision making to enable self-reconfigurability within a complex robotic system architecture. The fuzzy reasoning module incorporates multiple types of fuzzy inference models that passively monitor the constituent sub-systems for any anomalous changes. A response is generated in retrospect of this monitoring process that is sent to an Ontology-based rational agent in order to perform system reconfiguration. A reconfiguration routine is generated to maintain optimal performance within such complex architectures. The current research work will apply the proposed framework to the problem of autonomous visual navigation of unmanned ground vehicles. An increase in system performance is observed every time a reconfiguration routine is triggered. Experimental analysis is carried out using real-world data, concluding that the proposed system concept gives superior performance against non-reconfigurable robotic frameworks.

1 Introduction

Robotics and complex autonomous systems are becoming increasingly pervasive in almost every field of life [6, 17]. The evolution of robotic research and real-world application domains has resulted in a great number of sophisticated robotic

A. Shaukat (✉) · G. Burroughes · Y. Gao
Faculty of Engineering and Physical Sciences, Surrey Space Centre,
University of Surrey, Guildford GU2 7XH, UK
e-mail: a.shaukat@surrey.ac.uk

G. Burroughes
e-mail: g.burroughes@surrey.ac.uk

Y. Gao
e-mail: yang.gao@surrey.ac.uk

Y. Bi et al. (eds.), *Intelligent Systems and Applications*,
Studies in Computational Intelligence 650, DOI 10.1007/978-3-319-33386-1_7

architectures that are prone to mechanical, software and sensory failures. Although manual servicing is regularly carried out, it can become extremely time-consuming, expensive, risky [17] and even infeasible (such as for Martian rovers) in certain situations. As such there is a requirement to develop generic architectures of robotic systems that can enable self-reconfiguration of critical application software or hardware units in response to any environmental perturbation, performance deterioration and system malfunction. Hardware-based reconfiguration may include physical sensors and manipulators while software-based reconfiguration would allow the use of the most optimal software programs that is appropriate for a given situation. Rational agent-based architectures perform reconfiguration of higher-level goals, actions and plans [**?**]. A summary of the different types of reconfiguration that may take place in a robotic architecture is presented in Fig. 1.

This article addresses the problem of reconfiguration in complex robotic systems by using a multi-layered framework that builds upon the model introduced in [16]. The architecture presented in [16] incorporates the agent as a part of the reconfiguration layer. The proposed model adopts the rational agent as a separate layer, hence extending it to a three-layered architecture, thus further allowing the reconfiguration of higher level goals, actions and plans (for example multi-agent systems). This work will seek to establish the advantage of the proposed reconfigurable robotic framework by performing a number of quantitative and qualitative assessments.

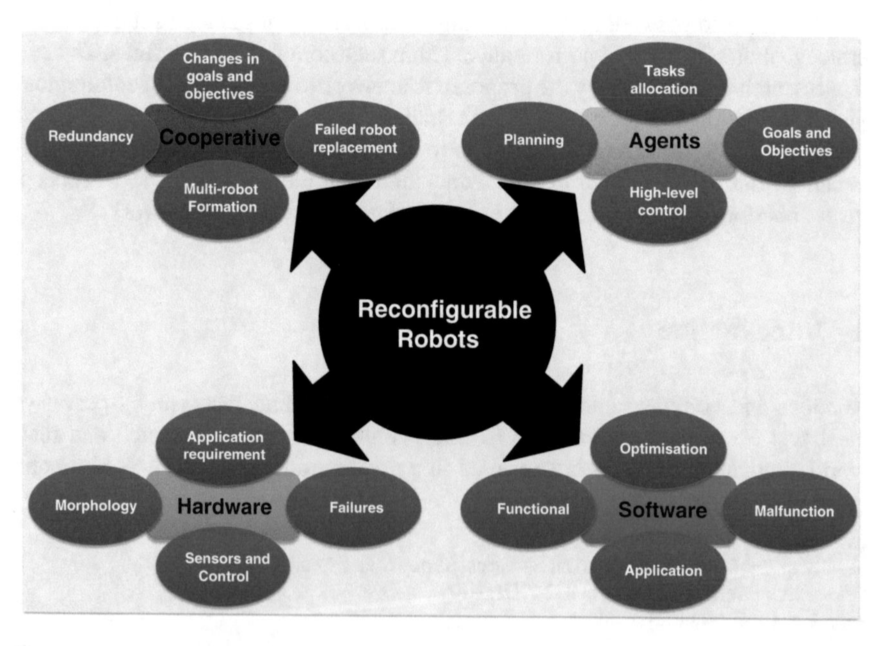

Fig. 1 Types of reconfiguration in robotic frameworks

2 Background

2.1 *Reconfigurable Systems*

In literature, the field of self-reconfigurable robotic systems has traditionally focused on the design and development of techniques and algorithms that can process reconfiguration of the morphological characteristics of modular systems [21]. However, there is an increasing understanding that the concept of reconfiguration in autonomous systems and robots expands beyond variation in geometry, shape, and hardware composition [12]. Many contemporary robotic architectures follow an isolated engineered system approach in comparison to many contemporary autonomous systems, which employ a *hybrid* agent architecture with many of the hardware and software sub-systems supervised by a *Rational Agent* [5]. Their highly modular and distributed architectures enable them with software and hardware reconfiguration within their capabilities in response to novel situations; whereas, the agent makes *rational* high-level decisions, define goals and perform planning/re-planning (refer to Fig. 1) [5]. While these advantages have yet to be fully utilised in the robotics field [21], there are a number of relevant examples in literature that have sought to develop techniques that can enable reconfiguration with limitations according to their specific applications.

Systems that can perform reconfiguration of controllers based on errors or specific performance requirements have also been proposed literature, known as *Plug & Play* control, and the controllers described as polymorphic [5]. In software architectures, there are examples of self-managing distributed processing systems, which can adapt to stochastic changes such as; *Autonomic Computing* [7], *Autonomic Control Loops* [5], *Self-CHOP, and Self-Organising Component Based Software* [18], MAPE-K [5] etc. For a few agent-based complex robotic systems, a number of multi-agent based reconfigurable architectures have also been proposed in literature [2]; however, certain limitations do exist in terms of generality.

2.2 *Fuzzy Inference Systems*

Fuzzy inference systems (FIS) are used in robotic systems to incorporate intelligent reasoning in situations where physical sensors and actuators have an intrinsic notion of uncertainty. Such systems have been widely used for practical control architectures. FIS have also proven to be very useful in logic-based decision-making and discerning patterns from low-level sensory inputs [16] and top-down reasoning in multi-layered hierarchical architectures that incorporate machine vision and learning [14, 20]. FIS use logic applicable to fuzzy sets with degrees of membership, which are formulated in terms of membership functions valued in the real unit interval [0, 1]. These membership functions can either be singular values, intervals or even sets of intervals. Machine learning techniques (such as, *neural networks*) are commonly used for optimisation and development of these membership functions. However, such methods rely on the scale of training data and the number of adjusted parameters [16]; therefore, they may not be suitable for complex robotic architectures with multiple levels of data abstraction and rational decision making. To avoid such problems, membership function definitions and parameter setting can also be performed using empirical knowledge of the robot sensory and control settings. Such knowledge can be extracted using *Ontology*.

2.3 Ontology

For a system to perform reconfiguration autonomously the system must be self-aware. In other words, the system must have a usable System knowledge representation of its internal subsystems, if it is going to be able attempt any autonomic characteristics. To accomplish this a Ontological model can be used.

An Ontology is a formal description of the concepts and relationships that can formally exist in a domain. Ontologies are often equated with taxonomic hierarchies of classes, class definitions, and the subsumption relation, but ontologies are not limited to these forms. An Ontology can be described as a Description Logic built on top of a Description Logic, where a Description Logic is any decidable fragment of FOL, that models concepts, roles and individuals, and their relationships [11].

The most prominent robot ontology research is the newly formed IEEE-RAS working group entitled Ontologies for Robotics and Automation (ORA) [13]; however, nothing is concrete as of yet. The goal of this working group is to develop a standard ontology and associated methodology for knowledge representation and reasoning in robotics and automation, together with the representation of concepts in an initial set of application domains.

Another example of an emerging robotics domain ontology is Intelligent Systems Ontology [13]. The purpose of this ontology is to develop a common, implementation-independent, extensible knowledge source for researchers and developers in the intelligent vehicle community. It provides a standard set of domain concepts, along with their attributes and inter-relations allow for knowledge capture and reuse, facilitate systems specification, design, and integration. It is based on a service based description akin to web service's WSDL [4], hence the use of a OWL-S [10] as the base language. It similarities to WSDL are profound, but what it does have is an ability to describe complex agents. What it does not have, is an ability to describe the physical environment; however, other ontologies can be appended to this ontology to make a more domain complete ontology.

3 A Reconfigurable Robotic Framework

The proposed multi-layered self-reconfigurable robotic architecture is illustrated in Fig. 2. It is a three-layered architecture; an *Agent*, a *Reconfiguration*, and an *Application* layer.

Briefly:

- The *Application Layer* contains the traditional elements of a robotic architecture, and doesn't contain any of the elements required for self-reconfiguration.

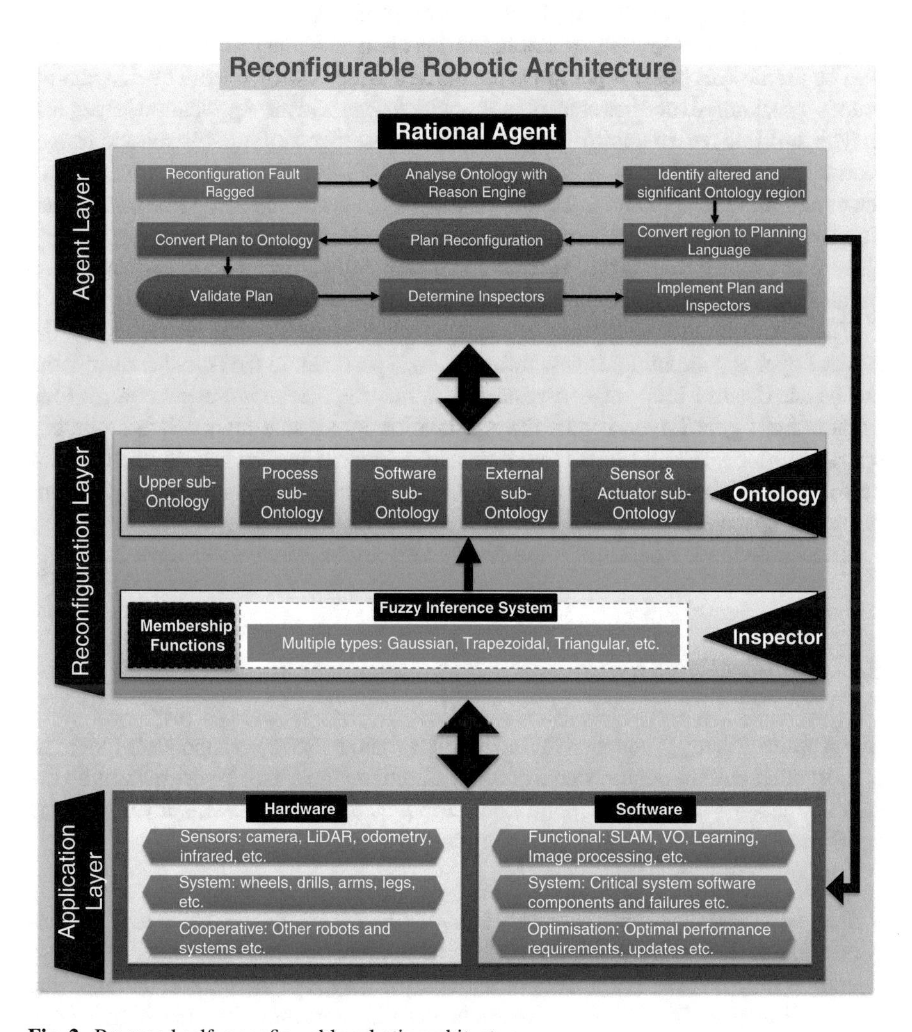

Fig. 2 Proposed self-reconfigurable robotic architecture

- The *Reconfiguration Layer* contains the elements of the system required for self-monitoring and self-awareness, and thus monitors the Application Layer's performance.
- The *Agent Layer* contains the Agent that computes Self-Reconfiguration for the system using the knowledge in the Reconfiguration Layer.

The layer of abstraction are ordered in this manner to: maximise modularity, which in turn allows the layers to be adapted and changed more easily; not require the Application Layer components to be integrated deeply with reconfiguration elements, which limits the number of faults/errors that Reconfiguration and Agent Layer unintentionally inject to the Application Layer; and it allows the Application Layer to be easily divorced from the Reconfiguration elements if they are no longer desired.

The Frameworks basic operations breaks into three control loops. The first control loop is the standard robotics control loops, which occurs in the Application Layer, this maybe multiple nested control loops like any non-self-reconfigurable robotic framework. The next control loop is in the Reconfiguration Layer, this separate control loop monitors the application layer, updating its internal knowledge of the system. The Reconfiguration Layer's control loop, if required, triggers the next control loop. The Final Control loop is the Agent control loop, which occurs in the Agent Layer. In the Agent control loop the required reconfiguration is calculated.

The Reconfiguration Control loop can carrying on monitoring system, while the Agent Layer is computing. If new information is pertinent to the Agent control loop, the Agent Control loop may be restarted. When the Reconfiguration control loop triggers the Agent Layer, it puts the Application into a safe mode. Safe mode is a operation mode, which limits the amount of damage a system can do itself, whilst performing sub-optimally. It is analogous to the system Safety modes of operations in robotic space-craft.

The following is a series of more detailed discussions in to these three layers.

3.1 Application Layer

The Application layer contains all the hardware and software components that exist in most other autonomous systems or robotic architectures. This layer contains all the classic elements of a non-reconfigurable robotics architecture. This layer contains no elements necessary for self-reconfiguration.

The hardware components may be:

- ***Sensors***: are various types of exteroceptive and interoceptive sensors such as, Cameras, LiDAR, Touch Sensors, Encoders, GPS, etc. For example, Sensor Reconfiguration may involve the process of replacing one type of sensor with another due to malfunction, fallible inputs or precision requirements from the higher level reconfiguration layer.
- ***System***: components are adherent to the system architecture and form an intrinsic part of some mechanism, such as, wheels, grippers, manipulators, etc. In this case,

reconfiguration may be required by the system course of actions, for example, orientation of the wheels to perform a specific type of manoeuvre.

- ***Cooperative scenarios***: are situations consisting of multiple robots working towards common mission goals, where a reconfiguration may be required when one of the robots either stops working, or the rational agent generates a new plan that may require an alternative formation of the robots.

Similarly, the software components can be:

- ***Functional components***: are application software and programs that use sensor data to enable autonomous capabilities, such as visual navigation from camera images, map generation and localisation from LiDAR, etc. Most of these algorithms are intrinsically stochastic, and their performance can be affected by spurious data. In such cases specific software can be reconfigured with more robust algorithms with better or more optimal performance for the scenario presented.
- ***System***: components are critical system software units that run and execute application software, for example, operating systems, boot-loaders, safety monitoring systems, etc. Any failure in this category can cause a critical paralysis of the whole system; however, a reconfigurable architecture can avoid this well in advance if certain anomalies are detected by the higher levels.

Under normal conditions, the system would operate without any anomalies and the application layer performance may not significantly differ from that of an engineered robotic system. However, within this framework, any changes that occur in the application layer are monitored and analysed. Any anomalies that are identified can then be dealt with as deemed appropriate by the Reconfiguration Layer.

3.2 Reconfiguration Layer

The Reconfiguration layer operates at a higher-level than the Application layer within the hierarchy of abstraction, which contains the necessary elements for building a knowledge of the changes in the Application layer and enabling the Rational Agent to trigger a reconfiguration. This from an Autonomic point of view would be the Self-Aware, Self-Monitoring, and Self-Analysing element of the system.

The principal components of this layer are: the Inspector, and the Ontology. The following are two individual discussions into these principal components.

3.2.1 Inspector

An Inspector is the lowest layer in the hierarchy of the Reconfiguration Layer, which passively monitors the Application Layer for *Reconfiguration Anomalies*, via FIS. The Inspector monitors the Application Layer for Reconfiguration Anomalies.

Reconfiguration Anomalies are things that deviate from what is standard, normal, or expected, which may affect an application optimal operation. These anomalies can be defined as:

A ***Self-Reconfiguration System Failure*** (SR-Failure) is an event that occurs when the configuration deviates from optimal strategy. These are distinct and adjoint from standard software failures, because a non-optimal configuration is not the same as the Fault.

A ***Self-Reconfiguration System Error*** (SR-Error) is that part of the system state that may cause a subsequent SR-Failure: a SR-Failure occurs when an SR-Error reaches the service interface and alters the service.

A ***Self-Reconfiguration System Fault*** (SR-Fault) is the adjudged or hypothesized cause of an SR-Error.

SR-Faults can include environmental, software, and hardware changes. Furthermore, these definitions limit SR-Faults to individual components. Whilst, SR-Failures can propagate through a system with these definitions, a SR-Fault is their point or points of origin.

To monitor for SR-Faults, outputs from the application layer are translated into fuzzy confidence values that provides a *semantic* notion of performance to the *Ontology*. To date, the proposed framework has implemented fuzzy inference based on *Mamdani's* method [9] due to its simple structure of *min–max* operations and high degree of success rate in many complex control architectures. However other types can be easily integrated in the Inspector. This allows for generic monitoring of low-level systems within the proposed framework, making it appropriate for any type of complex robotic system.

The FIS system uses multiple fuzzy variables that associate confidence values with the inputs from the Application layer; for example, these can be *errors* and *processing time* of an *extended kalman filter* within a simultaneous localisation and mapping system [1]. FIS subsumes conjunctive operators (T-norms; min, prod etc.) and disjunctive operators (T-conorms; max, sum, etc.) as well as hybrid operators (combinations of the previous operators). In particular, the T-norm (triangular norm) is generally a continuous function, $[0, 1] \times [0, 1] \Rightarrow [0, 1]$. Membership function definition depends upon the appropriate modelling parameters of the various units in the Application layer, and they can either be *gaussian*, *trapezoids* or *triangular*. A fuzzy rule-base combines the input variables in order to compute the rule strength. The antecedents within the fuzzy rule-base use the fuzzy operators to compute a single membership value for each rule. The outputs from multiple rules in the fuzzy rule-base are aggregated into a single fuzzy set, which is generally performed using the *max* operator. The fuzzy logic inference system allows the use of multiple input fuzzy variables, for example *processing time* and *error* to infer decisions as the consequence of the a priori set of fuzzy logic rules. The final output from the fuzzy inference system is a single confidence value instead that is used by the Ontology.

Consider an example of a fuzzy variable associated with the computation time characterised as,

$$\textit{compute\_time} = \langle \alpha, \mathbb{U}_{\alpha}, \mathbb{R}(\alpha) \rangle, \quad (1)$$

where α is the computation time required for processing within the Application layer, $\mathbb{U}_{\alpha}$ is the universe of discourse, and $\mathbb{R}(\alpha)$ is the fuzzy membership of α. Two piecewise linear continuous fuzzy membership functions, i.e., '$\text{comp\_time}_{\text{low}}(\alpha)$' and '$\text{comp\_time}_{\text{high}}(\alpha)$', can be associated with α, as defined in [16].

Similarly, a fuzzy variable associated with the estimation error can be characterised as follows,

$$\textit{estimation\_error} = \langle \beta, \mathbb{U}_{\beta}, \mathbb{R}(\beta) \rangle, \quad (2)$$

where β the error in state estimation of a localisation technique, $\mathbb{U}_{\beta}$ is the universe of discourse, and $\mathbb{R}(\beta)$ is the fuzzy membership of β. Similar to the previous case, two piecewise linear continuous fuzzy membership functions, i.e., '$\text{post\_error}_{\text{low}}(\beta)$' and '$\text{post\_error}_{\text{high}}(\beta)$', are associated with the input error from the Navigation system β [16].

The output variable defines the performance of the system in the Application layer on the *aggregated* confidence value computed from an inference technique applied to the fuzzy input variables,

$$\textit{system\_performance} = \langle \gamma, \mathbb{U}_{\gamma}, \mathbb{R}(\gamma) \rangle. \quad (3)$$

Two piecewise fuzzy membership functions, i.e., '$\text{system\_performance}_{\text{bad}}(\gamma)$' and '$\text{system\_performance}_{\text{good}}(\gamma)$' associated with the system performance output provide a confidence measure on how "*good*" or "*bad*" the lower-level SLAM system is performing [16].

A fuzzy rule-base combines the fuzzy input variables to compute the rule strength. The antecedents within the fuzzy rule-base use the fuzzy operators *AND* as t-norm and *OR* as t-conorm in order to compute a single membership value for each rule, for example:

- *if* $\text{comp\_time}_{\text{low}}(\alpha)$ *AND* $\text{post\_error}_{\text{low}}(\beta)$ *THEN* $\text{system\_performance}_{\text{good}}(\gamma)$,
- *if* $\text{comp\_time}_{\text{high}}(\alpha)$ *OR* $\text{post\_error}_{\text{high}}(\beta)$ *THEN* $\text{system\_performance}_{\text{bad}}(\gamma)$,

An implication operator is used to truncate the consequent's membership function output. The outputs from multiple rules within the fuzzy rule-base are aggregated into a single fuzzy set (refer to Fig. 3a for a graphical representation of the process). The final output is a single value, which is the final stage of the inference process known as *defuzzification* [14, 16, 20].

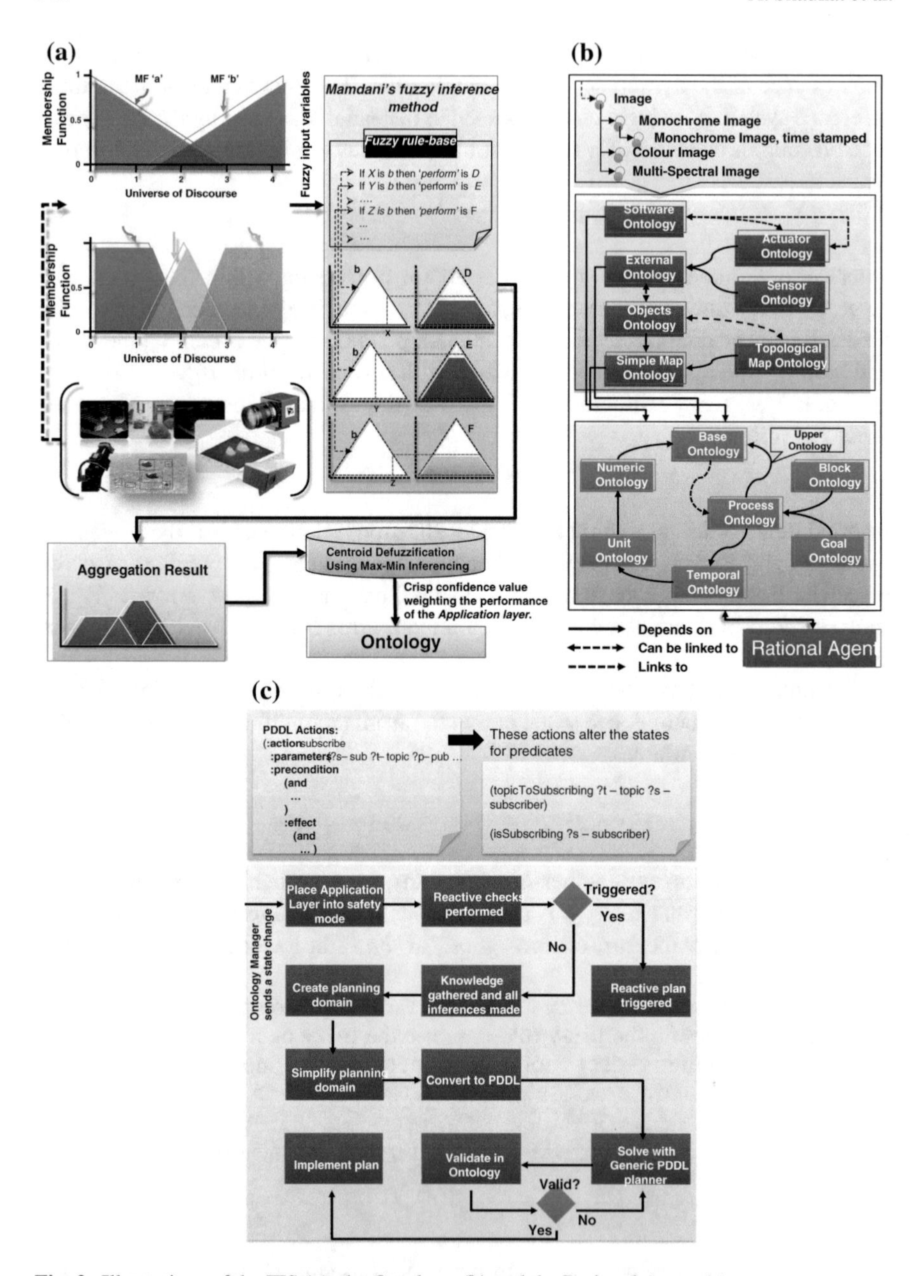

Fig. 3 Illustrations of the FIS (**a**), the Ontology (**b**) and the Rational Agent (**c**)

3.2.2 Ontology

The Ontology is the upper layer in the hierarchy of the reconfiguration layer, which contains all information used for a successful self-reconfiguration of the robotic system. The Ontology has been designed as a Modular Ontology with an Upper Ontology. This splitting of the Ontology into different modules increases extensibility, usability and readability of the Ontology. Furthermore, modularity adds the ability to perform validity checking and inference gathering on sub-ontologies, because there are fewer assertions and instances, it is more efficient.

The primary modules of the Ontology are:

- *Upper Ontology* forms the base syntax and logic for the description of goals and capabilities of modules, and is further sub-divided into essential modules like numerics, timings etc.
- *Process sub-ontology* describes the processes of autonomous software
- *Software sub-ontology* describes inner working and inter-communications of autonomous software
- *External sub-ontology* describes external environments
- *Sensor and Actuator sub-ontology* are both self-explanatory, and are dependent on the software and external sub-ontology

The Capabilities of Software and Hardware shall be represented in an extended IOPE format. The Input and Output (IO) are the communication input channels. Each channel connects to their appropriate service grounding, having a communication type, message type, rates, and their communications properties, they also have their necessity defined (i.e., is the communication input necessary for completion). The message formats and communication type exist in a subsumption hierarchy. The subsumption hierarchy has increasingly complex and specific properties and logics. The Precondition and Effect describe, in an extensible subsumptive description logic, the precondition for a service and the effect of implementing stated service. These conditions affect both the Environment and internal conditions. Then, there are non-functional properties, that can be used for quality of service metrics. Additionally, most configurations are encoded in states, so that each configuration represents a different IOPE. Continuous configurations parameters are encoded to states via parametric equations, they should be used not to make large changes, but mere parameter changes (e.g. rate). If continuous configuration parameter do make large changes to the functional operation of a service, it should be abstracted to another discontinuous parameter and another continuous parameter.

The Process logic is encoded as a single control loop of the autonomous software, multiple control loops can be described as a hierarchy, similar to a Hierarchical task network. The non-functional environmental and internal parameter condition change based on their separate conditions. For example, the battery usage for a service is given as a conservative estimate for complete standard cycle loop, battery usage is then verified at the lower abstraction layer, like the standard action planner and scheduler. The process is then logically modelled by a reduced version of Computational Tree Logic (CTL) translated to First order logic [19].

3.3 Agent Layer

The (*Rational*) Agent layer analyses the Ontology and performs planning/re-planning, followed by the *Validation* and *Verification* process, and finally implementation. The Rational Agent is the main source of triggering a self-reconfiguration routine in the lower layers. The basic process of the Rational Agent is outlined in the system diagram, Fig. 3c. The self-reconfiguration control loop is started by the *Inspector* in the *Reconfiguration layer* every time the system develops a fault or detects an anomalous behaviour.

The Rational agent triggers reconfiguration. The Rational agent takes a tiered view of the world. A subsumptive three-tier model is applied to the planning problem to reduce complexity and consequently reduce individual planning subspaces, and rendering the process of action planning more efficient [8]. Similarly, the planning problem for self-reconfiguration is not choreographing every action of the system, but the self-reconfiguration system adds another level to the subsumptive architecture that deals with a reconfiguration of the lower levels. Thus, the planning problem is to find a configuration of the system which allows the lower levels to perform their intended action.

The Rational agent has both reactive and deliberative elements, and is initialised by a change in the Ontology. Once the Rational agent has been informed by the Ontology Manager about a change, it will go through a series of steps in the attempt to rectify or optimise the system. For the purposes of speed, initial steps of the Rational agent are all reactive. The Reactive component has pre-calculated criteria, which when triggered react with a corresponding pre-calculated plan. The pre-calculated plans and criteria are prepared by a deliberative fault (or world-change) injection analysis, which calculates the most likely changes to the system. If none of the reactive criteria are fulfilled the Rational agent will use its deliberative techniques to devise an alternate plan.

To deal with the intrinsic complexity the Rational agent uses many techniques to mitigate the inherent complexity to remain efficient. The Deliberative component will first place the application layer in a *safety mode*, which halts all future operations until the safety of the system can be evaluated by the Rational agent. The Deliberative component then gathers all relevant information to the reconfiguration process and convert it into a form which can be used to generate the reconfiguration plan. While this could be achieved via first order logic in the Ontology's inference engine, directed planning algorithms with heuristics are more suited for this long chain style planning problem. Thus, the planning problem is converted into a generic planning language (i.e., PDDL2.1), then a generic Planner is used to find a self-reconfiguration plan. Using generic Planners allows for fast optimisation of planner algorithm and heuristic choice. This plan is then verified in the Ontology and implemented by the Rational agent.

The planning problem in the case of the self-reconfiguration of the application layer has an initial state that includes the (initial) world state. The goal of the planning problem includes the final world state. The possible actions in the planning domain are the services and the connection of services in which the appropriate communication link is used. Services follow the form of input, output, precondition, and effect (IOPE). The input and output refer to their communication requirements (e.g. publishing rate), and the precondition and effect refer to systems world state. The precondition and effect may include energy requirements, context requirements, other world state changes, and service requirements, e.g. whether a camera has previously been connected to the application layer. For all parameter configurations of a service there are different defined actions. The service can have discontinuous and continuous attributes and parameters. The services are assumed to perform with the most conservative scenario in respect to usable resources. This allows the other three levels of the hierarchical execution structure to be confident in being able complete their goals. The service can have any number of logical preconditions and restrictions on them. Then the solution to this planning problem is a set of services, their parameter configurations, and their connections.

To generate the PDDL, first order logic rules are used, which allows for very high level of extensibility since new rules for PDDL creation can be introduced easily. To minimise the planning domain sent to the PDDL planner, multiple space size reduction techniques are used. Using the knowledge that the plan will primarily focus on connecting services and then service configurations, a planning domain can be pruned for unreachable services that would not be apparent to generic algorithms as efficiently.

4 Experimental Scenarios

To demonstrate and prove the robustness of the Self-Reconfiguring Robotic Framework against unforeseen changes in the system, this paper presents two experiments incorporating software and hardware based self-reconfiguration scenarios. In the following, these individual scenarios will be discussed with their respective results followed by a discussion into the significance of these experimental findings.

The experimental workstation used for running the FIS comprises a quad-core Intel(R) Xeon(R) X5482 CPU (3.20 GHz) running Linux (Ubuntu 12.04, 64-bit architecture). Workstation used for the Agent, Application Layer, and the Ontology also had similar specifications.

4.1 SLAM in Unstructured Environments

The aim of the experiment is to test the ability of the Self-Reconfiguring Robotic Framework to optimise the system on the basis of changing system performance, which can alter because of changing environments and anomalous operations of the Application layer. The scenario in this experiment is a Planetary rover using a monocular camera based navigation system. The navigation system implemented in the Application Layer is the PM-SLAM localisation system introduced in [1].

PM-SLAM (Planetary monocular simultaneous localization and mapping) is a highly modular, monocular SLAM system for use in planetary exploration. PM-SLAM's modular structure allows multiple techniques to be implemented, such as, distinct SLAM filters, and visual feature tracking techniques. For example, multiple types of vision-based features may be implemented, SURF features or hybrid of semantic blob-based [1, 15] feature tracking, both having their own performance characteristics for different environments.

A visual SLAM system is susceptible to accumulated posterior state estimation error, and accruing computation time over multiple iterations [1]. The Self-Reconfiguring Robotic Framework attempts to maintain an optimal level of computational load and error over the course of operation, through self-reconfiguration of the subcomponents of PM-SLAM. For the purpose of monitoring the Application layer, this paper implements a FIS based Inspector. This Inspector monitors the predicted localisation error and computation loop time to generate a confidence value that quantifies the performance of the localisation system. The predicted localisation error for this experiment is derived from the covariances estimated by the SLAM filter. This Confidence measure is used to update the Ontology, which is then used by the Agent Layer to compute, and if necessary implement a new optimal configuration of the Application Layer.

A subset of the images generated by the European Space Agency field trial in the Atacama Desert, Chile in 2012 using the SEEKER rover platform is used for this experimental scenario [1].

4.1.1 Experimental Setup

PM-SLAM and the other Application Layer modules have been implemented in C++ using ROS and OpenCV. The Agent Layer is implemented in JAVA using Apache's JENA. The reactive component of the Agent Layer has been not implemented for this test, as it does not aid in demonstrating the capabilities and efficiency of the Agent Layer. The Ontology is implemented in OWL (Web Ontology Language), as it is the most widely used and supported Ontology Language, with a wealth of mature tools like Protégé and Apache JENA. The FIS Inspector has been implemented in MATLAB. Both the Agent Layer and the Ontology have been implemented as generic systems such that they can be adapted for various other types of robotic applications.

4.1.2 Experimental Results

The application layer takes two different configurations during the experimental process. An initial configuration of the application layer utilising SURF-based feature tracking, a depth perception module, and an EKF SLAM filter and a reconfigured application layer after the FIS Inspector confidence has dropped below acceptable levels (as determined by the Rational Agent). This configuration utilises a hybrid semantic blob and SURF feature tracking technique and an EKF SLAM filter [1].

As discussed in Sect. 3.2.1, the FIS generates confidence values using the inputs (*error* and *computation time*) from the SLAM system, which are added to the Ontology and used by the Rational agent to examine the system performance over the course of operation. The tests are run on-line over the images from the dataset; however, a reconfiguration routine is triggered by the Rational agent responding to the change in the Ontology as soon as the system performance confidence value goes below a specified threshold (0.5 in the current case). Referring to Fig. 4b, a significant reduction in the computation time is observed for the proposed reconfigurable system. In contrast, the engineered system continues its operation without any reconfiguration and therefore the computation time continues to increase. Similarly, Fig. 4c presents the behaviour of the system's posterior error over multiple iterations, which is marginally increased following the reconfiguration. This is because of the sudden change in the type and cardinality of the visual features used by the SLAM filter. However, this is deemed by the system as a suitable compromise to maximise the over performance of the system. This is also observed for the FIS-based system performance measure as shown in Fig. 4a, where the confidence value increases following the reconfiguration in the proposed system, while it continues to decrease for the engineered system.

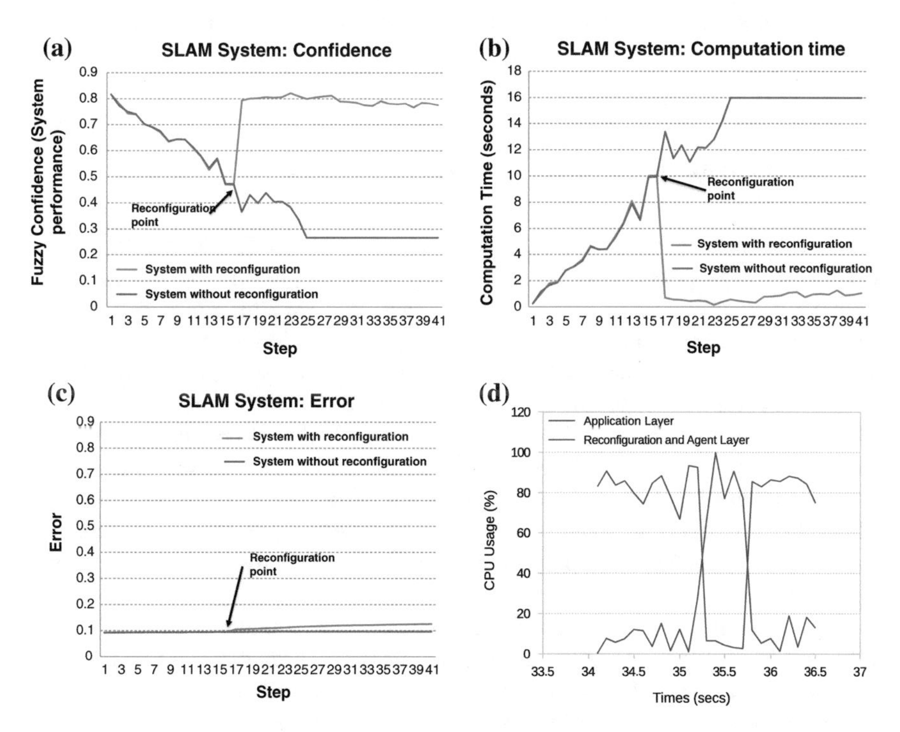

Fig. 4 FIS performance-based confidence output (**a**), computation time (**b**), weighted error (**c**) and CPU usage (**d**) for the proposed self-reconfigurable system against an engineered system in the SLAM scenario

As Fig. 4d shows the reconfiguration process takes approximately 0.46 s. The average slam loop time before the reconfiguration is approximately 5.65 s; therefore, the reconfiguration time is comparatively fast compared to the applications layer process. Similarly, the computation usage of the Application layer and Reconfiguration Layer, when they are in active full use, is of a similar amount. Moreover, the computational usage of the reconfiguration layer whilst not in actively calculating a reconfiguration plan is comparatively minimal.

4.1.3 Discussion

From the results presented in Sect. 4.1.2, it is clear that the Self-Reconfiguring Robotic Framework is capable of optimising the Application Layer, in that an improvement in performance is observed following a reconfiguration, as compared to a standard engineered system that does not allow for any reconfiguration. In particular, it demonstrate the system ability to reconfigure complex autonomous software at run-time, while maintaining continuity of service. Furthermore, it is clear that the system can self-optimise a system in such a way that includes compromising other systems, to maximise the systems overall performance. Moreover, the Self-Reconfiguring Framework demonstrates an ability to perform reconfiguration relatively efficiently.

4.2 Hardware in Complex Systems

The aim of this experiment is to test the ability of the Self-Reconfiguring Robotic Framework to respond to degrading or faulty hardware, to improve the continuity of operations of the system and optimise the systems overall performance. Similar, to the previous experiment a rover navigation system will be the scenario, with a camera that will be artificial impaired as to simulate degrading hardware or environmental conditions that may affect camera lens and cause unwanted artefacts in the input images. A common factor for noisy visual inputs is granulated dust particles. Such type of noisy conditions can be simulated by adding "*salt-and-pepper*" noise to images [1], hence simulating problem with one of the camera hardware. The Self-Reconfiguring Robotic Framework is expected to maintain continuity in the system, optimising the Application Layer, and eventually switching to another camera entirely.

Similar to the last system the Application Layer is monitored by a FIS based Inspector, which uses *entropy* as a statistical measure of randomness that can characterise the texture of the input image, and computation loop time to generate a confidence value for the performance of the Application layer system. This value is used to update the Ontology, which is then used by the Agent Layer to compute, and if necessary implement a new optimal configuration of the application layer.

In order to prove that the Self-Reconfiguring Robotic Framework can cope with such hardware problems, the framework should demonstrate a reconfiguration at run-time to provide continuity of service, even after a significant event in contrary to a traditionally engineered system, which may suffer a catastrophic failure.

4.2.1 Experimental Setup

The experimental setup is the same as for the previous scenario in Sect. 4.1.1. However, unlike the previous scenario the second camera (right) in the stereo pair from the SEEKER dataset, is used to simulate an auxiliary camera. Test images from left and right cameras are subsampled by adding *"salt-and-pepper"* noise over a varying scale of 0–100 % with a step size of 1 for the left camera (simulating very fast deterioration), and a step size of 0.1 for the right camera (simulating negligible damage).

4.2.2 Experimental Results

The application layer takes two different configurations as in the previous experiment, i.e., the initial configuration where the left camera is initially used by subsystem, followed by a reconfiguration process triggered by the Rational Agent switching the subsystem to the alternative right camera, which is still in working condition compared to the left camera.

Referring to Fig. 5, the Self-Reconfiguring Robotic Framework is capable of optimising the Application Layer, by switching to an alternative camera as soon as the fuzzy confidence measure goes below the specified threshold. This results in an improvement in performance of the system. Comparative analysis in this figure shows that a standard engineered system without any reconfiguration continues to use images from the faulty camera.

The FIS generated confidence values depend upon the entropy value and *computation time* from the SLAM subsystem, which are added to the Ontology and used by the Rational agent to check system performance. The tests are run on-line over the images from the sub-sampled dataset with simulated noise. A reconfiguration is performed by the Rational agent responding to the change in the Ontology on the basis of the system performance confidence value. Referring to Fig. 5b, the computation time is significantly reduced, while an improvement is observed in the entropy measure. This is a result of switching the faulty camera (left camera) to an alternative one that is still in working order (right camera). The engineered system continues its operation without any reconfiguration and therefore the computation time and the entropy measure continue to deteriorate over time. Similarly, the FIS-based system performance measure as shown in Fig. 5a, increases following the reconfiguration in the proposed system, while it continues to decrease without any reconfiguration.

As Fig. 5d shows the reconfiguration process takes approximately 0.47 s. The average slam loop time before the reconfiguration is approximately 3.61 s; therefore,

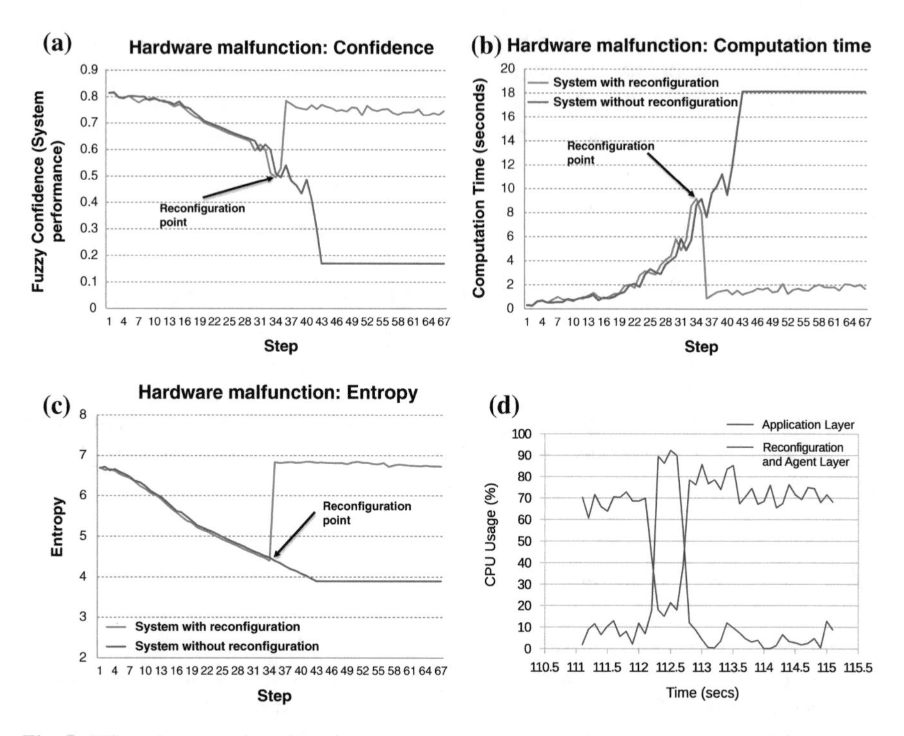

Fig. 5 FIS performance-based confidence output (**a**), computation time (**b**), entropy (**c**) and CPU usage (**d**) for the proposed self-reconfigurable system against an engineered system in the faulty hardware scenario

the reconfiguration time is comparatively fast compared to the applications layer process. The average time post reconfiguration is 1.80 s, which is comparatively slow compared to the Reconfiguration process. Similarly, the computation usage of the Application layer and Reconfiguration Layer, when they are in active full use, is of a similar amount. Moreover, the computational usage of the reconfiguration layer whilst not actively calculating a reconfiguration plan is comparatively minimal.

4.2.3 Discussion

From the results presented in Sect. 4.2.2, it is clear that the Self-Reconfiguring Robotic Framework is capable of optimising the Application Layer, in that an improvement in performance is observed following a reconfiguration, as compared to a standard engineered system that does not allow for any reconfiguration. In particular, the framework can reconfigure hardware for autonomous systems at run-time.

Furthermore, it demonstrates the system ability to cope with inferred measures of the systems performance. Moreover, the Self-Reconfiguring Framework demonstrates an ability to perform reconfiguration relatively efficiently.

5 Conclusion

This paper introduced a Self-reconfiguring Robotic framework in the form of a hierarchy combining fuzzy reasoning with ontology-based deliberative decision making to enable self-reconfigurability within a complex robotic system and increase the fault tolerance and robustness to unforeseen changes in the system and the environment. The framework is designed to be generic and therefore allows to be implemented in a wide variety of robotic applications. An Ontological model is used for system knowledge representation to enable it to be self-aware of its constituent subsystems. A fuzzy inference system quantifies the performance of the lower-level subcomponents within the Application layer in terms of fuzzy confidence measures which are intuitive for human operators that may be monitoring the system. At the highest level of the system hierarchy; a rational-agent constantly checks the underlying levels of the system for anomalies or inconsistent behaviour that can potentially cause fatal damage to itself, harm to the environment or failure to achieve desired objectives and trigger a reconfiguration in the Application layer.

Two different experimental setups were used to test the proposed system and its advantages against a system that does not allow reconfiguration in any situation. The setups involved real-world robotic hardware, such as, autonomous rovers and vision sensors that were used to generate the test and analysis data. The outcome of these experiments established that the proposed Self-Reconfiguring Robotic Framework either maintains or improves the performance of the over all system in challenging operating conditions. The layered design of the system, has proven to be an effective method for abstracting the reconfiguration of a complex system, without significantly influencing the underlying system (i.e., application layer) during normal operations.

In future work, an extension to the work presented in this paper could include inspectors that monitor the application layer using other logical inferencing techniques to discover various performance characteristics such as software faults for reconfiguration.

Acknowledgments The work presented here was supported by the U.K. Engineering and Physical Sciences Research Council under Grant EP/J011916/1 [Reconfigurable Autonomy] and Grant EP/I000992/1 [MILES]. The authors are very grateful to the Autonomous Systems Group, RAL Space, for providing the SEEKER dataset used for the experiments. No new data was created in the study.

References

1. Bajpai, A., Burroughes, G., Shaukat, A., Gao, Y.: Planetary monocular simultaneous localization and mapping. J. Field Robot. **33**(2), 229–242 (2015)
2. Bojinov, H., Casal, A., Hogg, T.: Multiagent control of self-reconfigurable robots. Artif. Intell. **142**(2), 99–120 (2002). International Conference on MultiAgent Systems 2000
3. Guy, B., Yang, G.: Ontology-based self-reconfiguring guidance, navigation, and control for planetary rovers. J. Aeros. Inform. Syst. doi:10.2514/1.I010378 (2016)

4. Christensen, E., Curbera, F., Meredith, G., Weerawarana, S., et al.: Web services description language (wsdl) 1.1 (2001)
5. Dennis, L.A., Fisher, M., Aitken, J.M., Veres, S.M., Gao, Y., Shaukat, A., Burroughes, G.: Reconfigurable autonomy. KI-Künstliche Intelligenz **28**(3), 199–207 (2014)
6. Gao, Y., Frame, T.E.D., Pitcher, C.: Peircing the extraterrestrial surface: integrated robotic drill for planetary exploration. Robot. Autom. Mag. IEEE **22**(1), 45–53 (2015)
7. Huebscher, M.C., McCann, J.A.: A survey of autonomic computing—degrees, models, and applications. ACM Comput. Surv. **40**(3), 7:1–7:28 (2008)
8. Knight, R., Fisher, F., Estlin, T., Engelhardt, B., Chien, S.: Balancing deliberation and reaction, planning and execution for space robotic applications. In: Proceedings. 2001 IEEE/RSJ International Conference on Intelligent Robots and Systems, 2001, vol. 4, pp. 2131–2139. IEEE (2001)
9. Mamdani, E.H., Assilian, S.: An experiment in linguistic synthesis with a fuzzy logic controller. Int. J. Hum.-Comput. Stud. **51**(2), 135–147 (1999)
10. Martin, D., Burstein, M., Hobbs, J., Lassila, O., McDermott, D., McIlraith, S., Narayanan, S., Paolucci, M., Parsia, B., Payne, T., et al.: Owl-s: Semantic markup for web services. W3C member submission, 22:2007–04 (2004)
11. McGuinness, D.L., Van Harmelen, F., et al.: Owl web ontology language overview. W3C Recomm. **10**(10), 2004 (2004)
12. Rouff, C., Hinchey, M., Rash, J., Truszkowski, W., Sterritt, R.: Autonomicity of nasa missions. In: Proceedings Second International Conference on Autonomic Computing, 2005. ICAC 2005, pp. 387–388. IEEE (2005)
13. Schlenoff, C., Prestes, E., Madhavan, R., Goncalves, P., Li, H., Balakirsky, S., Kramer, T., Miguelanez, E.: An ieee standard ontology for robotics and automation. In: IEEE/RSJ International Conference on Intelligent Robots and Systems (IROS), 2012, pp. 1337–1342 (2012)
14. Shaukat, A., Gilbert, A., Windridge, D., Bowden, R.: Meeting in the middle: a top-down and bottom-up approach to detect pedestrians. In: 21st International Conference on Pattern Recognition (ICPR), 2012, pp. 874–877 (2012)
15. Shaukat, A., Spiteri, C., Gao, Y., Al-Milli, S., Bajpai, A.: Quasi-thematic features detection and tracking for future rover long-distance autonomous navigation. In: 12th Symposium on Advanced Space Technologies in Robotics and Automation, Noordwijk, the Netherlands, European Space Agency, ESTEC (2013)
16. Shaukat, A., Burroughes, G., Gao, Y.: Self-reconfigurable robotics architecture utilising fuzzy and deliberative reasoning. SAI Intell. Syst. Conf. **2015**, 258–266 (2015)
17. Shaukat, A., Gao, Y., Kuo, J.A., Bowen, B.A., Mort, P.E.: Visual classification of waste material for nuclear decommissioning. Robot. Auton. Syst. **75**, Part B, 365–378 (2016)
18. Sterritt, R., Hinchey, M.: Spaace iv: self-properties for an autonomous & autonomic computing environment—part iv a newish hope. In: Seventh IEEE International Conference and Workshops on Engineering of Autonomic and Autonomous Systems (EASe), 2010, pp. 119–125 (2010)
19. Vakili, A., Day, N.A.: Reducing ctl-live model checking to first-order logic validity checking. In: Proceedings of the 14th Conference on Formal Methods in Computer-Aided Design, pp. 215–218. FMCAD Inc (2014)
20. Windridge, D., Felsberg, M., Shaukat, A.: A framework for hierarchical perception-action learning utilizing fuzzy reasoning. IEEE Trans. Cybern. **43**(1), 155–169 (2013)
21. Yim, M., Shen, W.-M., Salemi, B., Rus, D., Moll, M., Lipson, H., Klavins, E., Chirikjian, G.S.: Modular self-reconfigurable robot systems [grand challenges of robotics]. IEEE Robot. Autom. Mag. **14**(1), 43–52 (2007)

Design and Optimization of Permanent Magnet Based Adhesion Module for Robots Climbing on Reinforced Concrete Surfaces

M.D. Omar Faruq Howlader and Tariq Pervez Sattar

Abstract In this chapter, the detailed design of a novel adhesion mechanism is described for robots climbing on concrete structures. The aim is to deliver a low-power and sustainable adhesion technique for wall climbing robots to gain access to test sites on large concrete structures which may be located in hazardous industrial environments. A small, mobile prototype robot with on-board force sensor was built which exhibited 360° of manoeuvrability on a 50 × 50 mm meshed reinforcement bars test rig with maximum adhesion force of 108 N at 35 mm air gap. The proposed adhesion module consists of three N42 grade neodymium magnets arranged in a unique arrangement on a flux concentrator. Finite Element Analysis (FEA) is used to study the effect of design parameters such as the distance between the magnets, thickness and material of the flux concentrator, use of two concentrators, etc. Using two modules with minimum distance between them showed an increase of 82 N in adhesion force compared to a single module system with higher force-to-weight ratio of 4.36. An adhesion force of 127.53 N was measured on a real vertical concrete column with 30 mm concrete cover. The simulation and experimental results prove that the proposed magnetic adhesion mechanism can generate sufficient adhesion force for the climbing robot to operate on vertical reinforced concrete structures.

1 Introduction

Non Destructive Testing (NDT) plays a vital role in ensuring the safety and integrity of capital assets. NDT techniques have found application in the regular inspection of large safety critical infrastructures to determine their structural integrity and to plan maintenance and outage schedules. The applications range from large welded steel structures [1] (such as petrochemical steel storage tanks, ship hulls, pressure vessels, etc.) to composite structures (such as wind blades and modern aircraft wings and fuselage) and to concrete structures (such as cooling towers of power plant, bridge columns, dams, nuclear power plants buildings, etc.). While concrete can offer

M.D.O. Faruq Howlader (✉) · T.P. Sattar
Robotic and NDT Research Centre, London South Bank University, London, UK
e-mail: howladem@lsbu.ac.uk

Y. Bi et al. (eds.), *Intelligent Systems and Applications*,
Studies in Computational Intelligence 650, DOI 10.1007/978-3-319-33386-1_8

higher strength and durability compared to other building materials [2], atmospheric moisture and chlorides cause the concrete structures to lose their strength. Therefore, from a civil engineering perspective, NDT plays a critical role in detecting induced structural faults like reinforcement bar corrosion, cracking, and delamination of the concrete surface [3].

Currently, the NDT of concrete structures is mostly performed manually. The primary challenge of carrying out the procedure on a vertical concrete structure is to gain access to the inspection area which is normally located at a high altitude. The NDT of these vertical concrete structures must be addressed by ensuring the safety of operators, increasing the inspection speeds and hence decreasing the time taken to inspect them, and improving the quality of NDT measurements which suffer as a result of operator fatigue when performing NDT on large surfaces. The inspection speeds and cost reductions can be dramatically improved by using climbing robots to gain access to a test site by eliminating the need to erect scaffolding or make lengthy preparations to abseil down or use suspended platforms. The quality of NDT data can be improved by using automated robotic deployment of NDT sensors thereby eliminating human factors such as fatigue or inattention. As a result, climbing robots in the field of vertical structures are of increasing importance for inspections and maintenance tasks. A considerable amount of research has been devoted to develop various types of experimental prototypes for industrial inspection tasks with the application mostly dictating the design aspects of climbing robots [4, 5].

This chapter focuses on the development of a novel wall-climbing robot to carry out the Non-Destructive Testing (NDT) of safety critical reinforced concrete structures using permanent magnets to adhere to the structure. Permanent magnet systems do not require the supply of energy and hence have an inherent advantage for climbing robots.

The arrangement of multiple permanent magnets on the back of a flux concentrator provides a unique mechanism to concentrate and magnify the magnetic energy to couple with reinforcement bars (rebars) buried under the concrete surface. The research aim is to optimize the design parameters to provide increased payload capacity.

The chapter consists of eight sections. Section 2 gives an overview of the working environment of the mobile robot. Section 3 presents a review of the state-of-the-art of climbing robots. Section 4 specifies the design requirements of the climbing robot for successful deployment while Sect. 5 details the simulations carried out using Finite Element Analysis (FEA). Development of the prototype climbing robot is presented in Sect. 6 and the experimental results to validate the design are discussed in Sect. 7. The chapter concludes with a brief summary and recommendations in Sect. 8.

2 Current Practices to Access Vertical Structures

The current practice is to carry out manual NDT of large, vertical, safety critical structures and the biggest task in the performance of manual NDT is to first obtain access to test sites. This is mostly obtained by constructing scaffolding. The scaf-

Fig. 1 Manual NDT of weld lines on cargo container ships

foldings used are of two main types, fixed type and abseiling type. A fixed type scaffolding can offer stable access of the target area to the NDT operator. However, this type of scaffolding is expensive and time-consuming to erect and disassemble. The abseiling type is most commonly used for inspection purposes and is less expensive and time consuming than its fixed counterpart. But it is more vulnerable to wind gusts, bad weather and accidents [6]. A case study to show the problems of erecting a fixed scaffolding to inspect a 300 foot tall chimney is described in [7]. Figure 1 shows a scenario of manual NDT of approximately 0.5 Km of vertical and horizontal weld lines during the construction of new-build cargo container ships by erecting scaffolding consisting of wood planks and ropes.

However, for more complex structures such as bridge columns, decks, and concrete storage tanks etc. scaffolding is not a feasible option. In those cases, access devices like cranes and gondolas are deployed to the desired position. Sometimes professional climbers, called 'spider workers', are involved in the inspection if the structure cannot be accessed otherwise.

Other than the complexity, safety of the NDT operators and the cost of preparations are the two main issues that must be considered. Here the operators deal with hazardous working environments with possibly limited manoeuvrability [8]. More often than not, it's the working environment that causes the most concern rather than the altitude. In the nuclear industry for example, the spider-workers are exposed to radioactive air while inspecting concrete reactor tanks. Even though the nuclear regulatory commission has determined 50 mSv of radiation exposure in a year as a safe limit [9], it is the cyclic exposure to radiation that proves fatal. In 2015, a report in the British Medical Journal stated that, a total of 17, 957 workers in the nuclear industry from France, the United Kingdom, and the United States died from solid cancers caused by low level but cyclic radioactive exposure in the last 72 years [10].

Likewise, in the chemical industry, the operators have to deal with high temperatures as a boiler tube is shut down for inspection, it requires 2–3 days before it reaches a safe level [11]. In addition, in many of these inspection tasks, the freedom of movement is very restricted and the operator mobility is very low with the result that the NDT inspection is of poor quality. Therefore, interest in the development of climbing robots has grown significantly as robotic solutions can offer quick access to tall structures and remote test sites leading to increased operational efficiency, reduced cost and greater health and safety protection to human operators.

3 Climbing Robots—A Design Review

The workplace of climbing robots is on the vertical plane, therefore, it requires adhesion mechanisms to carry its payload along different types of vertical surface. The robot is required to support its own body weight as well as carry a given payload while climbing. A review of climbing methods based on locomotion and adhesion mechanisms is presented.

3.1 Adhesion Types

Permanent magnets have been used as the adhesive method when designing robots that are required to climb ferrous surfaces for a long period. The "WCR System" [12] is an example of a wall climbing robot developed by a team from Cambridge University. The application of the robot was to label the scale of steel oil storage tanks. Later a number of robots were developed and patented based on the same principle. A non-contact magnetic adhesion mechanism proposed in [13] implemented arrays of magnets under the chassis of the robot as shown in Fig. 2a. A constant gap of 5 mm between the magnet and the wall surface was maintained. The effect of multiple magnet layers and the distance between the magnets on the total adhesion force are not investigated in that study. Another permanent magnet based tracked robot presented in [14] investigated the effect of air gap between multiple magnets and the steel surface. The robot consisted of an aluminium frame and track wheels with permanent magnets located on the steel channels. The track wheels and the magnet attachment are shown in Fig. 2b. The effect of distance between the magnets, and the effect of magnet dimension on the overall adhesion force have not been investigated.

Another robot [15] was developed based on the magnetic wheel system. In this case, the wheel was constructed of magnetic material and the effect of different wheel rims on the distribution of magnetic flux lines was investigated on both planer and curved surfaces. The robot can achieve 60 N of maximum adhesion force while climbing a ferrous surface.

The most frequent approach to adhesion on a surface is to implement suction and pneumatic adhesion. These systems are mainly used for non-ferromagnetic, non-porous, relatively smooth surfaces, and can be applied to a large suction area to increase the adhesion force. Examples of robots using suction are a climbing robot for inspecting nuclear power plants [16], a robot using tracked wheel mechanism with suction pads [17] and two other four limbed robots with suction pads attached

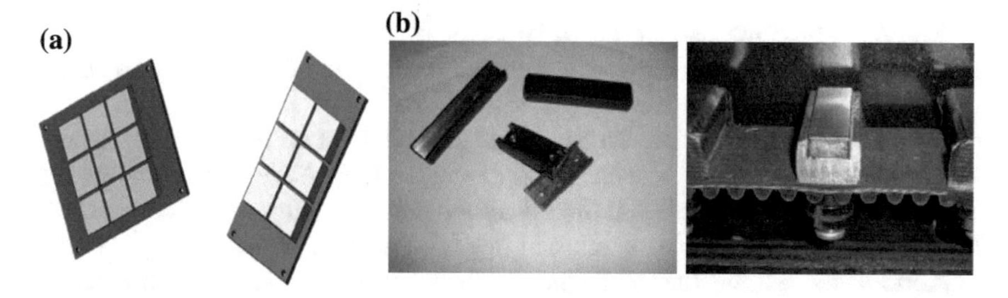

Fig. 2 **a** Arrays of rare earth permanent magnets on steel plates as proposed in [13]; **b** Tracked wheels and magnet spacing as in [14]

[18]. The initial limitation of the suction mechanism is that it requires time to develop enough vacuum to generate sufficient adhesion force and as a result, the speed of the robots using suction is relatively low [19]. To solve this problem a vacuum rotor package called Vortex generates a low pressure zone enclosed by a chamber [20]. The high speed rotor causes the air to be accelerated toward the outer perimeter of the rotor and away from the centre radially. The resulting air exhaust is directed towards the rear of the robot and creates an adhesion towards the climbing surface. However, the resultant adhesion force from vortex technology is not enough to support a reasonable payload. Moreover, as the gap between the climbing surface and the chamber is required to be absolutely minimized, robots using vortex technology cannot offer floor-to-wall transitions.

Researchers have also investigated the application of biological agents in robot design. The Lemur II is an example of such design [21]. It can climb vertical rough surfaces with the help of four limbs. Another example of this design is the RiSE robot which imitates the movement of a Hexapod [22] using a special type of micro spine attached to its six legs. A more advanced robot based on the same design principle is CLIBO [23] that uses 16 motors for high manoeuvrability. A robot design derived from human climbing behaviour was developed named the ROCR [24]. It consists of two modules and a tail that thrusts sideways and propels the robot upwards on a rough vertical surface.

3.2 *Locomotion Types*

For locomotion, wheel-type, track-type, wire and ligament based locomotion mechanisms are used to build robots for inspecting complex shaped structures with high mobility. The benefit of a wheel-type mechanism is that it can move and turn flexibly with a small contact surface between the wheels and the wall. Therefore, the robot's energy-use ratio is low. However, wheeled robots with suction type adhesion need to maintain a fixed minimum air gap to sustain the adhesion and this creates problems with loss of adhesion when climbing uneven surfaces. Also, these wheels cannot overcome larger steps, so they are exposed to slip effects.

The track-type mechanism has a larger contact area and can generate higher attraction force and it can move well in one direction but finds it is hard to change direction since it has only two degrees of freedom.

Wires and rail locomotion can be used to move a robot up vertical tall structures for some fixed applications like maintenance and cleaning. A tile-wall robotic system [25] for the NDT of concrete buildings is a good example of this sort of climber. This system consists of a portable module carrying NDT devices, with a ground and a roof platform connected together by a conveyor belt. The advantage of this mechanism is that the control system only has to position the NDT sensors at the point required and the additional payload is carried and held by the conveyor belt. Even though this system allows much simpler robot dynamics, the system is ineffective in the case of very tall structures as very long conveyor belts need to be installed between the roof and the ground.

The suction cup option is the most widely used method for non-magnetic surface climbing robots in comparison to other adhesion mechanism reported in the literature. Traditionally, vacuum pumps, often in the form of suction cups, are used as vacuum producers. They can produce high vacuum and large suction force. However, when suction cups and vacuum pumps are applied to a wall-climbing robot, other pneumatic components such as solenoid valves and regulating valves are also needed [23] which increase the weight of the robot especially when high flexibility of locomotion is required. Adhesion principles of vortex and electro adhesion are still in the development stage and not completely understood. As a result, few practical products have been developed for reinforced concrete surfaces, though wall-climbing robots have been researched for many years.

4 Design Requirements

The permanent magnet adhesion climbing robot discussed in this chapter is designed to deliver a solution to deploy a variety of NDT equipment on a robotic platform. A specific set of requirements can be postulated suitable for this task.

- Climbing capability—The robot must have a climbing capability on vertical reinforced concrete surfaces of up to 50 m high and curved surfaces with a minimum diameter of 10 m. This kind of structure is common for concrete buildings in nuclear power plants.
- Magnetic adhesion penetration depth—Concrete is a brittle material that can absorb strong compression forces but breaks at very low tensile forces [26]. Cracking and delamination can occur on non-reinforced concrete structures if excessive force is induced by external loading as shown in Fig. 3. To counter the excessive tensile force, concrete structures are reinforced with strong materials like steel reinforcement bars (rebars). Nuclear and other safety critical structures are reinforced with the dense meshing of rebars using special techniques to distribute the concrete as it is poured [27].

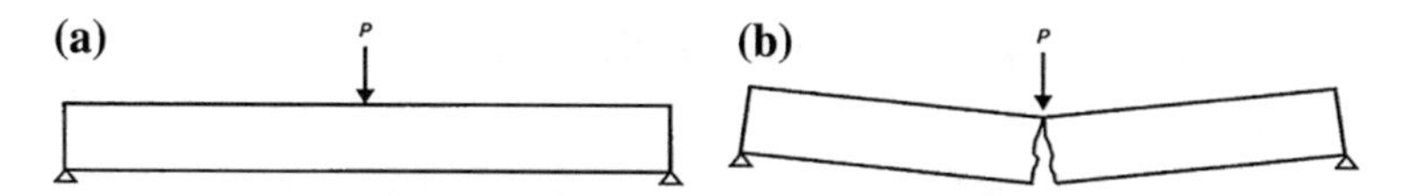

Fig. 3 Crack occurrence on a concrete slab **a** Bending Force < Tensile threshold force, **b** Bending Force > Tensile threshold [26]

As steel rebars are highly ferromagnetic, therefore magnetic coupling with the buried rebars can deliver a simple, low-energy adhesion mechanism. The main challenge is to shape and concentrate magnetic flux to penetrate the concrete cover as deeply as possible to gain sufficient magnetic adhesion. The nominal concrete cover for different environmental exposures is regulated by the BS 850 and Eurocode and is determined by concrete quality and intended working life of the structures [28]. To meet these standards, the proposed magnetic adhesion module is required to ascend on vertical nuclear buildings with an NDT payload by generating sufficient adhesion force while magnetically coupling with rebars at 45 mm depth. The following requirements must be satisfied:

- Dynamic force—A dynamic force analysis can determine the stability of the robot while climbing on vertical or inclined surfaces. This will also define the minimum adhesion force required to avoid sliding and roll-over failure. If the robot's weight is G, the wheels friction coefficient is μ, h is the height of the centre of gravity from the climbing surface, L is the distance between the front and back wheel, robot mass acceleration is m and a respectively then according to [29], adhesion force, F_a required to avoid sliding is

$$\mathrm{Fa} \geq (\mathrm{G} + \mathrm{ma})/\mu \tag{1}$$

And for roll-over avoidance,

$$\mathrm{Fa} \geq \frac{\mathrm{G} \times \mathrm{h}}{\mathrm{L}} \tag{2}$$

- Payload capacity—A variety of NDT sensors such as Ground Penetrating Radar (GPR), thermo graphic sensors, cover meter, Wenner probe etc. are used for concrete inspection. Depending on the application of the robot and the type of sensors used, a payload capacity of maximum 5 kg is required for safe operation during robotic inspection. Moreover, the robot weight should be kept to a minimum. In case of any adhesion failure, the impact of a heavy robot would cause damage to the plant equipment and surroundings which must be avoided.
- Reliability and safety—The safety and reliability of the robot must be ensured by robust hardware and mechanical design. A practical solution would secure the mobile robot with a safety harness with provision of power and NDT and control data transmission with an umbilical cable.

Ultimately, a strong and secure adhesion mechanism is the pinnacle of design requirements. The adhesion mechanism defines the system reliability and payload capacity.

5 Parametric Studies of the Adhesion Module

The proposed adhesion module is comprised of several magnets and a flux concentrator called the 'yoke'. The adhesion module, air gap and the steel rebar constitute a magnetic circuit as shown in Fig. 4. In order to achieve a target adhesion force, parametric studies of the adhesion module's design criteria such as material and thickness of the yoke, effect of rebar mesh on adhesion and use of multiple yokes have been carried out using Finite Element Analysis (FEA) software Comsol Multiphysics. The main properties of the materials used in the simulations are listed in Table 1.

The primary aim is to achieve maximum adhesion force to support the payload by using a minimum number of magnets. The ratio of the adhesion force and net weight of the system, η determines the module's performance by achieving maximum adhesion force F_a from the minimum net weight of the system W. Therefore,

$$\eta = Fa/W \tag{3}$$

5.1 *The Influence of Distance Between Magnets*

The initial dimensions used are magnet length, $M_l = 50$ mm, width, $M_w = 50$ mm, thickness, $M_t = 12$ mm, rebar diameter, $R_d = 12$ mm, yoke thickness, $Y_t = 10$ mm; concrete cover, $C_c = 35$ mm and the distance between the magnets is chosen as variable. With the distance varied from 10 mm to 150 mm, the adhesion force increases

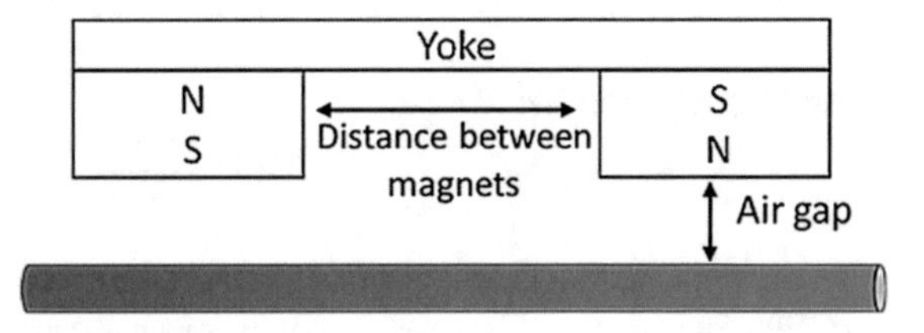

Fig. 4 Block diagram of the magnetic circuit

Table 1 Material properties used for Comsol simulations

Properties	Value
Magnetic induction intensity B_r (T)	1.31
Coercive force H_{cb} (KA/m)	915
Intrinsic coercive force H_{ci} (KA/m)	955
Magnetic energy product HB (KJ/m^3)	318
Relative permeability (μ_r)	1.05
Relative permeability of yoke (μ_r)	5000
Relative permeability of concrete surface (μr)	1
Relative permeability of steel rebar (μr)	1500

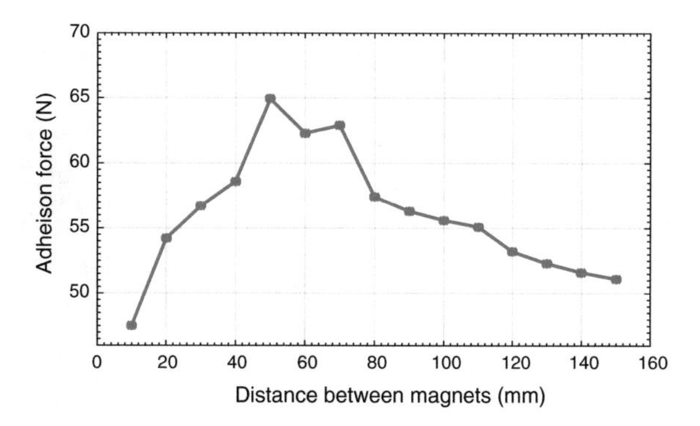

Fig. 5 Simulated adhesion force at different distances between magnets

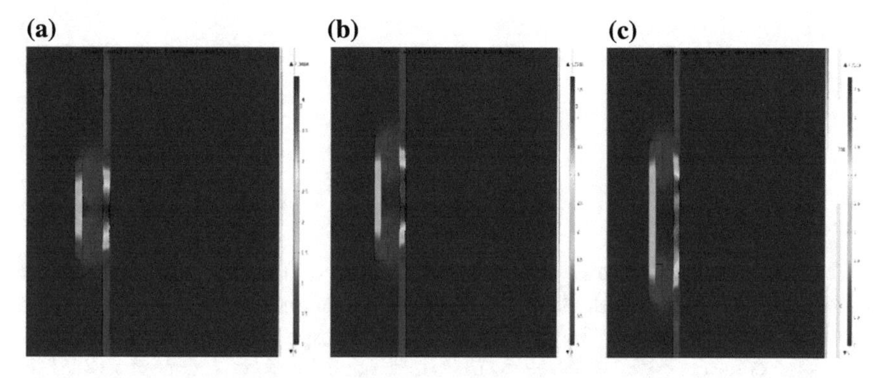

Fig. 6 Magnetic flux concentration norm at different distance between magnets; **a** Distance: 10 mm, **b** Distance: 50 mm, **c** Distance: 150 mm

rapidly at the beginning from 47.5 N to 64.93 N with the maximum at 50 mm distance as shown in Fig. 5. With the two magnets close to each other, the resultant magnetic attraction area is small. The biggest attraction area is indicated by the red hotspots at 50 mm in Fig. 6b which indicates the optimum distance between the magnets for a maximum adhesion force. At distances greater than 50 mm, the magnetic flux density along the rebar length starts to decrease and there is a gradual fall in the resultant adhesion force to 51.38 N at 150 mm distance between magnets.

5.2 *Magnetic Circuit Coupling Optimization*

Building the adhesion module with multiple magnets influences the magnetic flux distribution and the adhesion force. Previously, one magnetic circuit travelling from North Pole of one magnet to the South Pole of the other magnet was built using two magnets. To increase the number of magnetic circuits using additional magnets, two other arrangements as shown in Fig. 7 are investigated here.

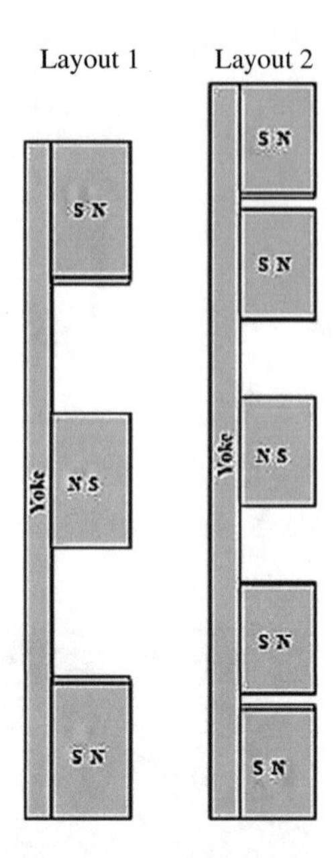

Fig. 7 Proposed magnetic circuit coupling layouts

It can be observed from Fig. 8a that, both layout 1 and layout 2 generates higher adhesion force than the method used in the previous section. At 10 mm concrete cover, resultant adhesion force is 312.57 N and 451.33 N respectively for layout 1 and 2. But as the concrete cover is increased to 20 mm, adhesion force reduces to 188.02 N and 248.17 N respectively i.e. a reduction of 39.84 % for layout 1 and 45 % for layout 2. Therefore, the rate at which adhesion force falls as the rebar distance increases is lower for layout 1 than layout 2. However, extra additional force generated by the layout 2 yields in higher values of η compared to the layout 1 which negates the additional weight of the extra magnets. The comparative values of η are shown in Fig. 8b.

Moreover, magnetic flux density norm shows that as two North Poles are located in close proximity at both ends of the adhesion module in case of layout 2, it does not concentrate magnetic flux as uniformly as layout 1. Therefore, layout 2 would be more applicable in cases where magnetic coupling with more than one layer of buried rebars is desired and layout 1 would be more suitable if the rebars are located at a lower depth. For the remainder of this chapter, layout 1 has been considered because of uniform distribution of magnetic flux.

5.3 *Effect of Yoke Thickness and Material*

The adhesion force generated by layout 1 without any yoke is 58.34 N while adding a 10 mm thick iron yoke increases the force to 83.71 N for 35 mm of concrete cover.

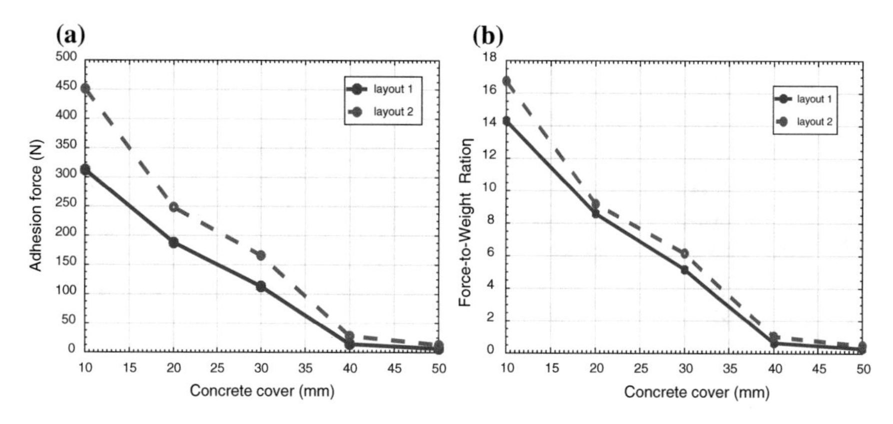

Fig. 8 **a** Comparison of adhesion force and **b** force-to-weight ratio between layout 1 and layout 2 for different concrete cover

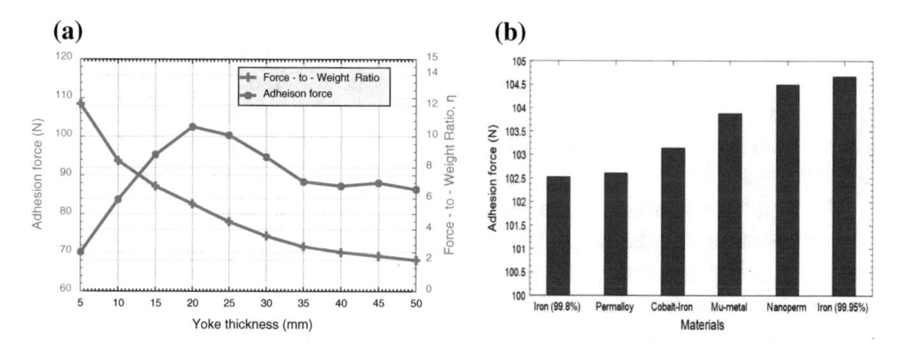

Fig. 9 Adhesion force comparison for: **a** different yoke thickness; **b** different materials

Therefore, the thickness of yoke plays a critical part to maximize adhesion force. To investigate this parameter further, models with varied yoke thickness from 5 mm to 50 mm are simulated. All other parameters are kept the same as in previous simulations. According to the results in Fig. 9a, adhesion force would increase significantly to 102.53 N for yoke thickness of 20 mm. Moreover, the magnetic flux leakage is lower in the yoke's opposite surface and more flux lines are concentrated toward the rebar shown. However, the adhesion force comes to a saturation point of approximately 85 N at 35 mm yoke thickness, where further increases in yoke thickness would not have any significant influence on the adhesion force but η value begins to decrease. The main reason for this is that when the yoke thickness reaches a certain value, the magnetic flux concentration is too small towards the rebar to influence the adhesion. Furthermore, adhesion force measurements for 20 mm thick yokes of different materials shows that the value of magnetic permeability has negligible influence on adhesion force as results are shown in Fig. 9b.

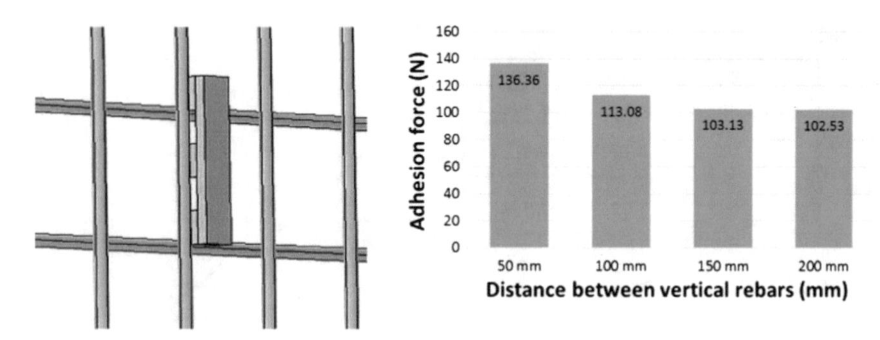

Fig. 10 Simulation setup of rebar mesh and resultant adhesion force at different mesh settings

5.4 Effect of Rebar Mesh

In practice, different spacing between the rebars is used depending on the concrete structure. The adhesion force for a rebar mesh with different spacing is presented here. In the simulation model, the distance between horizontal rebars is set to 150 mm which is the industry standard [28] and the distance between the vertical rebars is varied from 50 mm to 200 mm while keeping the concrete cover constant at 35 mm.

The simulated adhesion force measurements in Fig. 10 show that the adhesion force is increased as the distance between the vertical rebars is reduced. This is because in case of two rebars close together, magnetic flux can act on a larger area and generate bigger force compared to a single rebar. In case of safety critical infrastructures, the rebars are meshed at a distance of 50 mm for greater strength and longevity and this novel adhesion mechanism is much more suited to those environments.

5.5 Effect of Multiple Yokes

As structures are reinforced with a mesh of rebars therefore implementation of multiple yokes can increase the attraction area and ultimately generate higher adhesion force. To investigate this further, models are built using two layouts as in Fig. 11 and the adhesion force is measured for different distances between them. Rebar meshing of 50 mm × 50 mm is considered for this setup.

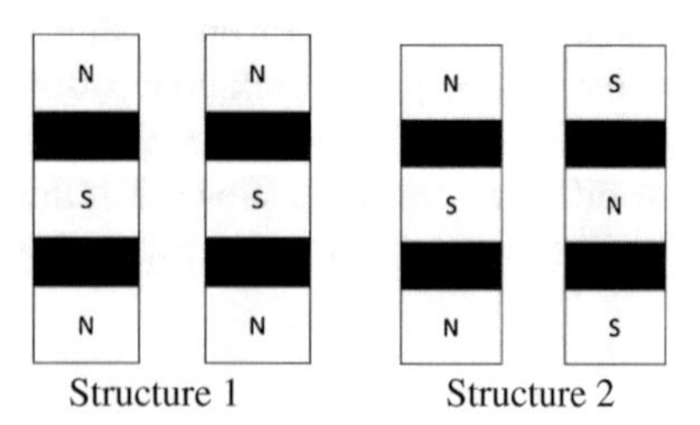

Fig. 11 Two proposed coupling structures using multiple yokes

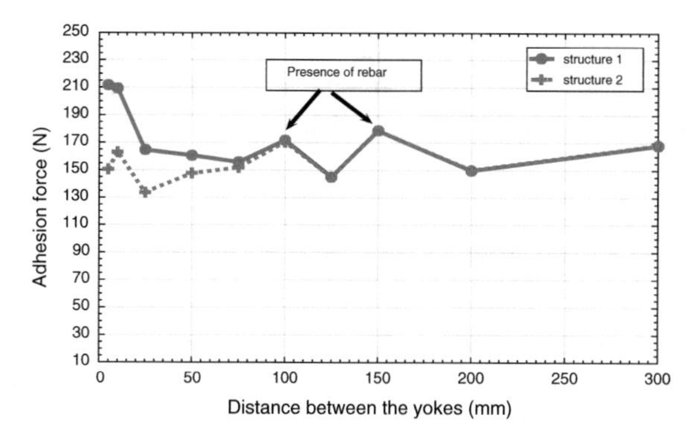

Fig. 12 Adhesion forces at different distances between the yokes for both structures

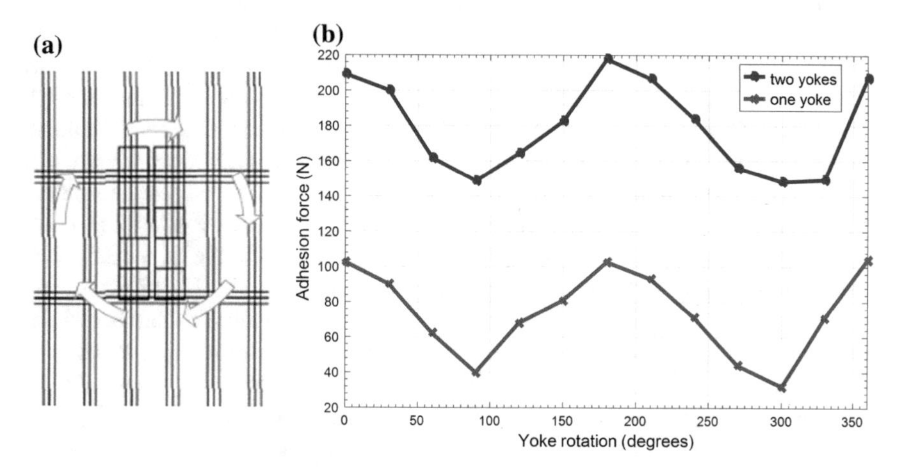

Fig. 13 Comparison of adhesion forces at different rotational positions of adhesion modules consisting of one and two yokes: **a** Simulation setup; **b** Resultant adhesion forces

Results in Fig. 12 show that, structure 1 provides higher adhesion force compared to structure 2 when the distance between yokes is minimized. At 5 mm distance, structure 1 generates 211.7 N of force compared to 150.3 N for structure 2. The adhesion force decreases as the distance is increased to 50 mm. At distances higher than 50 mm, both structures generate exactly the same amount of adhesion force as seen in Fig. 12. Moreover, an increase of adhesion force to 188 N is seen at 60 mm and 80 mm of distance between the yokes. At these two points, both of the yokes for structure 1 and 2 are located right in front of the vertical rebars.

Furthermore, dense meshing of the rebars can offer the feasibility of greater robot manoeuvrability on a vertical plane. To demonstrate that, the adhesion module arranged as structure 1 is rotated a full 360 degrees along its centre of gravity and the adhesion force is measured at 30° intervals. Figure 13b shows the results and it can be seen that the adhesion force is at its lowest value of 148 N when the yokes are rotated 90° and 270° from their initial upright position. From this, it can be deduced that the two yokes adhesion module can offer a robot the freedom to turn 360° and still maintain a minimum adhesion force of 140 N.

6 Prototype Design and Experimental Setup

To validate the simulation results and test performance of the adhesion modules, a magnetic adhesion system consisting of three N42 grade rare earth neodymium magnets were arranged as layout 1 and attached to a prototype climbing robot. The robot's dimension was 360 × 210 × 25 mm and each magnet's size was 50 × 50 × 12 mm with a 350 mm long yoke. Wheeled locomotion was chosen for higher velocity and manoeuvrability compared to other locomotion techniques [30]. Each wheel was independently driven using servo motors capable of delivering 2.16 Nm of torque at 30 rpm. The wheel's outer diameter was 63 mm and sufficient enough to keep the air gap between the magnet face and the concrete surface at 2 mm as a small gap increases the adhesion force significantly. The robot's net weight was 2.23 kg and 3.68 kg for a 5 mm and 15 mm thick yoked robot respectively. The robots required 46 N and 76 N of force for sliding avoidance (obtained by using Eqs. 1 and 2 when acceleration, a = 0.5 m/s^{-2} and wheel friction coefficient, μ = 0.5). The robot was equipped with four strain gauge based load cells capable of measuring up to 1500 N each. Load cells were affixed underneath the robot's chassis and the adhesion module was suspended freely in the loading platform using extensions which were secured in place to avoid displacement using additional screws. As the adhesion module presses toward the buried rebar, it applies force in the loading area of each load cell. The load cells produce a low-voltage milli-volt (mV) relative to the force applied. An operational amplifier HX711 was used to amplify the voltage and a AVR microcontroller based 10-bit analogue to digital module was used to convert the signals to numerical values and displayed on a computer screen using a Bluetooth data transmission system. A block diagram of the robot assembly and the prototype robot is shown in Fig. 14.

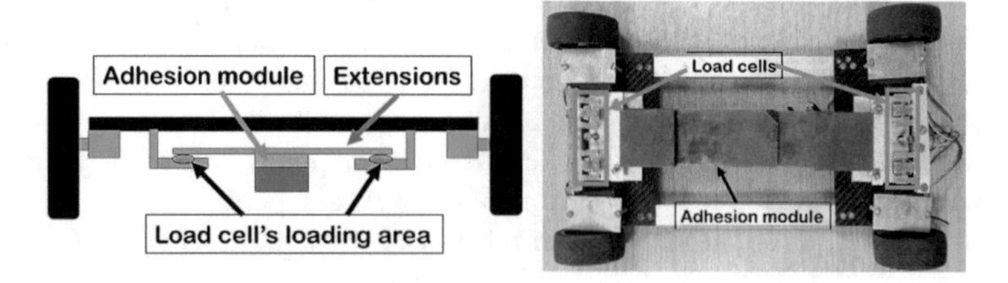

Fig. 14 Block diagram of the robot assembly and bottom view of the built prototype robot

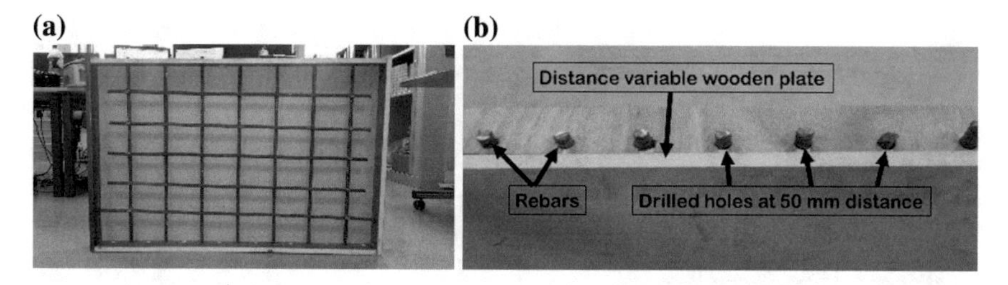

Fig. 15 Constructed experimental rig using wood frame and steel rebars: **a** Front view; **b** Top view

A test rig as in Fig. 15 was constructed to emulate a vertical reinforced concrete surface. 12 mm diameter holes were drilled along the length of a 1 m long and 70 cm tall wooden frame as the rebar holder. Distance between the rebars could be varied from 50 to 200 mm to investigate the effect of nearby rebars on adhesion force by slotting the rebars in through different holes. A 30 mm thick wooden plate simulating the concrete cover was also attached to the frame with screws which could be moved forward and backward using a slider to change the air gap.

7 Performance Validation of the Prototype Robot

To validate the performance of the prototype robot, experiments were carried out using the test rig at different rebar meshing and different air gaps. Measurements were taken at three different test points along the height of the test rig and the average value was considered. Force measurements were also recorded for rotational movements of the robot.

Firstly, the adhesion force was measured for 50 × 50 mm, 100 × 50 mm, 150 × 50 mm and 200 × 50 mm of rebar meshing with 30 mm air gap. The robot was positioned such that the adhesion module is aligned face-to-face with a vertical rebar. The robot's position was not changed while the rebar mesh settings were changed. Results shown in Fig. 16a suggest that when the rebars were densely meshed at 50 × 50 mm, highest adhesion force of 108.5 N could be achieved. In this case, the adhesion module could attract the front rebar as well as the adjacent vertical rebars located on its left and right side. As the distance increased to 100 mm, the module was only coupled with the front set of rebar and the next set of rebars were further apart to have any effect. At this stage, the adhesion force was 96.3 N. A further increase in the distance had no effect on the adhesion force which stayed at an average value of 95 N throughout.

For testing of rotational movements, the robot was rotated 360° along its centre of gravity and the resultant force was measured for different settings of rebar mesh. Figure 16b shows the respective results and according to that for 50 × 50 mm meshing, the adhesion force fell to a minimum value of 74 N at 60°, 120°, 240°, 330° of rotations. In these cases, the adhesion module was not aligned face-to-face with the

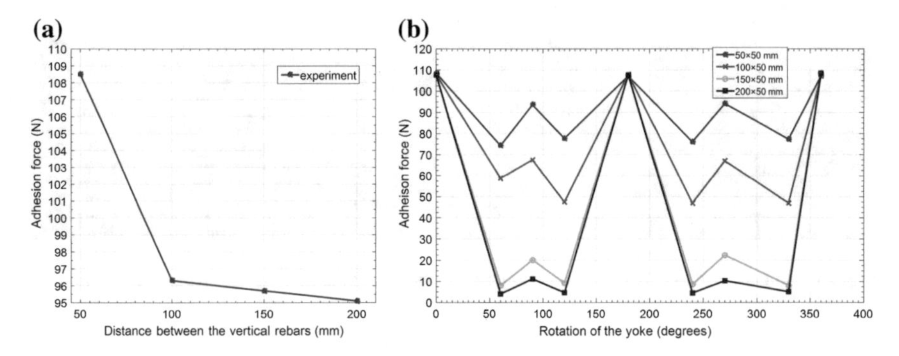

Fig. 16 Experimental results if the prototype: **a** Adhesion forces at different distances between the vertical rebars; **b** Adhesion forces at different rotations

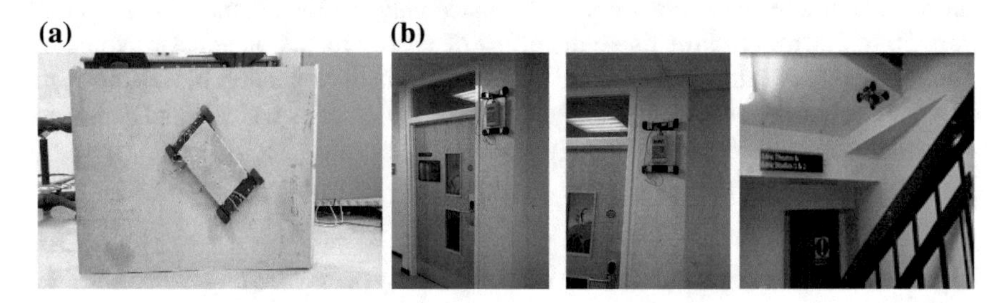

Fig. 17 **a** The prototype robot turning on the vertical test rig; **b** Climbing concrete column and ceiling with on board magnetic field sensor

rebar and as a result magnetic flux distribution was uneven. The same pattern was found for other setups. However, adhesion forces at 60°, 120°, 240°, 330° of rotation were decreasing with less dense meshing. The maximum adhesion force was at the same level of 108 N regardless of meshing settings while the module was aligned with a vertical rebar.

The robot was tested on the concrete column of a residential building. The rebars were reinforced in a 50 $\times$ 150 mm meshing at 30 mm depth. The concrete cover was measured using a concrete cover meter. The robot was carrying a Hall-Effect sensor as a payload that could identify the rebar location. The average adhesion force was measured to be 110.16 N with a maximum value of 127.53 N. The weight of the attached Hall-Effect sensor was 1.6 kg. Pictures of the robot climbing various surfaces are shown in Fig. 17. Adhesion force for different yoke thicknesses at a constant concrete cover of 30 mm also matched the same results achieved in simulations. Moreover, using thicker yokes resulted in higher adhesion forces as expected from simulations. However, increased weight of the yoke had a reverse relationship to the force-to-weight ratio. Figure 18 shows a gradual fall of η from 7.413 to 0.747 as the yoke thickness increased from 15 mm to 50 mm. Therefore, the yoke thickness of 15 mm should b an optimum design trade-off.

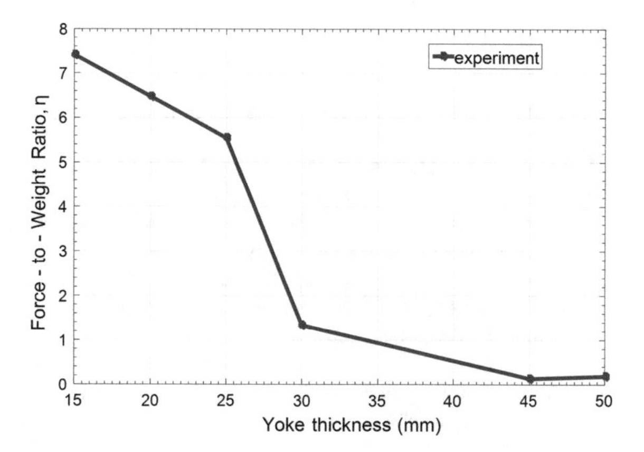

Fig. 18 Experimental values of force-to-weight ratios for different yokes used

8 Conclusion

Considering the workspace of the concrete wall climbing robot and safety requirements, a small size robot with high payload capacity has been presented in this chapter. Increased adhesion force has been achieved by arranging rare earth magnets on the back of a flux concentrator in a unique way and the robot's mechanical design ensures attachment of the magnet arrays at a constant distance from the surface while exhibiting small turning resistance. FEA based simulations have been carried out for parametric study of design variables. Distance between nearby magnets, yoke distance, number of yokes used were found to be the deciding factors of good performance. It was seen that by using two yokes, adhesion force could be maximized by 60 % compared to a single yoke system without compromising the force-to-weight ratio. To validate the simulation results, a prototype robot was built that was capable of displaying real time adhesion force values. Experiments on a test rig demonstrated the viability of 360° maneuverability of the robot on a reinforced rebar mesh surface. The minimum adhesion force was found to be 74.7 N on a 50 × 50 mm meshed surface when the robot is rotated around the vertical axis. This dense meshing was the only scenario where the robot was capable of completing a full 360° rotation. For other mesh settings, adhesion forces were critically low for full maneuvering. Though 12 mm diameter rebar was used for experiments, rebars used in the planned target application such as nuclear power plants have thicker rebars (up to 55 mm) and much denser meshing of the rebars making this method very feasible. Finally, 127.53 N of maximum adhesion force was measured with the robot climbing a real concrete column with 30 mm concrete cover while carrying 1.6 kg of additional payload. Higher grade (N52) of neodymium magnets could be used that will increase adhesion, while keeping the net system weight the same as N42 grade magnets. Considering the size, weight and high payload capacity of the robot, this is a practical solution for achieving zero-power permanent magnet based adhesion mechanisms for reinforced concrete environments.

References

1. Sattar, T., Hernando, E., Jianzhong, S.: Amphibious NDT robots. Int. J. Adv. Robot. Syst. **6**, 24 (2007)
2. Marceau, M., Michael, A., Vangeem, G.: Life Cycle Inventory of Portland Cement Concrete. Portland Cement Association, Skokie (2007)
3. Kosmatka, S., Kerkhoff, B., Panarese, W.: Design and Control of Concrete Mixtures. Portland Cement Association, Skokie (2008)
4. Sattar, T.P., Leon Rodriguez, H.E., Hussain, S.: Keynote paper: robotic non destructive testing. In: Proceedings of 26th International Conference on CAD/CAM, Robotics and Factories of the Future, Kuala Lampur (2011)
5. Garcia Ruiz, A., Sattar, T.P., Correa Sanz, M., Rodriguez-Filloy, B.S.: Inspection of floating platform mooring chains with a climbing robot. In: 17th International Conference on Climbing and Walking Robots and the Support Technologies for Mobile Machines, Poznan (2014)
6. Case Study on Oil and Gas: Comparison Between Rope Access and Scaffolding on 70 m Radio Tower Refurbishment. Megarme, Zirku Island (2011)
7. GSSI Case Study: A Tall Order: Scanning a 300-feet Chimney. Geophysical Survey Systems Inc, Salem (2009)
8. Zhiqiang, B., Yisheng, G., Shizhong, C., Haifei, Z., Hong, Z.: A miniature biped wall-climbing robot for inspection of magnetic metal surfaces. In: IEEE International Conference on Robotics and Biomimetics, Guangzhou (2012)
9. Bogue, R.: Robots in the nuclear industry: a review of technologies and applications. Ind. Robot: An Int. J. **38**(2), 113–118 (2011)
10. David, R., Elisabeth, C., Robert, D., Michael, G., Jacqueline, O.: Risk of cancer from occupational exposure to ionising radiation: retrospective cohort study of workers in France, The United Kingdom, and The United States (INWORKS). Br. Med. J. **351**, 5359 (2015)
11. Bayten, G.: Computerized Ultrasonic Inspection of LPD Vessel Construction Welds. Integrity NDT, Ankara (2012)
12. Zelihang, X., Peisun, M.: A wall climbing robot for labeling scale of oil tank's volume. Robot. Camb. J. **2**(2), 209–212 (2002)
13. Shang, J., Bridge, B., Sattar, T., Mondal, S., Brenner, A.: Development of a climbing robot for inspection of long weld lines. Ind. Robots Int. J. **35**(3), 217–223 (2008)
14. Kalra, P., Jason, G., Max, M.: A wall climbing robot for oil tank inspection. In: IEEE International Conference on Robotics and Biomimetics, Kunming (2006)
15. Zhang, Y., Dodd, T.: Design and optimization of magnetic wheel for wall and ceiling climbing robot. In: International Conference on Mechatronics and Automation, Xi'an (2010)
16. Luk, B., Collie, A., Cooke, D., Chen, S.: Walking and climbing service robots for safety inspection of nuclear reactor pressure vessels. In: Asia Pacific Conference on Risk Management and Safety, Hong Kong (2005)
17. Kim, H., Kim, D., Yang, H., Lee, K., Seo, K., Chang, D., Kim, J.: Development of a wall-climbing robot using a tracked wheel mechanism. J. Mech. Sci. Technol. **22**(8), 1940–1948 (2008)
18. Albagul, A., Asseni1, A., Khalifa, O.: Wall climbing robot: mechanical design and implementation. In: WSEAS International Conference on Circuits, Systems, Signal and Telecommunications, Mexico City (2011)
19. Manuel, F., Tenreiro, J.: A survey of technologies and applications for climbing robots locomotion and adhesion, ISBN: 978-953-307-030-8: InTech (2010)
20. Elliot, M., Morris, W., Xiao, J.: City-climber: a new generation of wall-climbing robots. In: IEEE International Conference on Robotics and Automation, Florida (2006)
21. Bretl, T., Rock, S., Latombe, J., Kennedy, B., Aghazarian, H.: Free-climbing with a multi-use robot. In: International Symposium on Experimental Robotics, Singapore City (2004)
22. Spenko, M., Haynes, G., Saunders, J., Cutkosky, M., Rizzi, A.: Biologically inspired climbing with hexapedal Robot. J. Field Robot. **25**, 4 (2008)

23. Avishi, S., Tomer, A., Amir, S.: Design and motion planning of an autonomous climbing robot with claws. J. Robot. Auton. Syst. **59**, 1008–1019 (2011)
24. Jensen-Segal, S., Virost, W., Provancher, R.: ROCR: energy efficient vertical wall climbing with a pendular two-link mass-shifting robot. In: International Conference on Robotics and Automation, California (2008)
25. Tso, S., Feng, T.: Robot assisted wall inspection for improved maintenance of high-rise buildings. In: International Symposium on Automation and Robotics in Construction, Eindhoven (2003)
26. Fanella, D.: Design of low-rise reinforced concrete buildings based on 2009 IBC. In: ASCE/SEI 7–05, ACI 318–08. International Code Council, Washington DC (2009)
27. Charles, R., James, C., Anthony, T.: Reinforced Concrete Designer's Handbook, 11th edn. Taylor and Francis Group ISBN: 9780419258308 (2007)
28. Design of Concrete Structures, General Rules. Eurocode 2, European Standard (2004)
29. Howlader, M., Sattar, T.: Novel adhesion mechanism and design parameters for concrete wall-climbing robot. In: SAI Intelligent Systems Conference, London (2015)
30. Schmidt, D., Berns, K.: Climbing robots for maintenance and inspections of vertical structures–a survey of design aspects and technologies. Robot. Auton. Syst. **61**(12), 1288–1305 (2013)

Implementation of PID, Bang–Bang and Backstepping Controllers on 3D Printed Ambidextrous Robot Hand

Mashood Mukhtar, Emre Akyürek, Tatiana Kalganova and Nicolas Lesne

Abstract Robot hands have attracted increasing research interest in recent years due to their high demand in industry and wide scope in number of applications. Almost all researches done on the robot hands were aimed at improving mechanical design, clever grasping at different angles, lifting and sensing of different objects. In this chapter, we presented the detail classification of control systems and reviewed the related work that has been done in the past. In particular, our focus was on control algorithms implemented on pneumatic systems using PID controller, Bang–bang controller and Backstepping controller. These controllers were tested on our uniquely designed ambidextrous robotic hand structure and results were compared to find the best controller to drive such devices. The five finger ambidextrous robot hand offers total of 13° of freedom (DOFs) and it can bend its fingers in both ways left and right offering full ambidextrous functionality by using only 18 pneumatic artificial muscles (PAMs).

1 Introduction

Robots play a key role in our world from providing daily services to industrial use. Their role is in high demand due to the fact they offer more productivity, safety at workplace and time reduction in task completion. Number of control schemes are found in the literature [1] to control such robot such as adaptive control [2], direct continuous-time adaptive control [3], neurofuzzy PID control [4] hierarchical control [5], slave-side control [6], intelligent controls [7], neural networks [8], just-in-time control [9], Bayesian probability [10], equilibrium-point control [11], fuzzy logic control [12], machine learning [13], evolutionary computation [14] and genetic algorithms [15], nonlinear optimal predictive con-trol [16], optimal control [17],

M. Mukhtar (✉) · E. Akyürek · T. Kalganova
Department of Electronic and Computer Engineering, Brunel University,
Kingston Lane, Uxbridge, London, UK
e-mail: Mashood.Mukhtar@brunel.ac.uk

N. Lesne
Department of System Engineering, ESIEE Paris, Noisy-le-Grand Cedex, France

Y. Bi et al. (eds.), *Intelligent Systems and Applications*,
Studies in Computational Intelligence 650, DOI 10.1007/978-3-319-33386-1_9

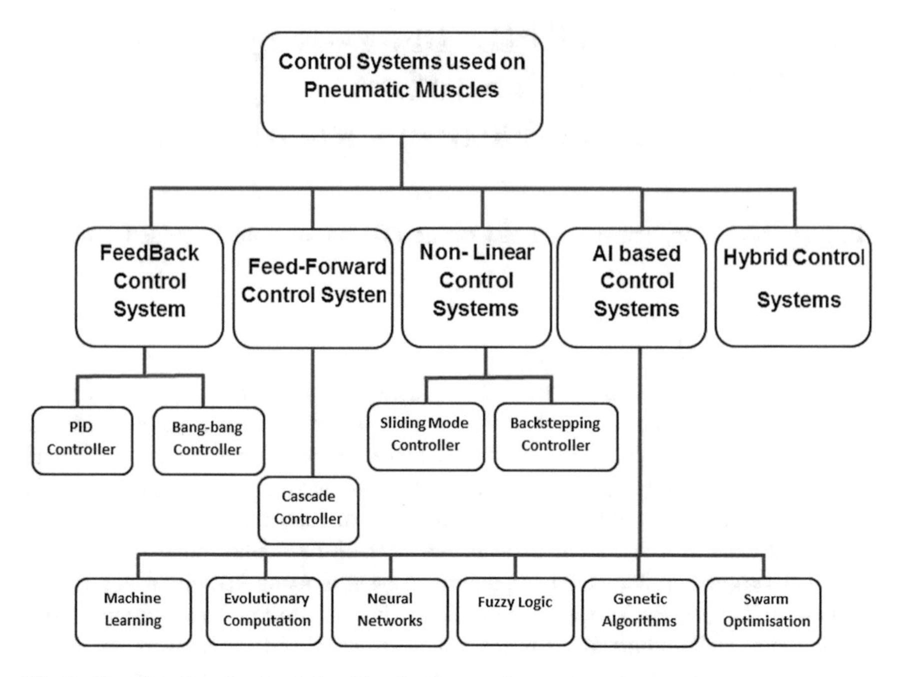

Fig. 1 Classification of control algorithm implemented on pneumatic systems

stochastic control [18], variable structure control [19], chattering-free robust variable structure control [20] and energy shaping control [21], gain scheduling model-based controller [22], sliding mode control [23], proxy sliding mode control [24], neuro-fuzzy/genetic control [25] but there are five types of control systems (Fig. 1) mainly used on pneumatic muscles.

Focus of our research was on the feedback control system and non-linear control systems. Feedback is a control system in which an output is used as a feed-back to adjust the performance of a system to meet expected output. Control systems with at least one non-linearity present in the system are called nonlinear control systems. In order to reach a desired output value, output of a processing unit (system to be controlled) is compared with the desired target and then feed-back is provided to the processing unit to make necessary changes to reach closer to desire output. The purpose of designing such system is to stabilise the system at certain target.

Tactical sensors are employed to investigate the grasping abilities of the ambidextrous hand. These sensors have been used several times in the past on the ro-bot hands driven by motors [26, 27] and robot hands driven by pneumatic artificial muscles [28, 29]. The automation of the ambidextrous robot hand and its interaction with objects are also augmented by implementing vision sensors on each side of the palm. Similar systems have already been developed in the past. For instance, the two-fingered robot hand discussed in [30] receives vision feedback from an omnidirectional camera that provides a visual hull of the object and allows the fingers to automatically adapt to its shape. The three-fingered robot hand developed by Ishikawa Watanabe Laboratory [31] is connected to a visual feedback at a rate of 1 KHz. Combined with its high-speed motorized system that allows a joint to rotate by 180° in 0.1 s, it allows the hand to interact dynamically with its environment, such as catching falling objects. A laser displacement sensor is also used for the two-fingered robot hand introduced in [32]. It measures the vertical slip displacement of the grasped object and allows the hand to adjust its grasp. In our research, the aim of the vision sensor is to detect objects close to the palms and to automatically trigger grasping algorithms. Once objects are detected by one of the vision sensors, grasping features of the ambidextrous robot hand are investigated using three different algorithms, which are proportional-integrative-derivative (PID), bang–bang and Backstepping controllers. Despite the nonlinear behavior of PAMs actuators [33], previous researches indicate that these three algorithms are suitable to control pneumatic systems.

The work presented in this chapter aim to validate the possibility of controlling a uniquely designed ambidextrous robot hand using PID controller, Bang–bang controller and Backstepping controller. The ambidextrous robot hand is a robotic device for which the specificity is to imitate either the movements of a right hand or a left hand. As it can be seen in Fig. 2, its fingers can bend in one way or another to include

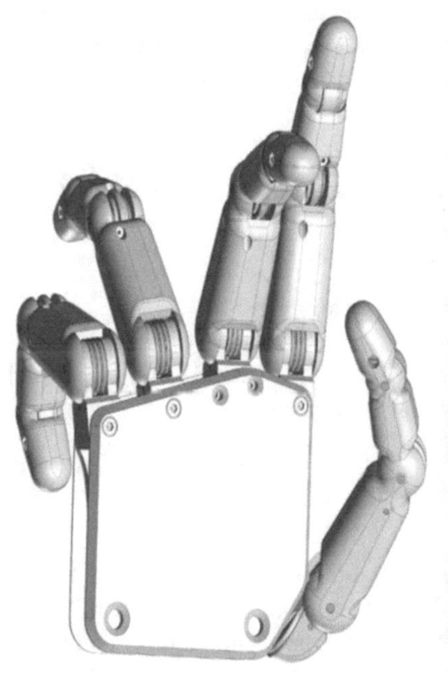

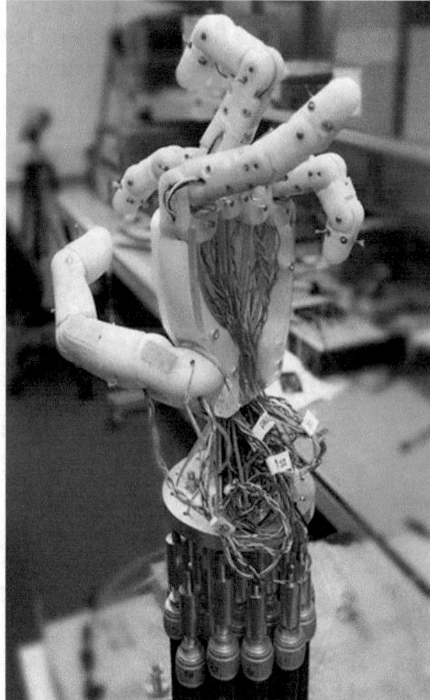

Fig. 2 Ambidextrous robot hand (Ambidextrous mode)

the mechanical behavior of two opposite hands in a single device. The Ambidextrous Robot Hand has a total of 13° of freedom (DOFs) and is actuated by 18 pneumatic artificial muscles (PAMs) [58].

2 Implementation of Controllers

2.1 PID Controller

PID controller is a combination of proportional, integral and derivative controller [34]. Proportional controller provides corrective force proportional to error pre-sent. Although it is useful for improving the response of stable systems but it comes with a steady-state error problem which is eliminated by adding integral control. Integral control restores the force that is proportional to the sum of all past errors, multiplied by time [35]

On one hand integral controller solves a steady-state error problem created by proportional controller but on the other hand integral feedback makes system overshoot. To overcome overshooting problem, derivative controller is used [36]. It slows the controller variable just before it reaches its destination. Derivatives controller is always used in combination with other controller and has no influence on the accuracy of a system. All these three control have their strengths and weakness and when combine together offers the maximum strength whilst minimizing weakness.

$$Output_{PID} = K_P \times e < t > +K_i \times \int_0^t e < t > \times dt + K_d \times \left(\frac{d}{dt} e < t > \right)$$

2.1.1 PID Controller Adapted to the Ambidextrous Hand

PID control loops were implemented on an ambidextrous robot hand using the parallel form of a PID controller, for which the equation is as follows:

$$u(t) = K_p e(t) + K_i \int_0^t e(\tau) d\tau + K_d \frac{d}{dt} e(t) \tag{1}$$

Output u (t) should be calculated exactly in the same way as described in Eq. (1) but this is not the case due to asymmetrical tendon routing of ambidextrous robot hand [37]. Mechanical specifications are taken into consideration before calculating the u (t). It is divided into three different outputs for the three PAMs driving each finger. These three outputs are $u_{pl}(t)$, $u_{mr}(t)$ and $u_{ml}(t)$, respectively attributed to the proximal left, medial right and medial left PAMs. The same notations are used for the gain constants. The adapted PID equation is defined as:

$$\begin{bmatrix} u_{pl}(t) \\ u_{mr}(t) \\ u_{ml}(t) \end{bmatrix} = \begin{bmatrix} K_{p_{pl}} & K_{i_{pl}} & K_d \\ K_{p_{mr}} & K_{i_{mr}} & K_d \\ K_{p_{ml}} & K_{i_{ml}} & K_d \end{bmatrix} \begin{bmatrix} e(t) \\ \int_0^t e(\tau)d\tau \\ \frac{d}{dt}e(t) \end{bmatrix} \quad (2)$$

To imitate the behaviour of human finger correctly, some of the pneumatic artificial muscles must contract slower than others. This prevents having medial and distal phalanges totally close when the proximal phalange is bending. Thus, experiments are done to collect PAMs' pressure variation according to the fingers' position. As the finger is made of one proximal phalange and two other phalanges for which the movement is coupled, the finger has nine extreme positions. These positions are numbered from 1 to 9. Position 1 refers to the proximal and medial/distal phalanges reaching both the maximum range on their left side. Position 2 refers to the proximal phalange reaching its maximum range on its left side, whereas the medial/distal phalanges are straight. Position 3 refers to the proximal phalange being on its maximum range on its left side, whereas the medial/distal phalanges are on their maximum range on the right side. Position 4 to 6 refers to the finger when the proximal phalange is straight. Last position, position 9, is when the proximal and medial/distal phalanges reach both the maximum range on their right side. The experiment is summarized in Table 1 and Fig. 3.

When the fingers' close from vertical to left position (position 5 to the average of positions 4 and 2), it is observed that the pressure of medial left PAM varies thrice less than the pressure of the proximal left PAM, whereas the pressure of the two medial PAMs vary from the same ratios but in opposite ways. Therefore, the proportional constant gains are consequently defined as:

Table 1 Pressure at ambidextrous robot hand's extreme positions

No. of position	Proximal left PAM (bars)	Medial left PAM (bars)	Medial right PAM (bars)
1	2.497	1.173	0.017
2	4.028	0.250	0.782
3	3.129	0.175	1.481
4	1.972	1.406	0.017
5	2.954	1.115	2.363
6	1.643	0.357	2.617
7	0.038	2.725	1.227
8	0.022	1.411	2.039
9	0.008	0.491	2.990

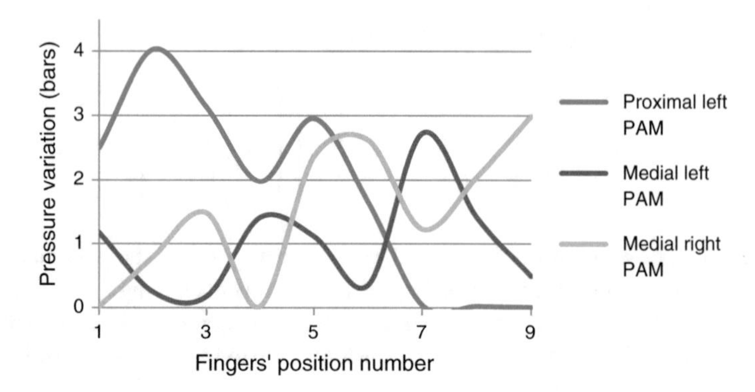

Fig. 3 Pressure variations at extreme finger positions

$$\begin{bmatrix} K_{p_{ml}} \\ K_{i_{ml}} \end{bmatrix} = \begin{bmatrix} K_{p_{pl}}/3 \\ K_{i_{pl}}/3 \end{bmatrix} = \begin{bmatrix} -K_{p_{mr}} \\ -K_{i_{mr}} \end{bmatrix} \tag{3}$$

When the fingers' close from vertical to right position (position 5 to the average of positions 6 and 8), it is observed that the pressure of medial left PAM varies 0.4 times slower than the pressure of the medial right PAM. Moreover, the pressure of the medial right PAM varies in an opposite way as the one of the proximal left PAM. Therefore, when the object interacts on the left side of the hand, equation can be written as follows:

$$\begin{bmatrix} K_{p_{mr}} \\ K_{i_{mr}} \end{bmatrix} = \begin{bmatrix} 0.4\ X\ K_{p_{ml}} \\ 0.4\ X\ K_{i_{ml}} \end{bmatrix} = \begin{bmatrix} -K_{p_{pl}} \\ -K_{i_{pl}} \end{bmatrix} \tag{4}$$

Same ratios are applied to the derivative gain constants when object interact on right side of the hand.

2.1.2 Results Obtained with the PID Controller

Using Eq. (2), PID control loops with identical gain constants are sent to the four fingers with a target of 1 N and an error margin of 0.05 N, whereas the thumb is assigned to a target of 12 N with an error margin of 0.5 N. A can of soft drink is brought close to the hand and data is collected every 0.05 s. The experiment result is illustrated in Fig. 4

From Fig. 4, it can be seen that grasping of a can of soft drink began at 0.15 approximately and it became more stable after 0.2 s. An overshoot has occurred on all fingers but stayed in limit of 0.05 N where system has automatically adjusted at the next collection. These small overshoots occurred because different parts of the fingers get into contact with the object before the object actually gets into contact with the force sensor. This results bending of fingers slower when phalanges touches the object. Since the can of soft drink does not deform itself, it can be deduced that

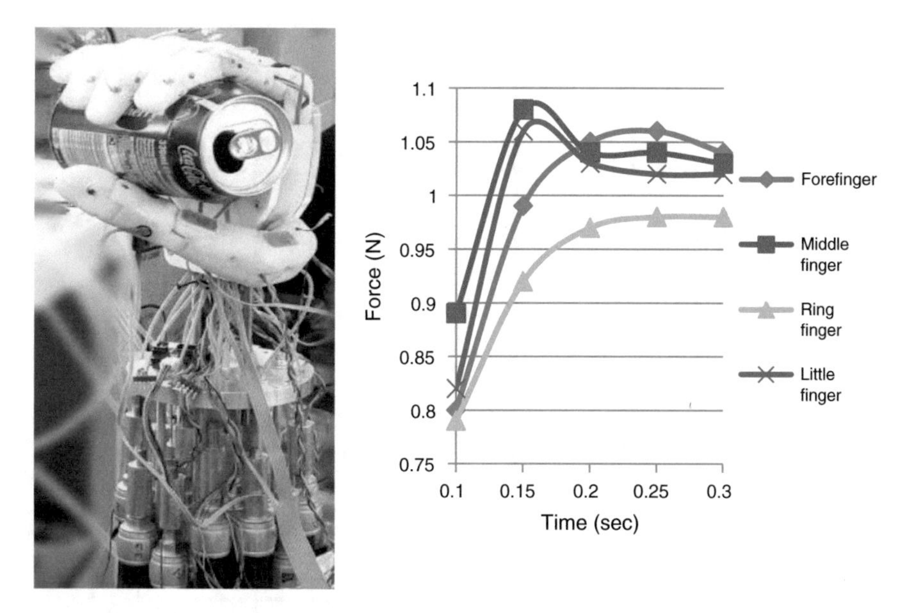

Fig. 4 The Ambidextrous robot hand holding a can (a grasping movement implemented with PID controllers) and a graph representing data collected after every 0.05 s

the grasping control is both fast and accurate when the Ambidextrous Hand is driven by PID loops.

2.1.3 Comparison with Other Grasping Algorithms Using PID

Mechanical and control features of the ambidextrous hand are compared with other hands in Table 2. The grasping algorithm designed by J.Y. Nagase et al. in [38] is the most accurate and the slowest engineered after 2010. As for fuzzy logic, it is observed that the movements are more precise but slower than the ones driven by PID and PD controls. It is also noticed that PD control is implemented more often than PID control to grab objects. This is explained because the grasping time is usually reached in a shorter delay than the full rotation of fingers from open to close positions. The Ambidextrous Hand has a grasping time and a maximum overshoot close to the best ones obtained with other robot hands. It can therefore successively grab objects despite its limited number of DOFs, and it is the only robotic model that can grab objects with an ambidextrous behavior, either as a left or a right hand, equally performing well.

Table 2 Comparison of various robot hands with our ambidextrous robot hand

Robot hand	# fingers	Type of actuators	# DOFs	Grasping algorithms	Anthropo-morphic positioning of fingers	Thumb opposite to other fingers during grasping	Grasping time (sec)	Max. overshoot or error (%)	Ambidexterity
Gifu Hand II [39]	5	Motors	16	PID	✓	✓	0.55[ab]	16[ab]	
High-speed hand [40]	3	Motors	8	PD		✓	0.05[a]	N/A	
ITU Hand [41]	2	SMAs	1[a]	N/A		✓	3.76	N/A	
I. Yamano and T. Maeno [42]	5	SMAs	20	N/A	✓	✓	N/A	6[a]	
L. Zollo et al. [43]	3	Motors	10[a]	PD	✓	✓	N/A	N/A	
S. Nishino et al. [29]	5	PAMs	13[a]	PID, Cascade	✓	✓	1.0[a]	6[a]	
D. Gunji et al. [32]	2	Motors	1	PD		✓	0.6[a]	N/A	
T. Yoshikawa [44]	2	Motors	4[a]	PID		✓	0.55[a]	22[a]	

(continued)

Table 2 (continued)

Robot hand	# fingers	Type of actuators	# DOFs	Grasping algorithms	Anthropo-morphic positioning of fingers	Thumb opposite to other fingers during grasping	Grasping time (sec)	Max. overshoot or error (%)	Ambidexterity
J.Y. Nagase et al. [45]	4	PAMs	N/A	Fuzzy logic	✓	✓	0.9[a]	9[ad]	
DLR Hand II [46]	4	Motors	16[a]	N/A	✓	✓	28.7	N/A	
DLR Hand [47]	5	Motors	19	Cascade	✓	✓	0.1[a]	N/A	
Shadow Hand [26]	5	Motors	20	PID	✓	✓	0.15[a]	N/A	
ACT Hand [48]	5	Motors	23	N/A	✓	✓	N/A	N/A	
T. Nuchkrua et al. [49]	3	PAMs	3	N/A	✓	✓	N/A	N/A	
TU Bionic Hand [50]	5	Motors	15	PID	✓		0.208	N/A	
DEXMART Hand [51]	5	Motors	20	NN	✓	✓	0.25[a]	25[ad]	
Ambidextrous Hand [52]	5	PAMs	13	PID	✓		0.20	8	✓

[a]Estimations made from curves, pictures or videos of the robot hands
[b]Experiments do not concern grasping but contact tasks
[c]Only for grasping
[d]For a target of 1 N ± 10 %, whereas the error can exceed 30 % for lower force targets

2.2 *Bang–Bang Controller*

Bang–bang controllers are used to switch between two states abruptly. They are widely used in systems that accept binary inputs. Bang–bang controller is not popular in robotics due to its shooting functions that are not smooth [53] and need regularisation [54]. Fuzzy logic [55, 56] and PID are usually combined with bang–bang controller to add flexibility and determine switching time of Bang–bang inputs [37]. Bram Vanderborght et al. in [57] controlled a bipedal robot actuated by pleated pneumatic artificial muscles using bang–bang pressure controller. Stephen M Cain et al. in [58], studied locomotor adaptation to powered ankle-foot orthoses using two different orthosis control methods. Dongjun Shin et al. in [59], proposed the concept of hybrid actuation for human friendly robot systems by using distributed macro-Mini control approach [60]. The hybrid actuation controller employ modified bang–bang controller to adjust the flow direction of pressure regulator. Zhang, Jia-Fan et al. in [61], presented a novel pneumatic muscle based actuator for wearable elbow exoskeleton. Hybrid fuzzy controller composed of bang–bang controller and fuzzy controller is used for torque control.

2.2.1 Bang–Bang Control Adapted to the Ambidextrous Hand

The bang–bang controller allows all fingers to grasp the object without taking any temporal parameters into account. A block diagram of Bang–bang controller is shown in Fig. 5. Execution of algorithm automatically stops when the target set against force is achieved. The controller also looks after any overshooting issue if it may arise. To compensate the absence of backward control, a further condition is implemented in addition to initial requirements. Since the force applied by the four fingers is controlled with less accuracy than with PID loops, the thumb must offset the possible excess of force to balance the grasping of the object. Therefore, a balancing equation is defined as:

$$F_{tmin} = W_o + \sum_{f=1}^{4} F_f \tag{5}$$

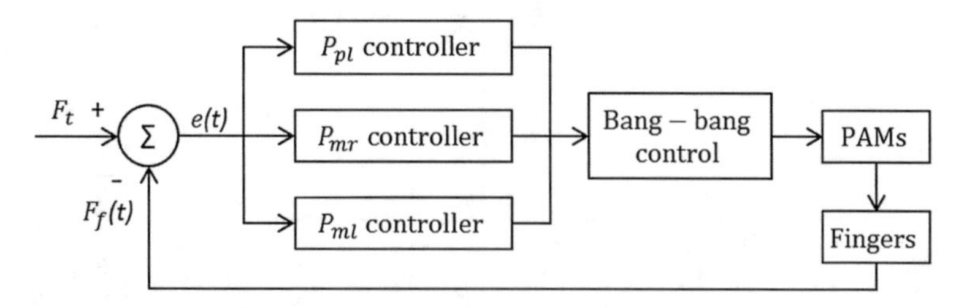

Fig. 5 Bang–bang loops driven by proportional controllers

2.2.2 Results Obtained with Bang–Bang Control

The phalanges must close with appropriate speed's ratios to tighten around objects when interacting with object. Results obtained from experiment using the bang–bang controller are provided in Fig. 6. This time, it is noticeable that the can of soft drink deforms itself when it is grasped on the left hand side. The graph obtained from the data collection of the left mode is provided in Fig. 6.

It is noted during experiment that speed of fingers does not vary when it touches the objects. That explains why we got higher curves in Fig. 6 than the previously obtained in Fig. 4. This makes bang–bang controller faster than PID loops. The bang–bang controllers also stop when the value of 1 N is overreached but, without predicting the approach to the setpoint, the process variables have huge overshoots. The overshoot is mainly visible for the middle finger, which overreaches the setpoint by more than 50 %. Even though backward control is not implemented in the bang–bang controller, it is seen the force applied by some fingers decreases after 0.2 s. This is due to the deformation of the can, which reduces the force applied on the fingers. It is also noticed that the force applied by some fingers increase after 0.25 s, whereas the force was decreasing between 0.2 and 0.25 s. This is due to thumb's adduction that varies from 7.45 to 15.30 N from 0.1 to 0.25 s. Even though the fingers do not tighten anymore around the object at this point, the adduction of the thumb applies an opposite force that increases the forces collected by the sensors. The increase is mainly visible for the forefinger and the middle finger, which are the closest ones from the thumb. Contrary to PID loops, it is seen in Fig. 6 that the force applied by

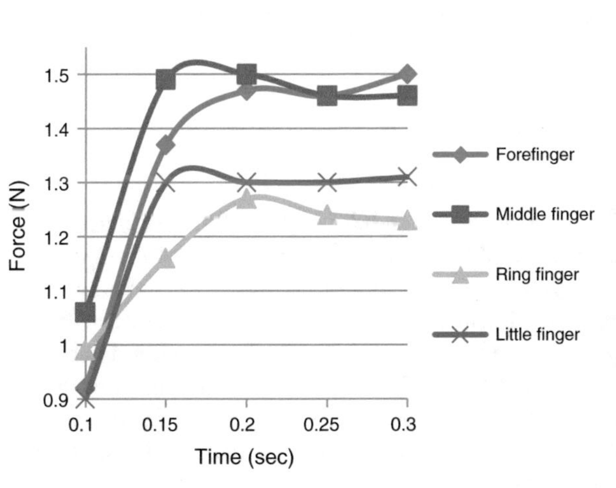

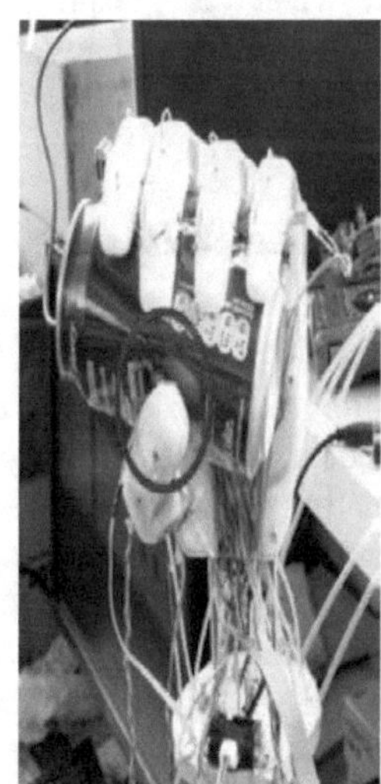

Fig. 6 Graph shows force against time of the four fingers while grasping a drink can with bang–bang controllers. The ambidextrous robot hand holding a can with a grasping movement implemented with bang–bang controllers. It is seen the can has deformed itself

some fingers may not change between 0.15 and 0.2 s, which indicates the grasping stability is reached faster with bang–bang controllers. The bang–bang controllers can consequently be applied for heavy objects, changing the setpoint of F_f defined in (5).

2.2.3 Comparison with Other Bang–Bang Controls

The features related to the bang–bang control implemented on the ambidextrous hands are compared to others in Table 3. For example, the bipedal walking robot discussed in [57] implemented bang–bang controller on robotic structure driven by pneumatic artificial muscles. and hand designed by Z. Xu et al. [62] is actuated by air cylinders instead of PAMs, its architecture is much closer to the one of the Ambidextrous Hand. Table 3 shows that bang–bang control is not usually implemented on complex structures, as the ambidextrous hand and the two legs discussed in [63] are the only architectures that exceed ten actuators. The ambidextrous hand is also the only robotic structure in Table 3 that has more than ten DOFs. The execution times are not very significant, given that the aims of the bang–bang controls are totally different from one project to other; it is nevertheless observed that the execution times never exceed 0.4 s, as bang–bang controls aim at making a system switch from one state to another as fast as possible. The system proves its efficiency for the walking robot introduced in [57], but provides a huge overshoot of 53 % when it is implemented on the Ambidextrous Hand. Consequently, despite the originality of the bang–bang control to grab objects, its implementation on an ambidextrous device and its grasping time of 0.15 s (25 % shorter than the one obtained with the PID control), the bang–bang control is not accurate enough to control the fingertips' force of the Ambidextrous Hand.

2.3 Backstepping Controller

Backstepping was offered as a mean to offer stabilise control with a recursive structure based on derivative control and it was limited to nonlinear dynamical systems. Mohamed et al. *presented* a synthesis of a nonlinear controller to an electro pneumatic system *in* [66]. First, the nonlinear model of the electro pneumatic system was presented. It was transformed to be a nonlinear affine model and a coordinate transformation was then made possible by the implementation of the nonlinear controller. Two kinds of nonlinear control laws were developed to track the desired position and desired pressure. Experimental results were also presented and discussed. P. Carbonell et al. compared two techniques namely sliding mode control (SMC) and Backstepping control (BSC) in [67] and found out BSC is somewhere better than SMC in controlling a device. He further applied BSC coupled with fuzzy logic in [68]. In [69], a paralleled robot project is discussed which is controlled by BSC. Since literature review revealed no multifinger robot hand (actuated by PAMs) is

Table 3 Comparison of bang–bang controls' characteristics between the Ambidextrous Hand and other robotic models

Bang–bang control	Robotic structure	Type of actuators	Aim of the bang–bang control	Controller used as an outer loop	# DOFs	# actuators	Execution time (sec)	Error max. (%)	Ambidexterity
R. Van Ham et al. [57]	Modular part of a leg	PAMs	Pressure control	PID	N/A	2	0.2[a]	N/A	
B. Vanderborght et al. [64]	Two legs	PAMs	Generate a joint trajectory	PI	3	6	0.4[a]	4 % for pressure[a] 0.27 % for angle[a]	
B. Vanderborght et al. [65]	Two legs	PAMs	Generate a joint trajectory	PID	3	6	0.2[a]	3 % for pressure 4.5 % for angle[a]	
B. Vanderborght et al. [63]	Two legs	PAMs	Generate a joint trajectory	Delta-p	6	12	0.15[a]	17 % for pressure[a] 17.5 % for position[a]	
Z. Xu et al. [62]	Index of a hand	Air cylinders	Evaluation of speed capabilities	None	2[b]	4[b]	0.33	N/A	
Ambidextrous Hand [37]	Hand	PAMs	Force control	Proportional	9[b]	14[b]	0.15	53 % for the force	✓

[a]Estimations are made from curves
[b]These numbers do not correspond with the actuators of the robotic structure but with the ones used for the bang–bang control

ever driven using BSC, research presented in this chapter validates the possibility of driving multifinger robot hand using BSC.

2.3.1 BSC Adapted to the Ambidextrous Hand

BSC consists in comparing the system's evolution to stabilizing functions. Derivative control is recursively applied until the fingers reach the conditions implemented in the control loops. First, the tracking error $e1_{(t)}$ of the BS approach is defined as:

$$\begin{bmatrix} e_1(t) \\ \dot{e}_1(t) \end{bmatrix} = \begin{bmatrix} F_t - F_f(t) \\ -\dot{F}_f(t) \end{bmatrix} \tag{6}$$

then evaluated using a first Lyapunov function defined as:

$$V_1(e_1) = \frac{1}{2}e_1^2(t) < F_{gmin} \tag{7}$$

$$\dot{V}_1(e_1) = e_1(t) * \dot{e}_1(t) = \dot{F}_f(t) * e_1(t) \tag{8}$$

The force provided by the hand is assumed not being strong enough as long as $e_1^2/2$ exceeds a minimum grasping force defined as F_{gmin}. In (8), it is noted that $\dot{e}_1(t) \neq 0$ as long as $F_f(t)$ keeps varying. Therefore, $\dot{V}_1(e_1)$ cannot be stabilized until the system stops moving. Thus, a stabilizing function is introduced. This stabilizing function is noted as a second error $e_2(t)$:

$$e_2(t) = k * \dot{e}_1(t) \tag{9}$$

with k a constant > 1. $e_2(t)$ indirectly depends on the speed, as the system cannot stabilized itself as long as the speed is varying. Consequently, both the speed of the system and $e_2(t)$ are equal to zero when one finger reaches a stable position, even if $F_f(t) \neq F_t$. k aims at increasing $\dot{e}_1(t)$ to anticipate the kinematic moment when $\dot{e}_1(t)$ becomes too low. In that case, the BSC must stop running as $F_f(t)$ is close to F_t. Both of the errors are considered in a second Lyapunov function:

$$V_2(e_1, e_2) = \frac{1}{2}(e_1^2(t) + e_2^2(t)) < F_s \tag{10}$$

$$\dot{V}_2(e_1, e_2) = e_1(t) * \dot{e}_1(t) + e_2(t) * \dot{e}_2(t) \tag{11}$$

where F_s refers to a stable force applied on the object. This second step allows stabilizing the system using derivative control. Using (9), (11) can be simplified as:

$$\dot{V}_2(e_1, e_2) = \dot{e}_1(t) * (e_1(t) + k\ddot{e}_1(t)) \tag{12}$$

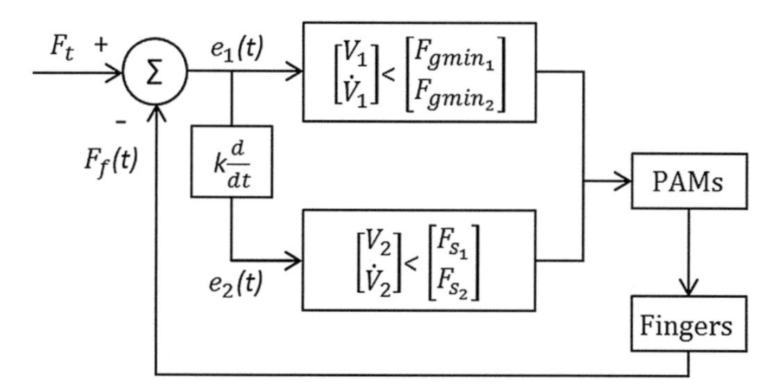

Fig. 7 Basic Backstepping controller is shown

The block diagram of backstepping process is illustrated in Fig. 7. According to the force feedback $F_f(t)$, the fingers' positions adapt themselves until the conditions of the Lyapunov functions $(V_1, \dot{V}_1)$ and $(V_2, \dot{V}_2)$ are reached.

2.3.2 Results Obtained with the BSC

The grasping features and the force against time graph is shown in Fig. 8. During the experiment, it was noticed that speed of finger tightening using backstepping control was much slower compared to PID controller and bang–bang controller. Since the

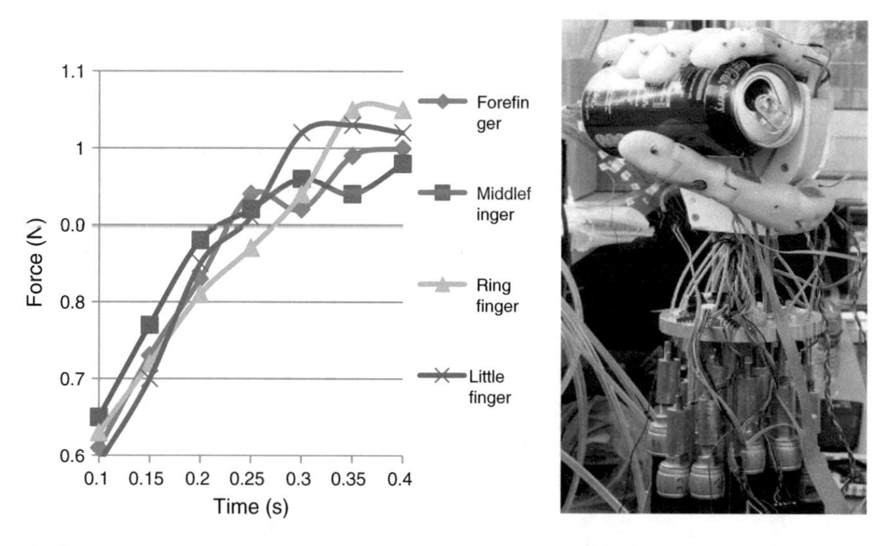

Fig. 8 Graph shows force against time of the four fingers while ambidextrous hand grasping a can of soft drinks using backstepping controllers

Table 4 Comparison of BSCs' characteristics between the Ambidextrous Hand and other robotic mode

BSC	Robotic structure	Type of actuators	Aim of the BSC	Algorithms to which it is combined	# DOFs	# actuators	Execution time (sec)	Error max. (in %)	Ambidexterity
C.-Y Su and Y. Stepanenko [70]	Manipulator	Motors	Trajectory control	None	N/A	N/A	0.3[a]	10[a]	
P. Carbonell et al. [67]	N/A, single PAM	PAM	PAM's length control	None	N/A	1	2.4[a]	8[a]	
P. Carbonell et al. [68]	N/A, single PAM	PAM	PAM's length control	Fuzzy logic	N/A	1	7[a]	1[a]	
D. Nganya-Kouya et al. [74]	Manipulator	N/A	Force and position control	None	4	N/A	9[a]	N/A	
Lotfazar et al. A. [75]	Manipulator	Motors	Trajectory control	None	5	5	2[a]	N/A	
S.-H. Wen S.-H. Wen and B. Mao. [76]	Manipulators	N/A	Force and position control	NN	N/A	N/A	1.9[ab]	9[ab]	

(continued)

Table 4 (continued)

BSC	Robotic structure	Type of actuators	Aim of the BSC	Algorithms to which it is combined	# DOFs	# actuators	Execution time (sec)	Error max. (in %)	Ambidexterity
H. Aschemann and D. Schindele [71]	Parallel robot	PAMs	Position control	None	2	4	0.3[a]	4[a]	
M.R. Soltanpour and M.M. Fateh [77]	Manipulator	Motors	Trajectory control	SMC	2	N/A	7.6[ab]	$<10^{-3}$[b]	
X. Liu and A. Liadis [78]	Parallel robot	Motors	Position control	Fuzzy logic	2	2	N/A	N/A	
L. Qin et al. [72]	Arm	N/A	Trajectory control	SMC	6	N/A	0.45[a]	2[a]	
H. Aschemann and D. Schindele [73]	Axis	PAMs	Position control	None	1	2	4[a]	1.4[a]	
Ambidextrous Hand, 2014 [79]	Hand	PAMs	Force control	None	9[c]	14[c]	0.37	4	✓

[a]Estimations are made from curve
[b]Results are obtained through a simulation
[c]Four DOFs and four actuators are unused for the BSC

backstepping controller's target is based on force feedback and speed stability, it offers greater flexibility than PID and bang–bang but takes longer to stabilise. Finger provided enough force to grab the can of soft drink at 0.30 s, but it is seen the system carries on moving until 0.40 s. Therefore it was deduced that BSC takes longer to stabilise. The force collected for the thumb at the end of the experiment is 13.10 N, which is a value close to the one obtained with the PID control. It can also be noted that the fingers' speed is slower using BSC, as none of the sensors collect more than 0.80 N after 0.15 s. The higher speeds of the PID and bang–bang controllers are respectively explained because of the integrative term and the lack of derivative control.

2.3.3 Comparison with Other BSCs

Backstepping control implemented on various hands is compared with the ambidextrous hand in Table 4. Some robotic structures actuated by motors are also included in the table as their number of DOFs is closer to the one of the ambidextrous hand and some of their BSCs are related to force control. BSCs are usually implemented on structures with less than five DOFs, such as manipulators, arms or parallel robots. The ambidextrous hand is the only robotic structure in Table 4 that has more than ten DOFs (BSC is only implemented on nine of them). It is quite clear that the number of DOFs does not exceed two when the BSCs are implemented on structures driven by PAMs. The BSC of the ambidextrous hand has an execution time of about 0.37 s, which is one of the shortest of Table 4, with [70, 71] that have execution times estimated to 0.3 s, as well as [72] with an execution time of 0.45 s. However, the maximum error of 4 % for the ambidextrous hand's BSC is much higher than the ones obtained with the practical results introduced in [68, 72, 73]. The control algorithms introduced in [72] appears to be ones of the most efficient, as the arm has 6 DOFs and the BSC is both among the fastest and the most accurate. Nevertheless, the BSC of the ambidextrous hand is the only one that is implemented on an ambidextrous structure and which is used to grab objects. In conclusion, the implementation of the BSC on the ambidextrous hand is quite successful, as the obtained results are among the best in Table 4. Nevertheless, its grasping time of 0.37 s is 85 % slower than the grasping time obtained with PID control in, and also much slower than the grasping times of most of robot hands summarized in Table 2. Consequently, the BSC is not the best option to grab objects with the Ambidextrous Hand.

3 Conclusion

PID, Bang–bang and BSC are tested and their performance in controlling an ambidextrous robotic hand is analyzed. PID controller was found the best when applied as compare to Bang–bang and Backstepping control. Backstepping control technique was validated for the first time on an ambidextrous robot hand. By combining PID controllers and force sensors, this research proposes one of the cheapest solutions possible. In future, PID controller could be used with artificial intelligent controllers to further improve the controls.

Acknowledgments The authors would like to cordially thank Anthony Huynh, Luke Steele, Michal Simko, Luke Kavanagh and Alisdair Nimmo for their contributions in design of the mechanical structure of a hand, and without whom the research introduced in this paper would not have been possible.

References

1. Andrikopoulos, G., Nikolakopoulos, G., Arvanitakis, I., Manesis, S.: Piecewise affine modeling and constrained optimal control for a pneumatic artificial muscle. IEEE Trans. Ind. Electron. **61**(2), 904–916 (2014)
2. Ortega, R., Spong, M.: Adaptive control. Elsevier, New York (1989)
3. Lilly, J.H.: Adaptive tracking for pneumatic muscle actuators in bicep and tricep configurations. IEEE Trans. Neural Syst. Rehabil. **11**(3), 333–339 (2003)
4. Chan, S.W., Lilly, J.H., Repperger, D.W., Berlin, J.E.: Fuzzy PD+I learning control for a pneumatic muscle. In: Proceedings of the 12th IEEE International Conference on FUZZ System, pp. 278–283 (2003)
5. Albus, J.S., Barbera, A.J., Nagel, R.N.: Theory and Practice of Hierarchcal control. (1980). www.robotictechnologyinc.com
6. Li, H., Kawashima, K., Tadano, K., Ganguly, S., Nakano, S.: Achieving haptic perception in forceps manipulator using pneumatic artificial muscle. IEEE/ASME Trans. Mechatron. **18**(1), 74–85 (2013)
7. Gupta, M.M.: Intelligent Control System. IEEE, New York (1996)
8. Omidvar, O., Elliott, D.L.: Neural Systems for Control. Elsevier (1997)
9. Kosaki, T., Sano, M.: Control of pneumatic artificial muscles with the just-in-time method based on a client-server architecture via the Internet. In: Proceedings of the IEEE International Conference on Automation Science and Engineering (CASE), pp. 980–985, Seoul, South Korea (2012)
10. Gelman, A., et al.: Bayesian Data Analysis. CRC press, Boca Raton (2013)
11. Ariga, Y., Pham, H.T.T., Uemura, M., Hirai, H., Miyazaki, F.: Novel equilibrium-point control of agonist-antagonist system with pneumatic artificial muscles. In: Proceedings of the IEEE ICRA, pp. 1470–1475. Saint Paul, MN, USA (2012)
12. Ross, T.J.: Fuzzy Logic with Engineering Applications. Wiley, New York (2009)
13. Domingos, P.: A few useful things to know about machine learning. Commun. ACM **55**(10), 78–87 (2012)
14. De Jong, K.: Evolutionary computation: a unified approach. In: Proceedings of the 2014 Conference Companion on Genetic and Evolutionary Computation Companion, pp. 281–296. ACM (2014)
15. Grefenstette, J.J.: Proceedings of the First International Conference on Genetic Algorithms and their Applications. Psychology Press (2014)
16. Reynolds, D.B., Repperger, D.W., Phillips, C.A., Bandry, G.: Modeling the dynamic characteristics of pneumatic muscle. Ann. Biomed. Eng **31**(3), 310–317 (2003)
17. Lewis, F.L., Vrabie, D., Syrmos, V.L.: Optimal Control. Wiley, New York (2012)
18. Åström, K.J.: Introduction to Stochastic Control Theory. Courier Dover Publications, USA (2012)
19. Repperger, D.W., Johnson, K.R., Phillips, C.A.: A VSC position tracking system involving a large scale pneumatic muscle actuator. In: Proceedings of the 37th IEEE Conference Decision Control, vol. 4, pp. 4302–4307 (1998)

20. Choi, T.-Y., Lee, J.-J.: Control of manipulator using pneumatic muscles for enhanced safety. IEEE Trans. Ind. Electron **57**(8), 2815–2825 (2010)
21. Navarro-Alarcon, D., Li, P., Yip, H.M.: Energy shaping control for robot manipulators in explicit force regulation tasks with elastic environments. In: Proceedings of the IEEE/RSJ International Conference Intelligent Robots and Systems (IROS), pp. 4222–4228 (2011)
22. Repperger, D.W., Phillips, C.A., Krier, M.: Controller design involvng gain scheduling for a large scale pneumatic muscle actuator. IEEE Int. Conf. Control Appl. **1**, 285–290 (1999)
23. Harald, A., Schindele, D.: Sliding-mode control of a high-speed linear axis driven by pneumatic muscle actuators. IEEE Trans. Ind. Electron. **55**(11), 3855–3864 (2008)
24. Van Damme, M., et al.: Proxy-based sliding mode control of a planar pneumatic manipulator. Int. J. Robot. Res. **28**(2), 266–284 (2009)
25. Chang, X., Lilly, J.H.: Fuzzy control for pneumatic muscle tracking via evolutionary tuning. Intell. Autom. Soft Comput. **9**(4), 227–244 (2003)
26. Shadow Dexterous Hand E1M3R, E1M3L. http://www.shadowrobot.com/products/dexterous-hand/
27. P. Srl.: The EH1 Milano Hand (2010). http://www.prensilia.com/index.php?q=en/node/41
28. S. R. C. L.: Shadow Dexterous Hand E1P1R, E1P1L (2013). http://www.shadowrobot.com/products/dexterous-hand/
29. Tsujiuchi, N., et al.: Development of pneumatic robot hand and construction of master–slave system. J. Syst. Design Dyn. **2**(6), 1306–1315 (2008)
30. Yoshikawa, T., Koeda, M., Fujimoto, H.: Shape recognition and optimal grasping of unknown objects by soft-fingered robotic hands with camera. Exp. Robot. Springer Tracts Adv. Robot. **54**, 537–546 (2009)
31. I. W. Laboratory: High-speed robot hand (2009). http://www.k2.t.u-tokyo.ac.jp/fusion/HighspeedHand/
32. Gunji, D., et al.: Grasping force control of multi-fingered robot hand based on slip detection using tactile sensor. In: Proceedings of the IEEE International Conference on Robotics and Automation (ICRA), pp. 2605–2610 (2008)
33. Chou, C.-P., Hannaford, B.: Measurement and modelling of McKibben pneumatic artificial muscles. IEEE Trans. Robot. Autom. **12**(1), 90–102 (1996)
34. Ang, K.H., Chong, G., Yun, L.: PID control system analysis, design, and technology. IEEE Trans. Control Syst. Technol. **13**(4), 559–576 (2005)
35. Hassan, M.Y., Kothapalli, G.: Comparison between neural network based PI and PID controllers. In: Proceedings of thre International Conference on Systems Signals and Devices (SSD), vol. 7, pp. 1–6 (2010)
36. Kristiansson, B., Lennartson, B.: Robust tuning of PI and PID controllers. IEEE Control Syst. **26**(1), 55–69 (2006)
37. Akyürek, E., Kalganova, T., Mukhtar, M., Paramonov, L., Steele, L., Simko, M., Kavanagh, L., Nimmo, A., Huynh, A., Stelarc, T.: Design and development of low cost 3D printed ambidextrous robotic hand driven by pneumatic muscles. Int. J. Eng. Tech. Res. (IJETR) **2**(10), 179–188 (2014)
38. Nagase, J.Y., Saga, N., Satoh, T., Suzumori, K.: Development and control of a multifingered robotic hand using a pneumatic tendon-driven actuator. J. Intell. Mater. Syst. Struct. **23**(3), 345–352 (2011)
39. Kawasaki, H., Komatsu, T., Uchiyama, K.: Dexterous anthropomorphic robot hand with distributed tactile sensor: gifu hand II. IEEE/ASME Trans. Mechatron. **7**(3), 296–303 (2002)
40. Namiki, A., Imai, Y., Ishikawa, M., Kaneko, M.: Development of a high-speed multifingered hand system and its application to catching. In: Proceedings of the 2003 IEEE/RSJ International Conference on Intelligent Robots and Systems (IROS 2003) (2003)
41. Dilibal, S., Tabanli, R.M., Dikicioglu, A.: Development of shape memory actuated ITU Robot Hand and its mine clearance compatibility. J. Mater. Process. Technol. **155**(156), 1390–1394 (2004)
42. Yamano, I., Maeno, T.: Five-fingered robot hand using ultrasonic motors and elastic ele-ments. In: Proceedings of the 2005 International Conference on Robotics and Automation, pp. 2673–2678 (2005)

43. Zollo, L., Roccella, S., Guglielmelli, E., Carrozza, M.C., Dario, P.: Biomechatronic design and control of an anthropomorphic artificial hand for prosthetic and robotic applications. IEEE/ASME Trans. Mechatron. **12**(4), 418–429 (2007)
44. Yoshikawa, T.: Multifingered robot hands: control for grasping and manipulation. Ann. Re-v. Control. **34**(2), 199–208 (2010)
45. Nagase, J.Y., Saga, N., Satoh, T., Suzumori, K.: Development and control of a multifingered robotic hand using a pneumatic tendon-driven actuator. J. Intell. Mater. Syst. Struct. **23**(3), 345–352 (2011)
46. Roa, M.A., Argus, M.J., Leidner, D., Borst, C., Hirzinger, G.: Power grasp planning for anthropomorphic robot hands. In: Proceedings of the IEEE International Conference on Robotics and Automation (ICRA), pp. 563–569 (2012)
47. Grebenstein, M., et al.: The hand of the DLR hand arm system: designed for interaction. Int. J. Robot. Res. **31**(13), 1531–1555 (2012)
48. Deshpande, A.D., et al.: Mechanisms of the anatomically correct testbed(ACT) hand. IEEE/ASME Trans. Mechatron. **18**, 238–250 (2013)
49. Nuchkrua, T., Leephakpreeda, T., Mekarporn, T.: Development of robot hand with pneumatic artificial muscle for rehabilitation application. In: Proceedings of the IEEE 7th International Conference on Nano/Molecular Medicine and Engineering (NANOMED), pp. 55–58 (2013)
50. Kakoty, N.M., Hazarika, S.M.: Local hand control for Tezpur University bionic hand grasp-ing, Air'13. In: Proceedings of Conference on Advances in Robotics, pp 1–7 (2013)
51. Palli, G., et al.: The DEXMART hand: mechatronic design and experimental evaluation of synergy-based controlfor human-like grasping. Int. J. Robot. Res. **33**(5), 799–824 (2014)
52. Mukhtar, M., Akyürek, E., Kalganova, T., Lesne, N.: Control of 3D printed ambidextrous robot hand actuated by pneumatic artificial muscles. In: Proceedings of the IEEE Sponsored SAI Intelligent Systems Conference, November 10–11, London (2015)
53. Silva, C., Trélat, E.: Smooth regularization of bang-bang optimal control problems. IEEE Trans. Autom. Control **55**(11), 2488–2499 (2010)
54. Bonnard, B., Caillau, J.B., Trélat, E.: Second order optimality conditions in the smooth case and applications in optimal control. ESAIM: Control Optim. Calc. Var. **13**(2), 207–236 (2007)
55. Nagi, F., Perumal, L., Nagi, J.: A new integrated fuzzy bang-bang relay control system. Mechatronics **19**(5), 748–760 (2009)
56. Nagi, F., Zulkarnain, A.T., Nagi, J.: Tuning fuzzy bang-bang relay controller for satellite attitude control system. Aerosp. Sci. Technol. **26**(1), 76–86 (2013)
57. Ham, R.V., Verrelst, B., Daerden, F., Lefeber, D.: Pressure control with on-off valves of pleated pneumatic artificial muscles in a modular one-dimensional rotational joint. In: Proceedings of the International Conference on Humanoid Robots, pp. 761–768 (2003)
58. Cain, S.M., Gordon, K.E., Ferris, D.P.: Locomotor adaptation to a powered ankle-foot orthosis depends on control method. Journal of NeuroEngineering and Rehabilitation. **4**, 1–13 (2007)
59. Shin, D., Sardellitti, I., Khatib, O.: A hybrid actuation approach for human-friendly robot design. IEEE Int. Conf. Robot. Autom. ICRA **2008**, 1747–1752 (2008)
60. Zinn, M., Khatib, O., Roth, B., Salisbury, J.K.: Towards a human centered intrinsically-safe robotic manipulator. In: Proceedings of the IARP/IEEERAS Joint Workshop Toulouse, France (2002)
61. Zhang, J.-F., Yang, C.-J., Chen, Y., Zhang, Y., Dong, Y.-M.: Modeling and control of a curved pneumatic muscle actuator for wearable elbow exoskeleton. Mechatronics **18**(8), 448–457 (2008)
62. Xu, Z., Kumar, V., Todorov, E.: A low-cost and modular, 20-DOF anthropomorphic robotic hand: design, actuation and modeling. In: Proceedings of the 13th IEEE-RAS International Conference on Humanoid Robots (Humanoids), pp. 378–375 (2013)
63. Vanderborght, B., et al.: Torque and compliance control of the pneumatic artificial muscles in the biped "Lucy". In: Proceedings of the 2006 IEEE International Conference on Robotics and Automation, pp. 842-847 (2006)
64. Vanderborght, B., Verrelst, B., Ham, R.V., Vermeulen, J., Lefeber, D.: Dynamic Control of a Bipedal Walking Robot actuated with Pneumatic Artificial Muscles. In: Proceedings of the 2005 IEEE International Conference on Robotics and Automation, pp. 1–6 (2005)

65. Vanderborght, B., Verrelst, B., Ham, R.V., Damme, M.V., Lefeber, D.: A pneumatic biped: experimental walking results and compliance adaptation experiments. In: Proceedings of the 5th IEEE-RAS International Conference on Humanoid Robots, pp. 44–49 (2005)
66. Mohamed, S., Xavier, B., Daniel, T.: Systematic control of an electropneumatic system: integrator backstepping and sliding mode control. IEEE Trans. Control Syst. Technol. **14**(5), 2–3 (2006)
67. Carbonell, P., Jiang, Z.P., Repperger, D.W.: Nonlinear control of a pneumatic muscle actuator: backstepping vs. sliding-mode. In: Proceedings of the IEEE International Conference on Control Applications(CCA '01), pp. 167–172 (2001)
68. Carbonell, P., Jiang, Z.P., Repperger, D.W.: A fuzzy backstepping controller for a pneumatic muscle actuator system. In: Proceedings of the IEEE International Symposium on Intelligent Control, (ISIC '01), pp. 353–358 (2001)
69. Soltanpour, M.R., Fateh, M.M.: Sliding mode robust control of robot manipulator in the task space by support of feedback linearization and backstepping control. World Appl. Sci. J. **6**(1), 70–76 (2009)
70. Su, C.-Y., Stepanenko, Y.: Backstepping-based hybrid adaptive control of robot manipulators incorporating actuator dynamics. Int. J. Adapt. Control Signal Process. **11**(4), 141–153 (1997)
71. Aschemann, H., Schindele, D.: Nonlinear model-based control of a parallel robot driven by pneumatic muscle actuators. In: Aschemann, H. (ed.) New approaches in Automation and Robotics. I-Tech Education and Publishing, Vienna (2008)
72. Qin, L., Liu, F., Liang, L.: The application of adaptive backstepping sliding mode for hybrid humanoid robot arm trajectory tracking control. Adv. Mech. Eng. **6**(307985), 1–9 (2014)
73. Aschemann, H., Schindele, D.: Comparison of model-based approaches to the compensation of hysteresis in the force characteristic of pneumatic muscles. IEEE Trans. Ind. Electron. **61**(7), 3620–3629 (2014)
74. Nganya-Kouya, D., Saad, M., Lamarche, L.: Backstepping adaptive hybrid force/position control for robotic manipulators. In: Proceedings of the 2002 American Control Conference, vol. 6,pp. 4595–4600 (2014)
75. Lotfazar, A., Eghtesad, M., Mohseni, M.: Integrator backstepping control of a 5 DOF robot manipulator with cascaded dynamics. IJE Trans. B: Appl. **16**(4), 373–383 (2003)
76. Wen, S.-H., Mao, B.: Hybrid force and position control of robotic manipulators using passivity backstepping neural networks. In: Advances in Neural Networks – ISNN 2007, vol. 4491, pp. 863–870 (2007)
77. Soltanpour, M.R., Fateh, M.M.: Sliding mode robust control of robot manipulator in the task space by support of feedback linearization and backstepping control. World Appl. Sci. J. **6**(1), 70–76 (2009)
78. Liu, X., Liadis, A.: Fuzzy adaptive backstepping control of a two degree of freedom parallel robot. Int. Robot. Appl. **7506**, 601–610 (2012)
79. Akyürek, E., Kalganova, T., Mukhtar, M., Paramonov, L., Steele, L., Simko, M., Kavanagh, L., Nimmo, A., Huynh, A., Stelarc, T: A novel design process of low cost 3D printed ambidextrous finger designed for an ambidextrous robotic hand. WSEAS Transactions on Circuits and Systems, ISSN / E-ISSN: 1109-2734 / 2224-266X, vol. 14, Art. #55, pp. 475–488 (2015)

Multiple Robots Task Allocation for Cleaning and Delivery

Seohyun Jeon, Minsu Jang, Daeha Lee, Young-Jo Cho, Jaehong Kim and Jaeyeon Lee

Abstract This paper presents a mathematical formulation of the problem of cleaning and delivery task in a large public space with multiple robots, along with a procedural solution based on task reallocation. The task in the cleaning problem is the cleaning zone. A group of robots are assigned to each cleaning zones according to the environmental parameters. Resource constraints make cleaning robots stop operation periodically, which can incur a mission failure or deterioration of the mission performance. In our solution approach, continuous operation is assured by replacing robots having resource problems with standby robots by task reallocation. Two resource constraints are considered in our formulation: the battery capacity and the garbage bin size. This paper describes and compares the performance of three task reallocation strategies: All-At-Once, Optimal-Vector, and Performance-Maximization. The performance measures include remaining garbage volume, cleaning quality, and cleaning time. Task allocation algorithms are tested by simulation in an area composed of 4 cleaning zones, and the Performance-Maximization strategy marked the best performance. Hence, a delivery task is added to the cleaning task. The delivery request operates as a new perturbation factor for the reallocation. The task allocation procedure for the delivery task includes the switching of tasks of the delivery robot itself as well as exchanging among cleaning robots to meet the balance of the cleaning performance. The experiment was conducted with 9 robots with the soft-

S. Jeon (✉) · M. Jang · D. Lee · Y.-J. Cho · J. Kim · J. Lee
Human Robot Interaction Research Team, Electronics and Telecommunications Research Institute (ETRI), 218 Gajeongro, Yuseonggu, Daejeon 34129, Korea
e-mail: happyseohyun@etri.re.kr

M. Jang
e-mail: minsu@etri.re.kr

D. Lee
e-mail: bigsum@etri.re.kr

Y.-J. Cho
e-mail: youngjo@etri.re.kr

J. Kim
e-mail: jhkim504@etri.re.kr

J. Lee
e-mail: leejy@etri.re.kr

Y. Bi et al. (eds.), *Intelligent Systems and Applications*,
Studies in Computational Intelligence 650, DOI 10.1007/978-3-319-33386-1_10

ware architecture that enables multi-functional of a robot and they performed both pseudo-clean and delivery task successfully.

1 Introduction

The problem of cleaning and delivery with multiple robots has been handled with various approaches. A common approach for cleaning with multiple robots is partitioning the space [9]. Ahmadi and Stone researched about partitioning the space for robots with different speed profile to minimize the overall cleaning time [2]. Once the space is partitioned, robots are distributed to their individual partitioned space to clean. Also, according to Ahmadi, the problem of cleaning a large public space can be defined as a *continuous area sweeping* task [2]. This means that the robot has to visit a place repeatedly rather than visiting once since the garbage is continuously piled on an open space. Similar application can be found for patrol and surveillance missions which sweeps the area continuously to detect the event of interest [7, 11].

Another approach for cleaning is a multi-robot coverage path planning within a single cleaning zone [10, 12, 17]. When multiple robots are assigned in a single space, paths of robots have to be generated, not to overlap with each other, to minimize the cleaning time. Also, a different approach for multi-robot cleaning is a formation control. If the cleaning target is impossible to be cleaned with one robot, a group of robots are coordinated to clean the target. Zahugi et al. researched about using a group of robots in a formation to clean the oil-spill [19].

Recently, Avidbots corp. introduced a autonomous cleaning robot system for a public space [1]. Two robots cooperate as a team for cleaning. Robots are able to

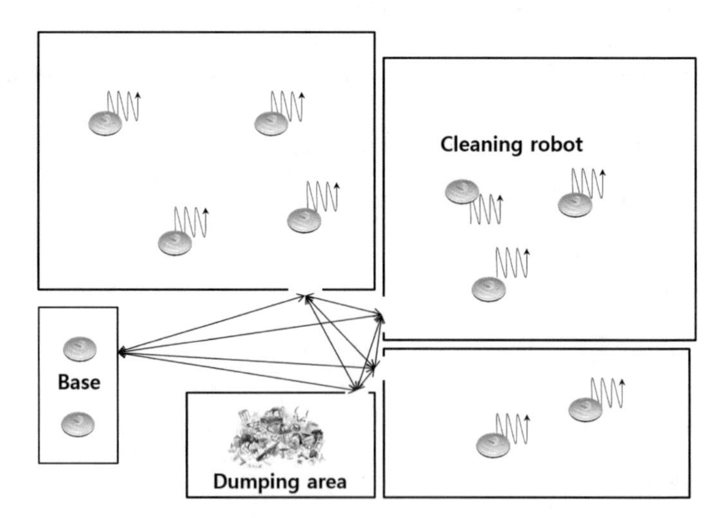

Fig. 1 Multi-robot cleaning as a team

autonomously localize, plan paths and sweep the floor. This system is just the beginning of the commercialized cooperative cleaning robot. More multi-robot systems will be used to clean a large public space in near future.

While operating cleaning robots, inoperative robots can be occurred for the reasons as discharge of battery or filled dust bin. Since such events as the battery and the bin can be solved after charging and emptying, these events are treated as normal inoperative situations. The advantage of using multiple robots stands out in this circumstance, that inoperative robots can be replaced with redundant robots.

Thus, the approach of this paper for multi-robot cleaning is task allocation. This approach is sending robots as a team to the cleaning zone and reallocate robots when an inoperative robot occurs (Fig. 1). The cleaning zone is derived from the topological map. Methods as doors detection or Voronoi diagram are used to generate the topological map from the 2D grid map [15, 18]. Each vertex of the topological map becomes the cleaning zone. Allocating robot as a team to the cleaning zones and planning the path of robots inside the cleaning zone reduces the computational complexity compared to the partitioning the space or planning paths as a whole workspace. Each vertex of the topological map contains cleaning information of the zone such as area or the garbage amount which affects the task allocation.

In addition, a delivery task can be added to the robot. A delivery task is one of the most demanding task for automation. The necessary technology for a delivery robot is not very far from the current stage. In fact, several robots are already commercialized in the field such as Aethon and Swisslog in the healthcare center, or Kiva systems in the warehouse. The key issue of the delivery task is the task switching within the multi-functional robot and load-balancing among all deployed robots in case of the absent (for delivery) robot. According to the multi-robot task allocation (MRTA) taxonomy, this problem is MT-MR-IA [13].

Centralized control of the multi-robot system is used to find an optimal allocation as in Fig. 2. The control system allocates tasks to robots and the robot autonomously interprets the task and performs the assigned task.

This paper describes the problem of cleaning and delivery with multiple robots in a large public space and proposes MRTA algorithms to allocate tasks to robots. The final goal of the solution is providing an allocation vector that represents the number of robots to be placed in the zone and selecting an appropriate robot for delivery while maintaining the cleaning performance. The mathematical formulation of the

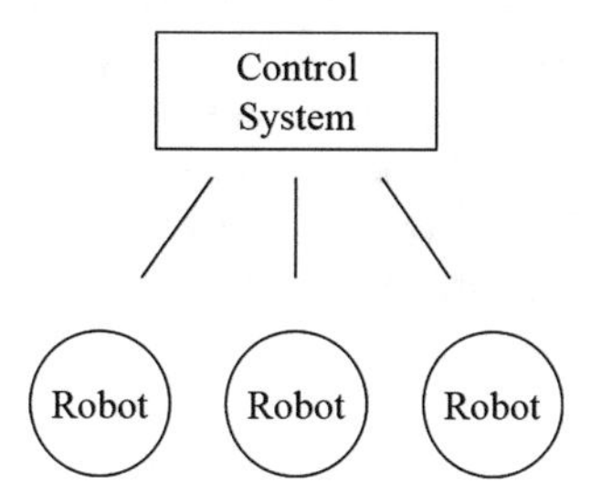

Fig. 2 Centralized multi-robot system

cleaning problem is described in Sect. 2 in which the task allocation strategies are introduced, and the simulated result and its analysis are discussed. The multi-robot delivery problem is explained in Sect. 3, followed by the conclusion in Sect. 4.

2 Multi-Robot Cleaning

2.1 Problem Description

In a large public area such as airport or department store, multiple robots are needed to clean the whole space. After the large space is divided into several cleaning zones, it needs to be decided how many robots are required to clean the single zone. If the garbage information of the environment and the specification of the cleaning robot is provided, the number of robot team for each cleaning zone can be decided. Thus, the object of the multi-robot cleaning problem is to find an appropriate team member of robots at each cleaning zones regarding the changing environment.

Since the battery discharges and the dust bin fills over time while the environment has constant inflow of the dust, the robot become absent to charge the battery and empty the bin. In this absent, robots become inoperative to conduct the assigned cleaning task. Inoperative robots have to be replaced to available robots to maintain the constant cleaning performance. Such replacement is defined as an allocation step. These constraints are assumed in this problem.

Constraints for the Inoperative Robot

Battery The robot has limited battery level that discharges linearly.
Dust bin The robot has limited capacity of the dust bin.

Goal The goal of this multi-robot cleaning problem is to find an allocation vector for each allocation step. The allocation vector is the number of robot in a team of a cleaning zone. The allocation vector at each reallocation step is denoted by $\mathbf{A_k}$.

$$\mathbf{A_k} = [r_k^1, r_k^2, \ldots, r_k^n] \tag{1}$$

The total number of working (cleaning) robots at kth step is r_k ($r_k = \sum_{i=1}^{n} r_k^i$). The number of standby robots at kth step is r_{ks}, and the number of inoperative robots is r_{ki}. Then, the available robots r_{ka} at kth step is the sum of standby robots and the working robots. And, this is the same to subtracting inoperative robots from the total number of robots ($r_{ka} = r_{ks} + r_k = r_t - r_{ki}$). When previously inoperative robot becomes ready after charging the battery or emptying the bin, the robot's status becomes standby robot.

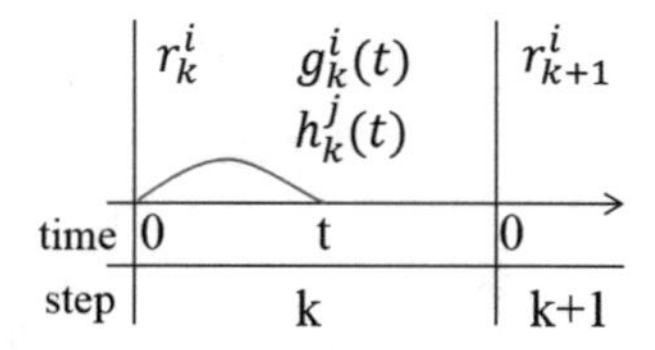

Fig. 3 The timeline used in the cleaning problem. k denotes the reallocation step, and t denotes the time passed after the reallocation. r_k^i and $g_k^i(t)$ is the number of assigned robots and the remaining garbage amount of ith zone, respectively, and $h_k^j(t)$ is the collected garbage amount of jth robot at time t of the step k

2.2 *Problem Formulation*

2.2.1 Timeline

To allocate robots at each zone, the environment and the robot should be modeled mathematically. To begin with, understanding timeline of the problem should be preceded. There are two types of timeline in this problem. One is the discrete time, and the other is the continuous time as in Fig. 3. Every reallocation steps is symbolized as k in the discrete time. The continuous time is used to show the time passed after reallocation, and it is symbolized as t. t is reset to 0 at every reallocation, and increases as time passes.

2.2.2 Robot

A set of r number of cleaning robots is represented by $R = \{R^j | j \in 1, \dots, r\}$, and each cleaning robot (R^j) has four attributes (A_R^j) and two parameters (P_R^j), and is defined as:

$$R^j =< A_R^j, P_R^j > \tag{2}$$

$$A_R^j = \{C, M, T_w, T_c\}, \quad P_R^j = (b_k^j(t), h_k^j(t)) \tag{3}$$

where C is the cleaning capability of a robot, which is the volume of the collected garbage by the robot per hour. M is the maximum capacity of the dust bin of a robot, T_w is the operating hours of a cleaning robot with the fully charged battery, and T_c is the time it takes to fully charge the empty battery. $b_k^j(t)$ is the battery level, and $h_k^j(t)$ is the collected garbage amount of jth robot at time t in the allocation step k. $h_k^j(0)$ becomes the initial amount of the collected garbage at the allocation step k.

$$h_k^j(t) = h_k^j(0) + \frac{h_k^j(t)}{dt} t \tag{4}$$

$$\frac{h_k^j(t)}{dt} = min\left(C, \frac{g_k^i(t)}{r_k^i}\right) = \begin{cases} C, & \frac{g_k^i(t)}{r_k^i} \geq C \\ \frac{g_k^i(t)}{r_k^i}, & \frac{g_k^i(t)}{r_k^i} < C \end{cases} \tag{5}$$

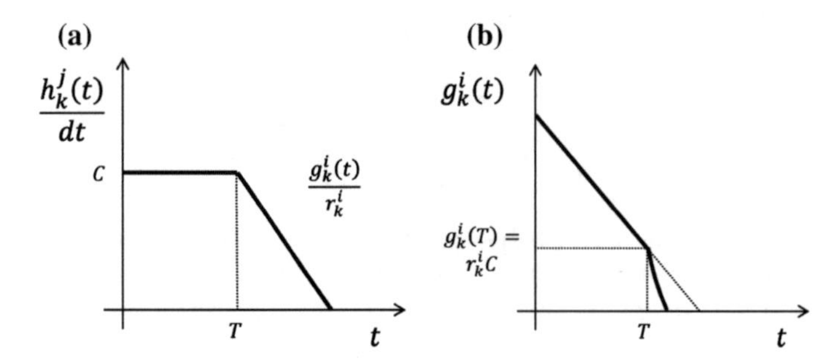

Fig. 4 If the remaining garbage of a cleaning zone is decreased below $r_k^i \cdot C$, the collecting garbage amount of a robot per hour is reduced to $\frac{g_k^i(t)}{r_k^i}$. In this case, $\frac{h_k^j(t)}{dt} = C$ at $t \leq T$, and $\frac{h_k^j(t)}{dt} = \frac{g_k^i(t)}{r_k^i}$ at $t > T$. **a** The collecting garbage amount of a robot. **b** The remaining garbage amount of a cleaning zone

Note that min$(\cdot)$ term in Eq. 5 is used to satisfy the condition that if there are enough garbage in the cleaning zone for robots which can sufficiently collect the garbage of their capability, then the collecting garbage amount of a robot is C. Otherwise, the collecting garbage amount per hour is reduced to the remaining garbage amount of the zone divided by the number of robots assigned to the zone ($\frac{g_k^i(t)}{r_k^i}$). Figure 4a shows that after the time T, the collecting garbage amount is reduced to $\frac{g_k^i(t)}{r_k^i}$ from C since there are fewer garbage as in Fig. 4b. Thus, the decrease rate of remaining garbage amount of the cleaning zone after time T shows second order curved line as in Fig. 4b.

Also, the collected garbage amount of a robot cannot exceed the maximum capacity of the dust bin mounted on the robot from the constraints for inoperative robots. ($h_k^j(t) \leq M$) If $h_k^j(t) = M$, the robot has to stop cleaning and return to empty the dust bin. Similarly, let B_M be the marginal value of the discharged battery of a robot, then, if $b_k^j(t) \leq B_M$, the robot has to return to the charging station.

2.2.3 Environment

A cleaning zone of the total workspace can autonomously be extracted and numbered from a grid map generated by a robot by the graph cut algorithm [5]. An example of a cleaning area can be seen in Fig. 5. Figure 5a shows the space is divided into 4 cleaning zones and an auxiliary zone. The auxiliary zone is a hallway connecting each cleaning zones in which exists a charging station and a dumping area. Corresponding topological map can be seen in Fig. 5b with 4 nodes. The edge of the topological map represents the connection and the distance to the entrance between nodes. The allocation is conducted based on these nodes of the topological map.

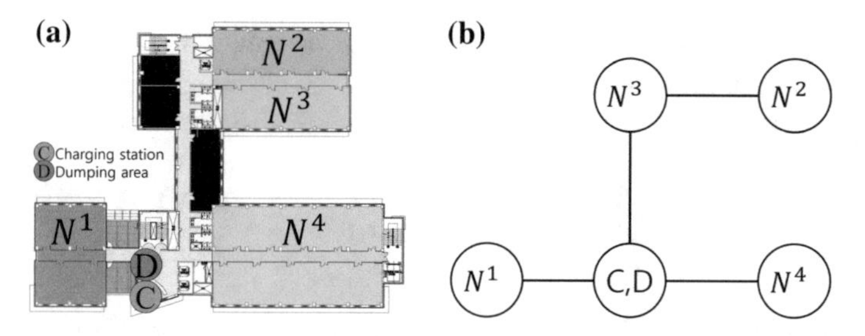

Fig. 5 An indoor 2D floor map and corresponding topological map of the cleaning area. **a** A 2D map with 4 cleaning zones. **b** Generated topological map from the 2D map

If there are n number of nodes $N = \{N^i | i \in 1, \ldots, n\}$, each node of the topological map contains two attributes (A^i_N) and parameters (P^i_N) about the cleaning zones and can be defined as:

$$N^i =< A^i_N, P^i_N > \tag{6}$$

$$A^i_N = \{A^i, \Delta^i\}, \quad P^i_N = (g^i_k(t), r^i_k) \tag{7}$$

where A^i is the area of the cleaning zone, Δ^i is the garbage increasing rate of the zone, g^i_k is the remaining garbage amount of ith node at time t in kth step, and r^i_k is the number of robots assigned for ith node at kth step.

There is various approaches to model the dirt distribution on the floor. Hess modeled the dirt distribution map of the floor with Poisson distribution of the grid-cell [8]. However, this paper assumed uniform distribution of the garbage that has linear increasing rate of (Δ^i).

$$g^i_k(t) = g^i_k(0) + A^i \Delta^i t - r^i_k \frac{h^j_k(t)}{dt} t \tag{8}$$

$$g_k(t) = \sum_{i=1}^{n} g^i_k(t) \tag{9}$$

In Eq. 8, $g^i_k(0)$ is the initial garbage volume amount in the ith node at the beginning of the step k. The second term $A^i \Delta^i t$ is the garbage volume occurrence in the ith node over time, while the last term indicates the total garbage volume collected by robots operating in the ith node in the duration of $t = [0, t]$.

Total number of robots Since there are limited number of robots, the user has to decide how many robots are needed at the workspace to clean continuously. If there are large amount of garbage occurring every time, it cannot be cleaned with small number of robots. Since the user cannot afford to buy infinite number of robots, there is a demand to know an optimal number of robots after comparing the cost and the cleaning performance. Hence, we want to maximize the performance of multiple robots with the limited number of robots. To achieve this, a task allocation algorithm is considered and this will be discussed in the Sect. 2.3.2.

This paper proposes the minimum number of robots required to clean the workspace. Given the information of robots and the cleaning space, the total minimum number of robots (r_t) required to clean the whole space can be defined as Eq. 10.

$$r_t = \sum_{i=1}^{n} \left\{ \left(\frac{A^i \Delta^i}{C} + \frac{g_0^i(0)}{C t_d} \right) \times \left(1 + \frac{T_w}{T_c} \right) \right\} \tag{10}$$

where t_d is a user defined parameter, which is the desired time of the space to be cleaned, and it is normally set to $t_d = 24\,\text{h}$.

2.3 *Task Allocation*

2.3.1 The Cleaning Procedure

The multi-robot system for cleaning a large public space is controlled as the following procedure.

1. Initialize parameters ($g_0^i(0), h_0^j(0)$).
2. Allocate robots ($\mathbf{A_k}$).
3. Perform cleaning.
4. Monitor parameters for the demand for reallocation ($b_k^j(t), h_k^j(t)$).
   If yes, go to step 2.

After initializing the environmental and robotic parameters, the control system decides the allocation vector ($\mathbf{A_k}$), that is the number of robots to be assigned at each cleaning zone. The allocation vector is derived by receiving the garbage amount ($g_k^i(t)$) and the number of available robots (r_{ka}) at the reallocation step. By applying allocation strategies, which will be explained in the following Sect. 2.3.2, the allocation vector is decided.

After deciding the number of robots, specific robots has to be selected to send to the target zone. Robots are selected based on the cost function (U_k^j). The cost function is a combination form of the distance to the target (d^j), remaining battery ($b_k^j(t)$) and the bin level ($M - h_k^j(t)$) as Eq. 11. Robots with minimum cost (U_k^*) will be assigned to the target zone. Since scales of parameters in the cost function are different, normalization is necessary. The cost function is a user-defined form, so weights of parameters can be defined by the user. Once the allocation vector is derived, the corresponding number of robots will move to the assigned zones and perform cleaning. The overall procedure can be seen in Fig. 6

$$U_k^j = f(d^j, b_k^j(t), M - h_k^j(t)) \tag{11}$$

$$U_k^* = \min_j(U_k^j) \tag{12}$$

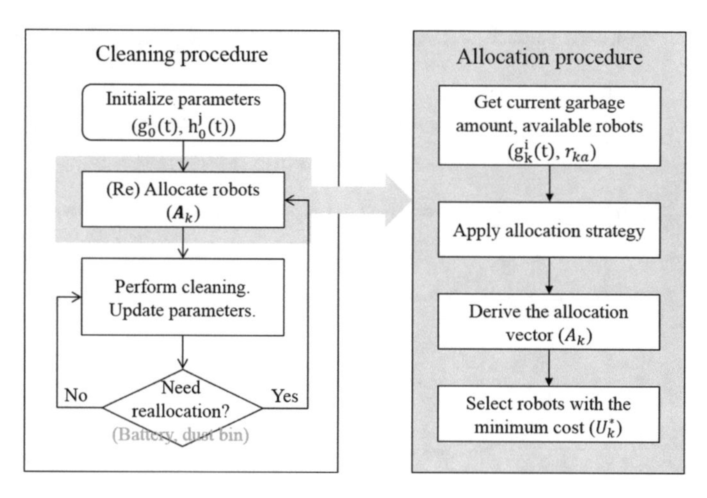

Fig. 6 Cleaning and its allocation procedure

When an inoperative robot (r_{ki}) occurs due to the battery discharge or full of the bin as discussed in Sect. 2.1, the robot cannot perform the cleaning task anymore, but has to return to the charging station or to the dumping area to resolve the situation. The control system monitors such encountered situation, and firstly searches for a substitute robot among standby robots (r_{ks}). If there is an available robot, the system replaces the inoperative robot to the standby robot. If not, the system has to decide whether to wait or to extract a robot from other cleaning zone, for persistent cleaning. This is decided by the task allocation strategy, and the reallocation result can be differed according to the strategy. This will be explained in the following section.

2.3.2 Task Allocation Strategy

Provided that environmental and robotic parameters are known, the control system decides the best allocation vector to clean the space since the performance of cleaning varies according to the initial allocation vector. Four different allocation vectors are tested as $\mathbf{A1_0}$, $\mathbf{A2_0}$, $\mathbf{A3_0}$, and $\mathbf{A4_0}$. The larger the number of allocation vector, the more robots are assigned at initial stage. The garbage parameters for the node 1, 2, 3 and 4 are set differently as in Table 1. It can be seen from the Fig. 7 that allocating many robots at initial stage does not always result in a good performance. Note that, even though the r_0 of $\mathbf{A4_0}$ is larger than $\mathbf{A3_0}$, the sum of remaining garbage of all nodes is not cleaned. (The line of the graph stays higher in $\mathbf{A4_0}$ than $\mathbf{A3_0}$.) This is because when initially many robots are allocated, there are not enough standby robots to substitute when former cleaning robots become inoperative and return to base for charging. In other words, intentionally keeping some standby robots initially will increase the performance of the cleaning. How many initial robots are needed to

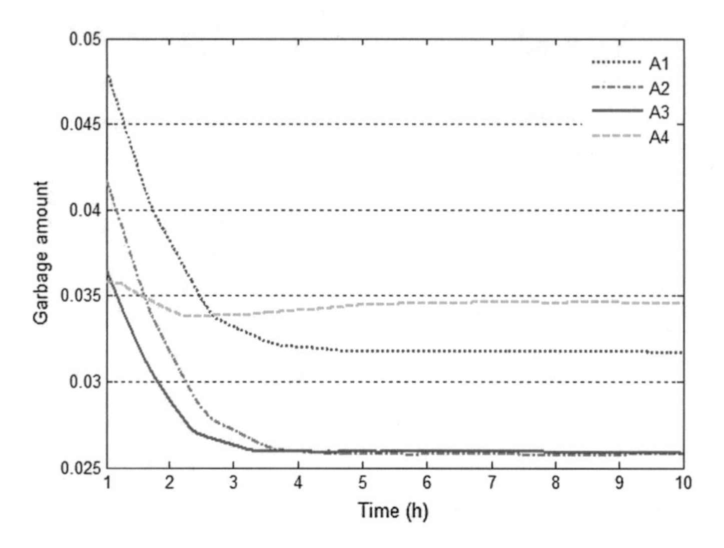

Fig. 7 The moving average of sum of the remaining garbage amount of all cleaning zones over time with different allocation strategies ($\mathbf{A1_0} = [2\ 3\ 3\ 6]$, $\mathbf{A2_0} = [3\ 4\ 4\ 7]$, $\mathbf{A3_0} = [4\ 5\ 5\ 8]$, $\mathbf{A4_0} = [5\ 6\ 6\ 9]$). The result of $\mathbf{A3_0}$ allocation shows the best performance with the least remaining garbage amount

be left for standby is decided by several allocation strategies. Considering this, three different allocation strategies are proposed as following.

All-at-once allocation (AOA)

This allocation method is relatively simple which is allocating all robots at once. All robot means robots that are available at the allocation moment. This allocation strategy does not leave standby robots. The number of robots for allocation at each node is as Eq. 13

$$r_k^{i*} = \frac{g_k^i(0)}{g_k(0)} \times r_{ka} \tag{13}$$

Optimal-vector allocation (OVA)

This method predefines the optimal number of robots for each cleaning zone and allocates robots only the number as defined even though there are remaining standby robots. This way standby robots is left and become substitutes when previously cleaning robots become inoperative. The allocation is as Eq. 14

$$r_k^{i*} = \frac{A^i \Delta^i}{C} + \frac{g_k^i(0)}{C t_d} + 1 \tag{14}$$

Performance-maximization allocation (PMA)

In this allocation, the number of robots at each cleaning zone is decided by the remaining garbage amount at the allocation step. This method tries to maximize the performance as Eq. 15.

First, by defining the base robot set ($r^i_{kb} = \frac{g^i_k(t)}{g_k(t)} \times r_{ks}$) and altering the number of robots from the base set by combinatorial search, the performance and the remaining garbage amount ($g_k(t)$) can be calculated. The r^i_k with the best performance ($\min(g_k(t))$) is finally selected. This is the same as maximizing the collecting garbage amount per robot ($\max(h^j_k(t))$).

$$r^{i*}_k = \min_{r^i_k}(g_k(t)) = \max_{r^i_k}(h^j_k(t)) \tag{15}$$

Due to several reasons, such as the longer charge time, the distance between cleaning zones, and the limited number of robots, often there are few standby robots when reallocation is required. In this case, the control system regenerates the appropriate allocation vector based on the number of available robots (r_{ka}) at the reallocation step and this can result the change of assigned cleaning zone.

2.4 *Simulation*

2.4.1 Experimental Setup

Proposed allocation strategies are tested in the simulation. Parameters of each cleaning zone in the simulation are set as Table 1 with the area, the garbage increasing rate, and the initial garbage amount. Homogeneous robots are used for cleaning, and their attributes are set as $C = 0.001\,\text{m}^3 = 1\,\text{L/h}$, $M = 0.004\,\text{m}^3 = 4\,\text{L}$, $T_w = 4\,\text{h}$, and $T_c = 6\,\text{h}$. ($A^j_R = \{0.001, 0.004, 4, 6\}$)

Based on the attributes of the cleaning zone as Table 1, the total number of robots (r_t) required for the space to be cleaned is 40. Therefore, 40 robots are used in the simulation. The garbage increasing rate (Δ^i) is assumed to be static and uniformly distributed within the cleaning zone. Initial garbage amount is set as Table 1, which is 5 hours passed from the zero amounts. The charging station and the dumping area is separated, so a robot has to visit two different places according to its status. The traverse distance among cleaning zones are not taken into account in the simulation since it does not affect the result of the performance comparison among strategies.

Table 1 Simulation environment

Cleaning zone	Area ($A^i (m^2)$)	Garbage increasing rate ($\Delta^i (m/h)$)	Initial garbage amount ($g^i_0(0)(m^3/h)$)
N^1	1,000	2×10^{-6}	0.01
N^2	1,500	1.5×10^{-6}	0.0125
N^3	2,000	1.5×10^{-6}	0.015
N^4	2,500	2×10^{-6}	0.025

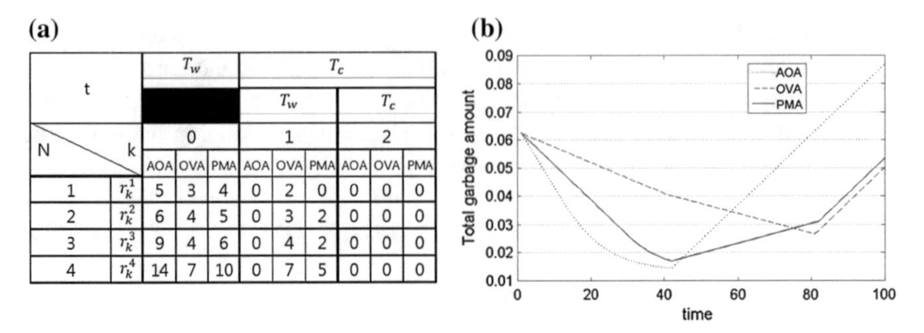

(a)

t		T_w			T_c					
					T_w			T_c		
N \ k		0			1			2		
		AOA	OVA	PMA	AOA	OVA	PMA	AOA	OVA	PMA
1	r_k^1	5	3	4	0	2	0	0	0	0
2	r_k^2	6	4	5	0	3	2	0	0	0
3	r_k^3	9	4	6	0	4	2	0	0	0
4	r_k^4	14	7	10	0	7	5	0	0	0

Fig. 8 Allocation of three strategies at first three allocation steps. **a** Allocation vector. The number of robots for each allocation strategies. **b** Garbage amount of the first allocation

2.4.2 Result and Analysis

The number of robots allocated to each cleaning zone varies according to the allocation strategy. The number of assigned robots at each cleaning zone and corresponding remaining garbage amount resulting from the allocation vector can be seen in Fig. 8. It shows the first allocation cycle of AOA, OVA, and PMA. The allocation cycle comes at every $(T_w + T_c)$ time. The allocation vector is repeated within the cycle for AOA and OVA, whereas PMA's changes since the garbage amount decreases and corresponding collecting garbage amount of a robot decreases below the capability (C). After first assigned group of cleaning robots returns for charging the battery or emptying the bin, the second group moves for cleaning at allocation step $k + 1$, and this is continued for the next cycle.

Figure 8b shows the first cycle of the remaining garbage of all cleaning zones at $k = 0, 1,$ and 2 for three strategies. Note that the remaining garbage of AOA increased dramatically since there are no cleaning robots at $k = 1$ and 2 while garbage is still increasing.

The overall cleaning performance of each allocation strategy is measured by three factors: The average boundary of the garbage amount, the cleaning quality, and the cleaning time.

During the simulation, the remaining garbage amount is increased and decreased within upper and lower boundaries. The lower the average between boundaries, the cleaner the area is. From Table 2, OVA and PMA is maintained cleaner than AOA.

Table 2 Cleaning performance comparison for three allocation strategies

Strategy	Average boundary of garbage	Cleaning quality	Cleaning time
AOA	0.0570	5.21×10^{-9}	∞
OVA	0.0228	1.09×10^{-9}	86.2 h
PMA	0.0228	0.87×10^{-9}	81.0 h

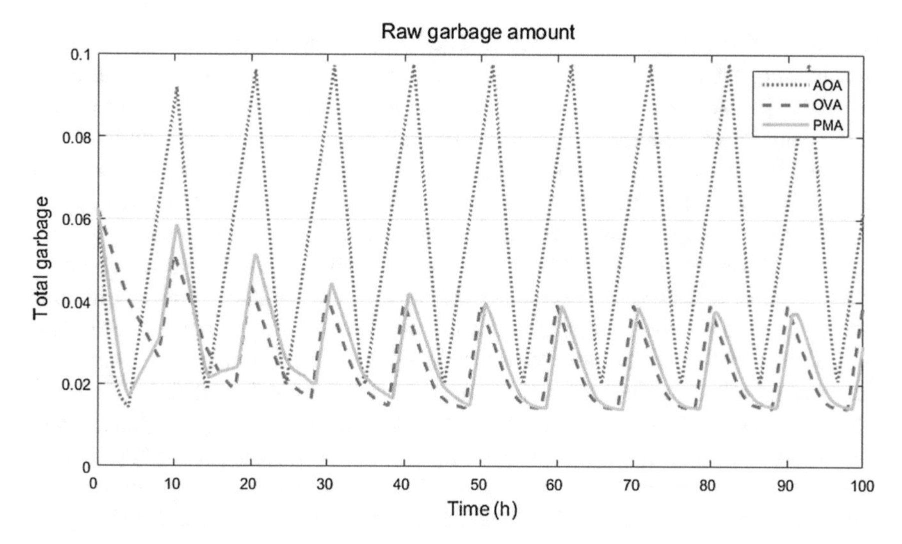

Fig. 9 The simulation result of the garbage amount of AOA, OVA, and PMA strategies

Although the average boundaries is low, if the difference of upper and lower boundaries is large, it can be said that the cleaning quality is low. It means the zone is cleaned but contaminated easily. This value is measured by calculating the deviation of two boundaries. The difference of the boundaries can be seen in Fig. 9 as saw-toothed shaped. From Table 2, the best cleaning quality is provided by PMA and it can be seen as orange solid line in Fig. 9.

The cleaning time is the time consumed to clean the garbage below the marginal amount. The marginal amount is set to $1.1 \times \Delta^i$ in this paper. Since the garbage increases continuously, there is no end of the cleaning, but, the first time that the garbage amount reaches the marginal amount is defined as the cleaning time. The calculated cleaning time of the PMA is slightly faster than OVA as in Table 2.

The simulation result shows that the initial allocation vector affects the cleaning performance, and leaving standby robots for replacement results in better performance as can be seen from PMA and OVA task allocation strategies.

3 Multi-Robot Delivery

3.1 Problem Description

Delivery is one of the highly demanding tasks for automation. For example, hospital staffs require frequent turn-arounds to deliver materials such as medicine, mails, and linens. Nurses walk maximum 10 Km a day, not only to care patients, but also, to get supplies [3]. To replace such mobility work inside a healthcare center, robots have been developed by Aethon and Swisslog to task over the delivery task [16]. Also, Kiva systems in the warehouse is one of the succeeded robotic system that used for delivering packages. They insisted that by improving the algorithms of coordinating robots, the number of robots required in the warehouse can be reduced to 5–10% [6]. Amazon is now extending their robotic application to deliver outside of the warehouse, directly to customers. From the aforementioned examples, it can be tested for robots to add a delivery task besides the cleaning task to maximize the usage. To achieve this, the robot should be multi-functional and the control system should select an appropriate robot for delivery while assigning the cleaning zone.

The performance of the delivery task can be evaluated by several indicators, in scopes of cost and time. The standard for carriers and logistics services providers categorized the key performance indicator into time, alert, security and efficiency [14]. The problem of robot for cleaning and delivery, in this paper, only considered the time and efficiency indicators since the experiment is conducted in the laboratory and the purpose of using robots are multi-functional which is not dedicated only for the logistics. Therefore, this problem disregards the alert and security problems. The time considered in this problem is waiting time of the pick-up position until a robot arrives after the delivery is requested. The efficiency is considered in scope of cleaning performance.

The delivery request is made with the simple input data of positions for pick-up and drop-off. Therefore, the problem of delivery is to find a nearest robot to the delivery position among cleaning robots while maintaining the cleaning performance.

3.2 *Task Allocation*

The reallocation due to the delivery request is similar to the battery and dust bin events in the cleaning problem. The task allocation procedure for delivery request is as in Fig. 10. When there is a request for a delivery, a nearest robot to the pick-up position is selected for the delivery. When delivery task is finished in the drop-off position, the robot returns to the cleaning task on the nearest cleaning zone from the finished position. To balance the cleaning performance, a robot from the drop-off position is selected to change the assigned zone with the delivery robot and is allocated to the new cleaning zone where the delivery robot has left. This way the cleaning performance is maintained continuously.

The delivery task while cleaning is conducted by following procedure:

1. A delivery is requested from the pick-up (PU) position to the drop-off (DO) position.
2. The task allocator selects a robot for the delivery and **reallocates** the cleaning tasks.
3. The selected robot moves to the PU position from the current position.
4. The robot arrives at the PU position and loads the package.
5. The robot moves to the DO position and unloads the package.
6. The robot returns to the cleaning state and perform cleaning on the newly assigned cleaning zone.

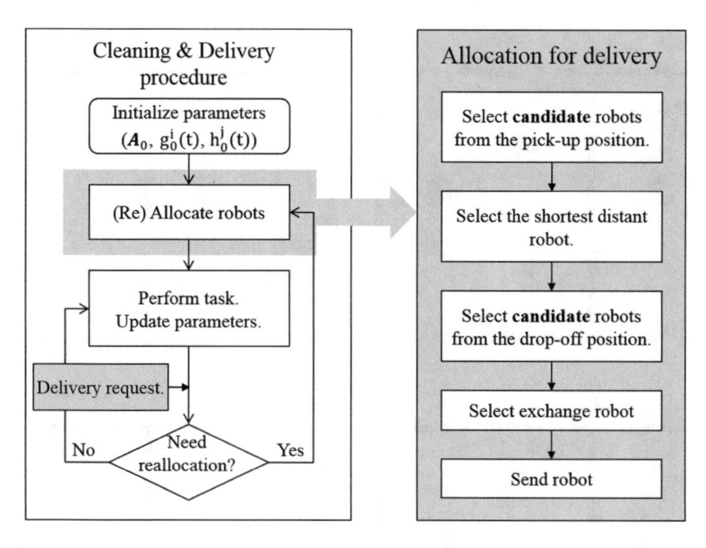

Fig. 10 The task allocation algorithm for cleaning and delivery

The task allocator considers the time efficiency, thus the cost function of the delivery task measures the distance (d).

$$\min_{r}(d) \tag{16}$$

Normally, the robot from the nearest cleaning zone that presents the minimum distance to the PU position is selected for the delivery. The absence of the delivery robot from the cleaning zone incur perturbation of previous allocation and becomes another factor for the task reallocation. The key point of this strategy that differs from other problem is that the second procedure from above that the task allocator reallocates the cleaning zone of all robots to maintain the balance of the cleaning task. In this procedure, a robot might change its assigned cleaning zone from the previous assigned cleaning zone to the new zone where the delivery robot has left. This way, the delivery robot can save time of returning to the cleaning task compared with going to the original cleaning zone, while maintaining the overall cleaning performance. This scenario is depicted in Fig. 11.

3.3 *Experimental Setup*

3.3.1 Hardware Setting

The experiment for the cleaning and delivery mission is conducted with 9 robots. The robot used for the experiment is Kobuki as in Fig. 12. Although the base platform came from the robot cleaner, this research robot does not have broom for cleaning.

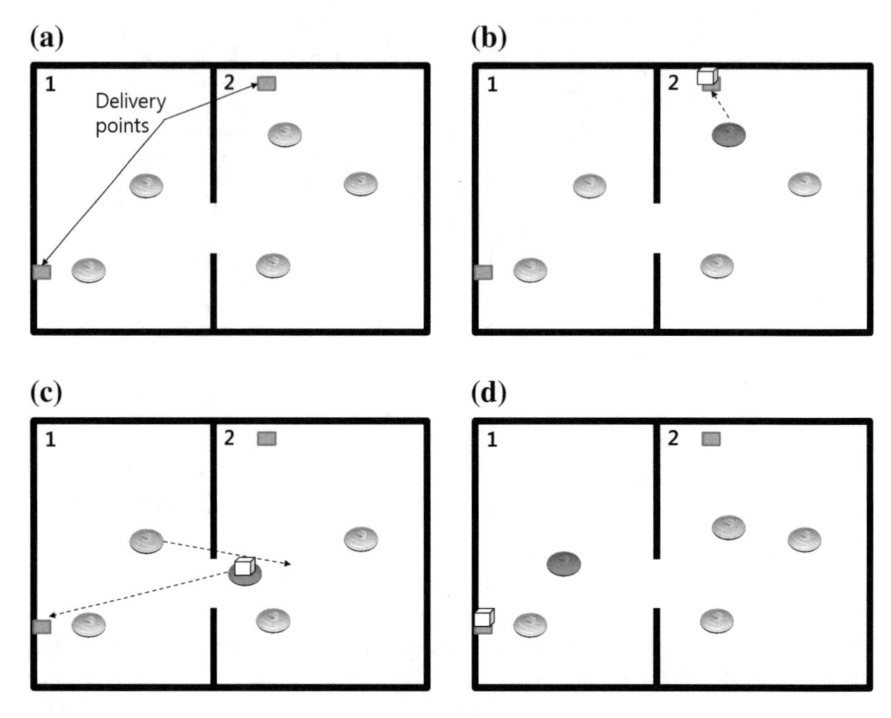

Fig. 11 Cleaning and delivery task scenario. **a** Initial allocation for cleaning. **b** A delivery request, and a (*red*) robot is selected for the delivery. **c** A (*green*) robot is selected for the substitute. **d** Two robot have changed their assigned cleaning zone after the delivery

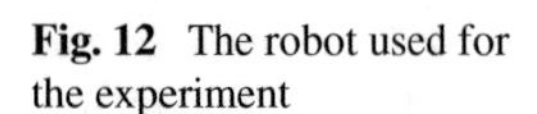

Fig. 12 The robot used for the experiment

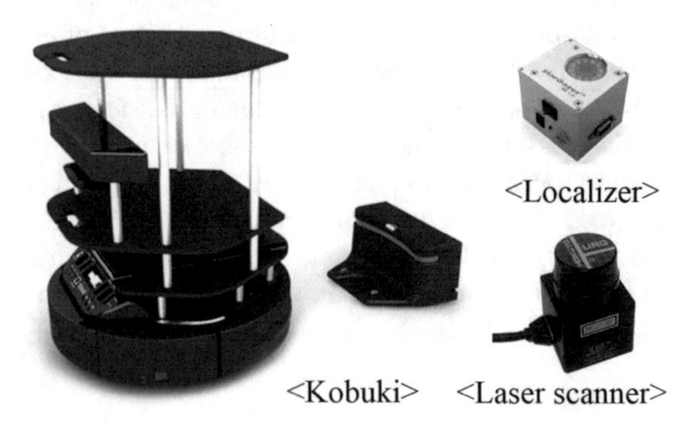

Since the only difference is the cleaning equipment, the movement of robot is programmed similar to the cleaning situation. Thus, it can be said that this robot performs pseudo-cleaning. The robot is mounted with a localizer which looks marker of the ceiling and a laser scanner for obstacle avoidance. A button for the user-interface of the delivery task and a light tower is used to indicate the status. All robots are connected via wifi to the Remote Operating Center (ROC).

3.3.2 Software Setting

The software architecture controlling robots and ROC is named as Cooperative Intelligent RObotic System (CIROS). CIROS is consisted of the Message processors, Context memory, Task controller, Individual task interpreter, Controller, Navigator, Robot platform, Monitor, and the Reporter as in Fig. 13.

When a mission is commanded from ROC, the Message processor receives the command, and the message is interpreted and matched from the Context Memory. If the message is a task, the Task controller receives the message and selects an appropriate task. Then, the Individual task interpreter takes action for the robot. If the command is cleaning, the Controller commands the robot to clean. And, if the command is delivery, the robot is controlled to follow the delivery action sequence. The uRON library is used for the Navigator which generates and follows the path with the localizer and avoids obstacle autonomously [4]. The final motion behavior of the command reaches to the hardware of the Robot platform to control sensor and actuator. All movements is monitored in the Monitor, and after matching it to the Context memory, it sends out the status of robot to ROC from the Reporter.

3.4 *Result and Analysis*

Nine robots are deployed in 3 zones on the $16\,\text{m} \times 6\,\text{m}$ space for the laboratory experiment as in Fig. 14. Seven robots are initially deployed for cleaning and 2 robots

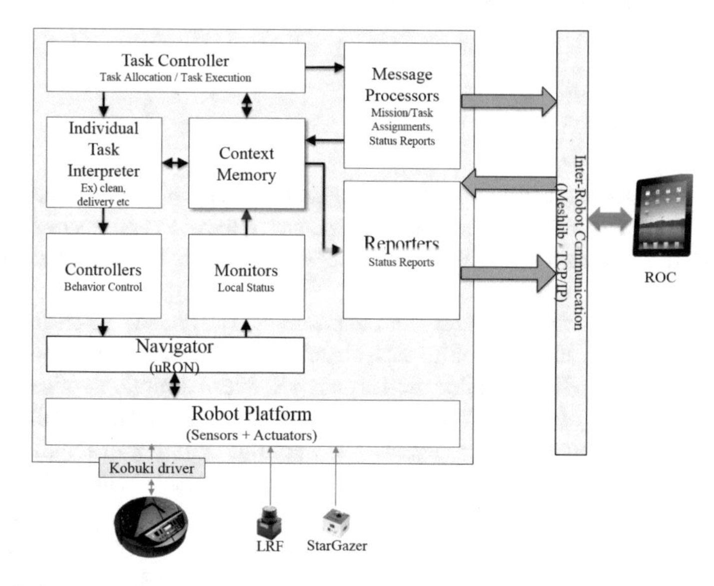

Fig. 13 Software architecture of a robot

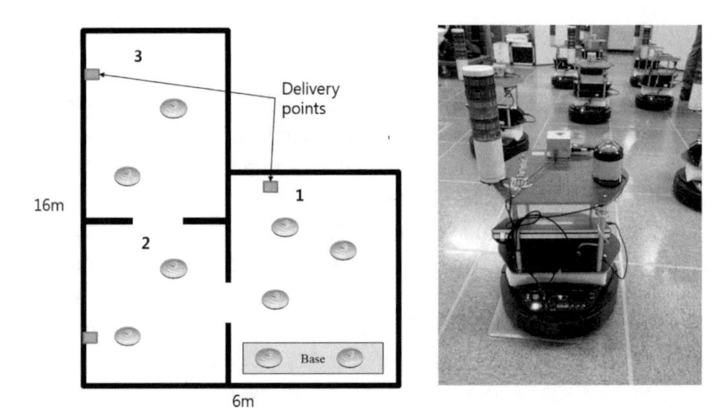

Fig. 14 Real application of multiple robots for cleaning and delivery mission

remained at the base for standby. Deployed robots move through the path generated by the boustrophedon method mimicking cleaning robot. Delivery positions are localized off-line and the delivery request is randomly called on-line. When a delivery request is made from zone 1 to zone 2, the nearest robot to the delivery position from zone 1 is selected for the delivery and moves to the pick-up position while a robot from zone 2 is selected to change the cleaning zone to 1. After loading the package, the delivery robot successfully moves to the drop-off position at zone 2 following the delivery task sequence while a robot from zone 2 is moved to zone 1 to fill out the absence. After the delivery is finished, the delivery robot stays in zone 2 for cleaning, and the exchange of robots was performed successfully (Fig. 14).

4 Conclusion

This paper describes the problem of cleaning and delivery mission in a large public space with the multi-robot system based on the task allocation approach. The multi-robot system seeks for optimization, so it inevitably follows the centralized control system.

This paper first describes the problem of cleaning. The goal of the cleaning problem is to allocate a group of robots to each cleaning zone according to the robotic and environmental status. Every reallocation is occurred due to limitations of garbage bin and battery amount of a robot. The problem is formulated mathematically regarding the robotic action and environment. The minimum number of required robots for cleaning the given space is suggested. Cleaning procedure with reallocation is explained. Three allocation strategies are proposed and tested by simulation. The number of robots allocated to each cleaning zone varies according to the allocation strategy. The result is compared with respect to the remaining garbage, the cleaning quality, and the overall cleaning time. It can be concluded that the performance-maximization allocation (PMA) strategy performs better than the optimal-vector allocation (OVA) or the all-at-once allocation (AOA) for cleaning a large space. The PMA and OVA strategies have in common that they intentionally remain standby robots, and this leads to better performance since the cleaning task can be successfully carried over to standby robots when the former robots become inoperative to perform the task.

And, the problem complexity is increased by adding the delivery task to the cleaning task. The delivery request becomes another factor for the reallocation. The delivery robot is selected to minimize the waiting time. The selected delivery robot follows the pick-up and drop-off task sequence. To balance the absent of the delivery robot and minimize the return time to cleaning task, the control system reallocates the cleaning zone to robots, and an exchange among robots of the assigned zone can occur to maintain the cleaning performance. The experiment with 9 robots in the laboratory shows the allocation of switching between two tasks was successfully conducted.

The future direction of this work will be developing the allocation strategy while increasing the cleaning performance and analyzing the effect of switching robots for frequent delivery request.

Acknowledgments This work was supported by the Robot R&D program of MOTIE/KEIT. [10051155, The Development of Robot Based Logistics Systems Applicable to Hospital-wide Environment]

References

1. Ackerman, E.: Avidbots wants to automate commercial cleaning with robots. IEEE Spectr. (2014)
2. Ahmadi, M., Stone, P.: A multi-robot system for continuous area sweeping tasks. In: Proceedings of the IEEE International Conference on Robotics and Automation (ICRA), pp. 1724–1729 (2006)
3. Bardram, J.E., Bossen, C.: Mobility work: the spatial dimension of collaboration at a hospital. Comput. Support. Coop. Work (CSCW) **14**(2), 131–160 (2005)
4. Choi, S., Lee, J.Y., Yu, W.: uRON v1.5: A device-independent and reconfigurable robot navigation library. In: Proceedings of the IEEE Workshop on Advanced Robotics and its Social Impacts (ARSO), pp.170–175 (2010)
5. Choi, J., Choi, M., Nam, S.Y., Chung, W.K.: Autonomous topological modeling of a home environment and topological localization using a sonar grid map. Auton. Robots **30**(4), 351–368 (2011)
6. D'Andrea, R.: Guest editorial: a revolution in the warehouse: a retrospective on kiva systems and the grand challenges ahead. IEEE Trans. Autom. Sci. Eng. **49**, 638–639 (2012)
7. Elmaliach, Y., Agmon, N., Kaminka, G.A.: Multi-robot area patrol under frequency constraints. Ann. Math. Artif. Intell. **57**(3–4), 293–320 (2009)
8. Hess, J., Beinhofer, M., Kuhner, D., Ruchti, P., Burgard, W.: Poisson-driven dirt maps for efficient robot cleaning. In: Proceedings of the IEEE International Conference on Robotics and Automation (ICRA), pp.2245–2250 (2013)
9. Jager, M., Nebel, B.: Dynamic decentralized area partitioning for cooperating cleaning robots: In: Proceedings of the IEEE International Conference on Robotics and Automation (ICRA), pp.3577-3582 (2002)

10. Janchiv, A., Batsaikhan, D., Kim, B.S., Lee, W.G., Lee, S.G.: Time-efficient and complete coverage path planning based on flow networks for multi-robots. Int. J. Control Autom. Syst. (IJCAS), **11**(2), 369–376 (2013)
11. Kolling, A., Carpin, S.: Multi-robot surveillance: an improved algorithm for the graph-clear problem. In: Proceedings of the IEEE International Conference on Robotics and Automation (ICRA), pp. 2360–2365 (2008)
12. Kong, C.S., Peng, N.A., Rekleitis, I.: Distributed coverage with multi-robot system. In: Proceedings of the IEEE International Conference on Robotics and Automation (ICRA), pp. 2423–2429 (2006)
13. Korsah, G.A., Stentz, A., Dias, M.B.: A comprehensive taxonomy for multi-robot task allocation. Int. J. Robot. Res. (IJRR) **32**(12), 1495–1512 (2013)
14. Koscielniak, T., et. al.: Key performance indicators for carriers & logistics service providers. Odette International Limited, p. 8 (2007)
15. Lee, J.H., Choi, J.S., Lee, B.H., Lee, K.W.: Complete coverage path planning for cleaning task using multiple robots. In: Proceedings of the IEEE International Conference on Systems, Man and Cybernetics (SMC), pp. 3618–3622 (2009)
16. Niechwiadowicz, K., Khan, Z.: Robot based logistics system for hospitals-survey. In: IDT Workshop on Interesting Results in Computer Science and Engineering (2008)
17. Rekleitis, I., Lee-Shue, V., New, A.P., Choset, H.: Limited communication, multi-robot team based coverage. In: Proceedings of the IEEE International Conference on Robotics and Automation (ICRA), pp. 3462–3468 (2004)
18. Santos, F.N., Moreira, A.P.: Towards extraction of topological maps from2D and 3D occupancy grids. Lecture Notes in Computer Science (LNCS), Prog. Artif. Intell. **8154**, 307–318 (2013)
19. Zahugi, E.M.H., Shanta, M.M., Prasad, T.V.: Design of multi-robot system for cleaning up marine oil spill. Int. J. Adv. Inf. Technol. (IJAIT), **2**(4), 33–43 (2012)

Optimal Tuning of Multivariable Centralized Fractional Order PID Controller Using Bat Optimization and Harmony Search Algorithms for Two Interacting Conical Tank Process

S.K. Lakshmanaprabu and U. Sabura Banu

Abstract The control of multivariable interacting process is difficult because of the interaction effect between input output variables. In the proposed work, an attempt is made to design a Multivariable Centralized Fractional Order PID (MCFOPID) controller with the use of evolutionary optimization techniques. The Bat Optimization Algorithm (BOA) and Harmony Search algorithm (HS), the evolutionary optimization techniques are used for the tuning of the controller parameters. As the process is a Two-Input-Two-Output process, four FOPID controllers are required for the control of the two interacting conical tank process. Altogether 20 controller parameters need to be tuned. In a single run, all the 20 parameters are founding considering the interaction effect minimizing Integral Time Absolute Error (ITAE). The BOA, HS based MCFOPID controller is validated under tracking and disturbance rejection for minimum ITAE.

1 Introduction

Interacting conical tanks finds application in many industrial plants like petrochemical plants, pharmaceutical industries, waste water treatment plants, food and beverages industries, metallurgical industries, etc. Their shape assists in the quick discharge of fluids. The occurrence and importance of multivariable control system has been increased in process industries. The process with interaction between inputs and outputs are complex to control. MIMO systems can be controlled either using multiloop controller or multivariable controller. Even though the multiloop decentralized PID controller are the commonly used industrial control, the steps involved in PID tuning is tedious [1–3].

S.K. Lakshmanaprabu (✉) · U. Sabura Banu (✉)
Department of Electronics and Instrumentation Engineering,
B.S. Abdur Rahman University, Vandalur, Chennai 600048, India
e-mail: prabusk.leo@gmail.com

U. Sabura Banu
e-mail: sabura.banu@bsauniv.ac.in

Y. Bi et al. (eds.), *Intelligent Systems and Applications*,
Studies in Computational Intelligence 650, DOI 10.1007/978-3-319-33386-1_11

The interaction of multiloop PID controller can be eliminated by addition feed forward element called decoupler. The design of decoupler requires very accurate model. In real time, decoupler fails to reject the interaction completely, there will be a leakage effect in the controller loop [4]. In such a case multivariable controller is necessary, but multivariable control scheme is not easy as single PID controller for implementation and understand. This drawback of decoupler and multivariable control implementation is over come by centralized controllers to reject the interaction disturbance.

The simplified centralized PI controller scheme is designed based on the inversion matrix control principle. The centralized controller parameter is obtained from the steady state of gain matrix [5]. The control scheme of centralized PI controller schemes is changed to fractional order PID controller. The performance of PI/PID controller performance is increased in fractional order PID controller by two additional parameters.

Fractional-Order Proportional-Integral-Derivative (FOPID) controllers become popular in the last two decades and it is applied in wide field of both industrial and academic point of view due to rediscovering of fractional calculus by scientists and engineers [6]. This technique is applied in the area of control theory but mostly, 90 % of industrial application like process control, flight control, motor drivers, etc., still uses PID controllers for easy implementation with simple structure and robust performance. The flexibility of PID controller can be increased by additional two additional parameters. It has also been easily tuned for wide range of operating conditions in the process plants. In general, fractional order controllers are described by differential equations which deal with derivatives and integrals from non-integer orders. The controllers based on fractional order derivatives and integrals have been used in industrial applications in various fields like system identifications, power electronics, robotic controls and heat diffusion systems. The use of FOPID controller in industries is limited rather than the PID controller because the performance improvement of FOPID controller has not been fully analyzed especially the set point tracking, load disturbance rejection, control effect and noise rejection. As well as there is no other effective and robust tuning rules available yet for tuning FOPID controller. The design of FOPID controller is not as easy as standard integer-order PID controllers [7, 8]. Designing of FOPID controller is more complex task because of the selection of five controller parameters (K_p, K_d, K_i, λ, μ).

Tuning of five different parameters of FOPID controller is difficult and challenging task. Many tuning method for fractional order PID controller has been proposed [9–12]. Recently various Meta-heuristic algorithm is used to solve complex real time problem in science, Engineering, economics etc. [13–15]. Meta-heuristic algorithm has been inspired by nature. The biological evolutionary process and animal behavior are imitated and combined with randomness to form meta-heuristic algorithm. Many successful algorithms imitated from biological evolutionary process and animal behavior. The Famous Particle swarm optimization and Ant colony optimization techniques are inspired from species [16–18]. Similarly Genetic algorithm (GA), differential evolution (DE) are developed from biological evolutionary process [19, 20]. Several engineering problem will have many local optimum and one global optimum solution. In such case, hybrid optimization developed algorithm with other meta heuristic algorithm are applied. Simulated annealing algorithm will not get trapped in local optima [21]. This Simulated annealing used with genetic algorithm reduces the execution time. The Harmony search optimization algorithm is stochastic algorithm which is widely used in real time optimization problems [22]. The HS algorithm is successfully used in the field of pipe network design, structural engineering, groundwater management model, multiple dam system scheduling, energy dispatch problems, clustering, visual tracking problems, sound power level optimization, vehicle routing problems, robotics, power and energy problem, medical, and also in other problems in scientific and engineering problem. The HS algorithm incorporates many features of Genetic algorithm (GA). Many meta-heuristic algorithms are sensitive to initial values and it requires large amount of record memory and uses complex derivatives which makes computation complex. The HS algorithm is derivative free and it uses gradient search. This algorithm can be easily implemented for various engineering optimization. The main reason of easy adaptation is it does not require initial values of parameter to be optimized [23, 24]. The Bat optimization algorithm also widely used in many engineering nonlinear, multimodal problems. The BOA is inspired from the echo location behavior of bats which is easy to implement and flexible to solve many complex problems [25–27]. This BOA uses the advantages of PSO, HS algorithms. The BOA become PSO when the loudness $A_i = 0$ and pluse rate (r_i) fixed as 1. The bat properties of varying pulse emission rate and varying loudness are control the exploration and exploitation enhance the BOA searching. The main objective of this paper is to tune the controller gain and order of fractional order controller using global best optimization algorithm like and BOA and HS algorithm.

Section 2 discusses basic concept of fractional calculus, the conical tank process description is given in Sect. 3, Multi-loop fractional order PID controller tuning detailed in Sect. 4 and the Harmony search optimization, bat optimization algorithm techniques elaborated in Sects. 5, 6 and Result analysis and conclusion given in Sects. 7, 8.

2 Basic of Fractional Calculus

All the real systems are generally fractional in nature [28]. It is proven that fractional order model is more accurate and adequate than the integer order model.

The application of fractional calculus has begun to gain importance in control theory. In recent time, the fractional integrals and derivatives are used in the modeling and control system design. The differ-integral operator, ${}_aD_t^q$, is a combined differentiation-integration operator commonly used in fractional calculus. This operator is a notation for taking both the fractional derivative and the fractional integral in a single expression and is defined by

$$ {}_aD_t^q = \begin{cases} \frac{d^q}{dt^q} & q > 0 \\ 1 & q = 0 \\ \int_a^t (d\tau)^{-q} & q < 0 \end{cases} \tag{1} $$

where q is the fractional order which can be a complex number and a and t are the limits of the operation. There are some definitions for fractional derivatives. The commonly used definitions are Grunwald–Letnikov, Riemann–Liouville and Caputo definitions [7]. The Grunwald–Letnikov definition is given by

$$ {}_aD_t^q = \frac{d^q f(t)}{d(t-a)^q} = \lim_{N->\inf} \left[\frac{t-a}{N}\right]^{-q} \sum_{j=0}^{N-1} (-1)^j \left(\frac{\Gamma(q+1)}{j!\Gamma(q-j+1)}\right) f\left(t - j\left[\frac{t-a}{N}\right]\right) \tag{2} $$

The Riemann–Liouville definition is the simplest and easiest definition to use. This definition is given by

$$ {}_aD_t^q f(t) = \frac{d^q f(t)}{d(t-a)^q} = \frac{1}{\Gamma(n-q)} \frac{d^n}{dt^n} \int_0^t (t-\tau)^{n-q-1} f(\tau) d\tau \tag{3} $$

where n is the first integer which is not less than q i.e. $n - 1 \leq q < n$ and Γ is the Gamma function.

$$ \Gamma(z) = \int_0^{\inf} t^{z-1} e^{-t} dt \tag{4} $$

For functions f(t) having n continuous derivatives for $t \geq 0$ where $n - 1 \leq q < n$, the Grunwald–Letnikov and the Riemann–Liouville definitions are equivalent. The Laplace transforms of the Riemann–Liouville fractional integral and derivative are given as follows:

$$ L\{{}_aD_t^q f(t)\} = s^q F(s) - \sum_{k=0}^{n-1} s^k {}_0D_t^{q-k-1} f(0); \qquad n - 1 < q < n \tag{5} $$

The Riemann–Liouville fractional derivative appears unsuitable to be treated by the Laplace transform technique because it requires the knowledge of the non-integer order derivatives of the function at $t = 0$. This problem does not exist in the Caputo definition that is sometimes referred as smooth fractional derivative in literature. This definition of derivative is defined by

$$ {}_aD_t^q f(t) = \begin{cases} \frac{1}{\Gamma(m-q)} \int\limits_0^t \frac{f^{(m)}(\tau)}{(t-\tau)^{q+1-m}} d\tau \; ; & m-1 < q < m \\ \frac{d^m}{dt^m} f(t) \; ; & q = m \end{cases} \quad (6) $$

where 'm' is the first integer larger than q. It is found that the equations with Riemann–Liouville operators are equivalent to those with Caputo operators by homogeneous initial conditions assumption. The Laplace transform of the Caputo fractional derivative is

$$ L\left\{ {}_aD_t^q f(t) \right\} = s^q F(s) - \sum^{n-1} S^{q-k-1} f^{(k)}(0) \; ; n-1 < q < n \quad (7) $$

Contrary to the Laplace transform of the Riemann–Liouville fractional derivative, only integer order derivatives of function f are appeared in the Laplace transform of the Caputo fractional derivative. For zero initial conditions, previous equation reduces to

$$ L\{ {}_0D_t^q f(t) \} = s^q F(s) \quad (8) $$

The numerical simulation of a fractional differential equation is not simple as that of an ordinary differential equation. Since fractional order differential equations do not have exact analytic solutions, approximations and numerical techniques are used. The approximation method, Oustaloup filter is given by

$$ s^q = k \prod_{n=1}^{N} \frac{1 + \frac{s}{\omega_{zn}}}{1 + \frac{s}{\omega_{pn}}} \quad ; \quad q > 0 \quad (9) $$

The approximation is valid in the frequency range $[\omega_l, \omega_h]$; gain k is adjusted so that the approximation shall have unit gain at 1 rad/sec; the number of poles and zeros N is chosen beforehand (low values resulting in simpler approximations but also causing the appearance of a ripple in both gain and phase behaviours); frequencies of poles and zeros are given by

$$ \alpha = \left(\frac{\omega_h}{\omega_l} \right)^{\frac{q}{N}} \quad (10) $$

$$ \eta = \left(\frac{\omega_h}{\omega_l} \right)^{\frac{1-q}{N}} \quad (11) $$

where,

$$ \omega_{zn} = \omega_{p,n-1}\eta, \quad n = 2, \ldots, N $$
$$ \omega_{pn} = \omega_{z,n-1}\alpha, \quad n = 1, \ldots, N. $$

3 Two Interacting Conical Tank Process

The proposed system consists of two conical tanks which are in the shape of an inverted cone fabricated from a sheet metal. The height of the process tank is 50 cm. The top end diameter is 40 cm. The two tanks are connected through an interacting pipe with valve (HV_1). The interaction of process can be changed by position of this valve (HV_1). It has a reservoir to store water and this is supplied through the pumps to the tanks. Provisions for water inflow and outflow are provided at the top and bottom of the tank respectively. Gate valves, one at the outflow of the tank 1 and the other at the outflow of the tank 2 are connected to maintain the level of water in the tanks. Variable Speed pump work as actuator and it is used to discharge the water from reservoir tank to process tanks. The schematic diagram of two interacting conical tank process is shown in Fig. 1.

The speed of pump is directly proportional to the input voltage. It consists of differential pressure transmitter for measuring the bottom pressure created by water level and it gives height in terms of milliamps. The nominal values of the parameters and variables are tabulated in Table 1.

3.1 *Mathematical Modeling*

The mathematical model of the process has been derived from the mass balance equation.

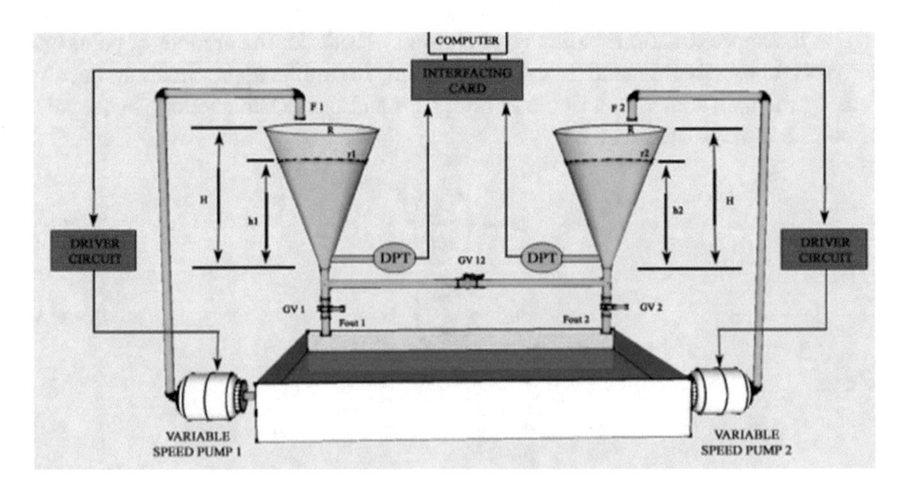

Fig. 1 Schematic diagram of two interacting conical tank process

Table 1 Nominal values of parameters used

Parameter	Description	Value
R,r	Top radius of conical tank	20 cm
H	Maximum height of Tank1, Tank2	50 cm
K_{pp1}, K_{pp2}	Pump gain	40 cm^3/V.sec
u_1, u_2	Maximum voltage applied to pumps	0–5 V
β_1	Valve co-efficient of MV1	0.30
β_{12}	Valve co-efficient of Mv12	0.9
β_2	Valve co-efficient of Mv2	0.35
a_1, a_{12}, a_2	Cross section area of pipe	1.227 cm^2

Mathematical model of the two interacting conical tank process is given by

$$\frac{dh_1}{dt} = \frac{K_{pp1}u_1 - \beta_1\, a_1\sqrt{2gh_1} - sign(h_1 - h_2)\,\beta_{12}\, a_{12}\sqrt{2g\,|h_1 - h_2|}}{\pi R_1^2 \frac{h_1^2}{H_1}} \tag{12}$$

$$\frac{dh_2}{dt} = \frac{K_{pp2}u_2 - \beta_2\, a_2\sqrt{2gh_2} + sign(h_1 - h_2)\,\beta_{12}\, a_{12}\sqrt{2g\,|h_1 - h_2|}}{\pi R_2^2 \frac{h_2^2}{H_2}} \tag{13}$$

The nominal values of the parameters and variables are tabulated in Table 1.

The open loop data was generated in the conical tank system by varying the applied voltage u_1 and u_2. The input output characteristics of system revealed the nonlinearity between input–output variables. Hence, the piecewise Linearization method is used and the operating points are found to develop a linearised model. The linearised model around operating points ($u_{1s} = 2.75$, $u_{1s} = 2.5$) is extracted.

$$G_p(s) = \begin{bmatrix} \dfrac{0.142s + 0.00619}{s^2 + 0.1367s + 0.00126} & \dfrac{0.003494}{s^2 + 0.135s + 0.00147} \\ \dfrac{0.003041}{s^2 + 0.137s + 0.0013} & \dfrac{0.126s + 0.00591}{s^2 + 0.1382s + 0.00128} \end{bmatrix}$$

4 Centralized FOPID Controller Design

In industrial control problems, multivariable system is decomposed and multi-loop controller is designed by considering most important loops. The control loop interaction is reduced by decoupler in multiloop controllers but decentralized controller fails to give reasonable response for the process with high interaction. A centralized controller is capable of eliminating the interaction effect by considering all the feedback output to compute each manipulated input. The controller output is depending on all the process output. The conventional centralized PI controller is designed by Davison's method which uses the steady state information of the system. The centralized controller scheme for multivariable (2 × 2 system) shown in Fig. 2.

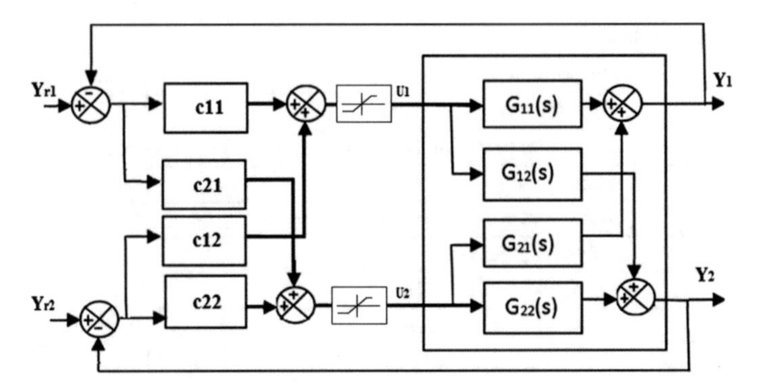

Fig. 2 Multivariable centralized FOPID control scheme

This scheme is adopted and new proposed fractional order PID controller is implemented [29].

The multivariable system is assumed to be of the form

$$Y(s) = G_p(s)U(s) \tag{14}$$

where,

Y is the output variables

U is the manipulated variables

$G_p(s)$ is the transfer function matrix of the system

The centralized controllers are given as

$$G_c(s) = K_c + K_I/s \tag{15}$$

where K_c and K_I are matrices of size nxn

$$K_c = [k_{c,ij}] \tag{16}$$

$$K_I = [k_{I,ij}/\tau_{ij}] \tag{17}$$

Here, $K_{c,ij}$ is the proportional gain matrix (K_c) & τ_{ij} is the integral time of the controllers.

4.1 Davison Method

Davison has proposed an empirical method of tuning a multivariable PI control system where it makes use of only the steady state gain matrix of the system for the design of centralized PI controllers [28]. The steady state gain matrix of the multivariable stable system can be obtained from transfer function matrix. The proportional gain K_C and integral gain K_I of centralized PI controller is given in Eqs. (18) and (19).

$$K_c = \delta\left[G_p(s=0)\right]^{-1} \tag{18}$$

$$K_I = \varepsilon\left[G_p(s=0)\right]^{-1} \tag{19}$$

Here $\left[G_p(s=0)\right]^{-1}$ is called the rough tuning parameters. The inverse of steady-state gain matrix is called as the rough tuning matrix.

4.2 Centralized FOPID Controller

The process with two input two output represented by

$$G(s) = \begin{bmatrix} g_{11}(s) & g_{12}(s) \\ g_{21}(s) & g_{22}(s) \end{bmatrix} \tag{20}$$

The multivariable controller scheme is given below

The Centralized Controller is,

$$G_c(s) = \begin{bmatrix} g_{c11}(s) & g_{c12}(s) \\ g_{c21}(s) & , g_{c22}(s) \end{bmatrix} \tag{21}$$

where,

$$g_{c11} = Kp_{11}\left(1 + \frac{1}{T_{11}S^{\lambda_{11}}} + T_{d11}S^{\mu_{11}}\right);\, g_{c12} = Kp_{12}\left(1 + \frac{1}{T_{i12}S^{\lambda_{12}}} + T_{d12}S^{\mu_{12}}\right)$$

$$g_{c21} = Kp_{21}\left(1 + \frac{1}{T_{i21}S^{\lambda_{21}}} + T_{d21}S^{\mu_{21}}\right);\, g_{c22} = Kp_{22}\left(1 + \frac{1}{T_{22}S^{\lambda_{22}}} + T_{d22}S^{\mu_{22}}\right)$$

Gc(s) is the control law of Fractional order PID controller where Kp_{ij} is the proportional gain, Ti_{ij} is integral time constant or reset time (mins/repeat), Td_{ij} derivative time, λ_{ij} integrator order and μ_{ij} derivative order.

The control law of Fractional order PID rewritten as

$$K_{ij}(s) = kp_{ij} + Ki_{ij}\cdot\frac{1}{s^{\lambda_{ij}}} + kd_{ij}\cdot s^{\mu_{ij}}\; i, j = \{1, 2, \ldots., n\} \tag{22}$$

where integral gain $ki_{ij} = kp_{ij}/Ti_{ij}$ and Derivative gain $kd_{ij} = kp_{ij}\cdot Td_{ij}$.

PID is most commonly used controller in many process industries. In past decades tuning of integer order PID controllers tuned by different methodologies such as auto tuning, self tuning and computational intelligence.

In the proposed method, Centralized Fractional order PID controller for MIMO interaction process is proposed. For the controller tuning purpose, interaction effect is considered as disturbances and the Fractional order PID control tuned optimally by harmony search Algorithm. The multi objective optimization function of harmony search algorithm is weighted percentage of ITAE of the two loops. The controller parameter range is selected from the conventional centralized PID controller design method.

5 Harmony Search Algorithm

Harmony search algorithm (HS) is a music-based metaheuristic algorithm is developed by zong woo Geem et al. in 2001. This HS algorithm is inspired by the jazz musician's improvisation process. The Hs algorithm first idealized by the following steps of musicians. When musicians composing a music, he or she follow the below steps, initially he play any famous music and then he try to improvise by adjusting pitch of musical instrument. The aim of musicians is to compose the pleasing harmony as determined by aesthetic standard; this process of achieving perfect state of harmony can be improved by adjusting pitch ranges. The musicians are like decision variables and they change musical instrument's pitch range and musical harmonics are improvised to make very good pleasant harmony to impress audience (objective function). At each try musician store the harmony in mind and improvise by tuning pitch, this refers to the iteration of optimization. The best composed pleasing harmony for optimal pitch ranges are considered as best pleasing harmony (final global objective value).

The harmony memory is related to best fit individuals in GA, the best harmonies (solution vector) are stored in harmony memory and usage of harmonies in harmony memory is defined by the harmony memory accepting rate *raccept* □ [0, 1].

The pitch adjusted can be made by pitch bandwidth b_{range} and a pitch adjusting rate rpa. The pitch solution x_{new} updated and stored in harmony memory after the adjusting action.

$$x_{new} = x_{old} + b_{range} * \beta \tag{23}$$

β is a random number generator in the range of $[-1, 1]$ and usually pitch adjusting rate *rpa* is used in the range $0.1 \sim 0.5$.

Randomization is also like pitch adjusting but it used for the global search of system to bring new global solution.

The probability of randomization is,

$$P_{random} = 1 - r_{accept} \tag{24}$$

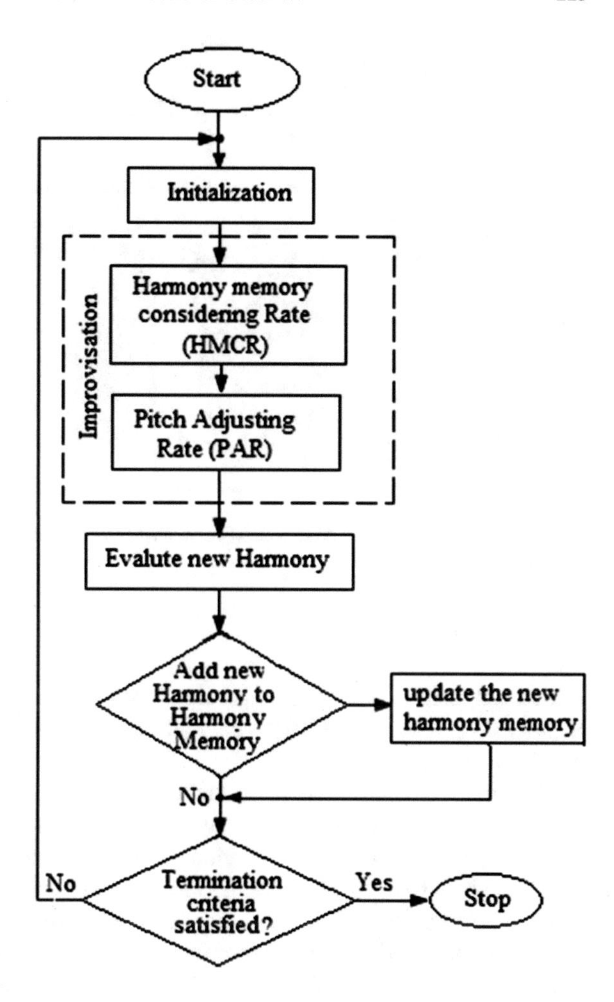

Fig. 3 Flow chart of HS algorithm (geem 2000)

The actual probability of pitch adjustment is

$$P_{pitch} = r_{accept} \times r_{pa} \tag{25}$$

The flow chart of harmony search is shown in Fig. 3.

The MCFOPID tuning parameter is obtained using harmony search algorithm technique which gives a better closed loop performance. The proposed BOA based centralised controller for TICT is shown in Fig. 4.

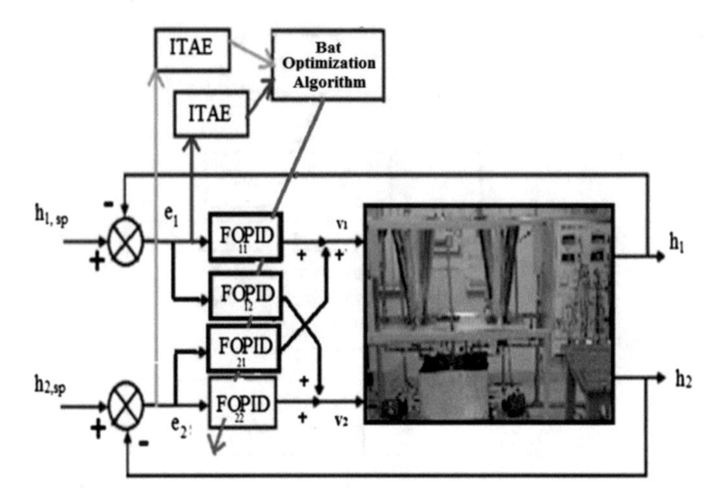

Fig. 4 Block diagram of the centralized Fractional Order PID controller tuned using BOA for TICT process

6 Bat Algorithm

The Bat algorithm is a Meta heuristic algorithm developed by xin-she yang. The Echolocation behavior of Micro bats are inspired and developed as a global optimization algorithm.

The micro bats have some special natural behavior to find the prey even in dark and the obstacle environment. The micro bat emits sound pulses with different frequencies and it receives back the echoes to identify the exact location of prey.

All the bats have some different kind of characteristics, hence some assumption made to develop Bat search algorithm. The first assumption is that all the bats have echolocation behavior and it has ability to find the distance between prey and its position. The second assumption is that, the bats fly randomly with velocity Vi at position Xi with the frequency f and loudness A0. The loudness varies based on the nearness of the prey. The loudness of bat decreases when the bat moves towards the prey and then the loudness rate is zero when it reaches the prey.

Procedure for Bat Algorithm

Step 1: Bat population Xi and velocity Vi (i = 1,2, …,n) is initiated for the controller. The pulse frequency range (fmin, fmax) and pulse rate ri and the loudness Ai are defined.
Step 2: The multi-objective optimal function are combined as a single objective function by providing 50 % weightage given to the ITAE of the first loop and 50 % weightage given to the ITAE of the second loop.
Step 3: New solutions are generated by adjusting frequency, updating velocities and loudness. These new location/solution Xi and velocity Vi at ‘t’ time is calculated by,

$$f_i = f_{min} + (f_{max} - f_{min})\,\delta \tag{26}$$

$$v_i^t = v_i^{t\text{-}1} + (x_i^t - x_{best}^t) f_i \tag{27}$$

The new solution,

$$x_i^t = x_i^{t\text{-}1} + v_i^t \tag{28}$$

where, the δ is a random vector from uniform distribution which is in the range of 0–1. x_{best}^t is the current global best location. The velocity increment of each bat based on the frequency (f_i) of each bat.
Step 4: A new solution is generated by flying randomly. Once the best solution is found, then a new solution is generated by local search.

$$x_{new} = x_{old} + \varepsilon\, L^t \tag{29}$$

where $^{\varepsilon}$ is the random number in the range of −1 to 1 and Lt is an average loudness of the entire bat at time't'.

If rand >ri; A solution is selected among the best solution; A local solution is generated around the selected best solution
Step 6: If rand <Ai and f(xi) < f(x+); The new solutions are accepted. When the bats moves towards best location/solution then the pulse emission rate ri increases and loudness Ai decreases. The loudness ranges is [Amin, A0] = [0, 1]. If the loudness reaches the value 0, it represents that the micro bats reaches the prey.

$$L_i^{t+1} = \mu L_i^t\,, \qquad r_i^{t+1} = r_i^0\left[1 - \exp(-\gamma t)\right] \tag{30}$$

where μ is a constant in the range of [0, 1] and γ is a positive constant. The value of μ, γ are fixed same to make optimization simple. So, the values are fixed as $\mu = \gamma = 0.9$.
Step 7: The bats are ranked at each iteration and the current best position x+, minimum objective function f(x+) are found. This is proceeded till the maximum number of iteration is reached or the sopping criteria met. The final iteration gives minimum function value and best optimized parameter.

The BOA is a minimization algorithm where it finds the optimal position to get the prey.

The integral performance indices like integral square error (ISE), integral absolute error (IAE), integral time absolute error (ITAE) are generally used cost function in controller design. The ISE penalizes small errors which leads fast response with small oscillation. The IAE yields a low overshoot, less oscillation but it gives slower response. The IAE, ISE are weight all the error uniformly which result in small overshoot and long settling time. Therefore time weighted IAE is used, it penalize more weight on the larger time error which lead to get faster settling time.

The multi objective function J is defined as,

$$J = \int_{0}^{\infty} \left[w_1.t \left| h_{1sp}(t) - h_1(t) \right| + w_2.t \left| h_{2sp}(t) - h_2(t) \right| \right] dt$$

$$J = \int_{0}^{\infty} [w_1.t\,|e_1(t)| + w_2.t\,|e_2(t)|]dt \tag{31}$$

Minimize J, subject to

$$K_{pij}^{min} \le K_{pij} \le K_{pij}^{max};\ k_i^{min} \le K_i \le K_i^{max};\ k_{d1}^{min} \le K_{d1} \le K_{d1}^{max};$$

$$\overset{min}{\lambda} \le \lambda \le \overset{max}{\lambda};\ \mu^{min} \le \mu \le \mu^{max};$$

where h_{1sp}, h_{2sp} are the setpoints of controller, w_1, w_2 are the weightage of sub objective functions.

7 Simulation Results and Discussion

In this section, the MCFOPID controller is applied on TICT process. The system was simulated using Matlab software and outaloup recursive approximation used in fractional order parts. The frequency range is fixed as $[10^{-4}, 10^{4}]$ and 5th order approximation is used [30].

The MCFOPID controller is designed for linearised process is Gp(s). This two input two output process have strong interaction between input and output pairs. The Inverse gain array of linearised process is calculated for find the initial guess of Kp, Ki controller values and Kd, integrator order λ derivative order μ range roughly fixed for optimization.

$$[G_p(0)]^{-1} = \begin{bmatrix} 0.2686 & -0.136 \\ -0.1346 & 0.2824 \end{bmatrix} \tag{32}$$

The optimization problem simplified by using Davison method, the controller parameter kp and Ki tuned using rough tuning parameter which is calculated by using Eqs. (18) and (19). In this study, complex centralized problem is simplified and control scheme is given below,

$$G_c(s) = \begin{bmatrix} 0.340*\delta + \frac{0.340*\varepsilon}{s^{\lambda}} + k_d s^{\mu} & 0.340*\delta + \frac{0.340*\varepsilon}{s^{\lambda}} + k_d s^{\mu} \\ 0.340*\delta + \frac{0.340*\varepsilon}{s^{\lambda}} + k_d s^{\mu} & 0.340*\delta + \frac{0.340*\varepsilon}{s^{\lambda}} + k_d s^{\mu} \end{bmatrix}$$

The tuning parameters δ, ε are generally in the range of 0 to 2 and k_d, λ, μ are considered same values for all FOPID controllers. Now, The objective function of system is, minimize J with subject to δ, ε, k_d, λ, μ.

The harmony algorithm based MCFOPID controller tuned using HS, the initial optimization parameters such as pitch bandwidth, pitch adjustment rate rpa, HMCR, PAR were fixed as 0.01, 0.2, 0.9, 0.3 respectively and optimized parameters bounds are fixed. The HMS is 30 and maximum no of iteration is 200. The δ, ε controller parameters ranges are selected from 0.1 to 2 and other parameters range is fixed, kd is 0–10; λ_{ij} integrator order bound is 0.1–1.5, μ_{ij} derivative order range is 0.01–1.5. The step response of closed loop system is used to find the optimal values of FOPID parameters. The equal weightage is given to sub objective function i.e., $w_1 = w_2 = 0.5$. After performing Harmony search, the FOPID controller parameters were found as follow,

$$G_{c-HS}(s) = \begin{bmatrix} 0.408 + \frac{0.153}{s^{0.97}} + 0.2s^{0.92} & -0.2508 - \frac{0.094}{s^{0.97}} + 0.2s^{0.92} \\ -0.2647 - \frac{0.993}{s^{0.97}} + 0.2s^{0.92} & 0.4734 + \frac{0.1775}{s^{0.97}} + 0.2s^{0.92} \end{bmatrix}$$

The BOA based MCFOPID controller also is designed using same procedure. The data for BOA algorithm is initialized and the lower and upper bounds of controller parameter is fixed. The population size if fixed as 100 and maximum iteration time is selected as 200. The BOA algorithm utilized to design the 5 MCFOPID controller parameters. The objective function weightage w_1, w_2 are fixed as 0.5/ The MCFOPID controller parameters evaluated by BOA is given below,

$$G_c(s) = \begin{bmatrix} 0.306 + \frac{0.068}{s^{0.9}} + 0.2s^{0.45} & -0.1881 - \frac{0.0418}{s^{0.9}} + 0.2s^{0.45} \\ -0.1985 - \frac{0.0441}{s^{0.9}} + 0.2s^{0.45} & -0.355 - \frac{0.0789}{s^{0.9}} + 0.2s^{0.45} \end{bmatrix}$$

7.1 Servo Response

Figures 5, 6 shows the servo responses of proposed controllers. Variable set point is given at an interval of 500 s and the controller has complete control over the process and the controller tracks the set point. The response settling time of BOA-MCFOPID is lesser than the HS-MCFOPID.

7.2 Regulatory Response

Figures 7, 8 shows the regulatory response of the proposed control scheme for two interacting conical tank process. Disturbance is given to the process at an interval of 500 s. By adjusting the manipulated variable, the process output is maintained constant irrespective of the disturbance applied within the range of ±10 %.

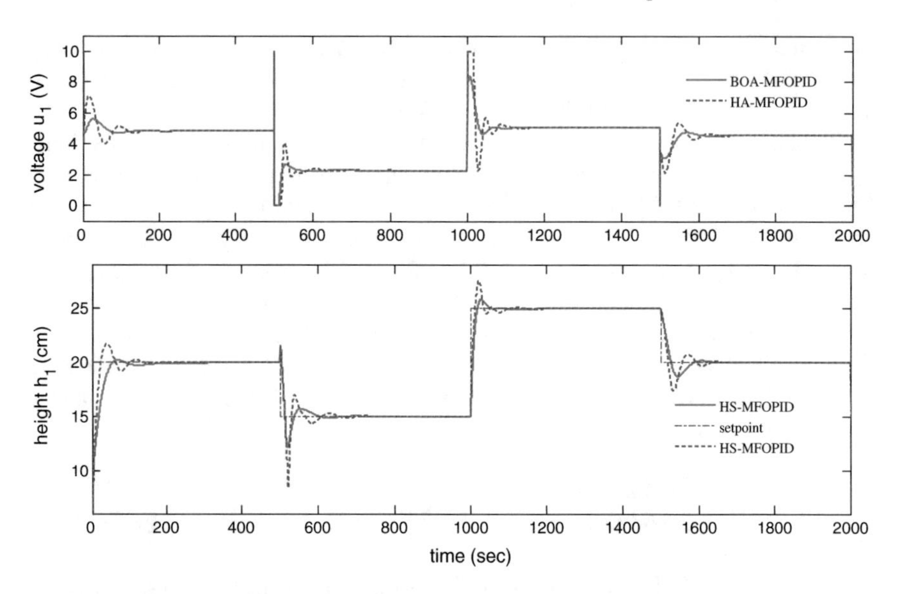

Fig. 5 Closed loop servo response of MCFOPID controller for TICT process (tank1)

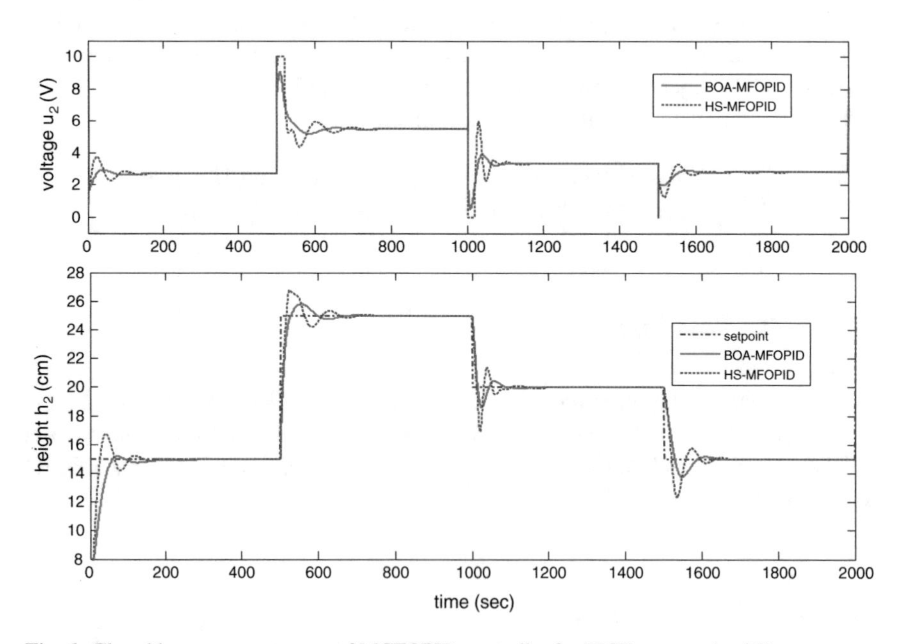

Fig. 6 Closed loop servo response of MCFOPID controller for TICT process (tank2)

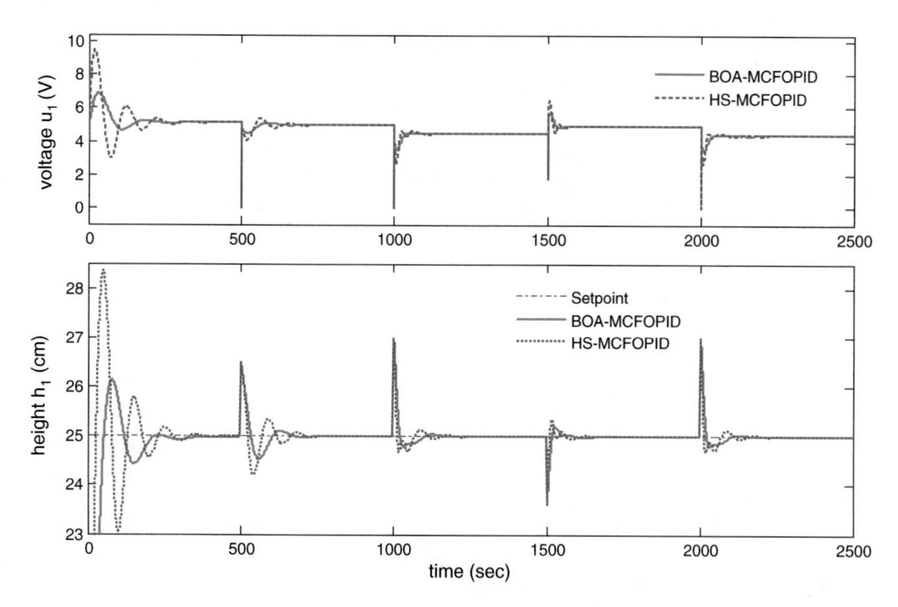

Fig. 7 Closed loop Regulatory Response of MCFOPID controller for TICT process (tank1)

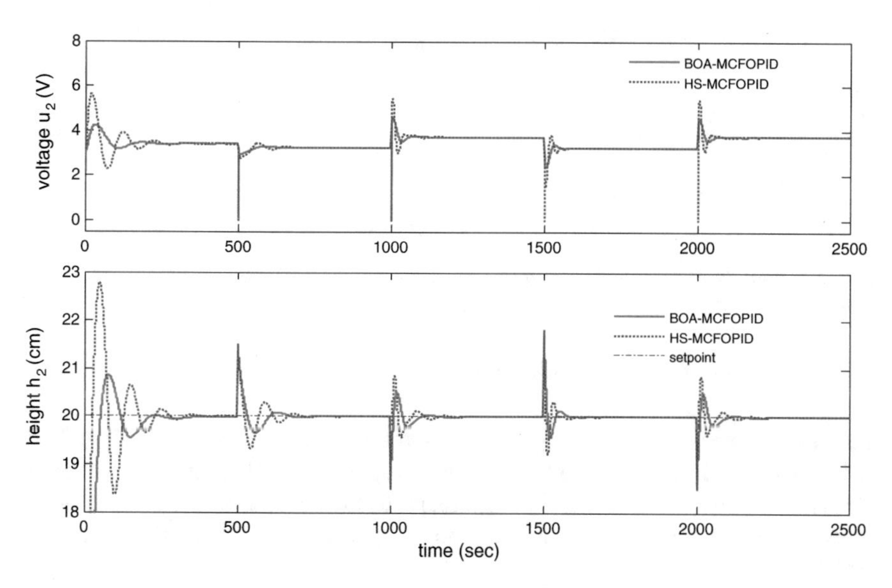

Fig. 8 Closed loop Regulatory Response of MCFOPID controller for TICT process

Table 2 Comparison of performance indices J under servo and servo-Regulatory problems

	HS-FOPID	BOA-MFOPID
	ITAE	ITAE
Servo	18920	17638
Regulatory	9821	8963
Servo Regulatory	27322	25831

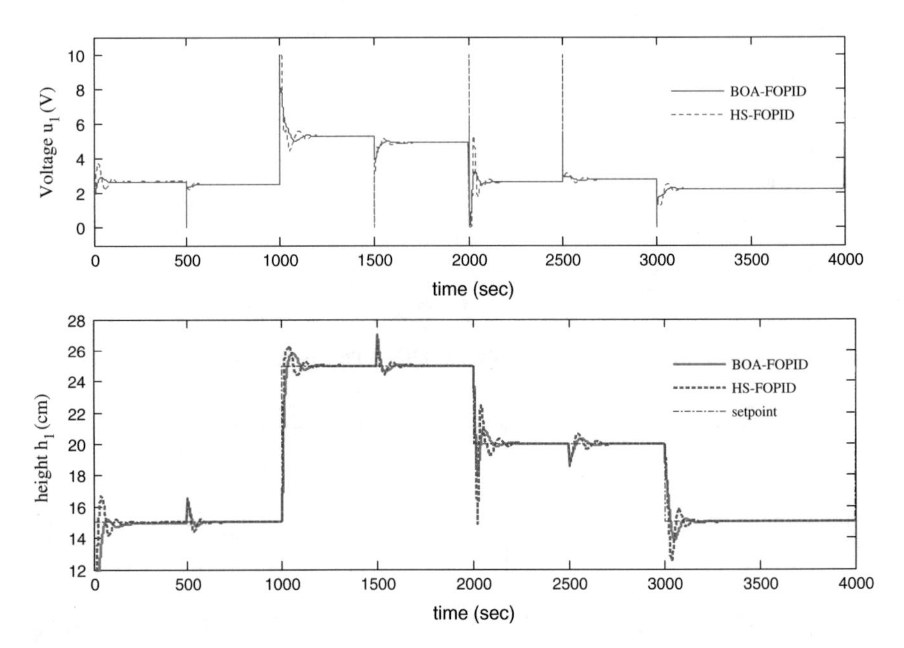

Fig. 9 Closed loop Servo and regulatory Response of MCFOPID controller for TICT process (tank1)

7.3 Servo Regulatory Response

Figures 9, 10 shows the servo regulatory response of the proposed controllers. Variable step inputs are applied and a disturbance is applied after the response reaches the steady state value. Even though both the set point and disturbance are applied simultaneously, the proposed controller scheme is capable of providing efficient control action.

The closed loop responses are shown in Figs. 5, 6, 7, 8, 9 and 10. All the responses showed that BOA-MFOPID provides faster and smoother response than the HS-MFOPID control scheme. The effect of external disturbance and interaction disturbance is rejection is better for BOA-MFOPID. The integral time absolute error is lesser for bat algorithm based FOPID controllers which an optimal global minimum. The performance of controllers are compared and tabulated in Table 2.

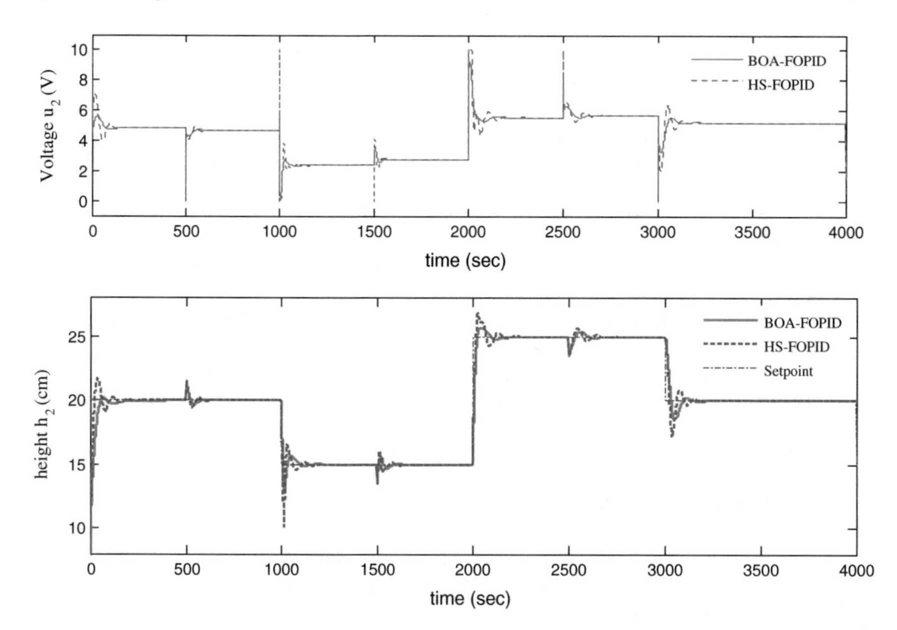

Fig. 10 Closed loop Servo regulatory Response of MCFOPID controller for TICT process (tank1)

8 Conclusion

In the proposed work, the conventional PID controller flexibility is increased by two more parameters and interaction effects between multi-loops are rejected by diagonal FOPID controller. The centralized multivariable fractional order PID controller is designed and optimally tuned using bat optimization algorithm and harmony search algorithm for two interacting conical tank process. The BOA, HS based MCFOPID controller is compared and it found that BOA based centralized FOPID controller provide better response with less ITAE and settling time. The proposed controller is validated for servo, regulatory and servo-regulatory problems and the result shows that the proposed BOA-MCFOPID scheme will result in a simple and optimal controller design of the multivariable interaction system.

References

1. Astrom, K.J. Johansson, K.H., Wang, Q-G.: Design of decoupled PI controllers for two-by-teo system, proceedings of IEEE Control Theory and Applications, **74**(1), 149 (2002) (special section on PID control)
2. Nordfeldt, P., Hagglund, T.: Decoupler and PID controller design of TITO systems. J. Process Control **16**, 923–93 (2006)
3. Nordfeldt, P., Hägglund, T.: Decoupler and PID controller design of TITO systems. J. Process Control **16**(9), 923–936 (2006)

4. Wang, Q.G., Huang, B., Guo, X.: Auto-tuning of TITO decoupling controllers from step tests. ISA Trans. **39**, 407–418 (2000)
5. Vijay Kumar, V., Rao, V.S.R., Chidambaram, M.: Centralized PI controller for interacting multivariable process by synthesis method. ISA Trans. **51**, 400–409 (2012)
6. Manabe, S.: The non-integer integral and its application to control systems. ETJ of Jpn **6**(3/4), 83–87 (1961)
7. Podlubny, I.: Fractional-order systems and controller. IEEE Trans. Autom. Control **44**(1), 208–214 (1999)
8. Padula, F., Visioli, A.: Tuning rules for optimal PID and fractional-order PID controllers. J. Process Control **21**, 69–81 (2011)
9. Valerio, D., Costa, J.: Tuning of fractional PID controllers with Ziegler–Nichols-type rules. Signal Process. **86**(10), 2771–2784 (2006)
10. Chen, Y.: A fractional order PID tuning algorithm for a class of fractional order plants, In: Proceedings of IEEE International Conference on Mechatronics & Automation (2005)
11. Monje, C.A., Vinagre, B.M., Feliu, V., Chen, Y.: Tuning and auto-tuning of fractional order controllers for industry applications. Control Eng. Practice **16**(7), 798–812 (2008)
12. Padula, F., Visioli, A.: Tuning rules for optimal PID and fractional order PID controller. J. Process Control **21**, 69–81 (2011)
13. Bianchi, L., Dorigo, M., Gambardella, L.M., Gutjahr, W.J.: A survey on metaheuristics for stochastic combinatorial optimization. Nat. Comput.: an Int. J. **8**(2), 239–287 (2009)
14. Yang, X.S.: Engineering Optimization: An Introduction with Metaheuristic Applications. Wiley, New York (2010)
15. Koziel, S., Yang, X.S.: Computational Optimization and Applications in Engineering and Industry. Springer, Germany (2011)
16. Zamani, M., Karimi-Ghartemani, M., Sadati, N., Parniani, M.: Design of a fractional order PID controller for an AVR using particle swarm optimization. Control Eng. Practice **17**(12), 1380–1387 (2009)
17. Binguland, Z., Karahan, O.: Tuning of fractional PID controllers using PSO algorithm for robot trajectory control. In: Proceedings of the IEEE International Conference on Mechatronics, Istanbul, Turkey (2011)
18. Zamani, M., Karimi-Ghartemani, M., Sadati, N., Parniani, M.: Design of fractional order PID controller for an AVR using particle swarm optimization. Control Eng. Practice **17**, 1380–1387 (2009)
19. Biswas, A., Das, S., Abraham, A.: SambartaDasgupta.: design of fractional order PIλDμ controller with improved differential evolution. Eng. Appl. Artif. Intell. **22**, 343–350 (2009)
20. Biswas, A., Das, S., Abraham, A., Dasgupta, S.: Design of fractional-order controllers with an improved differential evolution. Eng. Appl. Artif. Intell. **22**, 343–350 (2009)
21. Kirkpatrick, S., Gelatt, C.D., Vecchi., M.P.: Optimization by simulated annealing. Science **220**, 671–680 (1983)
22. Geem, Z.W., Kim, J.H., Loganathan, G.V.: A new heuristic optimization algorithm: harmony search. Simulation **76**(2), 60–68 (2001)
23. Lee, K.S., Geem, Z.W.: A new meta-heuristic algorithm for continuous engineering optimization: harmony search theory and practice. Comput. Methods Appl. Mech. Eng., Eng **194**, 3902–3933 (2004)
24. Geem, Z.W. (ed.): Music-Inspired Harmony Search Algorithm: Theory and Applications. Studies in ComputationalIntelligence. Springer, Berlin (2009)
25. Yang, X.S.: A new metaheuristic bat-inspired algorithm. Studies Comput. Intell. **284**, 65–74 (2010)
26. Yang, X.S., Gandomi, A.H.: Bat algorithm: a novel approach for global engineering optimization. Eng. Comput. **29**, 464–483 (2012)
27. Yang, X.-S.: Bat algorithm: literature review and applications. Int. J. Bio-Inspired Comput. **5**(3), 141–149 (2013)
28. Chen, YQ., Petra, I., Xue, D.: Fractional order control—a tutorial. In: Proceedings of American control conference. St. Louis (MO, USA): Hyatt Regency Riverfront, June 10–12 2009

29. Davison, E.J.: Multivariable tuning regulators: the feedforward and robust control of general servo-mechanism problem, IEEE Trans. Autom. Control, vol. 21, pp.35–47 (1976)
30. Moradi, M.: A genetic-multivariable fractional order PID control to multi-inputmulti-output processes. J. Process Control **24**, 336–343 (2014)
31. Monje, C.A., Chen, Y., Vinagre, B.M., Xue, D., Feliu-Batlle, V.: Fractional-order Sys-tems and Controls Fundamentals and Applications. Springer, London (2010)

Entity Configuration and Context-Aware reasoNer (CAN) Towards Enabling an Internet of Things Controller

Hasibur Rahman, Rahim Rahmani and Theo Kanter

Abstract The Internet of Things (IoT) paradigm has so far been investigating into designing and developing protocols and architectures to provide connectivity anytime and anywhere for anything. IoT is currently fast forwarding towards embracing a paradigm shift namely Internet of Everything (IoE) where making intelligent decisions and providing services remains a challenge. Context plays an integral role in reasoning the collected data and to provide context-aware services and is gaining growing attention in the IoT paradigm. To this end, a Context-Aware reasoNer (CAN) has been proposed and designed in this chapter. The proposed CAN is a generic enabler and is designed to provide services based on context reasoning. Discovering and filtering entities, i.e. entity configuration, become pivotal in analysing context reasoning to provide right services to right context entities at the right time. This chapter leverages the concept of entity configuration and CAN towards enabling an IoT controller. The chapter further demonstrates use cases and future research directions towards generic CAN development and facilitating context-aware services to IoE.

Keywords Context · Context-aware · Internet of things · Reasoning

1 Introduction

Context has been investigated in computer science for over last 50 years. Research into understanding context has introduced the notion context-aware applications. When an entity responds to changes in context, the entity is said to be context-aware. Context was initially utilized for location-based services only [1]. However, context offers more than only location. Context gained popularity with the introduction of mobile devices. Recently, context is gaining growing attention from Internet-of-Things (IoT) with the goal of providing services to enable a Connected World. The ultimate goal is to connect anything anywhere and anytime. This vision of Connected

H. Rahman (✉) · R. Rahmani · T. Kanter
Department of Computer and Systems Sciences (DSV), Stockholm University, Nod Buildning, 164 55 Kista, Sweden
e-mail: hasibur@dsv.su.se

Y. Bi et al. (eds.), *Intelligent Systems and Applications*,
Studies in Computational Intelligence 650, DOI 10.1007/978-3-319-33386-1_12

World has seen a rapid escalation of connected things. Hundreds of billions of things (devices) are expected to be penetrated in the market by 2020 [2, 3]. Such penetration is not limited to the scale of connected devices only but also expands in its scope such as agriculture, traffic management, environmental monitoring, home automation, healthcare, security surveillance, education, social activities, etc. [2]. This expansion leads to a paradigm shift from IoT to IoE (Internet of Everything). This paradigm shift gives rise to enormous amount of context information. Context information is "*any subset of information that can be used to characterize the situation of an entity as well as its relationship with other entities including the entity itself*" [4]. Furthermore, context awareness is "*the ability to realise an application or service over the situation of an entity or its context based relationships*" [4]. Here the entity refers to a physical device or a thing in IoT. Context-aware applications and services aim at provisioning context information to remote end nodes i.e. entities. This means context information needs to be accessed and shared among distributed entities. An entity (e.g. a connected device) that provisions the context information is thereby referred to as context entity. As the context entity escalates so does the context information. The rise of heterogeneous context information can be better utilized by managing via logical-clustering concept [5]. Logical-clustering topology enables to filter similar context information and can facilitate to providing better services by clustering [2].

To this date, IoT mostly focused on creating architectures and protocols to enable connectivity and scalability in the IoT domain [6]. The focus has now shifted to providing services to people by connecting everything by exploring the context information [3]. Figure 1 gives an idea about how IoT/IoE can provide services to people, for example Connected Home, Traffic Management, Smart Health, Environmental Monitoring, etc. There have been many approaches that are being proposed to broker context information such as SenseWeb [7] and IP Multimedia Subsystem platform [8]. However, these approaches employ centralized solution which is vulnerable to DoS attacks and configuration errors, and central point of failure [4]. This motivated researchers into developing distributed approaches such as MediaSense [6] and SCOPE [9]. MediaSense, offers an IoT architecture for connecting context entities and provisioning context information by extending Distributed Context eXchange Protocol (DCXP). DCXP enables real-time context information provisioning between distributed entities. DCXP was first introduced as context dissemination for Ambient Networks [10], and further adopted in [11] for distributed context support for ubiquitous mobile awareness services. DCXP is built upon five primitive messages such as REGISTER_UCI, RESOLVE_UCI, GET, SUBSCRIBE, and NOTIFY [10]. Context entity in DCXP protocol is designated as Universal Context Identifier (UCI). An entity such as sensor node along with its context information registers as UCI. When an entity along with its context information registers as UCI, other entity resolves the UCI to fetch the context information. The challenges such IoT platform faces are: configure (discover and further filter) the available entities and provide services to entities based on the context information.

A primary concern in IoT is to configure the entities that can be used for providing services. This corresponds to enabling entity configuration and entity reasoning.

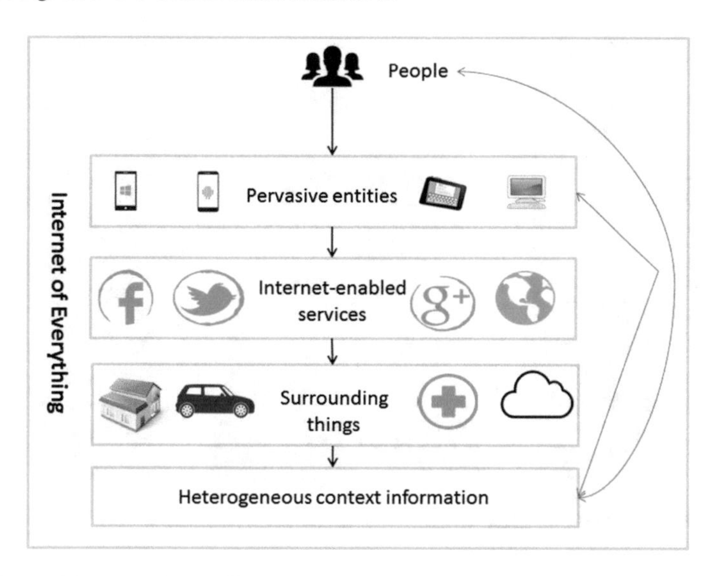

Fig. 1 Towards internet of everything

Entity configuration is about advertising the entities and further discovering newly joined entities. The discovered entities are not always of same kind and these entities are most likely interested in entities that are alike. For example, temperature sensors might not be interested in sensors that are being used for elderly care and vice versa. However, when a person, with a smartphone is interested in controlling her SmartHome and at the same wants to keep an eye on elderly person at home, might be interested in entities being used for both temperature sensing and elderly care. *Context* separates *who* needs *what* and *when*. This could be achieved by designing a reasoning service which could do the reasoning based on the context, i.e. a context-aware system. Furthermore, the upsurge of entities in IoT domain causes complexities in the network. This results in difficulties in managing the current and future IoT network. Entities are expected to join and leave the network frequently and it needs to analyse and adapt network-wide policies for adoption [12]. The above mentioned challenges can be countered by Software Defined Networking (SDN) [2, 13]. Martinez-Julia and Skarmeta proposed that controlling heterogeneous entities and complex networks such as those in IoT raises many challenges that can be resolved via SDN [13]. Existing and future IoT embraces heterogeneity; this heterogeneity is not limited to entities only but also includes the network in which entities immerse. There are many complexities and difficulties in connecting those heterogeneous entities and networks, but they could be countered by deploying SDN in IoT [13]. Figure 2 shows an example of SDN incorporation in IoT. The figure shows how two IoT entities communicate in SDN-enabled networks. Initial SDN proposal consisted of a single SDN controller only, but IoT requires another controller namely IoT controller as seen in Fig. 2. The IoT controller communicates with SDN controller to finalize connection and to start communication between entities [13]. MediaSense can be a

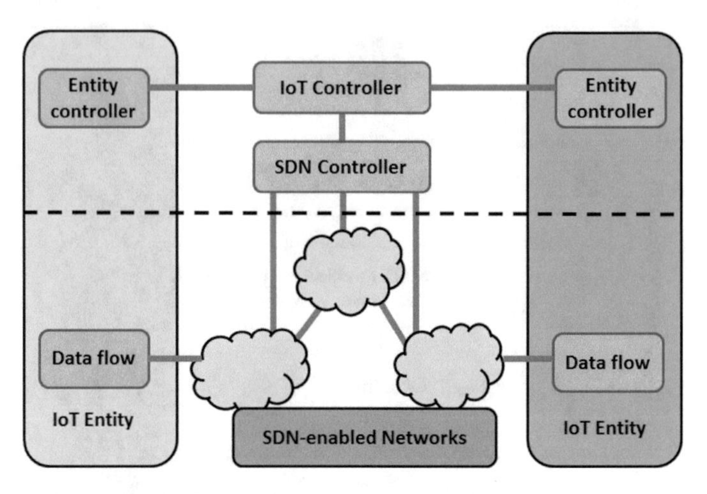

Fig. 2 SDN adoption in IoT [2, 13]

viable option for exploring as an IoT controller as shown earlier in [14]. The controller further requires going through few steps that includes entity joining, analysing and adapting, awareness etc. This work will look at creating a reasoning service for an IoT controller. The solution aims at designing a generic solution that can be explored by any IoT controller in the IoT domain.

By extending the earlier work on enabling distributed context entity discovery [15], this work designs and develops new algorithms to explore filtering for the entity configuration and further designs a novel reasoning service namely *Context-Aware reasoNer* (CAN). The *CAN* enables analysing, adapting and awareness in the MediaSense platform. These functions operate after an entity has joined the platform and before the platform has executed its internal operation. Figures 7 and 8 give an idea about how CAN can be explored towards enabling IoT controller. This work also reports simulation results on entity discovery and filtering, and services and future research directions of CAN.

2 Related Work

IoT so far has been concentrating on connecting things, i.e. any physical object that can be used to gather information (sensor) or can be operated on (actuator). Pervasive technologies such as RFID/Near Field Communication (NFC), short low power communication (6LoWPAN), high speed communication (LTE/4G), etc. make IoT a reality today [16]. This, commonly known as connectivity layer or also known as sensor layer, refers to real-time collection and processing of context information. IoT also necessitates connecting different type of networks into a single platform, in other words different networks need to be controlled by a single IoT controller. To counter this challenge, earlier research created MediaSense IoT platform [6] and further in [14] proposed to explore this platform as an IoT controller. MediaSense was implemented by extending the DCXP protocol with a simple mechanism of collecting and distributing context information among remote entities. These combination of MediaSense and DCXP offers novelty such as simple register/resolve protocol to sharing context information, real-time and fast communication between context entities, distributed framework capable of connecting any entity (both mobile and stationary) anywhere anytime.

2.1 Context Exchange Protocol

Context exchange protocol has not been research into extensively as yet, especially for sharing context information in IoT. Earlier works such as [17] discussed about context exchange protocol for context information sharing based on *request/response* in mobile environment, however, its implementation detail is unknown. Akin to this protocol's request/response mechanism, Distributed Context eXchange Protocol (DCXP) was proposed to share real-time context information by *register/resolve* mechanism and was initially built on for Ambient Networks project [10]. DCXP enables real-time context dissemination between remote end-nodes that are DCXP capable. DCXP further utilizes Universal Context Identifier (UCI) to register an entity and other entities resolve the UCI to fetch context information. UCI is a naming scheme similar to URI [10]. One of the advantages of utilizing UCI is that an entity can register several UCIs using DCXP [10, 14]. This is very useful especially when an entity acts as multi-purpose entity, for example, a smartphone can be used to collect information, control remote smart devices (thermostat, fridge, alarm, etc.), and at the same time it can also be used to share information. This smartphone can then be identified by registering different UCIs. Current DCXP consists of five primitives, Table 1 illustrates the primitives. The DCXP-capable entities that form a network are a Distributed Hash Table (DHT) overlay. All participating entities are located logically in a DHT ring. An entity with context information registers a UCI on the DCXP network by using the REGISTER_UCI; other entities resolve the UCI by using the RESOLVE_UCI. Depending on the need, entity can fetch instant context information or subscribe for future and continuous context information. The protocol is very simple to use but faces the challenge of discovering and filtering the entities registered as UCIs on the network. That means, currently when an entity wants to resolve a UCI, the resolving entity needs to know the UCI prior to resolving. This challenge was also raised in [14]. This work presents algorithms for entity configuration to enable entity discovery and filtering. This can be achieved by employing new primitive functions to the DCXP protocol [15].

Table 1 DCXP primitives messages [10, 15]

Message	Description
REGISTER_UCI	This message is used to register a particular UCI that holds context information in an entity
RESOLVE_UCI	This message needs to be invoked to fetch context information assocaited with the UCI
GET	Once the UCI is resolved at an entity, this message is used to get the context information
SUBSCRIBE	This enables an entity to subscribe to a particular context information and entity only gets updated when there is a new value
NOTIFY	Notifies the entity subscribes to a particular context information with the latest information

2.2 *IoT Architecture*

IoT has so far been prioritizing creating architectures to connect entities and to enable communication among the entities. Most of the architectures more or less extended from wireless sensor networks (WSNs) concept. e-SENSE is such an architecture, however, limited to neighbouring sensors in WSN [18]. It does not solve large-scale problem and lacks real-time functionality. Further, [19] has tabulated other architectures in IoT paradigm and shown that most of the architectures fail to provide real-time communication. MediaSense can counter this challenge and provides real-time communication and scalable context information sharing. MediaSense employs the DCXP protocol to share the context information. Figure 3 gives an idea how MediaSense works with DCXP. The detail information about MediaSense can be found in

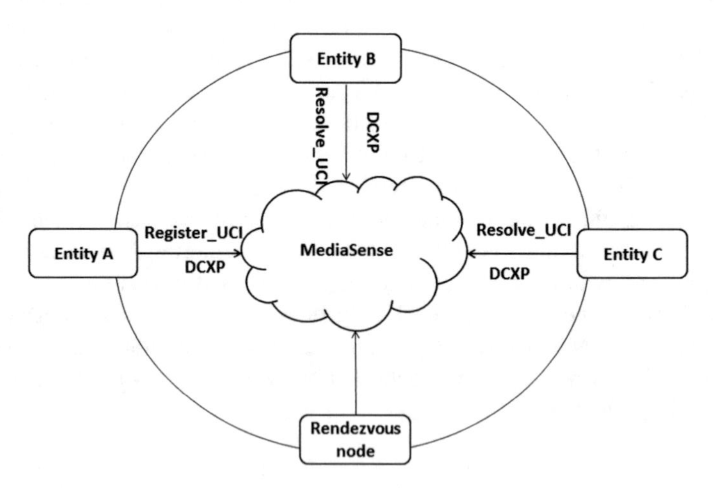

Fig. 3 MediaSense and DCXP [15]

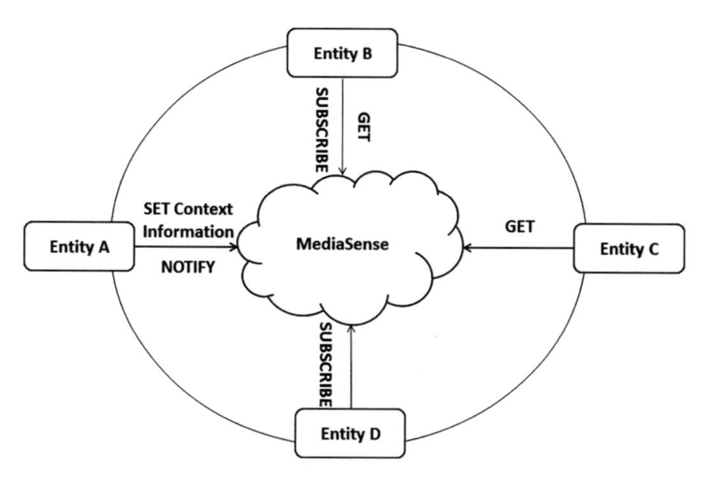

Fig. 4 MediaSense context information sharing [15]

[6]. MediaSense is a peer-to-peer (p2p) based context sharing IoT platform. As with most p2p system, MediaSense exploits a rendezvous host i.e. bootstrapping node to enable communication between nodes. The rendezvous node or the bootstrapping node must be initialized before any entity can join or leave the MediaSense platform. After registering a UCI, an entity needs to set the context information which other entities can get after resolving the UCI. This is further shown in Fig. 4.

2.3 Context-Aware Systems

A system that responds to context-changes is commonly known as a context-aware system. CybreMinder is one of the earliest such context-aware system that would allow users to send and receive notifications based on context [20]. However, its sole objective was to provide users a reminder tool. While other context-aware system such as CoBrA [19] and CaSP [19] provide context brokering for mobile solutions, SIM [19] and MidSen [19] are focused mainly for sensor information in smart home scenario and WSN respectively. Each of the context-aware system offers novel solutions to the respective domain as outlined. However, none of the solutions above provide the cutting-edge solution demanded by IoT as was also stressed in [19]. Furthermore, reasoning is one of the important aspects of IoT that allows collected data to provide more meaning and intelligence to the raw data. There have been many reasoning techniques available employing rule-engine and ontology based solutions, and combination of both has also been explored [21]. Most of the earlier researches only demonstrate those techniques in specific way for specific solution. However, as IoT incorporates heterogeneous context entities in heterogeneous domain such as connected home, healthcare, traffic management, environmental monitoring, agriculture, etc., this heterogeneity mandates to look into *multi-modal* reasoning approaches. The reasoner should also be able to provide required services to corresponding context entities. Vision of such a context-aware framework for distributed applications was demonstrated in [22].

3 Entity Configuration

This section illustrates the approach to configuring distributed context entity in the IoT. Here, entity configuration refers to discovering and filtering of context entities. To achieve the aforementioned objective, DCXP protocol is extended and its correctness is proven on a distributed, i.e. P2P based MediaSense platform. MediaSense already offers scalable Publish/Subscribe (PubSub) as shown in [23].

3.1 Entity Discovery

This section describes discovery of entities within the MediaSense platform. That means a context entity should be able to discover other context entities so that each entity can resolve UCIs and enable swift context sharing. Figure 5 shows how the approach would work and MediaSense with extended DCXP.

To accomplish the above goal, publish/subscribe approach is employed. The idea is to employ a UCI common to all entities and all the registered UCIs within MediaSense would be inserted as common UCI's context information. This common UCI, tagged as global_UCI, will be inserted when the rendezvous node is initiated (see Fig. 5), and whenever a new entity joins the platform will be subscribed to this global_UCI by default (see Fig. 5). To achieve this, a new primitive function named joinUCI has been implemented. This is illustrated in Fig. 5. This joinUCI primitive function now embeds other two primitive functions namely REGISTER_UCI and GET.

The primitive function, joinUCI, will enable discovering existing UCIs but in order to discover newly joined UCIs, a new primitive called DISCOVER has been implemented. DISCOVER will enable to fetching new UCIs after every t seconds. The interval, t, is implementation dependent. The algorithm for this primitive function is explained below.

3.1.1 Global_UCI

The algorithm first initiates the MediaSense rendezvous node. This node can be any entity that wants to join the MediaSense platform. The algorithms then checks if a global_uci exists or not; if global_uci does not exist then a global_uci is registered using MediaSense's publisher algorithm. However, if global_uci already exists then algorithm just renews the global_uci. MediaSense rendezvous node is run which

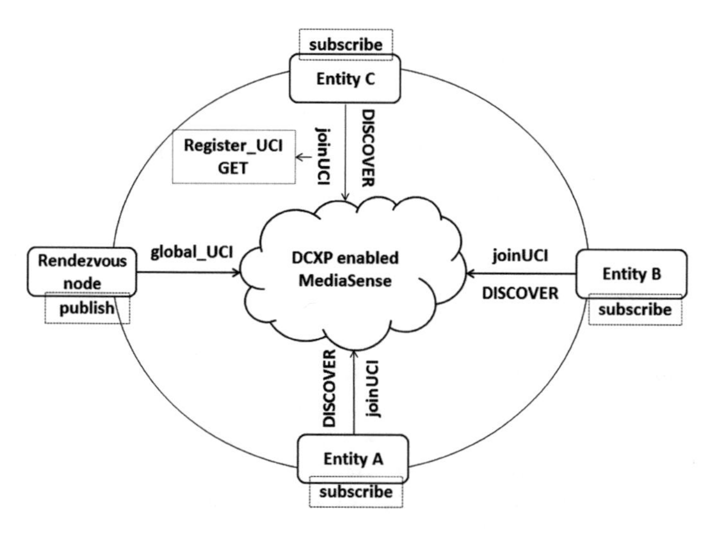

Fig. 5 MediaSense with extended DCXP [15]

is followed by starting the p2p communication. The IP address and bootport need to be passed as parameters to start the p2p communication. More on starting p2p communication is outlined in [6].

3.1.2 JoinUCI Primitive Function

MediaSense platform offers a single API called MediaSensePlatform which implements all the primitive functions offered by the DCXP. A new primitive function namely joinUCI is added which combines the two existing DCXP primitive functions (REGISTER_UCI and GET). The joinUCI algorithm first creates an instance of MediaSensePlatform and initializes with local network settings. The joinUCI is initialized and a UCI is declared for this context entity if the MediaSense rendezvous node is running. If this UCI is not listed on the global_UCI then the UCI is published on the global_UCI so that other entities can discover this UCI. If the UCI is already existent on the MediaSense platform's global_UCI then the UCI is registered by invoking REGISTER_UCI function. The algorithm then calls the DISCOVER function to synchronize with the system.

```
Algorithm 1. global_uci as publisher
begin
 initiate MediaSense_rendezvous_node
 if global_uci does not exist
   register global_uci (based on PubSub's publisher
algorithm)
   elseif
```

```
    renew global_uci
  endif
  run MediaSense_rendezvous_node
  start P2P Communication on the rendezvous node IP
address and bootport
end
```

Algorithm 2. joinUCI primitive function

```
begin
create an instance of MediaSensePlatform
initialize platform with network settings
while MediaSense rendezvous is running
  initialize MediaSensePlatform's joinUCI
  if the UCI is not listed on global_uci
    publish on global_uci by invoking MediaSense-
Platform's config method
    return the UCI's current configuration status
  elseif
    register the UCI on MediaSense platform
    publish on global_uci by invoking MediaSense-
Platform's config function
  endif
  synchronizes with the existing UCIs after every
T seconds
endwhile
end
```

3.1.3 DISCOVER

The joinUCI enables to put the UCI (trying to join) on the global_UCI and fetch existing (if any) UCIs. However, it does not cater for newly joined UCIs after this UCI has joined the system. To counter this, DISCOVER has been introduced. DISCOVER allows to synchronize with new UCIs after every t seconds. The algorithm first resolves the global_UCI based on the subscriber algorithm; the algorithm then fetches the subscribeable UCIs. The next step of this algorithm is to save the UCI (the UCI calling DISCOVER function) to the global_UCI and the algorithm then updates the global_UCI. Depending on the requirement global_UCI is either merged or renewed. The final step is to synchronize with the subscription.

3.2 *Entity Filtering*

The entity discovery described above returns all the available entities; this is helpful when an entity is interested in all the other entities, for example, in a SmartLiving scenario. In this scenario, the entity, i.e. a user with her smartphone, would want to retrieve all entities at her home designed for SmartLiving. These entities at home could be any appliances that can be connected to Internet, for example, temperature sensor, thermostat (actuator), lights, fridge, oven, TV, wearable devices, surveillance camera, etc. The same person with the same smartphone, i.e. entity could further be interested in other IoT applications other than SmartLiving, for example, retails information, traffic, parking (for private vehicles), sports, etc. Retails can further be divided into other different categories. Therefore, context in which an entity immerses is important to determine particular requirement of an entity at any moment of time. This can be achieved by exploiting idea such as logical-clustering [23], and it further necessitates designing a system that can respond to such context changes. This section will further design and portray algorithms to enable entity filtering. Figure 6 shows an example of entity filtering. Entity naming plays a role in filtering efficiently in this approach. Entity filtering is designed by employing a rule-based approach where entities are filtered if certain rules are met. The partial algorithm is shown below (Algorithm 3). As mentioned earlier, entities are identified as UCI in DCXP, so the naming of UCI needs to follow a standard naming scheme. After an entity has been named, there is a need to make decision if the entity requires discovering any other entities. For example, entities (sensors) that are placed at home for collecting information do not necessarily require discovering any entities at all. On the other hand, sensors that collaborate to carry out tasks together require discovering other similar sensors. Therefore, building a system that responds to different context is mandated.

Responding to context changes requires building a system that is context-aware and at the same time provides reasoning services. This is where the proposed CAN is useful (see Sect. 4). The combined approach of entity configuration and CAN enables creating an IoT controller. More about IoT controller and how it can be achieved can be found in [2, 14].

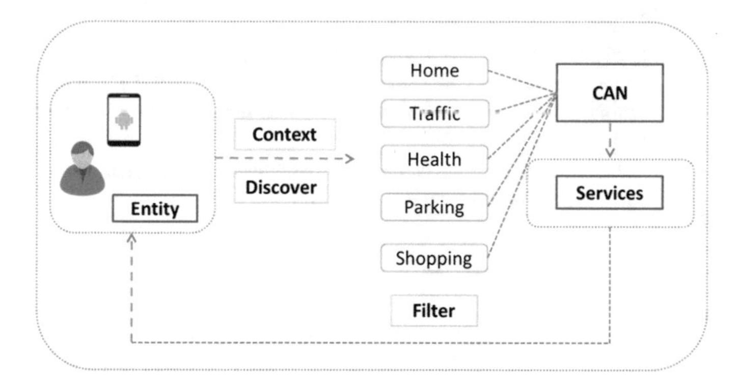

Fig. 6 Entity configuration and CAN

```
Algorithm 3. entity filtering
begin
invoke entity discover
while entity is not empty
  if the entity is not named
    name it
    invoke entity filtering
  elseif the entity is named
    determine context-naming
    find and match context (e.g. PubSub's subscribe
algorithm)
    return corresponding entities
  endif
endwhile
end
```

3.3 Evaluation

Table 2 shows the new primitive functions introduced to DCXP protocol. The earlier DCXP was evaluated for the time it needed to complete the primitive operations. Therefore, the extended DCXP has also been evaluated for the time requires completing the joinUCI and DISCOVER operations. Table 3 and 4 illustrate these respectively. Average measurements for several simulations are shown. From Table 3, it can be seen that joinUCI takes relatively more time compared to old DCXP. The extended DCXP's joinUCI combines two operations and those operations are now

Table 2 DCXP New primitives messages [15]

Message	Description
JoinUCI	This new functions now combines the old REGISTER_UCI and GET primitive functions, and enables context entity to join and retrieve context information at the same time
DISCOVER	This message needs to be invoked to discover existing UCIs and to synchronize with newly joined entities

Table 3 Time for joinUCI [15]

	REGISTER_UCI + GET	joinUCI
Sec	0.07	0.09

Table 4 DISCOVER [15]

UCIs	100	200
ms	293	450

called inside the MediaSensePlatform API. Furthermore, this function fetches existing UCIs at the time of UCI joining. These are responsible for the delay which is understandable. Nevertheless, this is still on par with the DCXP evaluation.

Table 4 shows the time required for DISCOVER function. The operation took 293 and 450 ms respectively for 100 and 200 UCIs. The time was measured after the UCIs were already registered. From the table, it can be observed that each UCI requires between 2.2 and 2.9 ms to be discovered. This is very significant in terms of fast discovery of context entity. However, joinUCI introduced a delay in the DCXP protocol. Nevertheless, discovering context entities has been a challenge for long time in DCXP-enabled MediaSense and IoT in general. This work successfully overcame this challenge and proposed new primitive operation for DCXP which now enables fast context entity discovery. The proposed algorithms can be extended to other p2p-based IoT platforms to enable entity discovery given that entity can be named and rendezvous node is available.

4 Context-Aware reasoNer (CAN)

Previous section portrayed the entity configuration, i.e. discovery and filtering of context entities in the IoT domain. As more and more entities are being penetrated in the IoT domain, this massive immersion of entities mandates to be managed by a capable controller [14]. This massive entities further immerse in heterogeneous networks. This can be countered by extending the SDN concept as shown in Fig. 2. The SDN relies on a single controller, for example an OpenFlow controller, however, the heterogeneity nature of the IoT further necessitates employing its own controller. This dedicated controller, i.e. IoT controller acts as a bridge between entity and SDN controller. In [14], it was communicated that MediaSense can be exploited as an IoT controller. Figure 7 shows this. The controller itself goes through many stages before it actually executes an entity. The stages include joining, analysing, adapting, executing and having awareness for each of the stages. The controller executes the entities based on the analysis, adaption and awareness. These three steps can be combined into a context-aware reasoning system, i.e. *Context-Aware reasoNer* (CAN). Figure 8 illustrates this. Context reasoning can further be divided mainly into two levels. Raw data is reasoned at the low level which in turn should be processed at the high level for extracting meaningful and intelligent context information and utilizing them for further purposes. This high level deduced context information, after reasoning is processed, can then be distributed among other entities. Furthermore, no single context-reasoner provides all the reasoning demanded by the IoT heterogeneity. This necessities to design a reasoner to accommodate IoT demands of a multi-modal context-reasoner. Some of the IoT applications where reasoner can be useful includes but not limited to: Connected Home, healthcare, SmartGrid, agriculture, transportation, retails, security, etc. In light of this, a multi-modal generic enabler CAN has been proposed for high level context information reasoning. Each

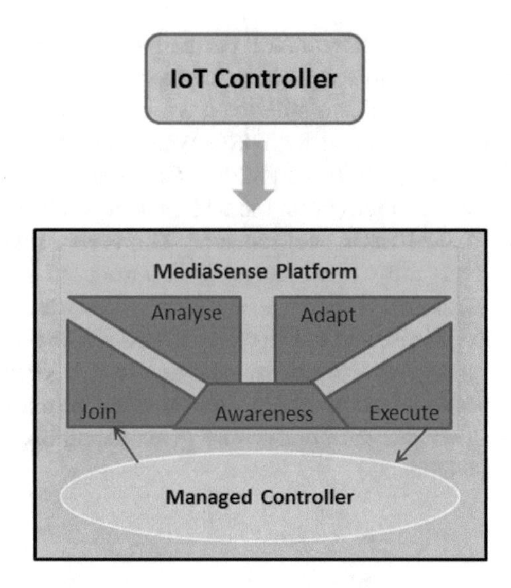

Fig. 7 MediaSense as an IoT controller

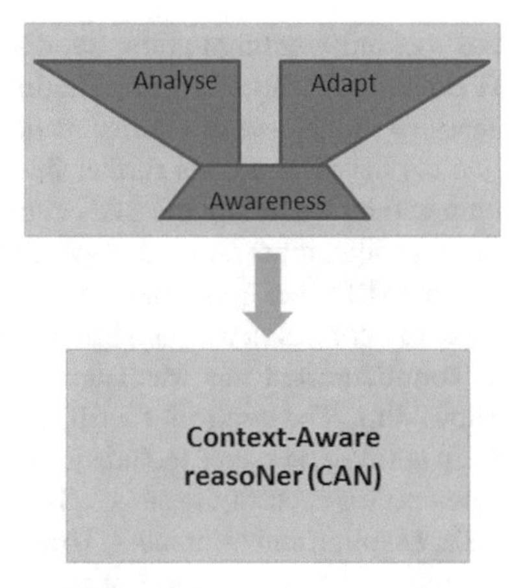

Fig. 8 Context-aware reasoNer (CAN)

step of the proposed CAN is described below. Entity configuration is pivotal in determining to provide right services to right context entity.

After an entity has joined a system, the CAN, part of IoT controller, first requires analysing based on the low level context information obtained. The context information at the low level can be processed by the entity controller (Fig. 2) which should include naming of the entity. CAN first analyses its name and based on its identification further analyses what kind of services it demands. The analysis can be done by employing either rule-based, ontology-based or mix of both. For a proof-of-concept, in the connected home example below, a rule-based reasoning analysis has been explored. This work did not focus on creating any new novel reasoning rule, emphasize was on creating a reasoner service which can provide required services based on the multi-modal reasoning. Hence, a rule-based reasoning for different applications is deemed sufficient. Following analysis, CAN adapts each entity according to its demands and provides the services. CAN adapts to any context changes in the entity thus provides the awareness. CAN then provides the services for each of the contexts to the entity. The following examples provide an insight to CAN. The section ends with an algorithm depicting the few of the rule-based reasonings for CAN. The algorithm is partial and only showcases some reasonings for the Connected Home scenario.

4.1 *Reasoner Services for Connected Home*

Take an example of a connected home where many Internet-enabled appliances are installed such as temperature sensors, thermostat, lighting, coffee machine, door lock, dish washer, etc. Figure 9 demonstrates a connected home scenario and what user might be interested in a connected home. A user opens a mobile application on her mobile phone to retrieve current status at her home and changes its status whenever required. Whenever user opens the mobile application, it first contacts the CAN and maps its identity (if already a registered entity) and returns the current status for the authorized home. The sensors and actuators that are connected with Internet and part of the home provide the raw data to the gateway at home (e.g. MediaSense on raspberry pi) and after collecting the data, gateway processes the raw data to context information and disseminates to the CAN via MediaSense. CAN

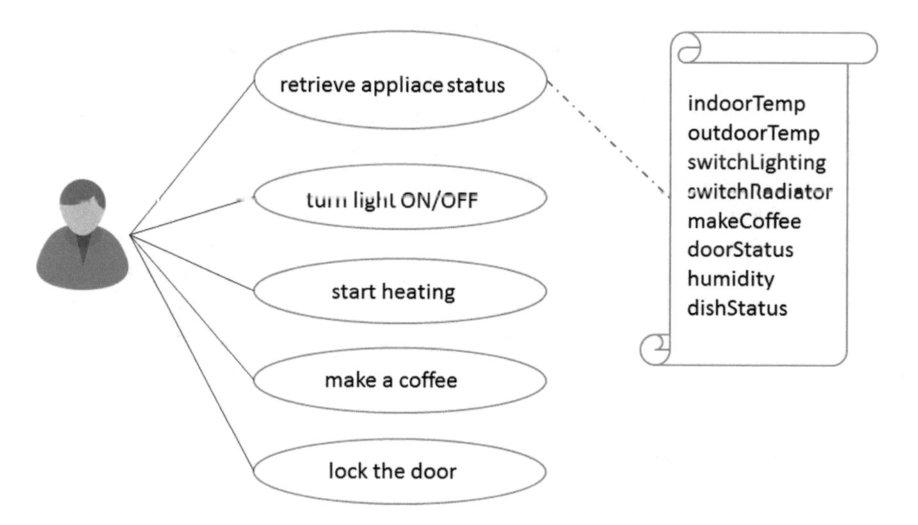

Fig. 9 Example CAN services for Connected Home

Table 5 *CAN services for Connected Home*

Services	Description
Status	Retrieves current status of a connected home. It communicates with a gateway that retrieves information from the connected home
Compare	Compares Time to Heat (TTH) and expected time of arrival, and alters thermostat radiator accordingly
Radiator	Turns the heating ON or OFF in a connected home
Light	Turns the lighting ON or OFF in a connected home
TTH	Calculates TTH based on current indoor temperature (T_{indoor}) and user's expected temperature (T_{home})
RegisterSH	Registers new connected home devices to the databases
Mapping	Maps user identity to available connected home ID and assigns its proper ID
Database	Saves connected home settings such as identification for each connected home devices, values for T_{home} and T_{away}, light status, connected home location, etc.

first filters out the appropriate entities according to the current context of the entity, in this particular scenario its home, and feeds the context information acquired from the home to user's mobile application. When user wants to changes the context at the home, for example turning on light or start heating, CAN forwards the context information to the MediaSense gateway which in turn alters the status of the entities based on the context.

To sum up, CAN provides different services based on context reasoning as shown in Table 5. For the connected home, as shown in Table 5, CAN provides user the required time to heat (TTH) based on the current indoor temperature and user's desired temperature at home. Furthermore, when TTH is less than 5 min, CAN sends a notification to the user if she is interested in turning on the light and/or would like to make coffee when she would reach home. If CAN can be connected to any external services which provides user with mobility service, i.e. expected time of arrival (ETA) from current location to home, for example as one provided by Google maps, then CAN compares the ETA and TTH. Based on the comparison if TTH is longer than ETA for example 60 or 90 min, it can advise the user to stop by a coffee shop before the radiator adjusts to the desired temperature. This service is very useful in cold countries where temperature in winter goes below freezing temperature. CAN also makes use of a database solution which allows to store user's settings (e.g. temperatures for home and away) and to register new users and devices for connected home. This connected home solution, ConnectedHome Reasoner, has already been developed and tested with a generic gateway solution with MediaSense [24].

4.2 Reasoner Services for Healthcare

Healthcare has become an attractive IoT application thanks to recent advancements in wearable devices. These wearable devices are becoming more and more realisable due to affordability and its ability to connect to the Internet. There have been many researches on this issue which is nowadays popularly known as Ambient Assisted Living (AAL) for elderly people. In this scenario, Internet-enabled devices can be used by both the caregiver and elderly people. Both can communicate and interact with the same devices via a gateway. IoT in healthcare allows to monitoring patient remotely and giving care remotely (either online or offline). With the context-aware system, it allows ageing people to remind about taking medication on time. CAN can provide the required context awareness and services to both caregiver and patients by discovering and filtering the respective identities. Figure 10 gives an insight to the communication between three patients and caregiver. The figure only demonstrates the communication and what CAN might provide, it does not show the IoT devices this example might consist of. Table 6 further highlights few of the CAN services to the healthcare application. Caregiver is a group of medical personnel taking care of a group of patients; the patients might have similar characteristics in terms of age, disease, treatments, etc. The caregiver can continuously or periodically monitor the patients via the retrieve patient status service. Based on the context of each of the patient, caregiver can issue prescription to the patients. Alike to connected home, healthcare also exploits the concept of gateway that continuously monitoring the IoT devices for the patients. When a patient is not active to the system or monitoring via camera at home fails, CAN can immediately notify the caregiver. CAN also allows both caregiver and patients to request and/or remind appointments. CAN also provides mechanism to forward health information to the caregiver via for example MediaSense gateway.

The above two examples are two of the probable IoT application scenarios that can be achieved via CAN. The examples gave an insight to the services, analysis,

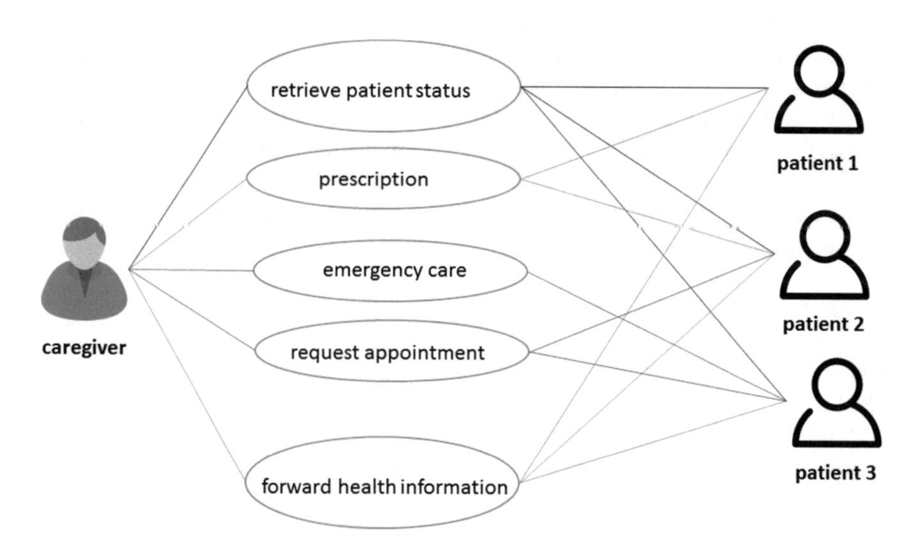

Fig. 10 Example CAN services for healthcare

Table 6 *CAN services for healthcare*

Services	Description
Status	Retrieves current status of patients currently under care of this caregiver. The caregiver can be a group of medical personnel
Prescription	Provides medication prescription to the concerned patient after analysing current status
Emergency care	Forwards an emergency request to the caregiver. It could be explicit request by the patient(s) or based on awareness in CAN of patient(s) inactivity
Request appointment	Reminds about scheduled appointment and further allows scheduling appointments
Forward health information	Forwards health information from patients to caregiver. Periodically or explicit forward
RegisterPatients	Registers new patients and their devices to the databases
Database	Saves information about patients, caregivers and the associated devices

and awareness that are provided by the proposed CAN. This shows that the proposed CAN provide multi-modal context-awareness that has been a challenge in IoT [19]. Furthermore, the concept can easily be adapted to other IoT applications such as traffic management, environmental monitoring, agriculture, etc. As mentioned earlier, context separates an application from other applications and awareness based on the context allows providing different services to different context entity in different time. This differentiation in IoT controller thereby in CAN is achieved via the entity configuration (discovering and then filtering the entities) as shown in Sect. 3. The entity configuration was part of the MediaSense platform and has already been designed and developed, the MediaSense as its current version allows connecting different IoT devices. MediaSense can be used as a gateway and it can run on resource-constrained devices such as on Raspberry pi as well as on Android platform. However, it lacked the functionality that now can be countered by employing the CAN concept. Following the development of CAN and integrating with MediaSense, it can now be used as an IoT controller as shown in Fig. 7. However, CAN has been designed to be a generic enabler, it can be used by any platform capable of sharing context information.

```
Algorithm 4. rule_based reasoning
begin
configure i.e. discover and filter context
while context is not empty
if context is connected home
  configure i.e. discover and filter context
  while context is not empty
  if authorized user
    if context is retrieve status
      return status about all entities at home
```

```
    elseif context is home trip
     ask for expected temperature
     calculate TTH
     compare TTH and ETA
     if TTH is greater than 90 minutes
     recommend for alternative activities
     if TTH is 5 minutes
     prompt to turn light on
     recommend to make a coffee
    elseif context is away trip
     switch radiator to away T_away
     turn all lights off
     if door is not locked
       lock the door
    elseif context is light
     turn on the lights
     endif
  endif
  endwhile
elseif context is healthcare
if context is caregiver
  do the reasoning and provide services to contexts
endif
if context is patients
  do the reasoning and provide services to contexts
endif
elseif context is traffic
  do the reasoning and provide services to contexts
elseif context is parking info
  do the reasoning and provide services to contexts
elseif … … … //any other IoT applications
endif
endwhile
end
```

5 Discussions and Future Work

This chapter aimed at showcasing approaches to enable entity configuration and further portrayed design of a novel generic reasoning approach namely CAN (Context-Aware reasoNer). Context has been very popular in mobile communication; context aware computing is now gaining tremendous attention in Internet-of-Things (IoT) paradigm. The growing attention is due to rapid growth of *things* (i.e. entities) deployment in the IoT and challenges these entities bring forth in providing services and making use of the collected data. Consequently, researchers have spent huge amount of time investigating protocols and IoT architectures to provide connectivity among the entities. Affordability and ability of entities to connect to the Internet made the vision of IoT a reality today. This rapid escalation in IoT deployment generates enormous amount of raw data. However, the focus of IoT has now been shifted to provide services based on the data collected from the enormous entities. This shift in focus eventually will allow experiencing a paradigm shift towards Internet of Everything (IoE). Providing context-aware services based on the collected and processed (e.g. reasoning) data to the entities is one of the goals and a challenge in IoE.

Prior to providing such context-aware services, configuring of those context entities is of utmost importance. In view of this, algorithms to enable such entity configuration for discovering and filtering entities have been designed and developed. The evaluation of algorithms suggests that these can provide fast and scalable entity configuration in the IoT domain. The correctness of these algorithms has been proven on a versatile IoT platform- namely MediaSense.

The proposed CAN aims at enabling context-aware services in the current and future IoT domain. CAN has been designed to be independent of any particular solution and is a generic solution which can be explored by any IoT architecture capable of providing communication between entities. Two out of many IoT applications have been showcased where reasoner for connected home has already been developed and further tested. CAN has achieved to provide different context-aware reasoning services to the context-entities in real-time. Mindful of future IoT heterogeneity, CAN has been designed to be multi-modal. It can be seen from Tables 5 and 6 that CAN provides different services to same entity or same service to different entity- thus proving the multi-modality. If early work on IoT is considered mostly on connectivity, the future IoT will focuse on providing intelligence to those connected entities. The proposed novel CAN attempts at providing such high-level intelligence to the connected entities in IoT. The feasibility of CAN has been demonstrated by illustrating two example use cases.

As the future IoT would require being controlled by a capable controller, i.e. an IoT controller, this can now be enabled by leveraging the two concepts. This leveraged concept is particularly useful for the high level reasoning of context. However, improvement of CAN can be achieved. CAN can be explored and developed for other IoT applications. Further intelligence can be provided by CAN by exploring machine learning techniques. Incorporating machine learning in CAN would allow providing further intelligent decisions based on collected data in the IoT domain. The reasoning that is used is only rule-based reasoning in this work, as CAN strives to be multi-modal so further work can be done in this area by looking into other reasoning techniques. Other reasoning techniques would allow CAN to be more intelligent, flexible and efficient.

References

1. Schmidt, A., Beigl, M., Gellersen, H.-W.: There is more to context than location. Computers & Graphics Journal, pp. 893–902. Elsevier, New York (1999)
2. Rahman, H.: Self-organizing logical-clustering topology for managing distributed context information, philosophy of licentiate thesis, Department of Computer and Systems Sciences, Stockholm University, September 2015
3. Clarke, R.Y.: Smart cities and the internet of everything. The Foundation for Delivering Next-Generation Citizen Services, white paper sponsored by Cisco, October 2013
4. Walters. J.: Distributed immersive participation—realising multi-criteria contextcentric relationships on an internet of things. Ph.D. thesis, Department of Computer and Systems Sciences, Stockholm University, November 2014
5. Rahmani. R., Rahman. H., and Kanter. T.: Context-based logical clustering of flow-sensors—exploiting hyperflow and hierarchical DHTs, In Proceeding(s) of 4th International Conference on Next Generation Information Technology, ICNIT, June 2013
6. Kanter. T., et al.: Mediasense—an internet of things platform for scalable and decentralized context sharing and control, In: Proceedings of the Seventh International Conference on Digital Telecommunications, ICDT 2012, pp. 27–32, April 2012
7. Grosky, W., Kansal, A., Nath, S., Liu, J., Zhao, F.: Senseweb: an infrastructure for shared sensing. Multimedia **14**, 8–13 (2007)
8. Gonzalo, C.: 3G IP Multimedia Subsystem (IMS): Merging the Internet and the Cellular, World. pp. 381 (2005)
9. Baloch, R.A., Crespi, N.: Addressing context dependency using profile context in overlay networks. In: Proceedings of 7th IEEE Consumer Communications and Networking Conference (CCNC), pp. 1–5. IEEE, January 2010
10. Swenson, M.: A distributed approach to context-aware networks—prototype system design, implementation, and evaluation, royal institute of technology, M.Sc thesis conducted at ERICSSON Research (2007)
11. Kanter, T., Pettersson, S., Forsstrom, S., Kardeby, V., Norling, R., Walters, J., Osterberg, P.: Distributed context support for ubiquitous mobile awareness services. In: Fourth International Conference on Communications and Networking in China. ChinaCOM, pp. 1–5, 26–28 Aug 2009
12. SDN defined by OpenNetworking Foundation, https://www.opennetworking.org/sdn-resources/sdn-definition. Accessed 18 May 2015
13. Martinez-Julia, P., Skarmeta, A.F.: White paper on:Lings to IPv6 with software defined networking (2014)
14. Rahman, H., Kanter, T., Rahmani, R.: Supporting self-organization with logical-clustering towards autonomic management of internet-of-things, (IJACSA). Int. J. Adv. Comput. Sci. Appl. **6**(2), 24–33 (2015)
15. Rahman, H., Kanter, T., Rahmani, R.: Enabling distributed context entity discovery for an internet-of-things platform. In: Proceedings of IEEE Intelligent Systems Conference, London, UK, November 2015
16. Feki, M.A., Kawsar, F., Boussard, M., Trappeniers, L.: The internet of things: the next technological revolution. IEEE Computer Society, IEEE New York (2013)
17. Tulla, J. M.: Context exchange between devices in mobile environment. M.Sc. thesis at Lappeenranta University of Technology, Finland (2008)
18. Lombriser, C., Marin-Perianu, M., Marin-Perianu, R., Roggen, D., Havinga, P., Troster, G.: Organizing context information processing in dynamic wireless sensor networks. In: Proceedings of ISSNIP, pp. 67–72, December 2007
19. Perera, C., Zaslavsky, A., Christen, P., Georgakopoulos, D.: Context aware computing for the internet of things: a survey, IEEE Communications Surveys & Tutorials (2013)
20. Dey, A.K., Abowd, G.D.: CyberMinder: a context-aware system for supporting reminders. In: Proceedings of the 2nd International Symposium on Handheld and Ubiquitous Computing (HUC2K), pp.172-186, Bristol, UK, 25–27 September 2000

21. Golbreich, C., Bierlaire, O., Dameron, O., Gibaud, B.: What reasoning support for Ontology and Rules? the brain anatomy case study. Galway, Ireland, In: OWL Experience and Directions, Nov 11–12 (2005)
22. Van Kranenburg, H., Bargh, M.S., Iacob, S., Peddemors, A.: A context management framework for supporting context aware distributed application. IEEE Commun. Mag. **44**(8), 67–74 (2006)
23. Rahman, H., Rahmani. R., Kanter, T.: Enabling scalable publish/subscribe for logical-clustering in crowdsourcing via mediasense. In: Proceedings of IEEE Science and Information (SAI) Conference, London, UK2014, August 27–29 2014
24. Rahman, H., Rahmani, R., Kanter, T., Persson, M., Amundin, S.: Reasoning service enabling smarthome automation at the edge of context networks. In: Proceedings of WorldCist'16—4th World Conference on Information Systems and Technologies (Springer), Recife, Brazil, March 2016

Track-Based Forecasting of Pedestrian Behavior by Polynomial Approximation and Multilayer Perceptrons

Michael Goldhammer, Sebastian Köhler, Konrad Doll and Bernhard Sick

Abstract We present an approach for predicting continuous pedestrian trajectories over a time horizon of 2.5 s by means of polynomial least-squares approximation and multilayer perceptron (MLP) artificial neural networks. The training data are gathered from 1075 real urban traffic scenes with uninstructed pedestrians including starting, stopping, walking and bending in. The polynomial approximation provides an extraction of the principal information of the underlying time series in the form of the polynomial coefficients. It is independent of sensor parameters such as cycle time and robust regarding noise. Approximation and prediction can be performed very efficiently. It only takes 35 μs on an Intel Core i7 CPU. Test results show 28 % lower prediction errors for starting scenes and 32 % for stopping scenes in comparison to applying a constant velocity movement model. Approaches based on MLP without polynomial input or Support Vector Regression (SVR) models as motion predictor are outperformed as well.

M. Goldhammer (✉) · S. Köhler · K. Doll
University of Applied Sciences Aschaffenburg, Würzburger Straße 45,
63743 Aschaffenburg, Germany
e-mail: michael.goldhammer@h-ab.de

S. Köhler
e-mail: sebastian.koehler@h-ab.de

K. Doll
e-mail: konrad.doll@h-ab.de

B. Sick
Intelligent Embedded Systems Lab, University of Kassel,
Wilhelmshöher Allee 73, 34121 Kassel, Germany
e-mail: bsick@uni-kassel.de

Y. Bi et al. (eds.), *Intelligent Systems and Applications*,
Studies in Computational Intelligence 650, DOI 10.1007/978-3-319-33386-1_13

1 Introduction

1.1 Motivation

In the World Health Organisation's last comprehensive status report on road safety, traffic deaths are listed as the eighth leading cause of death with annually more than 1.2 million global cases. Current estimates suggest the possibility that until 2030 traffic accidents will even become the fifth leading cause of death unless urgent action is taken [18]. Furthermore, for every recorded traffic fatality 4 permanent, 8 serious and 50 minor injuries are estimated with costs for the society of more than 100 billion Euro per year, only for Europe [4]. During the last decades large efforts were made to continually improve vehicle safety. Although 27 % of the victims are vulnerable road users (VRUs), e.g., pedestrians and cyclists, VRU protection remained an ongoing problem due to the absence of possibilities for effective passive or active safety mechanisms.

Only recently, vehicles get more often equipped with several types of sensors allowing them to perceive the local surrounding and thus offering advanced comfort and safety functionality to the driver. Common available examples of those "intelligent driving" applications are park distance control, lane assistant, traffic sign recognition or emergency brake assistant. This development offers a unique chance to address the great challenge of VRU safety effectively using early recognition of potentially critical situations to initiate active countermeasures at an early stage. However, mastering this task is complex: in a first processing step VRUs have to be detected and classified. Current state-of-the-art methods of pattern recognition and sensor data as well as image processing have to be performed in real-time generally using embedded hardware units in vehicles. Their computational power increases continuously. The latest developments lead to massive parallel processing by graphics processing units (GPUs) and hardware implementations of computationally complex algorithms using field programmable gate arrays (FPGAs), e.g., [13].

The second major step is understanding the current traffic scene based on information about the own (ego) vehicle, other road users, and the environment (road geometry, obstacles, etc.). The situation has to be continuously analyzed in order to detect critical situations. The criticality is defined by the potential for an accident and thus requires the prediction of the future behavior of potentially involved road users, in our case of the ego vehicle and the VRU. While the ego vehicle's behavior is relatively well known and predictable due to available on-board sensor data and existing models, the VRU has to be considered as an external system. The prediction of the VRU behavior has to be based only on external observations and on prior knowledge of his behavior. While many current approaches use the enhancement of the last movement state for this purpose (e.g., with Kalman filtering (KF) [1]), the VRU behavior is much more complex within the time ranges relevant for these applications. To gain and use the knowledge about VRU behavior for a more realistic movement prediction, models with higher complexity are needed. A new approach of movement prediction by polynomial approximation and supervised learning of multilayer perceptron (MLP) neural networks is in the focus of this publication. The subsequent step is using the gathered information for an adequate reaction of the intelligent vehicle to actively prevent the predicted accident. This may include the driver (information, warning) or it may happen in a completely autonomous way (breaking, evasive maneuver) if the timing constraints, the most important being the time to collision (TTC), fall into the range of human reaction times.

The state of the art in VRU motion modeling and path prediction is dominated by conventional movement models such as constant velocity (CV). As input information the measured position as well as further image and context information may be used. For a more detailed description of the state of the art we refer to own, preliminary work [10].

1.2 Our Contribution

The main contribution of our publication is a novel approach for self-learning trajectory prediction based on polynomial least-squares approximation and multilayer perceptron neural networks. Training and evaluation is done using trajectory data of uninstructed pedestrians in urban traffic scenarios whereby we assume that the method may also be applied to other VRU types. We focus on the prediction itself as major step, not a finished application in a vehicle. The method has the advantage not to be limited to certain types of movements, but it is able to handle all motion types included in training data. As self-learning predictor, it is independent of specific movement models since the network contains all required knowledge implicitly. The least-squares approximation of the discrete input track allows an extraction of the principal information of the current behavior of the pedestrian with the advantages of independence of sensor parameters such as cycle time and improved noise resistance. The output of the proposed method is a continuous position estimate up to a certain time horizon instead of only single trajectory points.

The article is structured as follows: In Sect. 2 the proposed path prediction approach is described. Section 3 outlines the methods used for evaluating the prediction quality while the according test results are set out in Sect. 4. A concluding summary is given in Sect. 5.

2 Methodology

In this section, we describe the usage of the measured trajectory information to predict the behavior of pedestrians represented by their future trajectory.

As the prediction is supposed to be invariant to the current global position and orientation of the pedestrian, the input data of the predictor is based on a time series of the absolute velocity $|v(t)|$ and the angular velocity $\omega(t)$ instead of directly on the global position measurements. The time series are approximated with multiple polynomials in sliding windows in a fixed position relative to the current time t_c in order to extract the polynomial coefficients every time cycle. They are serving as descriptor since they contain the principal information of the observed pedestrian behavior. The prediction output is also supposed to be invariant to the global position and orientation, so we use coefficients of polynomials describing the future velocity profile $v_{lon}(t)$, $v_{lat}(t)$ in the pedestrian's ego coordinate system for this purpose. As output, this representation shows better results than an output based on $|v(t)|$ and $\omega(t)$. The estimated future trajectory is rebuild from the predicted information by numerical integration and a retransformation into the global coordinate system.

The relation between the measured (input) and the future (output) pedestrian movement is established by a multi layer perceptron neural network model, which is capable of predicting all coefficients of the output polynomials based on those of the input polynomials within a single instance. As kernel-based comparison method we also evaluated Support Vector Regression. However, it requires one instance for each output value. The three consecutive steps of pedestrian tracking and data preprocessing, approximation with polynomials and prediction of polynomial coefficients of the future trajectory are described in the following Sects. 2.1–2.3.

2.1 Pedestrian Tracking and Data Preprocessing

The proposed method for pedestrian movement prediction is based on a short time tracking of the horizontal 2D position. The method is independent of the underlying sensor technology as long as it is capable of providing object positions in real world coordinates. Typical sensor setups are stereo cameras, lidar or radar sensors.

In our concrete setup we make use of a wide angle stereo camera system installed at a public urban intersection to generate pedestrian track data needed for training and testing of the self-learning algorithms. The field of view covers two crosswalks with pedestrian lights and two sidewalks of the crossing roads (see Fig. 1, left). The observed scenes provide a large variety of movement types such as straight walking, starting, stopping or bending in. The test site and the sensor setup are described in [9, 11].

The center of the pedestrian's head is serving as reference point for stereo triangulation and tracking. As the human gait can physically be described by the model of an inverted pendulum [17], the upper body and, in particular, the head indicates changes in motion very early (see Fig. 1, right). Furthermore, the center of the head can be recognized and located relatively stable from all directions by computer vision algorithms. This leads to a more robust trajectory measurement and potentially faster detection of upcoming movement changes compared to reference points which are based on the detection of whole persons, e.g., the center point of a detection window. The further major processing steps of the path prediction methodology described below are depicted in Fig. 2.

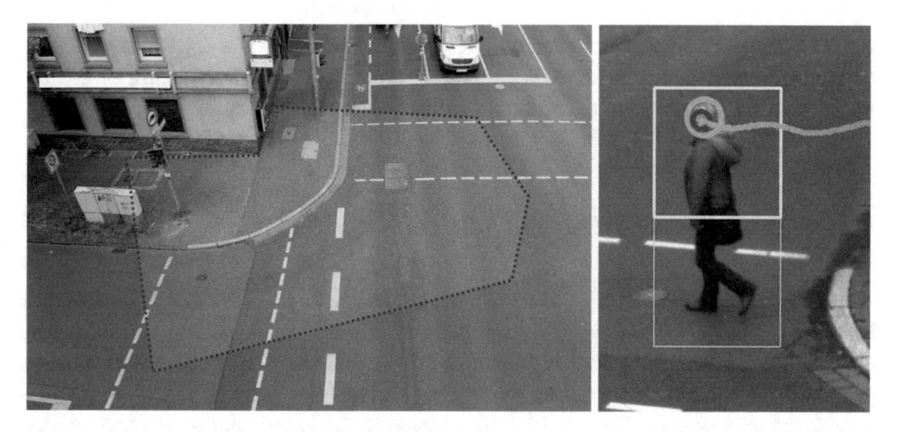

Fig. 1 *Left* field of view for 3D tracking by the wide angle stereo system at a public test intersection used to create the track database. *Right* optical people detection, head detection and -tracking to obtain pedestrian movement data

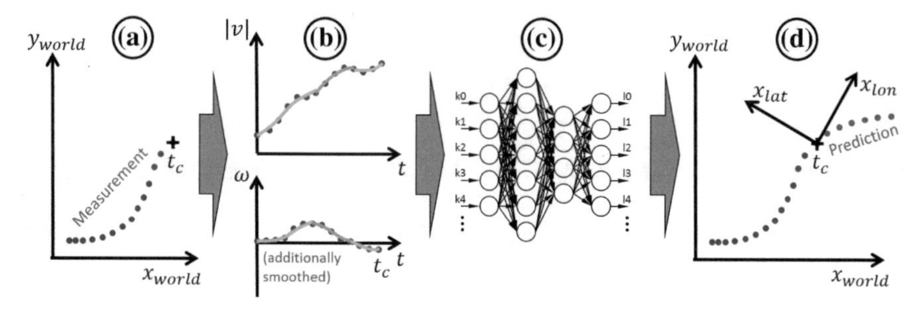

Fig. 2 Overview of the proposed path prediction method

In an own preliminary publication [10] we used the approach of transforming the measured past track into the pedestrian's ego coordinate system, in order to receive an input vector based on $v_{lon}(t)$ and $v_{lat}(t)$ independent of the global position x_{world}, y_{world} and orientation φ_{world}. This transformation has to be reapplied to the track at every time cycle since the pedestrian's position and orientation is continuously changing. In the approach described here, the transformation is substituted by an extraction of the absolute horizontal velocity $|v(t)|$ and the angular velocity $\omega(t)$. Those parameters are invariant to the current values of x_{world}, y_{world} and φ_{world}. Therefore, they have to be calculated only for the current time step while the preceding values remain the same. This sliding window behavior of the extracted time series has great advantages regarding the computational efficiency of the subsequent polynomial fitting, as the fast update algorithms of our *Fast Approximation Library* [7] can be applied. In return, $\omega(t)$ contains more noise compared to the previous approach and, therefore, an additional on-line exponential low-pass IIR filtering of the time series with

$$\omega_{sm,t} = \alpha \cdot \omega_t + (1 - \alpha) \cdot \omega_{sm,t-1} \tag{1}$$

is added with ω_t being the current, $\omega_{sm,t}$ the current filtered and $\omega_{sm,t-1}$ the preceding filtered angular velocity value. The smoothing factor α is included as additional input parameter to the optimization process for the prediction quality.

2.2 Approximation with Polynomials

The time series $|v(t)|$ and $\omega(t)$ are approximated with polynomials. This approximation is based on a least-squares error and orthogonal basis polynomials. The coefficients of the orthogonal expansion of the approximating polynomial serve as principal information sources as they represent the temporal development of the pedestrian's movement. The coefficients of the orthogonal expansion can be regarded as optimal estimators of average, slope, curve, change of curve, etc. of the time series in a considered time window [5, 6]. The approximating polynomial $f(t)$ is a linear combination of the basis polynomials $f_k(t)$

$$f(t) = \sum_{k=0}^{K} w_k \cdot f_k(t) \tag{2}$$

at a finite set of points in time $t_0, \ldots, t_N$. The objective is to solve the least squares problem

$$\min_{\mathbf{w}} \|\mathbf{F}\mathbf{w} - \mathbf{s}\|^2 , \tag{3}$$

where $\|\ldots\|$ is the Euclidean norm, $\mathbf{s}$ are the time series values (targets), $\mathbf{w}$ is the vector containing the coefficients of the respective polynomials and $\mathbf{F}$ is a matrix (the design matrix) of form

$$\mathbf{F} = \begin{pmatrix} f_0(t_0) & \ldots & f_K(t_0) \\ \vdots & \ddots & \vdots \\ f_0(t_N) & \ldots & f_K(t_N) \end{pmatrix} . \tag{4}$$

As the approximation has to be performed in a sliding window manner on our approach, the coefficients can be computed with extremely fast update algorithms if certain kinds of orthogonal basis polynomials, in our case discrete Legendre polynomials, are used. The approximation is capable of reducing the dimensionality of a feature vector serving as input for the subsequent self-learning predictor. At the same time the influence of measurement noise is reduced due to implicit data smoothing. A third advantage is that the coefficients can be calculated independent of sensor parameters such as the sampling rate. That is, a predictor could be trained with one and used with another sensor and sampling rate. Even changes of the sampling rate within one time series or handling of missing measurements are conceivable.

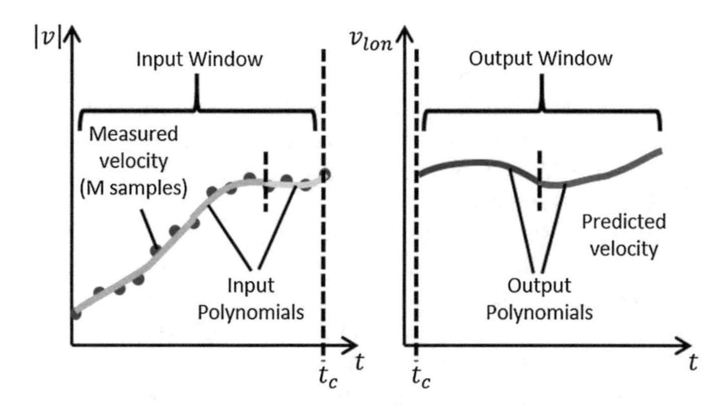

Fig. 3 Schematic representation of polynomial approximation for movement prediction for the case of $|v|$ (*left*, similar for ω, not shown). The coefficients of the fitted input window polynomials are used to estimate those of the v_{lon}, v_{lat} timeseries in the output window (*right*, only v_{lon} is shown). t_c indicates the current time step

In the proposed method we make use of multiple polynomials fitted in different temporal sub-windows together representing the overall input window. The example in Fig. 3 shows two consecutive input polynomials (left: green, orange) approximating the measured velocity (blue dots) where t_c indicates the current time step. The technique allows, e.g., to split the input window into separate sub-windows for short- and long-time observation as shown in this example. Another exemplary variant are multiple small input windows in order to get a closer approximation to periodic velocity variations, which occur within the human gait.

According to the procedure for the input time window the method also uses a polynomial representation for the predicted output information. However, contrary to the input, the 2D velocity in the pedestrian's ego system $v_{lon}(t)$, $v_{lat}(t)$ is taken as output time series as experiments show that even small prediction errors of $\omega(t)$ lead to relatively large errors when transforming back into estimated global positions. As the polynomial approximation for the output window has only to be performed to generate the training output coefficients but not during an on-line processing where the future track is generated from the predicted output coefficients, this does not yield a disadvantage regarding processing time.

For the training of the MLP-based prediction process polynomials are fit in a single or multiple defined output sub-windows. The polynomial coefficients are taken as targets for the MLP models in order to predict them based on the input window coefficients in an on-line mode. The output polynomials can thus be evaluated to obtain a continuous prediction of the future movement for the learned output time window. While overlapping sub-windows are possible as input only non-overlapping consecutive sub-windows are feasible for the output in order to get a unique position estimate for each future point in time. In the visualized example two consecutive output polynomials are predicted (right of t_c: violet, red). The output window in the example is also splitted into two sub-windows, here with constant size.

The number of polynomials, their temporal position and length, as well as the degree of each polynomial can be varied to optimize the prediction quality. The possible upper limit of the variation of polynomial degree K to get an unique approximation solution is $N-1$ where N is the number of measured velocity values within the considered time window. This limit is only relevant when short windows and low sampling rates are used at the same time.

The coefficients $k=0,\ldots,K$ are estimators of the average, slope, curve, change of curve, etc., of the polynomial. The underlying reference time unit is the cycle time $T_{cyc}=1/f_s$ of the time series, with f_s being the sampling rate of the source data. As we want the coefficients to be independent of the sampling rate, they are transformed to a reference time of 1 s by multiplying the scaling factor f_s^k with each coefficient. The transformation results in normalized coefficients with the physical units m/s, m/s^2, m/s^3, etc.

The variation of parameters leads to some special cases at the edges of the parameter space for the input sliding window configuration:

- One single polynomial with $K=0$: Corresponds to the average (absolute and angular) velocity of the regarded time window. Degree $K=1$ adds the average acceleration, and so on.
- The case of N polynomials with $K=0$ corresponds to the direct usage of the time series $|v(t)|$ and $\omega(t)$ as input pattern for the subsequent predictor. This is similar to the technique used for MLP prediction in [8].

2.3 Prediction of Polynomial Coefficients Using a Multilayer Perceptron

As predictor we use a feed-forward artificial neural network (ANN) in form of a multilayer perceptron (MLP, see e.g. [12]). A MLP is capable of predicting multiple output values at the same time what constitutes a great advantage when multiple time steps and dimensions shall be estimated in parallel (Fig. 2c). The network consists of neurons with the sigmoid activation function

$$f(x)=\frac{1-e^{-x}}{1+e^{-x}} \tag{5}$$

and is trained with the Resilient Backpropagation (RPROP) algorithm [16]. The normalization of the input data is done using the statistical z-transformation

$$z_i=\frac{x_i-\overline{x}}{s} \tag{6}$$

where x_i are the original input values, $\overline{x}$ the mean and s the standard deviation of training values per input dimension and z_i the z-transformed values. The size of the input layer is determined by the number of polynomials N_{Pol} and their degrees K_i with

$$N_{in} = 2 \cdot \sum_{i=0}^{N_{Pol}} (K_i + 1). \quad (7)$$

The size and number of hidden layers are variable and part of the optimization process. The network output consists of polynomial coefficients of the future trajectory represented in the ego velocity time series $v_{lon}(t)$ and $v_{lat}(t)$, which were determined for training by polynomial approximation. The output may also consist of multiple polynomials for separate time windows. To extract the trajectory estimation the output polynomials are evaluated for the required future points in time. From the current time on the predicted $v_{lon}(t)$ and $v_{lat}(t)$ are numerically integrated to obtain the positions $x_{lon}(t)$, $x_{lat}(t)$. Afterwards, the prediction is transformed back into the world coordinate space $x_{world}(t)$, $y_{world}(t)$ (see Fig. 2d).

3 Evaluation of Prediction Quality

To evaluate the quality of the proposed methods we regard a time window of $[-1, 0]$ s for the input and $]0, 2.5]$ s for the output of the predictor, relative to the time stamp of the current measurement. The older the input measurements are, the less influence they have on the result. Also, longer input windows require longer initialization time after the first detection of a pedestrian in a practical application, until the first prediction is available. Tests using MLP prediction based on direct input of the velocity measurements show that input windows larger than 1.5 s do not result in further improvements of the prediction quality on our data. With an input window of 1 s a prediction quality of over 97 % of the measured optimum at 1.5 s is reached. To make the prediction results of different configurations comparable we set the overall input window length to this value. The prediction horizon is set to 2.5 s since this value suffices for autonomous reactions of a vehicle as well as for effective driver warnings in advanced driver assistant system (ADAS) scenarios [15].

As major quality indicator we evaluate the average Euclidean error (AEE) from the predicted position to ground truth (GT). As ground truth the tracked head center position is used whereby the tracking in the stereo images and in world coordinates are manually inspected to avoid tracking errors and outliers. The AEE is defined for all P predictions in the test data with a specific prediction time step t_{pred} as

$$AEE(t_{pred}) = \frac{1}{P} \sum_{i=1}^{P} \sqrt{(x_{pred}(i) - x_{GT}(i))^2 + (y_{pred}(i) - y_{GT}(i))^2} \quad (8)$$

with $(x_{pred}; y_{pred})$ being the predicted and $(x_{GT}; y_{GT})$ being the ground truth positions. As the AEE is based on the Euclidean norm, the results are independent of the underlying coordinate system (here: global or ego coordinates). In order to evaluate the prediction quality for the longitudinal and lateral directions the average position error is also evaluated separately for these dimensions (AE_{lon} and AE_{lat}).

As the minimization of the *AEE* can only provide an optimal overall parameter setting for a specific prediction time step t_{pred} but not for the whole predicted output time window, another indicator is required as sole optimization target value in this case. Since the prediction error naturally rises with increasing t_{pred} an averaging of all *AEE* values would outweigh more recent prediction times. To avoid this problem we decided to consider the *AEEs* with regard to the respective t_{pred} and to calculate the weighted average of the AEEs as target value which has to be minimized:

$$ASAEE = \frac{1}{N} \sum_{i=1}^{N} \frac{AEE(t_{pred}(i))}{t_{pred}(i)}, \tag{9}$$

with N being the number of discrete subsequent prediction steps within the considered prediction horizon and *ASAEE* being the average specific *AEE*.

For our experimental studies we use a database of 1075 pedestrian tracks recorded with the sensor setup described in [9] at a public test intersection. All tracks have a sampling rate of 50 Hz and lengths between 4 and 10 s or 200–500 samples, respectively. They are divided into four types of movement: "Waiting", "Starting", "Walking" and "Stopping". Scenes of type "Waiting" are scenes with persons standing and usually waiting for the light signal to cross the road performing only small movements not exceeding a head velocity of 0.3 m/s. "Walking" consists of straight walking with constant or changing velocities as well as bending-in scenarios. "Starting" and "Stopping" include the corresponding motion transition with one second before and three seconds after. The track database is split into training (60 %, 643 tracks) and test data (40 %, 432 tracks) with equal split ratio for each included scene type (see Table 1). For the optimization of the individual stages of the prediction method the training data is further divided into 70 % training to 30 % test data. Due to the large number of variable parameters a *k*-fold cross-validation is not performed.

Table 1 Total number of trajectories sorted by training data, test data, and type of movement

Type	Training	Test	Total
Waiting	117	78	195
Starting	184	124	308
Walking	204	137	341
Stopping	138	93	231
All	643	432	1075

4 Results

In this section we present the tested parameter ranges and resulting prediction quality values based on the evaluation methods introduced in Sect. 3. As baseline method for comparison we use CV Kalman Filtering.

As the different phases of the proposed method (input polynomial approximation, MLP, output polynomial prediction) provide a large amount of parameters to minimize the overall prediction error, an extensive grid search in the entire parameter space is not feasible. Instead, the phases are optimized separately in alternating manner. The parameters of the separate stages are optimized with coarse-to-fine grid searches, if possible (MLP network structure), or by manual variation of parameters (the polynomial stages). The varied parameters include

- the exponential smoothing factors of the input time series,
- the number, temporal position, length and degrees of input and output polynomials,
- the number and size of hidden layers of the MLP,
- the parameters C, γ and ε of the alternatively used SVR method with RBF kernels.

4.1 Variation of Polynomial Parameters

In this section we evaluate the effect of changes in size, position, length, and degree of the input and output polynomials. The overall window sizes are thereby held constant at the values defined in Sect. 3: 1 s for the input and 2.5 s for the output window. Figure 4 shows the results of a variation of the polynomial degree K for four sample input configurations.

As base configuration for a comparison of different input configurations, we use $|v|$ and ω data directly (but normalized) as input for the MLP. This corresponds to an input layer size of 100 neurons in our sensor setup with 50 Hz and 1.0 s total input time. The configuration is equivalent to 50 polynomials per dimension with a degree of 0. For the following evaluations of the input structure, the structure of the MLP

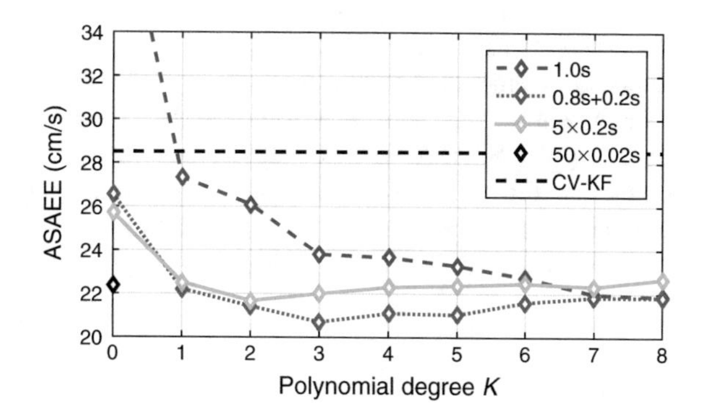

Fig. 4 Quality indicator *ASAEE* depending on input polynomial degree K, input sub-window numbers, positions, and sizes in comparison to the baseline technique of CV Kalman filtering (CV-KF). Lower *ASAEE* values are better. The examples only show an excerpt of the evaluated input configurations

and the output polynomials were kept constant at their optimized final configuration (two hidden layers with 20 and 8 neurons, 5 consecutive output polynomials with degree 2 and 0.5 s length). The given configuration leads to an *AEE* of 20.7 cm for 1.0 s and 71.0 cm for 2.5 s prediction time. The resulting *ASAEE* value is 22.4 cm/s (Fig. 4, single black marker at $K = 0$).

The evaluation for different input polynomial structures generally shows a trend to decreasing prediction errors with increasing polynomial degrees, according to the gain of usable information content for the neural network. The improvement slows down with higher degrees until a saturation plateau is reached, whereby a stronger splitting into more input polynomials leads to an earlier flattening of the *ASAEE* curve. With still higher polynomial degrees a slight increase of the prediction error is observable. In Fig. 4 the blue dashed line shows the configuration with a single polynomial covering the whole input time horizon of 1.0 s. Here, $K = 0$ describes the minimum tested input size using only the average two dimensional velocity. This already leads to an *ASAEE* of 41.8 cm/s. The step from $K = 0$ to $K = 1$ adds the values of mean acceleration to the input, which leads to a benefit of a 35 % lower *ASAEE* (27.3 cm/s) on our test data and already slightly outperforms CV Kalman filtering (28.5 cm/s, see Fig. 4, line "CV-KF"). Using a single polynomial the minimum error plateau is reached at degree $K = 8$ with 21.87 cm/s *ASAEE* according to an input layer size of 18. The green line represents the result for an input of five sequential input polynomials with the same window lengths of 0.2 s. In this case, polynomial degrees of 2 already suffice for the best prediction quality, but due to the higher number of sliding windows the input layer size grows to 30. The red dotted line represents an asymmetrical input configuration with a 0.8 s window followed by a 0.2 s window. The intention is to set an additional attention on short-time features in the second window, while the first covers the longer time distance up to 1.0 s. The configuration also reaches the quality of the previous configurations at a degree of 2 but already with a small input layer size of 12. The best result is reached at degree 3 and layer size 18 with an *ASAEE* of 20.7 cm/s. This input configuration will be used for further quality measurements presented in this work. Other tested input structures lead to similar results, e.g., 2×0.5 s, $0.4 + 0.3 + 0.2 + 0.1$ s, 10×0.1 s and configurations with overlapping sub-windows of 1.0 and 0.2 or 0.1 s. Altogether we can state that it is possible to use only 2 windows without reduction of the prediction quality.

The parameters of output polynomials have less influence on the prediction quality than the input parameters. The *ASAEE* also improves slightly with increasing polynomial degree but with a much earlier entering the plateau at degree 1–2 depending on window number and size. The overall *ASAEE* difference between a single and multiple output sub-windows is only approximately 2 % using a degree of 2. For special cases, e.g., the acceleration phase after an initial movement, the slight improvement is visible in velocity plots (see Fig. 5). For all further evaluations we choose an output of 5 consecutive polynomials of degree 2 and 0.5 s length each.

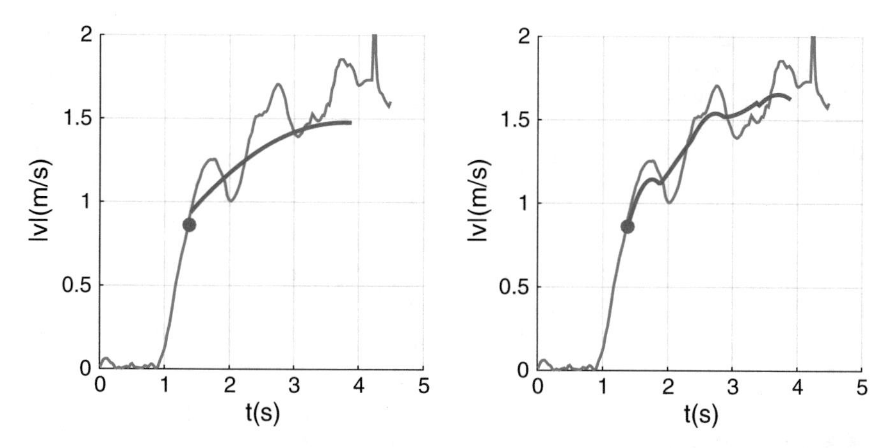

Fig. 5 Comparison of a sample prediction of a starting movement with two different output polynomial configurations in a velocity magnitude plot. The *blue line* represents the actual measurements over time, the *red dot* marks the current point in time, the *red line* represents the prediction. *Left* single polynomial with $K = 2$ and 2.5 s window length. *Right* 5 consecutive polynomials with $K = 2$ and 0.5 s window length each

4.2 Variation of Neural Network Structure

Though the size of the input and output layers is defined by the polynomial structure only size and number of hidden layers can be varied. We performed a coarse-to-fine grid search for network topologies with 0–3 fully connected hidden layers and numbers of 2–40 neurons per layer, including different layer sizes. The results show remarkable improvements for networks with two hidden layers in comparison to those with one or zero while an extension to three hidden layers shows no further advantages. The best result of a grid search with the $0.8 + 0.2$ s input configuration and degree 3 is given with 20 neurons in the first and 8 neurons in the second hidden layer (see Fig. 6). Similar network configurations with two hidden layers, e.g., using 4 instead of 8 neurons in the second hidden layer, lead to very similar results, in this case a slight increase of the *ASAEE* of 1.5 %.

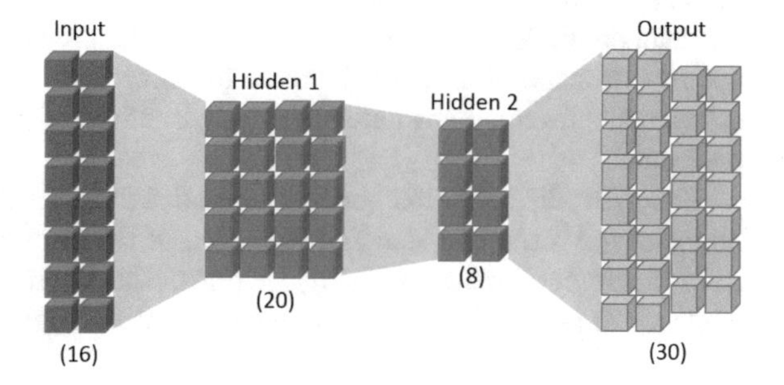

Fig. 6 Architecture of the finally used ANN. The best results were archived with a fully connected MLP with two hidden layers and 16–20-8–30 neurons

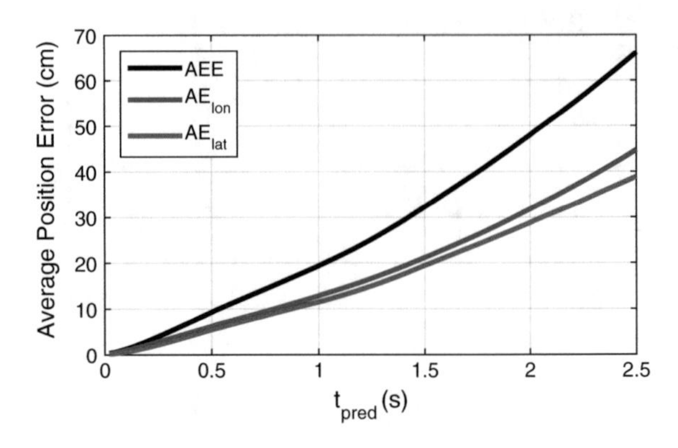

Fig. 7 Average Euclidean (AEE), average longitudinal (AE_{lon}) and lateral position errors (AE_{lat}) for the complete prediction horizon up to 2.5 s

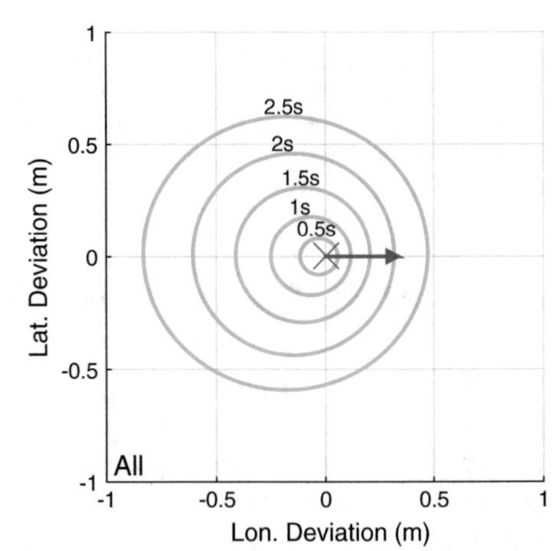

Fig. 8 Error ellipses for five specific prediction times from 0.5 to 2.5 s for all types of movement. The *red arrow* shows the direction of movement, the ellipses indicate the mean and standard deviation of the estimated position relative to the ground truth in the ego coordinate system of the pedestrian

4.3 Quality Depending on Prediction Horizon and Type of Movement

In order to compare the prediction error for different prediction times t_{pred} in this section we evaluate the AEE performance for several time steps. Figure 7 visualizes the change of the AEE over the regarded prediction time horizon of 2.5 s for all test data predictions. As one can observe the 2D prediction error increases slightly disproportionate with growing prediction time t_{pred}. The longitudinal and lateral components (average absolute errors AE_{lon} and AE_{lat}) show almost equal behavior over time.

The error ellipses for five specific prediction times are drawn in Fig. 8. Each ellipse indicates the mean and standard deviation of the estimated position to the ground truth in ego coordinates of the pedestrians. This overall evaluation shows a slight tendency of predicted velocities that are to low in longitudinal direction.

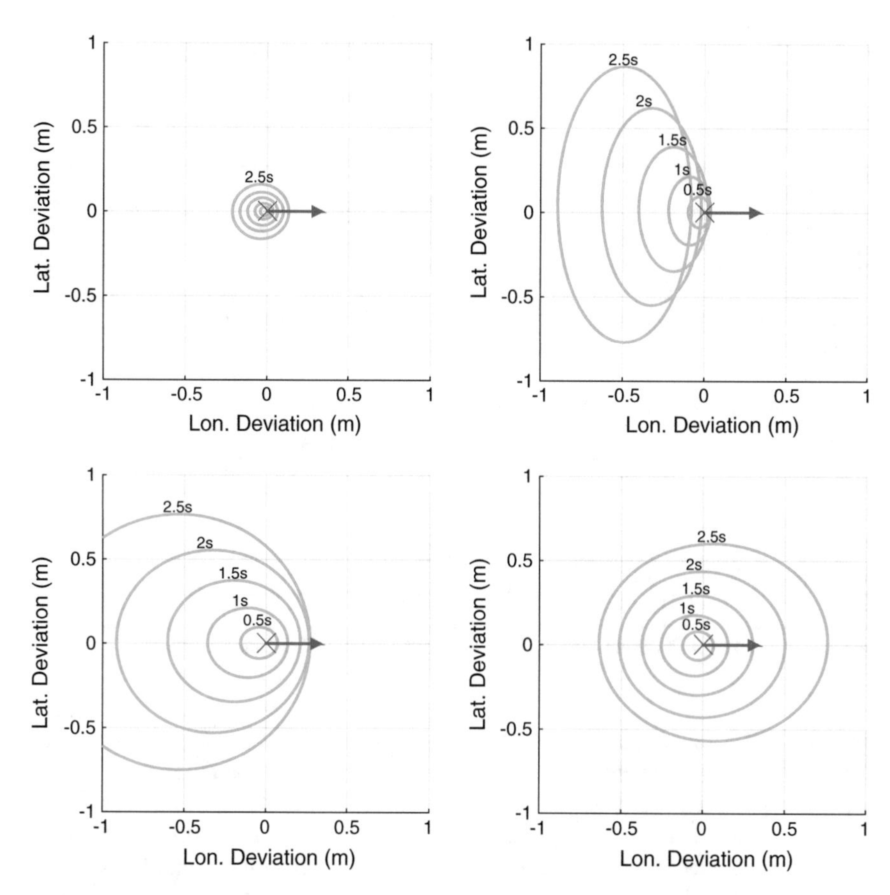

Fig. 9 Error ellipse plots for the four different types of movement labeled in the test data. Waiting (**a**), walking (**b**), starting (**c**), and stopping (**d**)

To investigate this aspect further a similar plot is generated for each of the four labeled movement types separately in Fig. 9. In Fig. 9a the result for people standing and waiting on the sidewalk shows relatively small errors, as expected. The error ellipses appear as concentric circles while the standard deviation even for the maximum prediction time of 2.5 s is just 1 cm. Regarding the evaluation of walking motion in Fig. 9b a shift of the means against the moving direction and a dominant lateral standard deviation are visible. This effect is mainly based on changes of walking direction generating a lateral deviation while the original movement vector in longitudinal direction gets shorter. Figure 9c shows the same plot for starting motions, where even more significant shifts of the mean prediction errors against the moving direction are visible. They arise from the fact that a prediction of the initial movement from a standing position over several seconds is a very difficult task, for machine vision as well as for human observers. The discrepancy between the prediction of an almost constant position and the person starting to move forward in the ground

Table 2 Quantitative prediction error results for the proposed polynomial-MLP method for the four investigated types of pedestrian movement and overall

Mov. type	AEE (1.0 s) (cm)	AEE (2.5 s) (cm)	ASAEE (cm/s)	Comp. to KF (%)
Overall	19.4	85.9	20.7	−27.4
Waiting	5.3	14.6	5.8	−0.8
Starting	26.6	97.6	28.6	−28.0
Walking	21.4	85.9	24.4	−23.1
Stopping	22.3	73.4	23.2	−32.0

In the right columns the percentage difference of *ASAEE* to CV Kalman Filter prediction is listed for comparison

Table 3 Quantitative errors of CV Kalman Filter prediction

Mov. type	AEE (1.0 s) (cm)	AEE (2.5 s) (cm)	ASAEE (cm/s)
Overall	27.9	95.0	28.5
Waiting	6.1	14.6	5.9
Starting	38.0	136.1	39.7
Walking	30.9	99.1	31.7
Stopping	33.2	122.8	34.1

The values are used as baseline for the evaluation of the proposed method in Table 2

truth data leads to this effect. The lateral errors show that estimating the movement direction for standing persons is also a challenging task for the algorithm. Ways to solve this problem could be the consideration of additional input information for prediction, such as the viewing direction of the pedestrian or the orientation of the road she or he intends to cross. In Fig. 9d the opposite effect is visible: For stopping persons the algorithm slightly tends to overestimate the velocity, such that a mean shift in direction of movement occurs. Since the stopping motions of people in traffic scenarios take generally more time than the starting motions (see [9]) the predictor has more time to react on appearing characteristic features.

Detailed numerical results for two prediction times 1.0 and 2.5 s as well as the *ASAEE* are given in Table 2. A comparison to the *ASAEE* values of the KF method taken as baseline (see Table 3) is shown in the right column (reduction of the *ASAEE* in percentage) and as bar plot in Fig. 10. The results show an overall improvement of the prediction result of 27.4 % compared to the Kalman Filter. Considering the individual movement types, the most benefit of our method occurs at starting (28.0 %) and stopping scenes (32.0 %) but also the prediction of walking scenes including bending is improved (23.6 %). The waiting scenarios show almost equal results for the proposed method and the Kalman Filter.

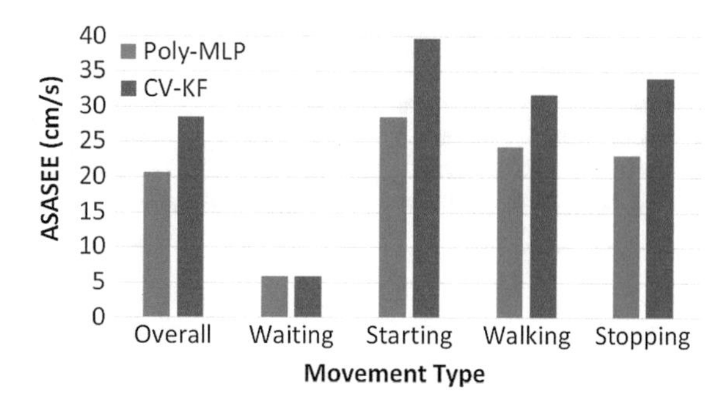

Fig. 10 Visual comparison of ASAEE values for the proposed polynomial-MLP and the baseline Kalman Filter method

4.4 *Early Recognition of Initiation of Gait*

As shown in the previous section the prediction of a starting motion for a standing pedestrian is a challenging task. At the same time this case is one of the most common and important ones in public traffic scenarios. Therefore we investigate the detection time and precision of the initial movement detection with the proposed method. The test data for this evaluation consists of 40 tracks of non-instructed pedestrians waiting at the roadside and then crossing the road. We define the moment of the heel-off of the first foot moving as reference time $t = 0$ s and labeled the associated time stamp by manually observing the video data. The initial movement is considered as detected successfully when the predicted position for $t_{pred} = 1.0$ s exceeds a specified Euclidean distance from the current position. This corresponds to a threshold for the absolute velocity averaged over the first second of the predicted time span. In order to vary the sensitivity of the motion detection this threshold can be shifted, what which is equivalent to moving the operating point of the detector. A lower velocity threshold accelerates the recognition of the starting motion but increases the probability for false alarms during the standing phase. For our evaluation we define already a single violation of the threshold during the standing phase of a scenario as false alarm (*FA*), while the correct detection during or after the labeled heel-off without prior false alarms counts as true positive (*TP*). The precision P over all scenes is defined as $P = TP/(TP + FA)$ and depends on the chosen threshold and, thus, on the detection time relative to heel-off.

The resulting relationship of precision and detection time is plotted in Fig. 11 (red line). The evaluation shows that already at the moment of heel-off a precision of 80 % is reached. This confirms the suitability of the head position tracking as input source for the neural network since 80 % of the initial movements are correctly detected before there is any movement of a foot. During the next 300 ms after heel-off the reachable precision exceeds 95 %. It should be mentioned that the predictor tested here is not optimized for this motion detection application, especially it is not designed to comply with the defined false alarm rate criteria of exceeding a low threshold during the waiting phase. The objective function in the training has been the goal to minimize the 2D position prediction error, not the binary movement state

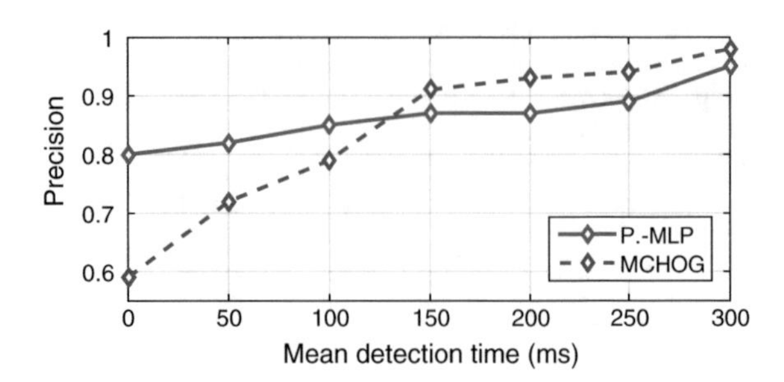

Fig. 11 Precision over mean relative detection time to heel-off for initial movement detection using the proposed prediction method (*red*) and the MCHOG method (*blue*)

(standing/walking). As comparison the same scenes are evaluated using the video-based initial movement detection method presented in [14] (MCHOG, blue dashed line). As one can see, our method shows advantages for the short prediction times up to 100–150 ms after heel-off while the MCHOG shows slightly higher precision rates afterwards.

4.5 Runtime Performance

The computation time of the used trajectory-based algorithms is very short compared to the cycle time of commonly used sensors and algorithms for pedestrian detection. Nevertheless, in this section the runtime performance of the proposed method is evaluated.

Using the multi-polynomial configuration evaluated in the previous section (two input and five output polynomials) the prediction from input to output track requires 35 μs on an Intel Core i7-3770 CPU with 3.4 GHz. So, under the preference of an available pedestrian detection and tracking system, the algorithms should operate on small embedded systems. Compared to the originally proposed method using v_{lon} and v_{lat} as prediction input [10] we could improve this value from 252 μs by a factor of 7, which is mainly due to faster preprocessing and exploiting the fast sliding window polynomial update functions for the $|v|$ and ω time series. The processing times of single steps are shown in Table 4.

Table 4 Processing time of the single modules, averaged over the entire test data

Module	Processing time (μs)
Track preprocessing	< 1
Polynomial fitting	14
Neural network prediction	5
Prediction reconstruction	15

Table 5 Comparison of *AEE* of MLP to SVR method for different prediction times

Pred. time (s)	0.5	1.0	1.5	2.0	2.5
AEE (cm) (MLP)	9.3	19.4	32.3	48.0	66.1
AEE (cm) (SVR)	9.4	19.9	33.4	49.5	68.6
Comparison (SVR: 100 %)	−1.1 %	−2.5 %	−3.3 %	−3.0 %	−3.6 %

The offline training of the predictor was performed on the same machine. It needed computational times between 20 s and 6 min for 643 tracks depending on the used polynomials and neural network topology.

4.6 Comparison to Support Vector Regression

For the purposes of comparison to a kernel-based alternative to MLP we also investigated Support Vector Regression (SVR, see [2]). Therefore, the MLP prediction module is substituted by an ε-SVR based on the LibSVM library [3] while the polynomial input remains unchanged.

We applied a radial basis function (RBF-) kernel and optimized the parameters C, γ, and ε by a coarse-to-fine grid search. Because SVR only allows a single output, we trained two instances to predict 2D positions for single prediction times in each case instead of a continuous estimation over a timespan based on the output polynomial coefficients. The resulting *AEE* values and a comparison to the above approach are set out in Table 5.

The evaluation shows a slight advantage of the MLP method. It yields smaller *AEE* values for all tested prediction horizons while the improvement rises with higher prediction times up to 3.6 % for 2.5 s. Besides the lower prediction errors a major benefit of the neural network is the capability to predict positions for a continuous future timespan with a single instance at the same time.

5 Conclusion

In this publication we proposed a method for the prediction of pedestrian trajectories by polynomial least-squares approximations in combination with MLP neural networks. The method is trained and tested on 1075 different tracks of pedestrians in real urban scenarios. Our implementation features the prediction of a continuous future trajectory for a time horizon of 2.5 s using camera-based head tracking data of the most recent time interval of 1.0 s as input. Due to the usage of a self-learning method as “implicit” movement model, a sole prediction technique is capable of handling different movement types crucial for traffic safety, e.g., starting and stopping. The polynomial approximation of the input tracks provides great flexibility as it allows for independence from sensor type and sampling rate. For a prediction time of 1 s our tests result in average Euclidean errors of 27 cm for starting and 22 cm for stopping scenes, and for 2.5 s in 98 and 86 cm, respectively. The proposed method outperforms the prediction quality of a CV model Kalman filter by 27.4 % over all test data, by 28 % for starting and by 32 % for stopping scenes.

Our future work will include the application of the method to bicyclists, who constitute another important proportion of VRUs. A vehicle-based implementation of the method requires an additional ego-motion compensation as the pedestrian track in global coordinates is used as input. Also, systems based on a forward-looking stereo camera will be investigated. As cars have a limited view on traffic scenarios and because not all intersections will be equipped with sensors, we envision a scenario where the protection of VRUs is realized in a cooperative way. The collective intelligence of cars, complemented by information from infrastructure (where available) and VRUs themselves (if equipped with intelligent devices such as smartphones) will be exploited to detect the intention of VRUs in a distributed way. This approach will not only provide an essential component for future traffic automation, it will also increase the safety of road users.

Acknowledgments This work partially results from the project AFUSS, supported by the German Federal Ministry of Education and Research (BMBF) under grant number 03FH021I3, and the project DeCoInt[2], supported by the German Research Foundation (DFG) within the priority program SPP 1835: “Kooperativ Interagierende Automobile”, grant numbers DO 1186/1-1 and SI 674/11-1.

References

1. Bar-Shalom, Y., Li, X.R., Kirubarajan, T.: Estimation with Applications to Tracking and Navigation: Theory Algorithms and Software. Wiley, New York (2001)
2. Chang, C.C., Lin, C.J.: Training ν-support vector regression: theory and algorithms. Neural Comput. **14**(8), 1959–1977 (2002)
3. Chang, C.C., Lin, C.J.: LIBSVM: a library for support vector machines. ACM Trans. Intell. Syst. Technol. **2**(3), 27:1–27:27 (2011). Software available at http://www.csie.ntu.edu.tw/~cjlin/libsvm. Accessed 9 Dec 2015
4. Euro NCAP: Euro NCAP 2020 Roadmap (2015). http://euroncap.blob.core.windows.net/media/16472/euro-ncap-2020-roadmap-rev1-march-2015.pdf. Accessed 9 Dec 2015
5. Fuchs, E., Gruber, T., Nitschke, J., Sick, B.: On-line motif detection in time series with Swift-Motif. Pattern Recognit. **42**(11), 3015–3031 (2009)
6. Fuchs, E., Gruber, T., Nitschke, J., Sick, B.: Online segmentation of time series based on polynomial least-squares approximations. IEEE Trans. Pattern Anal. Mach. Intell. **32**(12), 2232–2245 (2010)
7. Gensler, A., Gruber, T., Sick, B.: Fast approximation library. http://ies-research.de/Software. Accesed 17 Dec 2013
8. Goldhammer, M., Doll, K., Brunsmann, U., Gensler, A., Sick, B.: Pedestrian's trajectory forecast in public traffic with artificial neural networks. In: Proccedings of the 22nd International Conference on Pattern Recognition (ICPR), pp. 4110–4115 (2014)

9. Goldhammer, M., Hubert, A., Köhler, S., Zindler, K., Brunsmann, U., Doll, K., Sick, B.: Analysis on termination of pedestrians' gait at urban intersections. In: Proceedings of the IEEE 17th International Conference on Intelligent Transportation Systems (ITSC), pp. 1758–1763 (2014)
10. Goldhammer, M., Köhler, S., Doll, K., Sick, B.: Camera based pedestrian path prediction by means of polynomial least-squares approximation and multilayer perceptron neural networks. In: Proceedings of the SAI Intelligent Systems Conference, pp. 390–399 (2015)
11. Goldhammer, M., Strigel, E., Meissner, D., Brunsmann, U., Doll, K., Dietmayer, K.: Cooperative multi sensor network for traffic safety applications at intersections. In: Proceedings of the IEEE 15th International Conference onIntelligent Transportation Systems (ITSC), pp. 1178–1183 (2012)
12. Haykin, S.: Neural Networks and Learning Machines. Prentice Hall, New York (2009)
13. Kalarot, R., Morris, J.: Comparison of fpga and gpu implementations of real-time stereo vision. In: Proceedings of the 2010 IEEE Computer Society Conference on Computer Vision and Pattern Recognition Workshops (CVPRW), pp. 9–15 (2010)
14. Koehler, S., Goldhammer, M., Bauer, S., Zecha, S., Doll, K., Brunsmann, U., Dietmayer, K.: Stationary detection of the pedestrian's intention at intersections. Intell. Transp. Syst. Mag., IEEE **5**(4), 87–99 (2013)
15. Naujoks, F.: How Should I Inform my Driver? (2013). http://ko-fas.de/files/abschluss/ko-fas_c1_4_effective_advisory_warnings_based_on_cooperative_perception.pdf. Accessed 9 Dec 2015
16. Riedmiller, M., Braun, H.: A direct adaptive method for faster backpropagation learning: the RPROP algorithm. In: Proceedings of the IEEE International Conference on Neural Networks, pp. 586–591 (1993)
17. Winter, D.A.: Human balance and posture control during standing and walking. Gait Posture **3**(4), 193–214 (1995)
18. World Health Organization: Global Status Report on Road Safety 2013: Supporting a Decade of Action (2013). http://www.who.int/violence_injury_prevention/road_safety_status/2013/en. Accessed 9 Dec 2015

Assembly Assisted by Augmented Reality (A^3R)

Jun Okamoto Jr. and Anderson Nishihara

Abstract Traditionally, assembly instructions are written in the form of paper or digital manuals. These manuals contain descriptive text, photos or diagrams to guide the user through the assembly sequence from the beginning to the final state. To change this paradigm, an augmented reality system is proposed to guide users in assembly tasks. The system recognizes each part to be assembled through image processing techniques and guides the user through the assembly process with virtual graphic signs. The system checks whether the parts are properly assembled and alerts the user when the assembly has finished. Some assembly assisted by augmented reality systems use some kind of customized device, such as head mounted displays or markers to track camera position and to identify assembly parts. These two features restrict the spread of the technology whence, in this work, customized devices and markers to track and identify parts are not used and all the processing is executed on an embedded software in an off-the-shelf device without the need of communication with other computers to any kind of processing.

1 Introduction

In assembly tasks several elements need to be addressed such as the part's orientation and the assembly sequence. As a way to assist assembly tasks, several authors have proposed augmented reality systems [1–3]. They proposed that images augmented with instructions may be an alternative to the traditional paper manual and it would improve the assembly time. Implementations of augmented reality systems were compared to computer assisted instructions, paper manuals and tutorial by an expert. Although augmented reality (AR) systems did not improve assembly time compared to a tutorial by an expert [2], it was evidenced that assembly time can be improved

J. Okamoto Jr. (✉) · A. Nishihara
Department of Mechatronics and Mechanical Systems Engineering,
Escola Politécnica of the University of São Paulo, São Paulo, Brazil
e-mail: jokamoto@usp.br

A. Nishihara
e-mail: anderson.nishihara@usp.br

Y. Bi et al. (eds.), *Intelligent Systems and Applications*,
Studies in Computational Intelligence 650, DOI 10.1007/978-3-319-33386-1_14

in comparison to computer assisted instructions [1, 3] and paper manuals [1–3]. Yet, when an augmented reality system was used the number of errors in the assembly process was reduced [1, 3].

This research consists in development of an augmented reality system to assist the assembly process in real time as a replacement to conventional assembly manuals. The proposed system behaves as an interactive instruction manual where virtual models and graphic signs are overlaid on a video of the real scene so that the user is guided until the final assembled state is reached. To prove the concepts described here, a tablet is used between the user and the workspace in a way that a set of randomly placed assembly parts can be viewed on the display. From the video feed, the parts to be assembled are identified through image processing techniques and graphic signs show to the user where the piece should be assembled. Lastly, the system checks the assembly state and informs the user when the final state is reached.

The developed system has the following characteristics:

- Fiducial markers are not used: objects are detected by image processing and named from a database;
- The solution is known beforehand: the system is not for finding a solution to the assembly problem;
- The system does not teach the assembly process to the user, its purpose is to be a guide through the assembly process;
- The display is placed between the user and the assembly workspace and it is fixed during all the process;
- A general assembly guidance process is proposed and validated;
- The tablet implementation runs an embedded software, only resources from the tablet are used and no network connection is required.

This paper presents the results obtained by the object recognition subsystem and the assembly guidance process of the developed system that is being called Assembly Assisted by Augmented Reality (A^3R).

2 Augmented Reality in Assembly Tasks

One of the first efforts to build an AR tool was shown in [4] to guide the assembly of cable harnesses. Information derived from engineering design of parts and processes come to the factory floor in the form of templates, assembly guides, cable lists and location markings. Expenses and delays on manufacturing can be minimized if changes on engineering design of parts and processes can be quickly mirrored to templates, assembly guides, cable lists and location markings. To address this issue, an AR system composed of a wearable computer and a head mounted display (HMD) is demonstrated in [4]. Later, researchers demonstrated applications to guide printer maintenance [5] and car door-lock assembly [6].

Specifically on AR applications in assembly processes, we can analyze the research in this field under 4 main aspects which enables the AR system implementation: display, type of tracking system, user interaction and platform. Additionally, an assembly sequence representation that is needed to guide the user through the assembly process is also analyzed.

2.1 Display

Basically, there are two different approaches to show information to the user in AR systems: optical see-through and video see-through. In the optical see-through approach a half-transparent mirror is placed in front of the user's eyes. In this way the world can be seen though the half-transparent mirror and information is reflected on these half-transparent mirrors, thereby combining the real world and virtual information. Optical see-through displays do not present parallax effect, however, the combination of virtual elements by half-transparent mirror can reduce the brightness and contrast of virtual images. In the video see-through approach the user sees the world captured by cameras and information is superimposed on the digitized video. Video see-through approaches are cheaper and easier to be implemented. Besides, it is easier to add or remove elements from the display, as the world is already digitized. Still, as the camera and user's eyes have different field of views, video see-through approaches suffer from parallax effect. As for the placement of the display, some devices are placed on the user's head and the display is positioned in front of the eyes (HMD—head mounted displays), projectors can be used to display information on real objects (spatial displays) and some displays are placed on a hand's reach (hand-held displays).

On the literature, optical see-through [3, 7] and video see-through HMD [1, 7–13] are commonly used in AR systems as they enable the use of both user's hands on the assembly process. However, HMDs available on the market are bulky, expensive or requires connection to a computer.

Another approach is the use of spatial displays that use projectors to directly display information on real objects. This is the most integrated technology to the environment and several users can interact simultaneously. Projector-based AR systems implementations are shown in [14–16]. However, the only use of a projector as a spatial display to guide the user to a solution of a 3D puzzle is presented in [15]. Although spatial systems enable more immersion on the task than other alternatives, it requires a controlled environment illumination and preparation of workspace parameters to correctly superimpose information on assembly parts.

Smartphones, tablets and PDAs represent hand-held displays. A system composed of a smartphone and PC to guide an assembly process of a 3D puzzle is shown in [17]. While the PC made all the image processing, the smartphone was used to display images augmented with assembly information.

2.2 Tracking

When augmented reality is used only to show information to the user and it is not required to superimpose information aligned to real world, this is called augmented reality without context as demonstrated in [3, 8, 15, 18]. In this kind of AR, images, animations and texts are shown to the user on a display similarly to an electronic manual. The main difference between an electronic manual and AR without context is the faster access to assembly information without change of attention, as the information is in the same display as the real world seen by the user.

Augmented reality with context requires the knowledge of camera and object position. In some cases, camera and object position are not known beforehand, therefore, methods to extract this information are needed. These methods are often called tracking. The main challenge to tracking methods is the depth information which is lost when a camera captures the world. Some AR systems use a combination of camera and another sensor to acquire the depth information. Magnetic sensors [1], infrared sensor [19] and multiple cameras [20] have been used to extract the depth data.

Specifically on AR uses on assembly processes, the main approach to tracking is the application of computer vision techniques on images taken by a singular camera. Two different approaches using computer vision techniques are found on the literature: tracking with fiducial markers and marker-less tracking. Fiducial markers make the tracking and recognition more robust, however, it requires the preparation of the workspace. AR systems which use fiducial markers tracking approach are presented in [9–12, 16, 17, 21–23].

There are mainly two kinds of marker-less tracking: frame-to-frame and tracking by detection. Frame-to-frame methods estimate the current position by using information from previous frames. Tracking by detection estimates position by matching extracted features from images to features extracted from a known object model. An AR system that uses tracking by combining frame-to-frame and tracking by detection methods was shown in [24]. While frame-to-frame tracking is achieved by an implementation of a particle filter algorithm, tracking by detection is made by matching features from several 2D synthetic views of the 3D geometric features generated by moving a virtual camera along a sphere.

2.3 User Interaction

Some researchers used mouse and PC keyboard [9, 10, 16] or smartphone keyboard [17] to enable the AR system interaction, while others proposed interfaces with selectable panels by using a 3D stick tool [21], hand and gesture recognition methods [12] and haptic interfaces [20].

Two kinds of tools are used to interact in AR environments: computer vision tools and haptic tools. A system controlled by hand gestures is demonstrated in [10]. The detection of hands is made by color segmentation and virtual menus are opened by holding the hands on an activation area. A stick tool detected by image segmentation to interact with virtual panels is presented in [21]. Following the work in [21], the segmentation technique to recognize hands and gestures, respectively was improved in [25, 26].

An AR system with voice control was presented in [10, 27]. The voice control is made by a connection from the AR system with a stand-alone voice command system which makes the voice recognition of simple English commands like "next", "previous", "next phase" and "previous phase".

Yet, an AR system with a haptic device where users wear a device to manipulate virtual objects as they were real objects was demonstrated in [20], which enables a more integrated and natural way to interact with the virtual elements.

2.4 Platform

Several AR systems were implemented by a combination of PC and HMD/monitor [1, 3, 8, 10–16, 24, 27, 28]. The PC versatility, connection diversity which enables several devices connected at the same time, wide availability on market and processing power are the main reasons for choosing PCs as platforms for AR systems. Still, there are restrictions on connections to HMD, as a connection to a PC is required.

One main requirement on AR systems is near real time response. Manipulation of 3D virtual objects, tracking and object recognition require suitable processing power and memory. Due to this requirement, few approaches have been developed on mobile devices and markers for object recognition or tracking have been used to reduce the need of complex image processing algorithms. An AR system using a mobile phone as a client and a PC as a server on a client/server architecture is shown in [17]. A mobile phone is used to take photos of the workspace with markers and the images are sent to a server. The images are processed on the server and then the assembly information is sent back to the mobile phone. On the research presented in [24] a marker-less tracking subsystem for AR systems was described. Images at 640×480 resolution were used to estimate camera pose in hundreds of milliseconds on a PC equipped by Intel Core i7-860 CPU and 3GB of RAM. The development of an AR system with the same architecture as proposed in [17] was proposed in [24] to eliminate the need to move bulky hardware components so that the mobile device's computational requirements could be lowered. Up until this work, only one project implemented on a tablet [23] was found on the literature, mainly due to limitations on processing power and memory. Nowadays, mobile devices are becoming increasingly more powerful platforms. The iPad is a mobile device that stands out for its computing power, which can allow execution of image processing algorithms in real time and may be used for AR systems [29].

2.5 Assembly Sequence Representation

Some assembly sequence representations have been proposed in [30, 31]. Whenever an intelligent control system for assembly processes is being designed, a way to describe the assembly sequence representation is required. The choice of the representation impacts in the required memory to store the information and the difficult to derive the representation from an assembly process.

The following definitions are used for the assembly sequence representation:

- *Subassembly*—represents a set of joined parts, it can contain one or more parts
- *Assembly task*—is a connection between subassemblies
- *Assembly sequence*—ordered list of assembly tasks
- *Assembly state*—list of connected subassemblies
- *Assembly layout*—represents the part disposition in the space

The connection between solid parts forming an stable unit is a mechanical assembly. The assembly parts are connected by surface contact reducing the degrees of freedom. Subassemblies are sets of parts containing one or more joined parts. Whenever subassemblies are connected an assembly task is performed. A succession of an ordered list of tasks or assembly states is an assembly sequence. An assembly process starts with all parts separated and ends with all parts connected to form a whole object.

An assembly sequence representation named as Assembly Sequence Table (AST) was introduced in [31]. This representation is used to describe all possible solutions to an assembly problem for evaluation and selection of an assembly sequence. Table 1 shows an example of AST. The first column is the level number, the second column contains the assembly states representing all feasible subassemblies formed after the assembly task is performed. Assembly tasks represent the feasible subassemblies from previous level to be joined. At any level L the assembly states column have all the subassemblies with L number of parts. Hence at level 1 only subassemblies with 1 part on assembly states are listed, on level 2 there are 3 assembly states with sets containing 2 parts each and so on.

Table 1 AST example of parts identified by a, b, c and d

Level	Assembly states	Assembly tasks
1	{a}; {b}; {c}; {d}	–
2	{a, b}	{(a), (b)}
	{a, c}	{(a), (c)}
	{c, d}	{(c), (d)}
3	{a, b, c}	{(a, b), (c)};{(a, c), (b)}
4	{a, b, c, d}	{(a, b, c), (d)};{(a, b), (c, d)}

3 The Proposed System

The Assembly Assisted by Augmented Reality (A^{3}R) system was developed to run an embedded software on an off-the-shelf tablet with a camera, display, gyroscope and accelerometer sensors. It uses a video see-through approach to show the assembly space to the user and graphical signs to guide the user through the solution of the assembly process. Consequently, no customized device as an HMD is used and no connection to another computer is required. The parts to be assembled are identified by object recognition techniques in a way that markers are not used as an aid to the parts identification process and environment preparation is not needed. Instead of markers, models are used to identify the parts to be assembled.

On a setup phase, the proposed system loads the models and the assembly sequence representation from files or from the internet, as pictured on Fig. 1a. Models are composed by a label to identify the assembly part and edges. The edges are described by a label and sets of points, representing the beginning and the end of an edge. For each assembly part model, features are extracted and stored on a database. The assembly sequence is composed of a list of tasks and each task contains the sequence order and a set of connections. Each connection is composed by a set of labels which identifies the edges to be connected in order to join the subassemblies. A representation derived from [31] is used to represent the assembly sequence. The AST Table, described in [31], contains all possible solutions to the assembly process, whereas in the proposed system only one assembly solution is represented per assembly process, thus there is no need to represent the assembly state on each step. Only the assembly tasks column

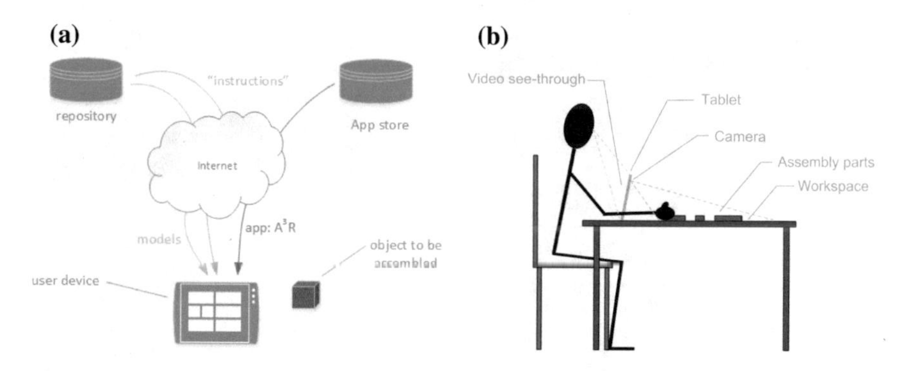

Fig. 1 Proposed system. **a** Models and instructions loading from the internet. **b** System setup

Table 2 Example of the proposed system assembly sequence representation

Sequence	Task
0	{I3}; {I0}
1	{T0, I2}
2	{W0, T6}; {W1, T5}

from the AST representation is used. An example of the assembly sequence representation is presented in Table 2. The Sequence column corresponds to the sequence order and the Task column represents the assembly task. An Assembly task contains sets of edges labels. On the Sequence 0, it is assumed that the first assembly part (I) is located at $(x, y) = (0, 0)$ on a cartesian base requiring only the edge that is on axis $x = 0$ and $y = 0$. Therefore, edges "I3" and"I0" are on $x = 0$ and $y = 0$ axis, respectively. On the Sequence 1, the Task column contains the connection between edge "T0" from the "T" assembly part and the edge "I2", from the "I" assembly part. Finally, on the Sequence 2, the Task column contains the connection between edge "W0" and "T6" and the connection between edges "W1" and "T5", from assembly parts "W" and "T", respectively. The assembly guidance is carried out by reading the table from the Sequence 0 to the last Sequence, stepping to the next Sequence whenever the connection between the subassemblies is detected.

Still on the setup phase, the intrinsic camera parameters and lens distortion are determined by a camera calibration process and stored on the database. Later, this information is used for the estimation of the object position and to correct the distortions caused by the lens on the camera image.

On the assembly guidance process phase, the tablet device is placed between the user and the assembly parts in a way that the user can see the assembly parts on the tablet's display in a video see-through display approach, as shown schematically on Fig. 1b.

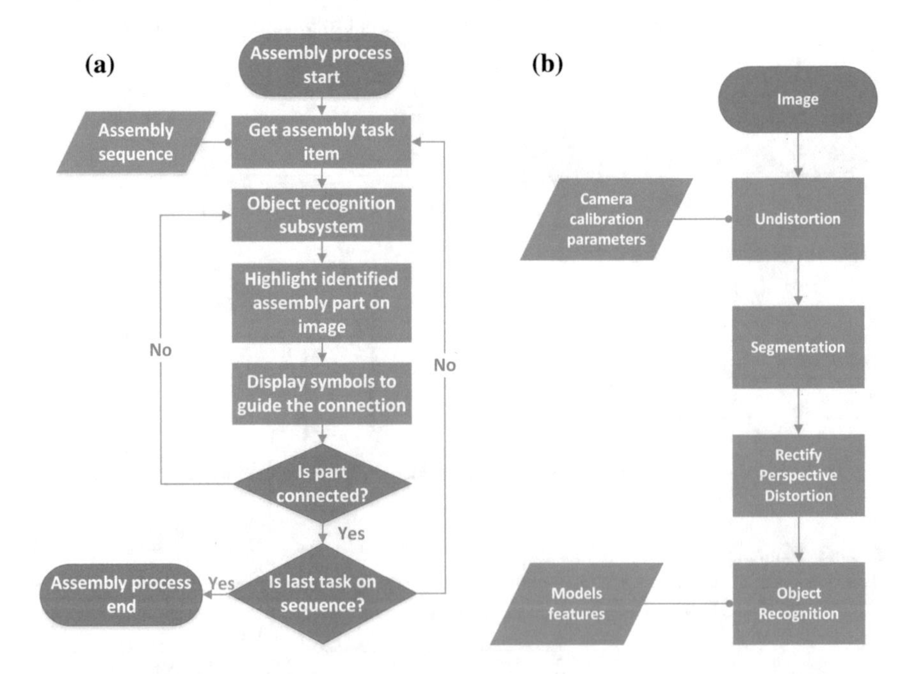

Fig. 2 System block diagram. **a** Assembly process diagram. **b** Object recognition subsystem

Figure 2a shows the assembly process algorithm. At first, the system retrieves an assembly task from the assembly sequence ordered list. If the assembly task is not the last one, the part associated with the assembly task is identified by the object recognition subsystem and is highlighted on the tablet's display to show to the user the part to move. Graphical signs are shown to the user indicating the following actions: flip the assembly part if necessary and connect the part to a subassembly. While a highlighted assembly part is not connected to the subassembly, the system continuously indicates to the user the assembly part and its placement for connection even while the user moves the part around the workspace, except in the case of occlusion that is not being taken care at this time. After the part is placed on the indicated location and the connection to the subassembly is detected, the system steps to the next assembly task. This loop is repeated until all parts are placed on assembly layout and the proper connections are validated, thus the assembly process is completed.

For each image frame, a sequence of image processing algorithms are executed on the object recognition subsystem, as shown on Fig. 2b. The first algorithm is executed to correct the lens distortion from the camera image by the use of the distortion lens parameters previously estimated on the setup phase. After the undistortion, objects on the workspace are segmented. Due to the tablet's camera position, segmented parts are rectified to remove the perspective distortion. Features are extracted from the segmented contours and then invariant features relative to rotation and scale are computed. On the object recognition algorithm, features are used to match the parts and the models.

4 Implementation

The implemented system uses an iPad Air 2 as platform with a video see-through approach. An 18 mm focal distance wide angle lens was placed on the tablet camera to broaden the field of view. Xcode was used to implement the software and the OpenCV library was used to develop the image processing algorithms. All implemented routines were written with Objective-C and C++ languages.

As a use case, the system was implemented to guide the assembly process of a pentomino puzzle [32]. This puzzle is composed of 12 pieces where each piece is the result of 5 square concatenated by their edges. Figure 3 shows the pentomino puzzle pieces and the given name of each piece that is used to identify each individual part and in the assembly sequence table. A common puzzle objective is to tile all the pieces in a rectangular box with an area of 6×10 squares. This particular configuration has 2,339 possible solutions. This application was selected because it represents a generic assembly process with known solutions where the user can place the pieces in well defined positions starting from a random distribution of pieces on the workspace.

On the setup phase, the camera calibration is performed using a method implemented in the OpenCV library [33]. Using the tablet's camera with a wide angle lens attached, twenty images were taken from a chessboard on arbitrary orientations to estimate the camera intrinsic matrix and lens distortion parameters. If the sum of the

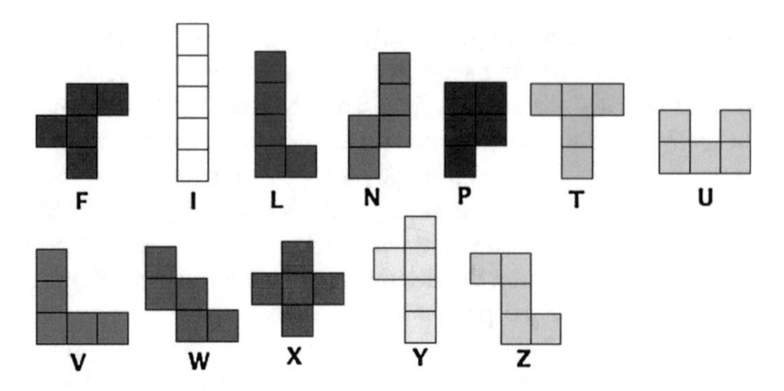

Fig. 3 Pentomino pieces identified by letters

squared distances between the observed projection points and the projected points (re-projection error) is less than 0.5, the camera intrinsic matrix and lens distortion parameters are stored for later use. If the estimated re-projection error is not less than 0.5, the calibration process is repeated. The correction (undistortion) of the camera lens distortion is made in two steps: computing the mapping of points in the distorted space to the undistorted space and interpolating the pixels on the undistorted image after the mapping. The mapping result is the same on all camera images with the same lens, therefore, for optimization proposes it is cached once computed and only the interpolation of pixels on the undistorted images is processed for each camera image.

Assembly parts images were segmented in four steps: (1) color segmentation is made by thresholding the pixels by hue, saturation and value; (2) a Canny Edge algorithm [34] is used to detect edges; (3) a 2×2 rectangular element size is used on a morphology closing process to close open borders on the detected edges; (4) a Border Following algorithm, as described by [35], is used to join the borders in vectors which represents the border points in a counter-clockwise order.

In the particular case of the 2D puzzle solution guidance, it is necessary to make the rectification of perspective distortion to compute assembly parts as a top–down view. Given that the pieces on puzzle are on the plane $Z = 0$, we can relate a point d on the image to a point D on the plane $Z = 0$:

$$sd = HD \tag{1}$$

where H is the homography matrix and s is an arbitrary scale.

The image points are computed as if the image was taken by a virtual camera placed on $C_v = (0, 0, h_v)_g$ on the global coordinate in O, as shown on Fig. 4. Using the (1) a point in the coordinate system O can be represented in the coordinate system C as:

$$s_c d_c = A \begin{bmatrix} -\cos(\theta) & 0 & h \cdot \frac{\sin^2(\theta)}{\cos(\theta)} \\ 0 & 1 & 0 \\ \sin(\theta) & 0 & h \cdot \sin(\theta) \end{bmatrix} D = H_c D \tag{2}$$

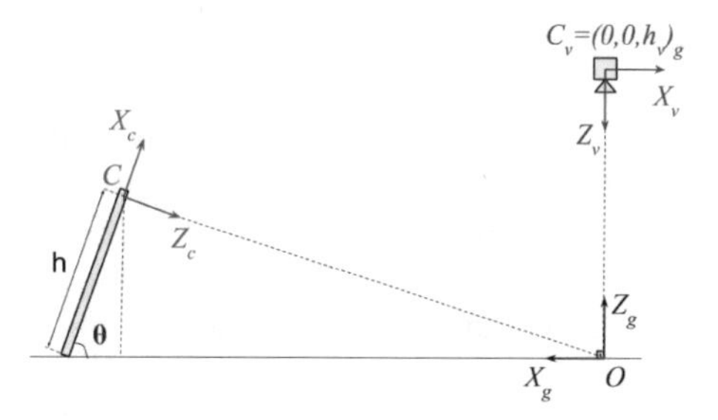

Fig. 4 Perspective effect removal: camera coordinate system *C*, global coordinate system *O* and virtual camera coordinate system *V*

and a point in the coordinate system *O* can be represented in the coordinate system *V* as:

$$s_v d_v = A \begin{bmatrix} \cos(\pi) & 0 & 0 \\ 0 & 1 & 0 \\ -\sin(\pi) & 0 & h_v \end{bmatrix} D = H_v D \tag{3}$$

where the d_c and d_v are the points in the images of the real camera and the virtual camera, respectively; *A* is the camera intrinsic matrix; *D* is the point on the plane of the assembly parts; H_x is the homography matrix of the real camera or virtual camera; θ is the device inclination angle; *h* is the distance between the camera and the device opposite side; h_v is the height of the virtual camera.

Relating the Eqs. (2) and (3), we can relate an image point to a point as taken by a virtual camera:

$$\alpha d_v = \begin{bmatrix} -1 & 0 & 0 \\ 0 & 1 & 0 \\ 0 & 0 & h_v \end{bmatrix} \begin{bmatrix} -\cos(\theta) & 0 & h \cdot \frac{\sin^2(\theta)}{\cos(\theta)} \\ 0 & 1 & 0 \\ \sin(\theta) & 0 & h \cdot \sin(\theta) \end{bmatrix}_c^{-1} d_c \tag{4}$$

or

$$\alpha d_v = H_v H_c^{-1} d_c \tag{5}$$

where α is an arbitrary constant scale.

The distance *h* of the camera and the opposite side of the tablet is known and we used the gyroscope and accelerometer sensor of the tablet to compute the inclination angle θ.

Following the rectification, we used the Ramer–Douglas–Peucker algorithm [36, 37] implemented on OpenCV to approximate the segmented contours to polygons, which compressed the edge points and reduced noise. Precision of 1 % of contour length was used on the algorithm.

The approximated contours of the segmented parts change if the assembly part is rotated or if they are farther or near from the camera. The parts on the image are recognized by matching to the models of a database, so the extraction of features that are invariant to translation, rotation and scale is necessary to make the comparison possible between segmented parts and a model in the database. The models in the database are loaded on the setup phase and they are described in more details later in this section. As presented in [38], the turning function is a description of a shape contour as a representation that is invariant to translation, rotation and scale. Suppose that a polygon is represented by a curve, then we can re-scale this curve in a way that its total length is equal to 1. The Turning Function $\phi(s)$ measures the change of the angle ϕ along the curve as a function of the curve length s. Following the curve's path to the counter-clockwise direction, the angle increases on counter-clockwise turns and decreases on clockwise turns, as the turns are accumulated along the curve. The turning function has the following features: it is translation invariant, scale invariant, the rotation is represented by a shift on accumulated turns and the choice of the initial starting point corresponds to a shift on the curve segment length.

The matching between two polygons is done by calculating a distance as discussed in [39]. The method used is the Polygonal Method. The choice of the initial starting point affects the turning functions, so it is necessary to compute the minimum distance from all possible turning functions computed by different initial starting points. The following metric is used to make the curve matching between two closed polygons A and B given their turning functions $\phi_1(s)$ and $\phi_2(s)$:

$$D(A, B) = \min_{\substack{\alpha \in \Re \\ u \in [0,1]}} \left[\int_0^1 (\phi_1(s) - \phi_2(s+u) + \alpha)^2 ds \right]^{\frac{1}{2}} \tag{6}$$

The optimal α is:

$$\alpha = \int_0^1 [\phi_1(s) - \phi_2(s)]\, ds - 2\pi u \tag{7}$$

where u is the choice of a starting point and α is the rotation difference between the two polygons.

The turning function is computed for each segmented object. Then we compute the minimum distance between the segmented object turning function and each of the model's turning function in the database by the use of Eq. (6). The minimum is reached when the best orientation between the curves and the starting point is selected and a matching is detected when the minimum distance to the model is less than a defined threshold of 0.8. Due to the nature of the pentomino pieces, some flipped pieces are not recognized as the same object because the turning function of a flipped piece is not the same depending on the piece's shape. So the turning function of the pentominoes pieces that presented distinct turning function on the flipped state were included. In this way some pentominoes pieces have two possible turning functions, as is the case of the "P" piece.

Each assembly part on the pentominoes puzzle has a unique shape, therefore it is necessary to treat repeated identified parts in the object recognition algorithm. If a model is matched to more than one contour, the segmented image object with the minimum distance to the model is held as a match to the model. Yet, due to noise on the captured image, the segmentation algorithm can detect this noise as a small contour. Elimination of this noise contour is made by comparing the arc length of contours matched to the same model. If the arc length of one contour is less than 20 % of the comparing contour, the contour with the minor arc length is discarded as a possible match to the model.

For the assembly guidance process the implemented system loads the assembly parts models and the solution description from files on the setup phase. The assembly parts models are loaded by a JSON formatted file which lists the assembly parts with the following information:

1. Assembly part label;
2. Set of connected edges, each edge is labeled and described by vertices.

On Fig. 5 a JSON formatted file with the description of a pentomino part model is shown. In this example, the pentomino piece “V” is described by a label and by an array of edges. The edges are described by a label “Vn” (with n between 0 and 5) and the model’s vertices. Each vertex is described by a point on a cartesian base. The models from the JSON formatted file are stored on a database and the model’s turning functions for object recognition are derived from them.

Figure 6 shows a JSON formatted file with a description of an assembly sequence. The assembly sequence is represented by a list of assembly tasks containing:

1. “Sequence”—the sequence order;
2. “Part”—the associated assembly part model to be moved on the task and;
3. “Connections”—the assembly connections. The assembly connections are represented by a set of connected edges.

An assembly layout is used for guidance on user interface. The assembly layout is a grid of cells displayed on the camera image to guide the connection of assembly parts by indicating the part location on the whole object. The location of an assembly part is shown by highlighting the cells to be occupied, determined by the assembly task. Computing which cells are to be highlighted on the assembly task is made by:

1. reading the models and solution description;
2. translating and rotating the assembly part on each assembly task for the assembly part placement on assembly layout;
3. estimating the cells to be highlighted on each assembly task.

The assembly part model rotation is computed by rotating the associated part on the assembly task until all connected edges are parallel. Translation is computed by estimating the distance from the connected edges, which are all parallel after rotation. Given the rotation and translation of the assembly part model the validity of the connection is verified. A connection is valid whenever the rotated and translated part do not overlap another subassembly. The systems steps through each assembly task until all the parts are verified to have valid connections. When this condition occurs and it is in the end of the assembly sequence, the object is considered to be completely assembled.

```
{   "Parts":
      [{
         "Label": "V",
         "Edges":
            [{
               "Label": "V0",
               "Vertices": [{"X": 0,"Y": 0},{"X": 0,"Y": 300 }]
             },
             {
               "Label": "V1",
               "Vertices": [{"X": 0,"Y": 300},{"X": 100,"Y": 300 }]
             },
             {
               "Label": "V2",
               "Vertices": [{"X": 100,"Y": 300},{"X": 100,"Y": 100 }]
             },
             {
               "Label": "V3",
               "Vertices": [{"X": 100,"Y": 100},{"X": 300,"Y": 100 }]
             },
             {
               "Label": "V4",
               "Vertices": [{"X": 300,"Y": 100},{"X": 300,"Y": 0 }]
             },
             {
               "Label": "V5",
               "Vertices": [{"X": 300,"Y": 0},{"X": 0,"Y": 0 }]
             },
            ]
       },
       ...
      ]
}
```

Fig. 5 Assembly part JSON example. In this example the "V" part of the pentomino puzzle is described on Parts array

```
{
   "AssemblySequence":[
     ...,
     {
       "Sequence": "2",
       "Part": "L",
       "Connections": [{"Edges":["L4","W5"]},{"Edges":["L5","W4"]}]
     },
     ...
   ]
}
```

Fig. 6 Assembly sequence JSON example

5 Results and Discussion

We processed 35 images with the 12 pentominoes pieces displaced on random orientation and position without occlusion to evaluate the segmentation and object recognition algorithms. From a total of 420 pieces, the algorithm segmented 413 pieces, thus, the segmentation ratio is 0.98.

Analyzing the pieces which could not be segmented, we noticed that the segmentation failed due to borders that where not fully closed. The morphology closing operation with one interaction of a 2×2 rectangular element size was not sufficient to close some borders and the border following algorithm accumulated the points of the border as an open contour. The segmentation result may be improved if a larger element is used on the morphology closing operation. By the other hand, larger elements may deform the shape, which makes the identification task more challenging.

For the object recognition we assumed that the pentominoes pieces are unique. On the distance computation between segmented contours and models turning functions only segmented contours turning function with distance lower than 0.8 are selected as candidates for identification. As the object recognition depends directly from the segmentation result, we only used successfully segmented parts for the object recognition. Then, from a total of 413 segmented parts, 407 were correctly identified. Figure 7 demonstrates an example of the identified parts and their labels. We noticed that some of the unrecognized puzzle parts were near the edge of camera image. They were affected by lens distortion correction that acts more aggressively on pixels far from the center of image, which deforms the piece shape.

It is important to evaluate the execution time of each algorithm as near real time performance is desired on AR systems. We placed pentomino pieces in a way that the camera could capture all the 12 pieces and executed the algorithms 500 times. Execution time for each process is summarized on Table 3. Segmentation execution time had a great significance on the whole process execution time. As assembly guidance algorithm is executed in parallel on the implemented system, the assembly guidance algorithm execution time do not affect the image processing and the object

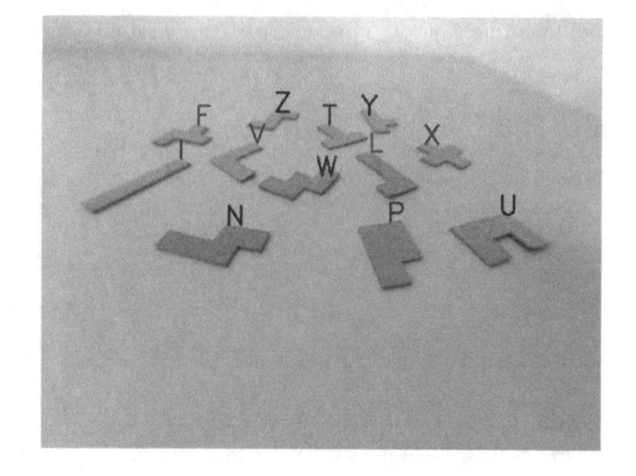

Fig. 7 Example of pentomino parts recognition assuming unique pentominoes

Table 3 Algorithms execution time

Algorithm	Time (ms)	
	Mean	Standard deviation
Undistortion	19.1	2.18
Segmentation	68.9	4.39
Perspective rectification	0.435	0.0706
Object recognition	20.7	1.71
Total	*109*	*8.35*

recognition execution time. Segmented contours turning function derivation and distance computation can be parallelized, therefore, the object recognition execution time can be improved by running part of the algorithm on a parallel process.

When an assembly part is required to be flipped, the system alternates between highlighting the current assembly part in white and green shapes signs, as shown in Fig. 8a, b, respectively, in order to indicate to the user the need for this action. The system produces an audible warning whenever the need for a flip action is detected. Figure 9 presents an intermediate state of the assembly guidance process. The assembly part to be moved on this sequence is highlighted and an arrow indicates to the user where the part must be moved to in order to connect to the subassembly. The highlighted cells near the arrow tip represent the placement location of the part for connection and the remaining filled cells represent already assembled and connected parts, or subassemblies, on previous steps. When the system detects that the part is placed on the right position and is connected to the subassembly, an audible confirmation is produced to inform the user and the system steps to the next assembly sequence. This process is repeated until the assembly process is detected as completed, as illustrated in Fig. 9b.

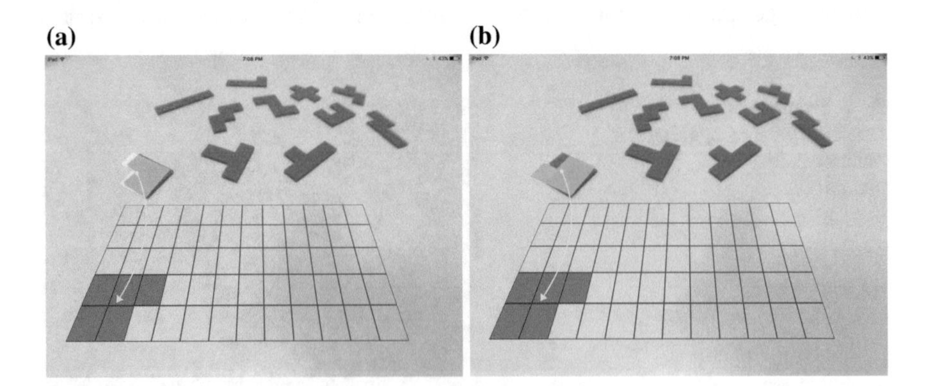

Fig. 8 Indication to the user of part to be flipped. **a** Alternating symbol for current detected part in *white*. **b** Alternating symbol for part in final flipped state in *green*

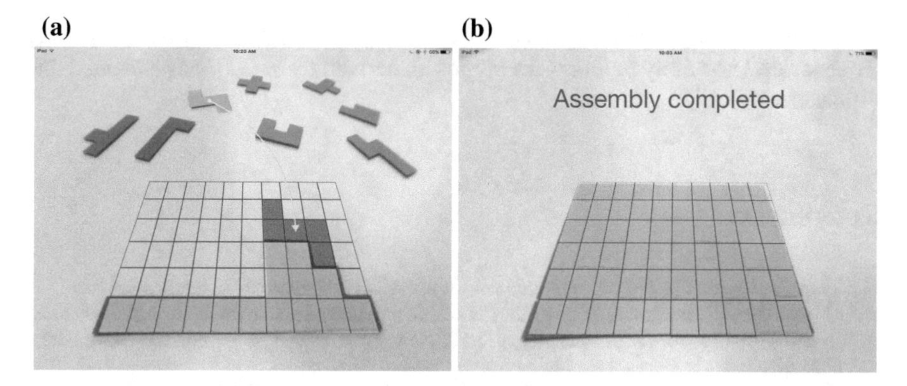

Fig. 9 Assembly process guidance viewed by the user. **a** Assembly process on an intermediate state. **b** Completed assembly process

6 Conclusion

This paper presented an AR system to guide an user through an assembly process. The system is based on image processing techniques to detect the parts to be assembled, to follow up the assembly process and uses graphical signs to inform the user of each step in the assembly sequence. Segmentation and object recognition are two important aspects of this project. Assembly parts segmentation is made by a combination of color segmentation, Canny Edge algorithm, morphology closing operation and border following algorithm. Segmentation rate of 0.98 has been achieved. Segmented assembly parts are identified by computing the turning function and the distance between the turning functions of segmented assembly parts and models. We made the assumption that all the 12 pentomino parts are present on the evaluated images and a filter algorithm have been used to reduce the likelihood of noise contour being incorrectly labeled as a pentomino part. We achieved a rate of 0.98 for object recognition with this process.

We evaluated the execution time of each image processing step: undistortion, perspective rectification, segmentation and object recognition. A mean of approximately 109 ms to the whole process execution time and an execution time of 20.7 ms for object recognition of 12 pentomino pieces were achieved in an iPad Air 2. We reckon that in about two generations of hardware upgrades real time performance will be possible.

The graphical interface guides the user through a series of movement operations such as flipping and connecting a part to a subassembly. Color symbols and audible confirmations and warnings give the necessary instructions and feedback to the user. The system has been implemented and tested with several puzzle solutions, from start point, where all parts are placed randomly in the workspace, until the full assembly is completed. General descriptions of the assembly model, assembly task, assembly sequence and assembly layout were proposed, derived from [31], implemented and validated.

References

1. Tang, A., Owen, C., Biocca, F., Mou, W.: Comparative effectiveness of augmented reality in object assembly. In: Proceedings of the SIGCHI Conference on Human Factors in Computing Systems, pp. 73–80. ACM Press, New York, New York, USA (2003). doi:10.1145/642625.642626
2. Wiedenmaier, S., Oehme, O., Schmidt, L., Luczak, H.: Augmented reality (AR) for assembly processes design and experimental evaluation. Int. J. Hum. Comput. Interact. **16**(3), 497–514 (2003)
3. Baird, K.M., Barfield, W.: Evaluating the effectiveness of augmented reality displays for a manual assembly task. Virtual Real. **4**(4), 250–259 (1999). doi:10.1007/BF01421808
4. Caudell, T., Mizell, D.: Augmented reality: an application of heads-up display technology to manual manufacturing processes. In: Proceedings of the Twenty-Fifth Hawaii International Conference on System Sciences, pp. 659–669 vol. 2 (1992). doi:10.1109/HICSS.1992.183317
5. Feiner, S., Macintyre, B., Seligmann, D.: Knowledge-based augmented reality. Commun. ACM **36**(7), 53–62 (1993). doi:10.1145/159544.159587
6. Reiners, D., Stricker, D., Klinker, G., Müller, S.: Augmented reality for construction tasks: Doorlock assembly. In: Proceedings of the IEEE and ACM IWAR (1998)
7. Odenthal, B., Mayer, M.P., Kabuß, W., Schlick, C.M.: A comparative study of head-mounted and table-mounted augmented vision systems for assembly error detection. Hum. Factors Ergon. Manuf. Serv. Ind. **24**(1), 105–123 (2014). doi:10.1002/hfm
8. Boud, A., Haniff, D., Baber, C., Steiner, S.: Virtual reality and augmented reality as a training tool for assembly tasks. In: 1999 IEEE International Conference on Information Visualization (Cat. No. PR00210), pp. 32–36. IEEE Comput. Soc (1999). doi:10.1109/IV.1999.781532
9. Zauner, J., Haller, M., Brandl, A., Hartman, W.: Authoring of a mixed reality assembly instructor for hierarchical structures. In: Proceedings of the Second IEEE and ACM International Symposium on Mixed and Augmented Reality, 2003, pp. 237–246. IEEE Comput. Soc (2003). doi:10.1109/ISMAR.2003.1240707
10. Salonen, T., Sääski, J., Hakkarainen, M., Kannetis, T., Perakakis, M., Siltanen, S., Potamianos, A., Korkalo, O., Woodward, C.: Demonstration of assembly work using augmented reality. In: Proceedings of the 6th ACM International Conference on Image and Video Retrieval - CIVR '07, pp. 120–123. ACM Press, New York, New York, USA (2007). doi:10.1145/1282280.1282301
11. Henderson, S., Feiner, S.: Exploring the benefits of augmented reality documentation for maintenance and repair. IEEE Trans. Vis. Comput. Graph. **17**(10), 1355–1368 (2011)
12. Wang, Z.B., Ong, S.K., Nee, A.Y.C.: Augmented reality aided interactive manual assembly design. Int. J. Adv. Manuf. Technol. **69**(5–8), 1311–1321 (2013). doi:10.1007/s00170-013-5091-x
13. Khuong, B.M., Kiyokawa, K., Miller, A., La Viola, J.J., Mashita, T., Takemura, H.: The effectiveness of an AR-based context-aware assembly support system in object assembly. In: 2014 IEEE Virtual Reality (VR), pp. 57–62 (2014). doi:10.1109/VR.2014.6802051
14. Chiang, H.K., Chou, Y.Y., Chang, L.C., Huang, C.Y., Kuo, F.L., Chen, H.W.: An augmented reality learning space for PC DIY. In: Proceedings of the 2nd Augmented Human International Conference, p. 12. ACM Press, New York, New York, USA (2011). doi:10.1145/1959826.1959838

15. Kitagawa, M., Yamamoto, T.: 3D puzzle guidance in augmented reality environment using a 3D desk surface projection. In: 2011 IEEE Symposium on 3D User Interfaces (3DUI), pp. 133–134. IEEE (2011). doi:10.1109/3DUI.2011.5759241
16. Fiorentino, M., Uva, A.E., Gattullo, M., Debernardis, S., Monno, G.: Augmented reality on large screen for interactive maintenance instructions. Comput. Ind. **65**(2), 270–278 (2014). doi:10.1016/j.compind.2013.11.004
17. Hakkarainen, M., Woodward, C., Billinghurst, M.: Augmented assembly using a mobile phone. In: 2008 7th IEEE/ACM International Symposium on Mixed and Augmented Reality pp. 167–168 (2008). doi:10.1109/ISMAR.2008.4637349
18. Nakanishi, M., Ozeki, M., Akasaka, T., Okada, Y.: Human factor requirements for applying augmented reality to manuals in actual work situations. In: 2007 IEEE International Conference on Systems, Man and Cybernetics, pp. 2650–2655 (2007). doi:10.1109/ICSMC.2007.4413588
19. Li, Y., Kameda, Y., Ohta, Y.: AR replay in a small workspace. In: 2013 23rd International Conference on Artificial Reality and Telexistence (ICAT), pp. 97–101. IEEE (2013). doi:10.1109/ICAT.2013.6728913
20. Murakami, K., Kiyama, R., Narumi, T., Tanikawa, T., Hirose, M.: Poster: A wearable augmented reality system with haptic feedback and its performance in virtual assembly tasks. In: 2013 IEEE Symposium on 3D User Interfaces (3DUI), pp. 161–162. IEEE (2013). doi:10.1109/3DUI.2013.6550228
21. Yuan, M.L., Ong, S.K., Nee, A.Y.C.: Augmented reality for assembly guidance using a virtual interactive tool. Int. J. Prod. Res. **46**(7), 1745–1767 (2008). doi:10.1080/00207540600972935
22. Novak-Marcincin, J., Barna, J., Janak, M., Novakova-Marcincinova, L., Torok, J.: Visualization of intelligent assembling process by augmented reality tools application. In: 2012 4th IEEE International Symposium on Logistics and Industrial Informatics, pp. 33–36. IEEE (2012). doi:10.1109/LINDI.2012.6319505
23. Serván, J., Mas, F., Menéndez, J., Ríos, J.: Assembly work instruction deployment using augmented reality. Key Eng. Mater. **502**, 25–30 (2012). doi:10.4028/www.scientific.net/KEM.502.25
24. Alvarez, H., Aguinaga, I., Borro, D.: Providing guidance for maintenance operations using automatic markerless augmented Reality system. In: 2011 10th IEEE International Symposium on Mixed and Augmented Reality pp. 181–190 (2011). doi:10.1109/ISMAR.2011.6092385
25. Wang, Z.B., Shen, Y., Ong, S.K., Nee, A.Y.C.: Assembly design and evaluation based on bare-hand interaction in an augmented reality environment. In: 2009 International Conference on CyberWorlds, pp. 21–28. IEEE (2009). doi:10.1109/CW.2009.15
26. Ong, S., Wang, Z.: Augmented assembly technologies based on 3D bare-hand interaction. CIRP Ann. - Manuf. Technol. **60**(1), 1–4 (2011). doi:10.1016/j.cirp.2011.03.001
27. Säääski, J., Salonen, T., Hakkarainen, M., Siltanen, S., Woodward, C., Lempiäinen, J.: Integration of design and assembly using augmented reality. Micro-Assem. Technol. Appl. **260**, 395–404 (2008). doi:10.1007/978-0-387-77405-3_39
28. Westerfield, G.: Intelligent Augmented Reality Training for Assembly and Maintenance. Master of science, University of Canterbury (2012)
29. Carmigniani, J., Furht, B., Anisetti, M., Ceravolo, P., Damiani, E., Ivkovic, M.: Augmented reality technologies, systems and applications. Multimed. Tools Appl. **51**(1), 341–377 (2010). doi:10.1007/s11042-010-0660-6
30. Homem de Mello, L.S., Sanderson, A.C.: Representations of mechanical assembly sequences. IEEE Trans. Robot. Autom. **7**(2), 211–227 (1991). doi:10.1109/70.75904
31. Gottipolu, R.B., Ghosh, K.: A simplified and efficient representation for evaluation and selection of assembly sequences. Comput. Ind. **50**(3), 251–264 (2003). doi:10.1016/S0166-3615(03)00015-0
32. Golomb, S.W.: Polyominoes: puzzles, patterns, problems, and packings, 2nd edn. Princeton University Press, Princeton (1996)
33. Zhang, Z.: Flexible camera calibration by viewing a plane from unknown orientations. In: Proceedings of the Seventh IEEE International Conference on Computer Vision **1**(c), 666–673 (1999). doi:10.1109/ICCV.1999.791289

34. Canny, J.: A computational approach to edge detection. IEEE Trans. Pattern Anal. Mach. Intell. **8**(6), 679–698 (1986)
35. Suzuki, S., Abe, K.: Topological structural analysis of digitized binary images by border following. Comput. Vis., Graph., Image Process. **30**(1), 32–46 (1985). doi:10.1016/0734-189X(85)90016-7
36. Ramer, U.: An iterative procedure for the polygonal approximation of plane curves (1972). doi:10.1016/S0146-664X(72)80017-0
37. Douglas, D.H., Peucker, T.K.: Algorithms for the reduction of the number of points required to represent a digitized line or its caricature. Cartogr.: Int. J. Geogr. Inf. Geovis. **10**(2), 112–122 (1973). doi:10.3138/FM57-6770-U75U-7727
38. Arkin, E.M., Chew, L.P., Huttenlocher, D.P., Kedem, K., Mitchell, J.S.B.: An efficiently computable metric for comparing polygonal shapes. IEEE Trans. Pattern Anal. Mach. Intell. **13**(3), 209–216 (1991). doi:10.1109/34.75509
39. Cakmakov, D., Celakoska, E.: Estimation of curve similarity using turning functions. Int. J. Appl. Math. **15**, 403–416 (2004)

Maximising Overlap Score in DNA Sequence Assembly Problem by Stochastic Diffusion Search

Fatimah Majid al-Rifaie and Mohammad Majid al-Rifaie

Abstract This paper introduces a novel study on the performance of Stochastic Diffusion Search (SDS)—a swarm intelligence algorithm—to address DNA sequence assembly problem. This is an NP-hard problem and one of the primary problems in computational molecular biology that requires optimisation methodologies to reconstruct the original DNA sequence. In this work, SDS algorithm is adapted for this purpose and several experiments are run in order to evaluate the performance of the presented technique over several frequently used benchmarks. Given the promising results of the newly proposed algorithm and its success in assembling the input fragments, its behaviour is further analysed, thus shedding light on the process through which the algorithm conducts the task. Additionally, the algorithm is applied to overlap score matrices which are generated from the raw input fragments; the algorithm optimises the overlap score matrices to find better results. In these experiments real-world data are used and the performance of SDS is compared with several other algorithms which are used by other researchers in the field, thus demonstrating its weaknesses and strengths in the experiments presented in the paper.

1 Introduction

Every cell in the body has a complete copy of about 3.2 billion[1] DNA base pairs or letters which build the human genome [22]. DNA has all the information necessary to build the whole living organism. Although the letters of the genetic alphabet

[1] In American English, 1 billion is equated to a thousand million (i.e. 1,000,000,000).

F. Majid al-Rifaie (✉)
Department of Computer Science, Kristianstad University,
S-291 88 Kristianstad, Sweden
e-mail: fatimah.al-rifaie0018@hkr.se

M. Majid al-Rifaie
Computing Department, Goldsmiths, University of London,
London SE14 6NW, UK
e-mail: m.majid@gold.ac.uk

Y. Bi et al. (eds.), *Intelligent Systems and Applications*,
Studies in Computational Intelligence 650, DOI 10.1007/978-3-319-33386-1_15

Adenine (A), Thymine (T), Cytosine (C) and Guanine (G) are meaningless on their own, they are joined into useful instructions in genes. It is interesting to note that more than 99 % of human's structure is genetically identical [21].

Imagine having several copies of the same book written in a language you cannot understand. Every page of each copy has been randomly cut into horizontal strip and a piece from one copy may overlap a piece from another copy. Assuming that some of strips are missing and some are splashed with ink, and maybe some of the books have random typos and error throughout, in different places. Try to arrange all the strips and assemble a single copy of the original book without any typos or errors. This process is similar to the important task of DNA sequencing.

In this work, novel applications of a swarm intelligence technique is introduced as a proof of principle. The swarm intelligence algorithm used is Stochastic Diffusion Search (SDS) which has a good potential to work in large search spaces and noisy environments. This algorithm is explained in the paper and its application to the problem is detailed.

This paper starts by presenting the swarm intelligence algorithm along with a simple example demonstrating its use. Then a brief introduction is given to the DNA assembly problem and the solutions offered so far using swarm intelligence techniques. Subsequently, some experiments are designed and the performance of SDS is investigated using various benchmarks and then its performance is contrasted against several other techniques. Finally, the reason behind using SDS is further elaborated and the difference between utilising SDS and Smith–Waterman algorithm is discussed. Additionally in the second set of experiments, initially SDS is shown to be generating the overlap score matrices (a task that is historically accomplished by Smith–Waterman algorithm); and then the details of using the SDS generated matrices to optimise the overlap scores are described. This is followed by a conclusion and directions for future research.

2 Swarm Intelligence

The paper is based on swarm intelligence which is one of the categories of artificial intelligence. Swarm intelligence is based on the study of behaviour of simple individuals (e.g. ant colonies, bird flocking, and honey bees, animal herding) that mimics the behaviour of swarms of social insects or animals [5]. More and more researches are interested in this field as swarm intelligence offers new ways of designing intelligence systems.

Among the successful examples of optimisation techniques inspired by swarm intelligence are: ant colony optimisation (inspired by foraging behaviour of real ant colonies) and particle swarm optimisation (inspired by bird flocking) [5]. In this work, Stochastic Diffusion Search (SDS) [1] algorithm is used. This algorithm also belongs to the category of swarm intelligence and is based on mimicking the foraging behaviour of one type of ants *Leptothorax acervorum*. More details about SDS are provided in Algorithm 1 and a simple example is presented next.

Algorithm 1 SDS algorithm

Initialisation phase: Allocate agents to random hypotheses in the search space

Until (all agents congregate on the best hypothesis)

- **Test phase**
  - Each agent evaluates its hypothesis
  - Each agent is classified into active or inactive
- **Diffusion phase**
  - Each inactive agent randomly chooses another agent to communicate with. If the inactive agent selects another inactive agent, no information will be transferred between the agents. Therefore the selecting agent should choose another hypothesis randomly. If the selected agent is active, the active agent communicate its hypothesis to the selecting agent

End

2.1 Search Example with SDS

In the following example the aim is to find a 4-letter model (Table 1) in a 32-letter search space (Table 2).

There are four agents; and a hypothesis identifies four adjacent letters in the search space (e.g. hypothesis '6' refers to D-N-A-F; hypothesis '17' refers to A-S-S-E, etc.). In the first step, each agent initially picks a random hypothesis from the search space (see Table 3). Assume that:

- The first agent points to the 27th entry of the search space and randomly picks one of the letters (e.g. the fourth one, (B): O B L **E**
- The second agent points to the 14th entry and randomly picks the first letter (E): **E** N T A

Table 1 MODEL

Index:	0	1	2	3
Model:	D	N	A	F

Table 2 Search space

Index:	0	1	2	3	4	5	6	7	8	9	10	11	12	13	14	15
Search space	T	H	I	S	I	S	D	N	A	F	R	A	G	M	E	N
Index:	16	17	18	19	20	21	22	23	24	25	26	27	28	29	30	31
Search space	T	A	S	S	E	M	B	L	Y	P	R	O	B	L	E	M

Table 3 Initialisation and iteration 1

Agent no	1	2	3	4	5
Hypothesis position	27	14	8	20	4
	OBL**E**	**E**NTA	A**F**RA	EM**B**L	I**S**DN
Letter picked	4th	1st	2nd	3rd	2nd
Status	×	×	×	×	×

- The third agent refers to the 8th entry in the search space and randomly picks the second letter (F): | A | **F** | R | A |
- The fourth agent goes the 20th entry and randomly picks the third letter (B): | E | M | **B** | L |
- The fifth agent refers to the 4th entry in the search space and randomly picks the second letter (S): | I | **S** | D | N |

The letters picked are compared to the corresponding letters in the model that is D-N-A-F (see Table 1). In this case:

- The fourth letter from the first agent (E) is compared against the fourth letter from the model (F) and because they are not the same, the agent is set inactive.
- For the second agent, the first letter (E) is compared with the first letter from the model (D) and because they are not the same, the agent is set inactive.
- For the third, fourth and fifth agents, letters 'F', 'B' and 'S' are compared against 'N', 'A' and 'N' from the model. Since none of the letters correspond to the letters in the model, the status of the agents are set inactive.

In the next step, each inactive agent chooses another agent and gets the same hypothesis if the selected agent is active. If the selected agent is inactive, the choosing agent generates a random hypothesis. Assume that the first agent selects the third one; since the third agent is inactive, the first agent chooses a new random hypothesis from the search space (e.g. 6). Figure 1 shows communication between agents.

The process is repeated for the other four agents. When the agents are inactive, they all choose new random hypotheses (see Table 4).

In Table 4, the first, third, fourth and fifth agents do not refer to their corresponding letter in the model, therefore they become inactive. The second agent, with hypothesis '6', chooses the second letter (N) and compares it with the second letter of the model (N). Since the letters are the same, the agent becomes active.

In this case, consider the following communication between the agents: (see Fig. 2)

- The third and fourth agents choose the second one
- The first agent chooses the third one

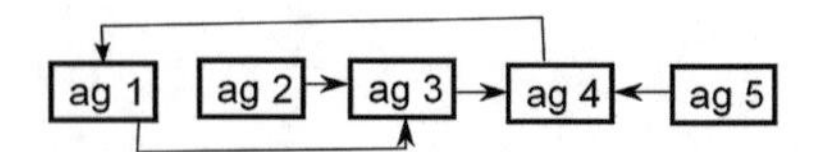

Fig. 1 Agent Communication 1

Table 4 Iteration 2

Agent no	1	2	3	4	5
Hypothesis position	1	6	22	12	17
	HISI	D**N**AF	BLY**P**	GM**E**N	AS**S**E
Letter picked	1st	2nd	4th	3rd	3rd
Status	×	√	×	×	×

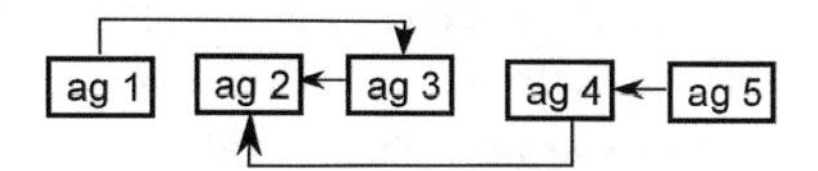

Fig. 2 Agents Communication 2

- The fifth agent chooses the fourth one

At this stage, the first and fifth agents, which chose the inactive third and fourth agents, have to choose other random hypotheses from the search space. However, agents three and four use the hypothesis of the active agent, two.

This process is repeated until all agents are active pointing to the location of the model inside the search. Depending on the problem, there are alternative termination strategies; for instance, in some cases, SDS algorithm is set to terminates only if all agents are active and refer to the same hypothesis.

The next section, provides a brief introduction to DNA assembly problem, stating the main phases and the major challenges faced by researchers in this field. This is followed by an overview of some of the algorithms that aimed to address the problem. Afterwards the experiments and results are reported.

3 Understanding DNA Assembly

There is no single solution available for NP-hard problems [22] and it is often not possible to find an extremely good algorithm that solves such problems [18].

In DNA assembly, a process is required to join the relevant fragments together. In other words, the overlapping fragments are to be assembled back into the original DNA sequence. Therefore, the goal of genome projects is to reconstruct the original genome sequence of an organism. To achieve the goal, DNA fragment assembly process is divided into three phases [6, 9]:

1. **Overlap Phase** is tasked to find the common sequence among the prefix of one sequence and suffix of another.
2. **Layout Phase** uses alignment strategies to determine the order of fragments based on high overlap scores and according to the level of similarity.
3. **Consensus Phase** assembles all fragments into the consensus sequence and omits the similar parts.

The quality of a consensus sequence is measured by the term coverage [6, 19]. Coverage is evaluated according to the following equation:

$$\text{Coverge} = \frac{\sum_{i=1}^{n} \text{length of fragment } i}{\text{target sequence length}} \tag{1}$$

where n is the number of fragments.

The higher the coverage, the higher the probability of covering original genome, the higher the correctness of the assembled parts, the fewer the number of the gaps, and the better the result [6, 11].

The Layout Phase is the most complex step due to the difficulty of finding the best overlap. This difficulty is caused by the following challenges [6, 11]:

- **Unknown orientation**: After the original sequence is divided into many fragments, the direction may change.
- **Base call errors:** substitution, insertion, and deletion errors are types of base call error. The errors happen because of experimental errors in the electrophoresis procedure that affects the finding of fragment overlaps.
- **Incomplete coverage:** It occurs when the algorithm cannot assemble a given fragments into one contig.
- **Repeated regions:** the problem occurs when some sequences are repeated two or more times in the DNA. None of the current assembly programs can solve the problem without an error [18].
- **Chimeras and contamination:** Chimeras arise when two fragments that are not adjacent, or overlapping on the target molecule, join together into one fragment. Contamination occurs due to the incomplete purification of the fragment from the vector DNA.

3.1 DNA Sequence Assembly and Swarm Intelligence

DNA Assembly problem is still open to a large extent because of the principal issue of "scaling up to real organism". Some of the swarm intelligence and evolutionary algorithms, such as genetic algorithms and ant colony optimisation have been used for the fragment assembly problem focusing on the overlap, layout and consensus approach [14].

In 1995, Rebecca Parsons and Johnson created performance improvements for a genetic algorithm applied to the DNA sequence assembly problem [17]. In 2003 Kim and Mohan used a new parallel hierarchical adaptive genetic algorithm. The method is reported as accurate and noise-tolerant compared to previous methods [10]. In the same year, Meksangsouy and Chaiyaratana proposed ant colony optimisation. The goal of the search was to find the right order and orientation of each fragment to create a consensus sequence [15]. In 2005 Fang proposed approach speeded up the searching process and maximised the similarity or overlaps between given fragments [7]. Alba and Luque presented several methods, including genetic algorithm, a CHC method, scatter search algorithm, and simulated annealing to solve accurately DNA Assembly problem in 2005 [12]. They also proposed a local search method named PALS in 2007 [2]. In 2008, Luque and Alba studied the behaviour of a hybrid heuristic algorithm that combines a heuristic, PALS, with a meta-heuristic, a genetic algorithm, achieving an assembler to find optimal solutions for large instances of this DNA assembly problem [3]. In 2010 Kubalik presented a method called Prototype Optimisation with Evolved Improvement Steps (POEMS). Also in the same year Minetti and Alba presented a paper about how noiseless and noisy instances of this problem are handled by three algorithms: problem aware local search, simulated annealing and genetic algorithms [16].

There are some other solutions that are proposed in 2011 for DNA sequence assembly problem using Particle Swarm Optimisation (PSO) with Shortest Position Value (SPV) rule [22]. In 2012 Firoz analysed and discussed the performance of two swarm intelligence based algorithms namely Artificial Bee Colony (ABC), and Queen Bee Evolution Based on Genetic Algorithm (QEGA) to solve the fragment assembly problem [8]. In 2013 Fernandez-Anaya et al. designed a nature inspired algorithm (PPSO + DE) based on Particle Swarm Optimisation and Differential Evolution [13].

4 Experiments and Results I

In order to understand the process through which SDS is adopted and adapted for DNA sequence assembly problem, a number of fragments are used in the experiments. The fragments are the input of the program and the program is responsible to assemble the fragments and create one long sequence. This is achieved by taking a fixed number of characters from the end of the first fragments and trying to find those characters in the other fragments using SDS algorithm. Once the other fragment is found, the two fragments are joined and the repeated part is deleted from one of the fragments. This will create a longer fragment. This process is repeated until all fragments are joined. The steps required for SDS to assemble a set of fragments are detailed in Algorithm 2.

In the experiments reported in this paper the agent size is empirically set to 100 and the model size for SDS is set to 50. Table 5, as proposed by Mallén-Fullerton et al. [14], shows the benchmarks used by SDS for DNA assembly.

Using the benchmarks provided, SDS algorithm assembles the entire sequences correctly. Table 6 shows the performance of SDS when assembling the nine aforementioned benchmarks. Each benchmark is assembled 50 times. As the table shows, the larger the coverage, the more SDS iterations it takes to fully assemble the datasets. While the number of overall algorithm cycles needed follow the same structure, there are some exception caused by the order of the fragments. Observing the sum of active agents over all the iterations and their consistent proximity (check the negligible dif-

Algorithm 2 DNA sequence assembly using SDS

Choose a model from the end of 1st fragment in the search space

While (true)

- Use SDS to search the model in the fragments
  - If no matching fragment is found
    - Choose the model from the beginning of the first fragment
    - Use SDS to search the model in the fragments
    - If no matching fragment is found
      Break
- Compile a list of fragments where the model is found
- Pick the fragment (j^{th}) with the maximum similarity (based on agents activity)
  - assemble fragments *i* and *j*.
  - Delete the jth fragment
  - Choose a new model from the end of assembled fragment

End While

Table 5 Benchmark datasets

Benchmark	Coverage	Mean fragment length	Number of fragments	Original sequence length
x60189 4	4	395	39	3,835
x60189 5	5	286	48	
x60189 6	6	343	66	
x60189 7	7	387	68	
m15421 5	5	398	127	10,089
m15421 6	6	350	173	
m15421 7	7	383	177	
j02459 7	7	405	352	20,000
bx842596 7	7	703	773	77,292

ference between the median and the mean, as well as the value of the standard deviation) shows the robustness of the technique.

The results shown in Table 6 indicate that three of the benchmarks ($m15421$ 6, $m15421$ 7 and $bx842596$ 7) are not assembled fully into one sequence. SDS has been able to assemble two large, accurate sequences from the fragments of each of these datasets which make up the whole dataset. However up to this point, given there were no similarities between the two resulting sequences, they are returned separately. Therefore, caution is taken and they are reported as not completely assembled.

Next, one of the benchmarks is chosen ($x60189$ 4) and the analysis are reported based on this benchmark. The results are compatible with the ones generated from the other benchmarks. Figure 3-left shows the level of agents activity at various stages of SDS assembling process, including both when a match is found and when a match is

Table 6 Summary of assembling four datasets

	Cycles	SDS Itrs	Sum of active agents				
			Median	Mean	Stdev	Max	Min
x60189 4	23	46,899	384,075	384,746	4,007	392,361	374,717
x60189 5	17	53,649	418,777	418,744	4,119	430,935	409,683
x60189 6	28	114,799	752,539	752,176	4,382	763,060	740,365
x60189 7	26	124,249	1,095,673	1,096,029	5,928	1,109,507	1,083,767
m15421 5	57	476,149	2,175,028	2,176,785	9,722	2,220,938	2,150,915
m15421 6	–	–	–	–	–	–	–
m15421 7	–	–	–	–	–	–	–
j02459 7	129	3,174,449	12,092,968	12,087,911	21,613	12,127,170	12,021,776
bx842596 7	–	–	–	–	–	–	–

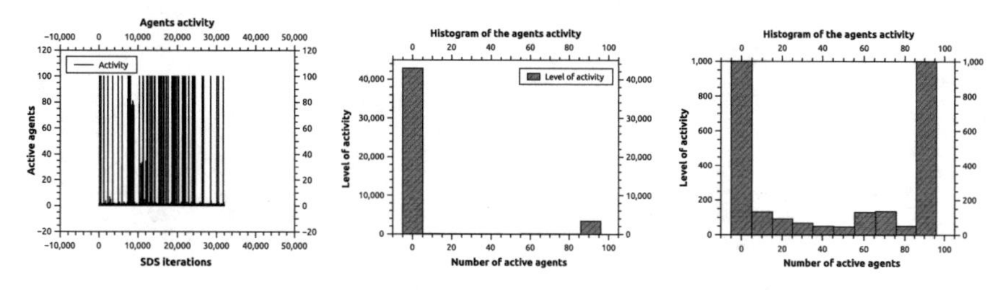

Fig. 3 *Left* activity of the agents in the fragments of $x60189$ 4; *middle* the histogram of the activity of the agents; *right* zooming to show the activity of agents between 0 and 100

not found in any given fragment. The activity of the agents is in the range [0, 100], however if less than the entire agent population (i.e. 100) are active, the agents' hypotheses are not taken into account for the assembling purpose; this ensures the presence of a full match. Reducing the 100 % accuracy would cater for a noisy environment which is one of the strengths of SDS algorithm.

To provide a better understanding, the histogram of the agents' activity is presented in Fig. 3-middle. This graph clearly shows that in most cases there is no high similarity between fragments (note that the similarity between fragments is evaluated by comparing the model to the fragments). However when there is a match (i.e. 100 % activity), the fragments are joined on the fly.

Figure 3-right provides a close-up view of the graph on its left and demonstrates that when there are no exact matches, some of the SDS agents could be activated; however if there are no full match, the activated agents eventually lose their active status in the consequent iterations when they choose a different micro-feature. This feature is particularly useful in a noisy environment whose complete analysis will be provided in an expanded future publication.

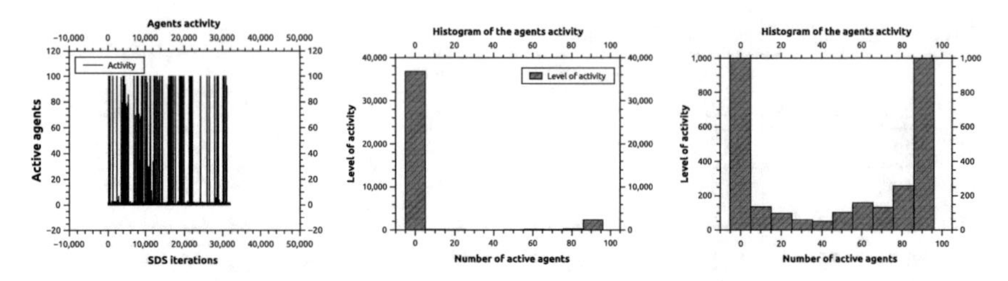

Fig. 4 *Left* activity of the agents in noisy set of fragments; *middle* histogram of the activity of the agents in noisy set of fragments; *right* zooming to show the activity of agents between 0 and 100

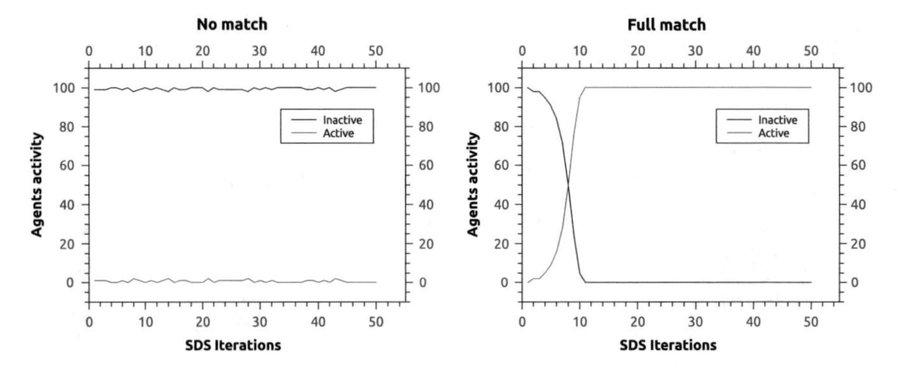

Fig. 5 *Left* absence of a match; *right* presence of a full match

In a similar experiment and in order to analyse the behaviour of the agents when some of the fragments are contaminated with noise, some noise (i.e. of type substitution) is added to all the fragments. Figure 4-left shows the activity of the agents and Fig. 4-middle and right illustrate the frequency of activity level at various iterations. Note that there are fewer number of iterations needed before SDS terminates (as at some point during the process, no match is found from either end of the growing sequence). However the proportion of agents activity between 0 and 100 is increasing with the presence of noise. Despite the fact that the entire fragment is contaminated with noise, SDS is able to produce more than half the length of the target sequence. Further research is required to improve this rate.

In order to illustrate the activity of the agents at each SDS iterations, the graphs in Fig. 5 are presented. In Fig. 5-left the activity of SDS agents are displayed when there is no match. As shown, most of the agents are inactive and very a few flicker from being active and then back to being inactive.

However, on the contrary to the lack of a match, when there is a full match, as shown in Fig. 5-right, soon after the start of the SDS iteration and through agents' communication and information exchange, the entire population becomes active and points to the right position, which is the position of the model within the fragment.

Table 7 Comparison with other techniques

	SDS	PALS	GA	PMA	CAPS	Phrap
x60189 4	1	1	1	1	1	1
x60189 5	1	1	1	1	1	1
x60189 6	1	1	–	1	1	1
x60189 7	1	1	1	1	1	1
m15421 5	1	1	6	1	2	1
m15421 6	2	NA	NA	NA	NA	NA
m15421 7	2	1	1	2	2	2
j02459 7	1	1	13	1	1	1
bx8425696 7	2	2	–	2	2	2

4.1 Comparison with Other Techniques

In another analysis, the performance of SDS is compared against a few other algorithms tasked with assembling the benchmarks. These algorithms, which are used in this context in the literature, are genetic algorithm (GA), a pattern matching algorithm (PMA), Problem Aware Local Search (PALS) and commercially available packages: CAP3 and Phrap. The algorithms are compared in terms of the final number of contigs assembled. Despite being in the early stages of its application in DNA assembly problem, SDS shows a competitive performance (see Table 7[2]). Other than an isolated case (m15421 7), where PALS and GA outperform SDS, in the rest of the cases (89 %), SDS either presents similar or better outcome. SDS is also tried on a benchmark (m15421 6) that is not attempted by the rest of the techniques. The accuracy of the assembled sequences is 100 %; in other words, whenever the accuracy is less than 100 %, the results are considered unsuccessful.

In these experiments, SDS deals with various issues common in DNA sequence assembly,[3] including but not limited to fragments with varying lengths, unknown orientation, incomplete coverage, repeated regions, chimeras and contamination, etc.

4.2 SDS Versus Smith–Waterman Algorithm I

Many DNA sequence assembly techniques use Smith–Waterman algorithm [20], which is a pairwise alignment method to create a similarity matrix between the fragments, therefore generating a complete picture of the entire available data before setting off to the overlapping and assembling stage. While Smith–Waterman

[2]The results of these algorithms, other than SDS, are borrowed from [2].

[3]These issues are explained in Sect. 3.

algorithm provides a precise and detailed account of the input data, it comes at the expense of being time consuming and computationally expensive [23].

Assuming there are n fragments, once the similarity between each pair is calculated using Smith–Waterman algorithm, an $n \times n$ matrix is created. The matrix is then used by other optimising algorithm to conduct the overlapping phase. The results of many of these algorithms are reported in [14].

In the experiments reported earlier in the paper, instead of using Smith–Waterman algorithm to calculate the similarities between fragments, SDS picks a model from a given fragment and aims to find the model in the rest of the fragments. Among the fragments containing the model, the one with the highest similarity is picked and assembled on the fly and then removed from the search space, thus reducing the subsequent computational cost.

On the contrary to many other swarm intelligence and evolutionary computation, SDS has been successful in assembling the benchmarks without using Smith–Waterman algorithm, therefore avoiding its time consuming and computational expensive nature. To understand the full picture of the process, further analysis is needed, among other things, to verify the impact left on the assembling process without accessing the very detailed information provided by Smith–Waterman algorithm.

5 Experiments and Results II

In the second set of experiments of this paper, SDS is tasked to generate overlap score matrices and then the same algorithm is used to optimise the generated matrices. In the next subsections, the process through which the overlap score matrices are generated are described, then the constrains in creating these matrices as well as choosing SDS hypothesis are presented. Afterwards SDS is applied to the benchmarks in the dataset and the results are reported.

5.1 Generating Matrices of Overlap Score by SDS

In this section, SDS is shown to be generating the overlap score matrices. As in the previous set of experiments, an already commonly used dataset [14] is used. More details about the reason behind using SDS (instead of the commonly used Smith–Waterman algorithm) for generating the overlap score matrices are provided in Sect. 5.5.

In the meantime, the steps through which the overlap scores between fragments are computed using SDS are described below:

1. Take a model from the end of the first fragment, and search each fragment one by one to check if other fragments have overlaps with the first fragment. For instance,

if fragment 1 and fragment 5 have 100 bases in common, this value is stored in the matrix as $(1, 5) = 100$.

2. After going through all the fragments, another model is taken from the beginning of the first fragment and again the program searches all the fragments. If the program finds a matching fragment, it stores the value of the overlap score in the matrix.
3. Then the orientation and bases of the first fragment is changed (i.e. A replaces T, T replaces A, C replaces G and G replaces C. Then the orientation is reversed).
4. Steps 1–3 are repeated for all other fragments (in the second cycle, the second fragment will be verified, and in the third the third fragment, until all fragments are checked, their matches are found and their overlap scores are stored in the correct entry of the matrix).

5.2 *Constraints in Overlap Score Matrices and Hypotheses Choice*

Before preceding to the experiments, some important points about the input matrices, overlap scores and the constraints and rules are listed below:

- The matrix always has symmetric overlap score.
- The matrix always has the overlap score of zero on the diagonal line.
- The dimension of the matrix is equal to the number of fragments in the dataset.
- For computing the overlap score of a dataset (that for instance) has 40 fragments, the program should select 40 indices from the matrix. In this work, each one of these elements is called the index hypothesis.
- For choosing the hypothesis from the matrix, the program should follow some rules and consider some constraints.

The following constraints should be considered when choosing the hypotheses (also see Fig. 6):

- Assume each hypothesis has one row and one column. The row and column should not be equal. In other words, hypothesis should not be chosen from the diagonal line of the matrix as the values on the diagonal line are always zero.

Fig. 6 Matrix Constraints (The entries highlighted in green are the valid elements (coordinates) of the hypothesis, and the entries in *red* are the invalid ones)

	0	1	2	3	4	5
0	0	10	50	300	65	0
1	10	0	44	72	100	0
2	50	44	0	35	84	0
3	300	72	35	0	26	0
4	65	100	84	26	0	0
5	0	0	0	0	0	0

- Two different hypotheses should not be on the same row.
- Two different hypotheses should not be on the same column.

Having mentioned the rules and constraints, the next section presents the experiments designed for this paper, demonstrating the performance of the proposed algorithm. Then the results are reported along with comparisons against other techniques.

5.3 Applying SDS on Overlap Score Matrices

As mentioned before, SDS has the three phases of initialisation, test and diffusion. In this part, these phases are explained in detail and it is shown how SDS is applied to the matrices in order to find the optimum solution. Thus, on the contrary to the previous set of experiments where the inputs were DNA bases, in this section the inputs data are overlap score matrix.

5.3.1 Initialisation Phase

The initialisation phase of SDS algorithm should adhere to the rules and constraints described above. Also it is important to note the following problem-dependant issues:

- Every fragment (except the first and last one) is a prefix of one fragment and suffix of another fragment.
- The first fragment in a contig has no prefix, and the last fragment has no suffix.
- The fragments cannot be appended to itself.

During the implementation stage of the initialisation phase, a hypothesis (i.e. member of matrix) is generated for every agent with respect to the constraints.

For generating the hypothesis, the following steps should be taken:

1. Create a list and populate the cells with values from $[0, n - 1]$ where n is the size of any sides of the matrix.
2. Initialise the first value in X with zero (see Fig. 7).
3. Randomly choose the Y value from the list created. According to Fig. 7, 2 is chosen randomly. Now the row and column (X, Y) of the first hypothesis are assigned.
4. In the next step, the Y value of the previous hypothesis will be assigned as the X value of the second hypothesis. Then that number will be removed from the array (i.e. 2) which was created in the first step. Again the Y value of the second hypothesis will be selected randomly.
5. Finally when all the elements of the array are removed, zero is assigned as the Y value of the last hypothesis. Note that the algorithm is prevented from choosing zero as the Y value during generating the hypotheses until it reaches to the last hypothesis. As stated earlier, the reason behind picking the value of ‘0’ is that

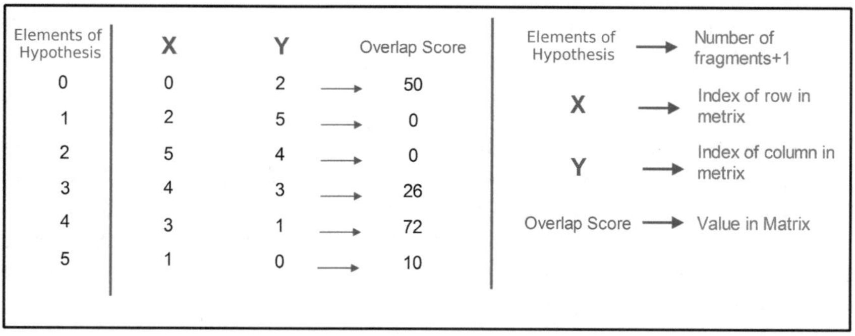

Elements of Hypothesis	X	Y		Overlap Score
0	0	2	→	50
1	2	5	→	0
2	5	4	→	0
3	4	3	→	26
4	3	1	→	72
5	1	0	→	10

Elements of Hypothesis	→	Number of fragments+1
X	→	Index of row in metrix
Y	→	Index of column in metrix
Overlap Score	→	Value in Matrix

Fig. 7 Sample Hypothesis of an agent (based on the matrix in Fig. 6)

the first and the last fragments in every sequence only have suffix and prefix respectively.

6. All these elements form one SDS hypothesis for the first agent.

After the initialisation phase, every agent has a SDS hypothesis which consists of n coordinates from the overlap score matrix. Therefore there will be an overlap score associated with every agent (e.g. the overlap score of the first agent is $50 + 0 + 0 + 26 + 72 + 10 = 158$). This will pave the way for comparing the agents against one another.

5.3.2 Test Phase

The test phase is explained by taking agent '0' as the starting point to determine whether it should be active or inactive.

- Agent 0 is compared with a random agent
- If agent 0 has a higher overlap score than the randomly selected agent, it will be active
- If agent 0 has a lower overlap score than the randomly selected agent, it will be inactive
- This continues until all agents are labelled as either active or inactive

5.3.3 Diffusion Phase

This phase is similar to the original description of the SDS's diffusion search and works as follows:

- Each inactive agent selects an agent randomly
- If the selected agent is active, the hypothesis (which contains all the x and y coordinates) will be copied from the active to the inactive agent.

- If the selected agent is not active, again, another *valid* hypothesis will have to be generated for that inactive agent.

This technique shows a good initial outcome however it does not return an optimum solution since the agents communicate with each other with limited information (the hypotheses that are chosen randomly) and there is not sufficient comparison with the "outside worlds" (all other possible hypotheses that can be generated from the matrix are not considered). Therefore, to improve the program, in the next section, another important phase (i.e. Reform Phase) is added. This phase complements the three previously discussed phases.

Prior to explaining the Reform phase, it is important to discuss what happens if one of the elements of the hypothesis moves from one position to another.

5.3.4 Hypothesis Sliding and the Reform Phase

In the following part, the steps are shown as to how to exchange the values of the coordinates of hypothesis in each agent.

Figure 8 shows a sample matrix representing a dataset with 8 fragments. Therefore each agent has a set of 8 coordinates belonging to the hypothesis. The details of the figure are explained below:

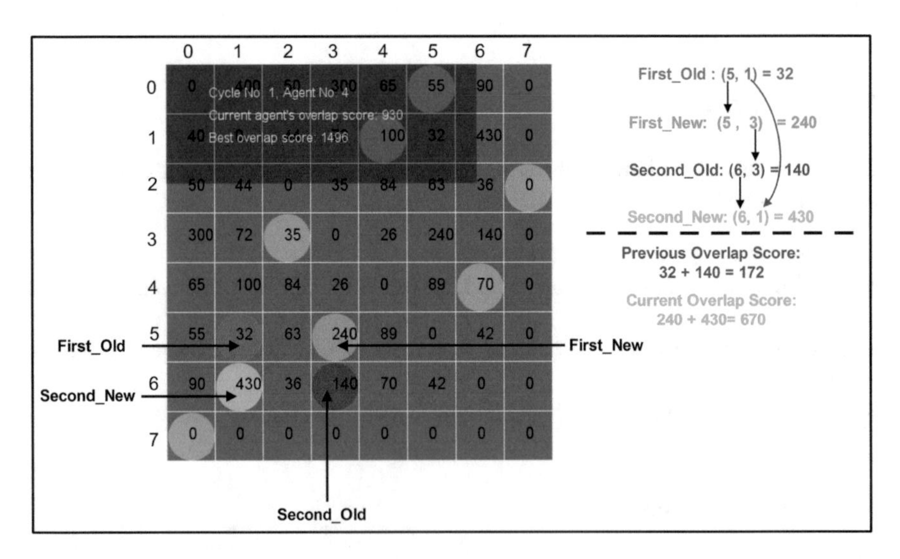

Fig. 8 Hypothesis Sliding (This figure shows the elements of a sample agent hypothesis. The hypothesis' elements are highlighted in *grey* as well as the one in *red*. When the *red* elements of the hypothesis are updated and the new coordinates are shown in *green*. Note that the number of hypothesis elements are 8)

- The grey circles are the hypotheses in the current agent.
- The red circles are the old hypotheses that are replaced by the new ones
- The green circles are the new hypotheses that are replaced with the old ones.

The below steps, describe how the Reform Phase works on the input matrix:

1. Initially the minimum hypothesis (with the smallest overlap score) other than zero, which is called 'First Old' is picked. Assume that the hypothesis (5, 1) which has the minimum value is picked.
2. The aim is to have the minimum hypothesis exchanged with an index in the matrix which has a value (overlap score) more than the minimum value. This index is called 'First New'.
   - The position of this index (i.e. First New) should be selected from the same row of where the minimum overlap score (First Old) is located. In other words, 'First Old' and 'First New' are in the same row. In order to do that, a column is chosen randomly. Assuming the fourth column is selected, the index of 'First New' will be (5, 3).
3. Next, the program should find a hypothesis that is called 'Second Old'. The steps below show how this is accomplished:
   - The 'Second Old' has the same column as the 'First New'. According to the previous stage, the column of 'First New' was 3, therefore the column of 'Second Old' will become 3.
   - In order to find the row, the algorithm checks which hypothesis in the current agent has column 3. Then it picks the row of that hypothesis. Suppose the row of that hypothesis is 6. Thus the index of 'Second Old' hypothesis becomes (6, 3).
   - As mentioned before, in every agents hypothesis, every column is used once, and every row is used once too. In other words, no two items of the hypothesis should be in the same column or the same row.
4. The next tasks is to find the position of the new coordinate in the original matrix. This coordinate is called the 'Second New', which takes the row of the 'Second Old' and the column of the 'First Old'.

5.4 Results

This section presents the results of optimising the overlap score on the real dataset. These experiments are conducted on the following datasets from Table 5: x60189_4, x60189_5, x60189_6, x60189_7.

In the experiments reported in this paper the agent size is empirically set to 100 and the model size for SDS is set to 50. Table 5, as proposed by Mallén-Fullerton et al. [14], shows the benchmarks used by SDS for DNA assembly.

Table 8 Collection of results for the commonly used 4 benchmark datasets along with SDS results

Benchmark	SDS	LKH	PPSO+DE	QEGA	PALS	SAX	POEMS
x60189 4	11,322	11,478	11,478	11,478	11,478	11,478	11,478
x60189 5	13,930	14,161	13,642	14,027	14,021	14,027	–
x60189 6	15,160	18,301	18,301	18,266	18,301	18,301	–
x60189 7	19,728	21,271	20,921	21,208	21,210	21,268	21,261

As part of the results reported, the performance of SDS is compared against a few other algorithms used for computing the overlap scores. These algorithms, which are used in this context in the literature, are Prototype Optimisation with Evolved Improvement Steps (POEMS), Problem aware local Search (PALS), Queen Bee Evolution Based on Genetic Algorithm (QEGA), Particle Swarm Optimisation and Differential Evolution, which is called (PPSO+DE), SAX and Lin-Kernighan (LKH). The algorithms are compared in terms of the total overlap score (see Table 8[4]) and the results show that the preliminary investigation of SDS behaviour demonstrates a competitive performance.

As explained before, the starting point in the reform phase is the minimum (smallest) element of the hypothesis. The algorithm aims to find a larger value to replace the minimum element. The question is: should the program search the entire matrix environment to find a bigger value (global search) or should it search just, for instance, the row or the column where the minimum elements are sitting in (local or neighbourhood search). At the moment the paper has explored the neighbourhood search and global search is the topic of an ongoing research.

5.5 *SDS Versus Smith–Waterman Algorithm II*

In the dataset used [14] for the experiments reported, the file 'matrix conservative' contains matrices that store the overlap score amongst all fragments for each benchmark in the dataset. The overlap scores are computed by the Smith–Waterman algorithm. Smith–Waterman algorithm takes into account the overlap of each fragment with every other fragments individually, therefore, while providing a comprehensive picture, it is computationally expensive. In other words, Smith and Waterman which was proposed in 1981, is a dynamic programming local sequence alignment algorithm that is used for this purpose. This algorithm's most common setting are 1 for a match, 3 for a mismatch, and 2 for a gap. This algorithm must of run on the entire possible combination of fragment pairs taking both orientations into account (regular and reverse compliment); this is necessary due to the unknown orientation

[4]The results of these algorithms, other than SDS, are borrowed from [14].

of every input fragment. Upon finding the maximum overlap by the algorithm, the actual overlap needs to be calculated. Once complete, the final values are inserted into the overlap score matrix, which details the overlap between any two pair of fragments. As stated before, this is a time consuming and computational expensive task, with the computational complexity of $O(n^2)$. As stated in [14], even if there are only 500 fragments of approximately 100 bases long, around 2,500,000,000 Smith–Waterman elements would be required to calculate the overlap matrix.

In order to explore an alternative less time consuming approach, SDS algorithm is tasked to accomplish the creation of the overlap score matrices.

SDS ignores the non-significant overlap. This is achieved by assigning the minimum overlap score by setting the model size. When the overlap between two fragments is less than the model size, the overlap score value of the fragments is set to zero in matrix. For instance, if model size is 50 and the overlap score between fragments 1 and 4 is 40, the value in entry $(1, 4)$ in the matrix will be zero. The advantage of using this technique (i.e. ignoring non-significant overlaps) is speeding up the search. Additionally the matrix could be compressed easily by removing the zero values.

6 Conclusions

Since DNA fragment assembly problem is NP-hard, it is difficult to find optimal solutions. The increasing presence of biological data and the requirements to study and understand them closely lead researchers and scientists in this field to use computational approaches. This work has shown how DNA fragment assembly problem can be addressed with meta-heuristics. This paper presents Stochastic Diffusion Search (SDS), which belongs to the extended family of swarm intelligence algorithms, in the context of DNA fragment assembly problem. An initial study into the behaviour of SDS is provided, offering an analysis into the agents' activity using several benchmarks. The results are promising as they demonstrate how the activity of the agents shed light into the way agents interact and eventually finalise the assembling process. Additionally it is shown that the level of agents' activity provides a measure of similarity between fragments, thus allowing more similar fragments to be joined in the assembling process.

Subsequently, SDS algorithm is used to optimise the overlap score of the input overlap score matrices. In this optimisation task, input matrices with overlap scores of fragments is given as input to the system and the adapted SDS algorithm is responsible for finding the optimum overlap score in order to assembly the fragments. Taking into account the initial attempt of using this algorithm, the results are close to the those of other researchers.

As part of the future research, this algorithm will be compared against other evolutionary computation techniques used in this field; also larger datasets with more complex features are to be used, and more research is needed in order to theoretically determine the two values (population size and model size) of the SDS parameters.

Additionally, CPU time and memory usage will be taken into account for all the comparisons to provide a more comprehensive account on the performance of the proposed algorithm.

References

1. al-Rifaie, M.M., Bishop, M.: Stochastic diffusion search review. In: Paladyn, Journal of Behavioral Robotics, vol. 4(3), pp. 155–173. Springer, Heidelberg (2013)
2. Alba, E., Luque, G.: A new local search algorithm for the dna fragment assembly problem. In: Evolutionary Computation in Combinatorial Optimization, pp. 1–12. Springer, Heidelberg (2007)
3. Alba, E., Luque, G.: A hybrid genetic algorithm for the dna fragment assembly problem. In: Recent Advances in Evolutionary Computation for Combinatorial Optimization, pp. 101–112. Springer, Heidelberg (2008)
4. Bishop, J.: Stochastic searching networks. In: Proceedings of the 1st IEEE Conference on Artificial Neural Networks. pp. 329–331. London, UK (1989)
5. Blum, C., Li, X.: Swarm Intelligence in Optimization. Springer, Berlin (2008)
6. Cotta, C., Fernández, A., Gallardo, J., Luque, G., Alba, E.: Metaheuristics in bioinformatics: DNA sequencing and reconstruction. In: Optimization Techniques for Solving Complex Problems, pp. 265–286 (2009)
7. Fang, S.C., Wang, Y., Zhong, J.: A genetic algorithm approach to solving dna fragment assembly problem. J. Comput. Theor. Nanosci. **2**(4), 499–505 (2005)
8. Firoz, J.S., Rahman, M.S., Saha, T.K.: Bee algorithms for solving dna fragment assembly problem with noisy and noiseless data. In: Proceedings of the Fourteenth International Conference on Genetic and Evolutionary Computation Conference. pp. 201–208. ACM, New York (2012)
9. Huang, K.W., Chen, J.L., Yang, C.S., Tsai, C.W.: A memetic particle swarm optimization algorithm for solving the dna fragment assembly problem. Neural Comput. Appl. pp. 1–12 (2014)
10. Kim, K., Mohan, C.K.: Parallel hierarchical adaptive genetic algorithm for fragment assembly. In: The 2003 Congress on IEEE Evolutionary Computation CEC'03, vol. 1, pp. 600–607. (2003)
11. Li, L., Khuri, S.: A comparison of dna fragment assembly algorithms. METMBS **4**, 329–335 (2004)
12. Luque, G., Alba, E.: Metaheuristics for the DNA fragment assembly problem. Int. J. Comput. Intell. Res. **1**, 98–108 (2005)
13. Mallén-Fullerton, G.M., Fernández-Anaya, G.: DNA fragment assembly using optimization. In: IEEE Congress on Evolutionary Computation (CEC), pp. 1570–1577 (2013)
14. Mallén-Fullerton, G.M., Hughes, J.A., Houghten, S., Fernández-Anaya, G.: Benchmark datasets for the dna fragment assembly problem. Int. J. Bio-Inspir. Comput. **5**(6), 384–394 (2013)
15. Meksangsouy, P., Chaiyaratana, N.: DNA fragment assembly using an ant colony system algorithm. In: The 2003 Congress on IEEE Evolutionary Computation CEC'03, vol. 3, pp. 1756–1763 (2003)
16. Minetti, G., Alba, E.: Metaheuristic assemblers of dna strands: noiseless and noisy cases. In: IEEE Congress on Evolutionary Computation (CEC), pp. 1–8 (2010)
17. Parsons, R., Johnson, M.E.: DNA sequence assembly and genetic algorithms-new results and puzzling insights. In: ISMB, pp. 277–284 (1995)
18. Pevzner, P.: Computational molecular biology: an algorithmic approach. MIT press, Cambridge (2000)
19. Setubal, J.C., Meidanis, J., Setubal-Meidanis.: Introduction to Computational Molecular Biology. PWS, Boston (1997)

20. Smith, T.F., Waterman, M.S.: Identification of common molecular subsequences. J. Mol. Biol. **147**(1), 195–197 (1981)
21. U.S. National Library of Medicine: Cells and DNA. what is DNA? http://ghr.nlm.nih.gov/handbook/basics/dna. Accessed 06 Jan 2015
22. Verma, R.S., Singh, V., Kumar, S.: DNA sequence assembly using particle swarm optimization. Int. J. Comput. Appl. **28** (2011)
23. Wieds, G.: Bioinformatics explained: Blast versus smith-waterman. CLCBio. http://www.clcbio.com/index.php. (2007)

A Comparative Analysis of Detecting Symmetries in Toroidal Topology

Mohammad Ali Javaheri Javid, Wajdi Alghamdi, Robert Zimmer and Mohammad Majid al-Rifaie

Abstract In late 1940s and with the introduction of cellular automata, various types of problems in computer science and other multidisciplinary fields have started utilising this new technique. The generative capabilities of cellular automata have been used for simulating various natural, physical and chemical phenomena. Aside from these applications, the lattice grid of cellular automata has been providing a by-product interface to generate graphical patterns for digital art creation. One notable aspect of cellular automata is symmetry, detecting of which is often a difficult task and computationally expensive. This paper uses a swarm intelligence algorithm—Stochastic Diffusion Search—to extend and generalise previous works and detect partial symmetries in cellular automata generated patterns. The newly proposed technique tailored to address the spatially-independent symmetry problem is also capable of identifying the absolute point of symmetry (where symmetry holds from all perspectives) in a given pattern. Therefore, along with partially symmetric areas, the centre of symmetry is highlighted through the convergence of the agents of the swarm intelligence algorithm. Additionally this paper proposes the use of *entropy* and *information gain* measure as a complementary tool in order to offer insight into the structure of the input cellular automata generated images. It is shown that using these technique provides a comprehensive picture about both the structure of the images as well as the presence of any complete or spatially-independent symmetries. These technique are potentially applicable in the domain of aesthetic evaluation where symmetry is one of the measures.

M.A. Javaheri Javid (✉) · W. Alghamdi · R. Zimmer · M.M. al-Rifaie
Department of Computing, Goldsmiths, University of London,
London SE14 6NW, UK
e-mail: m.javaheri@gold.ac.uk

W. Alghamdi
e-mail: walgh001@gold.ac.uk

R. Zimmer
e-mail: r.zimmer@gold.ac.uk

M.M. al-Rifaie
e-mail: m.majid@gold.ac.uk

Y. Bi et al. (eds.), *Intelligent Systems and Applications*,
Studies in Computational Intelligence 650, DOI 10.1007/978-3-319-33386-1_16

1 Introduction

Creating aesthetically pleasing images has been investigated by many researches in the context of evolutionary computing, including the Bimorphs of Dawkins [16], Mutator of Latham [41], and Virtual Creatures of Sims [39]. Although some impressive results have been achieved, there still remains problems with the aesthetic selection. According to [28], first, the subjective comparison process, even for a small number of phenotypes, is slow and forms a bottleneck in the evolutionary process. Human users would take hours to evaluate many successive generations that in an automated system could be performed in a matter of seconds. Secondly, genotype-phenotype mappings are often not linear or uniform. That is, a minor change in genotype may produce a radical change in phenotype. Such non-uniformities are particularly common in tree or graph based genotype representations such as in evolutionary programming, where changes to nodes can have a radical effect on the resultant phenotype. In this study we approach the problem in the framework of dynamical systems and define a criterion for aesthetic selection in terms of its association with symmetry. The association of aesthetics and symmetry has been investigated from different points of view.

In this work, a brief account on cellular automata is presented, followed by a section on symmetry and its significance in aesthetics. Subsequently, the newly proposed measure of information gain is described and the main experiments related to this measure are included. Then a swarm intelligence algorithm—Stochastic Diffusion Search (SDS)—is explained, highlighting its main features, including its unique partial function evaluation aspect. Afterwards, the application of the algorithm in detecting points of symmetry is detailed, illustrating the performance of the method proposed. At last but not least, a comparison is provided between information gain and Stochastic Diffusion Search emphasising their strength and weaknesses. The paper is then concluded with a summary of the finding and the suggested future research.

2 Cellular Automata

Cellular automata (CA) are one of the early bio-inspired systems invented by von Neumann and Ulam in the late 1940s to study the logic of self-reproduction in a material independent framework. CA are known for their capabilities in exhibiting complex behaviour from iterative application of simple rules. In this section mathematical formulations of 2D CA are provided which are specific for the purpose of this paper and for the rest of the paper all the notions will be referring to these formulations.

Definition 1 A deterministic finite automaton is formally defined [26] as a quintuple of $\mathscr{M}$ such that:

$$\mathscr{M} = \langle Q, \Sigma, \delta, q_0, F \rangle \tag{1}$$

1. Q is a finite set of *states*,
2. Σ is a finite set of symbols as *input alphabet*,
3. $\delta : Q \times \Sigma \mapsto Q$ is the *state transition function*,
4. $q_0 \in Q$ is the *start* or *initial state*,
5. $F \subseteq Q$ is a set of *accepting* or *final states*.

The state transaction function δ determines the transitions from one state to another state. It takes two arguments as $q \in Q$ and an input symbol $a \in \Sigma$ then maps them to a state $q_1 \in Q$ (i.e., $\delta(q, a) = q_1$).

Definition 2 A lattice (L) is a regular tiling of a space by a unit cell.

The *Euclidean* plane is considered so the lattice L is over Z^2. Lattices can have *square*, *hexagonal* or *triangle* for their unit cells. A lattice can be *infinite* with *open boundary conditions* or *finite* with *periodic boundary conditions*. A finite lattice with periodic boundary conditions where the opposite borders (up and down with left and right) are connected, forms a virtual torus shape (Fig. 1).

Definition 3 A cellular automaton is a regular tiling of a lattice with uniform deterministic finite state automata.

A cellular automaton $\mathscr{A}$ is specified by a quadruple $\langle L, S, N, f \rangle$ where:

1. L is a finite square lattice of cells (i, j).
2. $S = \{1, 2, \ldots, k\}$ is set of states. Each cell (i, j) in L has a state $s \in S$.
3. N is neighbourhood, as specified by a set of lattice vectors $\{e_a\}$, $a = 1, 2, \ldots, N$. The neighbourhood of cell $r = (i, j)$ is $\{r + e_1, r + e_2, \ldots, r + e_N\}$. A a cell is considered to be in its own neighbourhood so that one of $\{e_a\}$ is the zero vector $(0, 0)$. With an economy of notation, the cells in the neighbourhood of (i, j) can be numbered from 1 to N; the neighbourhood states of (i, j) can therefore be denoted $(s_1, s_2, \ldots, s_N)$. Periodic boundary conditions are applied at the edges of the lattice so that complete neighbourhoods exist for every cell in L.

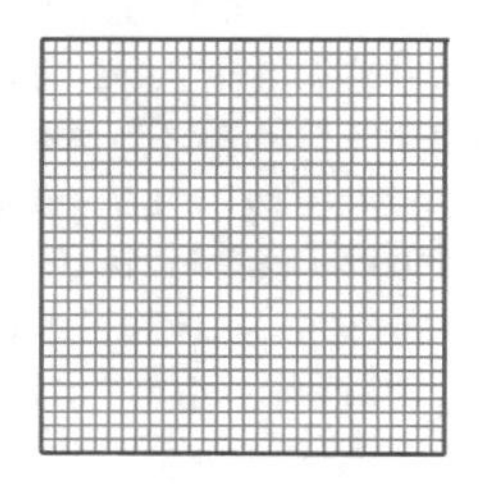

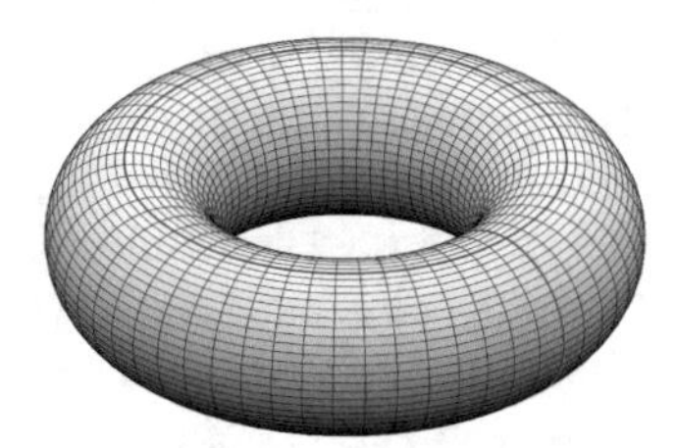

Fig. 1 The formation of virtual torus shape in a lattice with periodic boundary conditions

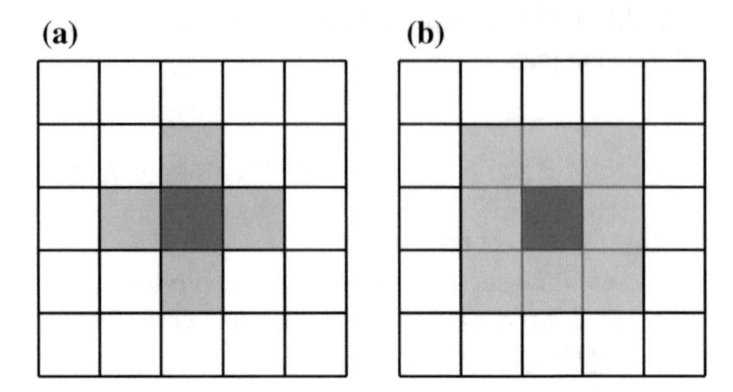

Fig. 2 von Neumann (**a**) and Moore (**b**) neighbourhood templates

4. f is the update rule. f computes the state $s_1(t+1)$ of a given cell from the states $(s_1, s_2, \ldots, s_N)$ of cells in its neighbourhood: $s_1(t+1) = f(s_1, s_2, \ldots, s_N)$. A quiescent state s_q satisfies $f(s_q, s_q, \ldots, s_q) = s_q$.

Remark 1 There are two common neighbourhoods; a five-cell von Neumann neighbourhood $\{(0,0), (\pm 1, 0), (0, \pm 1)\}$ and a nine-cell Moore neighbourhood:

$$\{(0,0), (\pm 1, 0), (0, \pm 1), (\pm 1, \pm 1)\}.$$

The collection of states for all cells in L is known as a configuration C. The global rule F maps the whole automaton forward in time; it is the synchronous application of f to each cell. The behaviour of a particular $\mathscr{A}$ is the sequence $c^0, c^1, c^2, \ldots, c^{T-1}$, where c^0 is the initial configuration (IC) at $t = 0$ (Fig. 2).

CA behaviour is sensitive to the IC and to L, S, N and f. The behaviour is generally nonlinear and sometimes very complex; no single mathematical analysis can describe, or even estimate, the behaviour of an arbitrary automaton. The vast size of the rule space, and the fact that this rule space is unstructured, mean that knowledge of the behaviour a particular cellular automaton, or even of a set of automata, gives no insight into the behaviour of any other CA. In the lack of any practical model to predict the behaviour of a cellular automaton, the only feasible method is to run simulations.

3 Symmetry and Aesthetic

Symmetry or having proportionality and balance is an important element of aesthetics. The association of aesthetics and symmetry has been investigated extensively in the literature. A study to investigate the effect of symmetry on interface judgements, and relationship between a higher symmetry value and aesthetic appeal for the basic imagery, showed that subjects preferred symmetric over non-symmetric images [9]. Further studies found that if symmetry is present in the face or the body, an individual is judged as being relatively more attractive and if the body is asymmetric the face is rated unattractive, even if the person doing the rating never sees the body [18, 35].

Symmetry plays a crucial role in theories of perception and is even considered a fundamental structuring principle of cognition [25]. In the Gestalt school of psychology things [objects] are affected by where they are and by what surrounds them... so that things [objects] are better described as more than the sum of their parts [10].

The Gestalt principles emphasise the holistic nature of perception where recognition is inferred, during visual perception, more by the properties of an image as a whole, rather than its individual parts [21]. Thus, during the recognition process elements in an image are grouped from parts to whole based on Gestalt principles of perception such as proximity, parallelism, closure, symmetry, and continuation [32]. In particular, symmetric objects are more readily perceived [14]. It is not surprising that we humans find sensory delight in symmetry, given the world in which we evolved. In our world the animals that have interested us and our ancestors (as prey, menace, or mate) are overwhelming symmetric along at least one axis [34].

3.1 *Evolutionary and Computational Approaches*

Evolutionary psychologists examine physical appearances like as symmetry, and perceived level of aesthetics as an indirect measure in mate selection [30, 31]. In this view symmetrical faces are examined as more attractive faces. In other words symmetry is positively linked with both psychological and physiological health indicators [36]. In geometry symmetrical shapes are produced by applying four operations of translations, rotations, reflections, and glide reflections. However developing computational methods which generate symmetrical patterns is still a challenge since it has to connect abstract mathematics with the noisy, imperfect, real world; and few computational tools exist for dealing with real-world symmetries [27]. Applying evolutionary algorithms to produce symmetrical forms leaves the formulation of fitness functions, which generate and select symmetrical phenotypes, to be addressed. Lewis describes two strategies in evolutionary algorithms approach for generating and selecting symmetrical forms: “A common approach is to hope for properties like symmetry to gradually emerge by selecting for them. Another strategy is to build in symmetry functions which sometimes activate, appearing suddenly. However this leads to a lack of control, as offspring resulting from slight mutations (i.e., small steps in the solution space) bear little resemblance to their ancestors [24]”.

There are several algorithms designed to detect exact symmetry in images; amongst which, and in the case of a collection of n points in the plane, Atallah describes an algorithm for enumerating all axes of symmetry under reflection of a planar shape [7]. In another work, Wolter et al. give exact algorithms, based on string matching, for the detection of symmetries of point clouds, polygons, and polyhedra [44]. These algorithms are often impractical due to their sensitivity to noise and high computational expense, because of their restricted nature to exact symmetries.

In terms of approximate symmetry there are two broad categories: the first defines approximate symmetry by an infimum of a continuous distance function quantifying how similar is a shape to its transformed version. One of the works introduced in this category is reported in [45]. The second approach aims to address the computational complexity issue by translating the search into a proxy domain, realising that the set of admissible symmetries is sparse in the transformation space; among the examples of this type of work are [29, 33, 40]. Other variations of symmetry detection have being proposed that fall within the broad above-mentioned categories (e.g. [23] which addresses reflection symmetry detection problem or [46] focusing on boolean functions, which is based on *information theory* [38] that detects symmetry by comparison of two-variable cofactors).

4 Entropy and Information Gain Measure

The *information theory* was an attempt to address a reliable communication over an unreliable channel [37]. Entropy is the core of this theory [15]. Let $\mathscr{X}$ be discrete alphabet, X a discrete random variable, $x \in \mathscr{X}$ a particular value of X and $P(x)$ the probability of x. Then the entropy, $H(X)$, is:

$$H(X) = -\sum_{x \in \mathscr{X}} P(x) \log_2 P(x) \tag{2}$$

The quantity H is the average uncertainty in bits, $\log_2(\frac{1}{p})$ associated with X. Entropy can also be interpreted as the average amount of information needed to describe X. The value of entropy is always non-negative and reaches its maximum for the uniform distribution, $\log_2(|\mathscr{X}|)$:

$$0 \leqslant H \leqslant \log_2(|\mathscr{X}|) \tag{3}$$

The lower bound of relation (3) corresponds to a deterministic variable (no uncertainty) and the upper bound corresponds to a maximum uncertainty associated with a random variable. Despite the dominance of entropy as a measure of order and complexity, it fails to reflect on structural characteristics of 2D patterns. The main reason for this drawback is that it only reflects on the distribution of the symbols, and not on their spatial arrangements.

Considering our intuitive perception of complexity and structural characteristics of 2D patterns, a complexity measure must be bounded by two extreme points of complete order and disorder. It is reasonable to assume that *regular structures*, *irregular structures* and *structureless* patterns lie along between these extremes, as illustrated in Fig. 3.

$$\textit{order} \xleftarrow{\text{regular structure}|\text{irregular structure}|\text{structureless}} \textit{disorder}$$

Fig. 3 The spectrum of spatial complexity

A complete regular structure is a pattern of high symmetry, an irregular structure is a pattern with some sort of structure but not as regular as a fully symmetrical pattern and finally a structureless pattern is a random arrangement of elements [20].

A measure introduced in [6, 8, 42] and known as *information gain*, has been proposed as a means of characterising the complexity of dynamical systems and of 2D patterns. It measures the amount of information gained in bits when specifying the value, x, of a random variable X given knowledge of the value, y, of another random variable Y,

$$G_{x,y} = -\log_2 P(x|y). \tag{4}$$

$P(x|y)$ is the conditional probability of a state x conditioned on the state y. Then the *mean information gain* (MIG), $\overline{G}_{X,Y}$, is the average amount of information gain from the description of the all possible states of Y:

$$\overline{G}_{X,Y} = \sum_{x,y} P(x, y) G_{x,y} = -\sum_{x,y} P(x, y) \log_2 P(x|y) \tag{5}$$

where $P(x, y)$ is the joint probability, prob$(X = x, Y = y)$. $\overline{G}$ is also known as the conditional entropy, $H(X|Y)$ [15]. Conditional entropy is the reduction in uncertainty of the joint distribution of X and Y given knowledge of Y, $H(X|Y) = H(X, Y) - H(Y)$. The lower and upper bounds of $\overline{G}_{X,Y}$ are

$$0 \leqslant \overline{G}_{X,Y} \leqslant \log_2 |\mathscr{X}|. \tag{6}$$

Definition 4 A structural complexity measure G, of a cellular automaton configuration is the sum of the mean information gains of cells having homogeneous/heterogeneous neighbouring cells over 2D lattice.

For a cellular automaton configuration, $\overline{G}$ can be calculated by considering the distribution of cell states over pairs of cells r, s,

$$\overline{G}_{r,s} = -\sum_{s_r, s_s} P(s_r, s_s) \log_2 P(s_r, s_s) \tag{7}$$

where s_r, s_s are the states at r and s. Since $|\mathscr{S}| = N$, $\overline{G}_{r,s}$ is a value in $[0, N]$.

The relative positions for non-edge cells are given by matrix M:

$$M = \begin{bmatrix} (i-1,j+1) & (i,j+1) & (i+1,j+1) \\ (i-1,j) & (i,j) & (i+1,j) \\ (i-1,j-1) & (i,j-1) & (i+1,j-1) \end{bmatrix}. \tag{8}$$

Correlations between cells on opposing lattice edges are not considered. The result of this edge condition is that $G_{i+1,j}$ is not necessarily equal to $\overline{G}_{i-1,j}$.

In addition the differences between the horizontal (vertical) and two diagonal mean information rates reveal left/right (up/down), primary and secondary orientation of 2D patterns.

So the sequence of generated configurations by a multi-state 2D cellular automaton can be analysed by the differences between the vertical $(i, j \pm 1)$, horizontal $(i \pm 1, j)$, primary diagonal (P_d) and secondary diagonal (S_d) mean information gains by

$$\Delta\overline{G}_{i,j\pm1} = |\overline{G}_{i,j+1} - \overline{G}_{i,j-1}| \tag{9}$$

$$\Delta\overline{G}_{i\pm1,j} = |\overline{G}_{i-1,j} - \overline{G}_{i+1,j}| \tag{10}$$

$$\Delta\overline{G}_{P_d} = |\overline{G}_{i-1,j+1} - \overline{G}_{i+1,j-1}| \tag{11}$$

$$\Delta\overline{G}_{S_d} = |\overline{G}_{i+1,j+1} - \overline{G}_{i-1,j-1}| \tag{12}$$

The advantages of $\overline{G}$ over H in discriminating structurally different patterns is illustrated in Fig. 4 where the 4-state 2D CA configurations with the complete

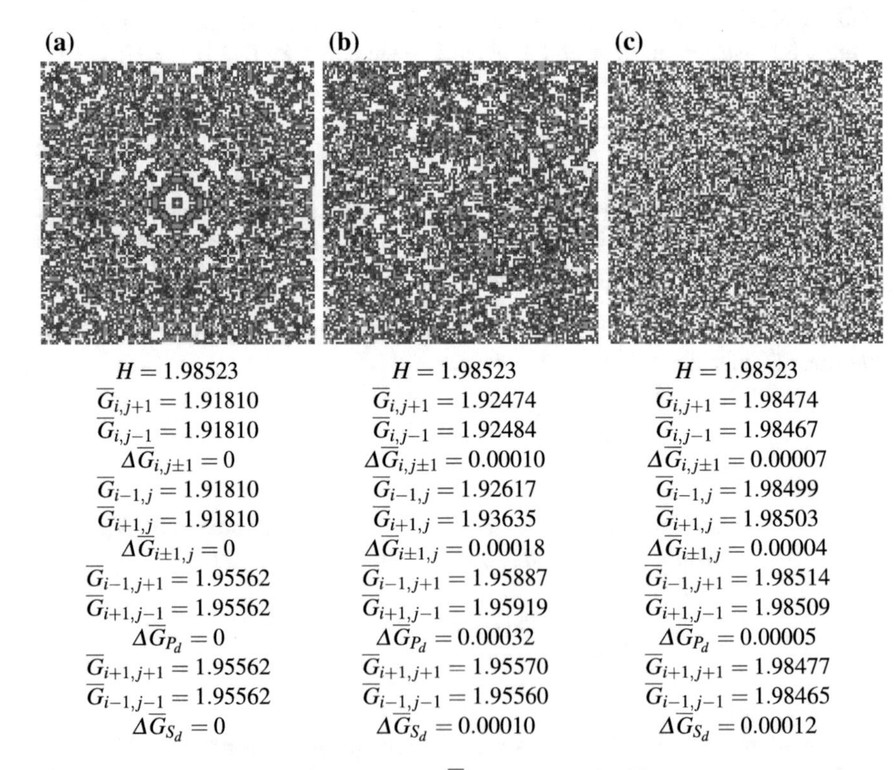

Fig. 4 The comparison of H with measures of $\overline{G}_{i,j}$ for structurally different 4-state CA configurations

symmetrical (Fig. 4a), the partially structured (Fig. 4b) and the structureless and random (Fig. 4c) patterns are evaluated. As it is evident, the measures of H are identical for structurally different patterns, however, the measure of $\overline{G}s$ and $\Delta\overline{G}s$ are reflecting not only the complexity of patterns but their spatial arrangements (i.e. symmetries) too. Figure 4 clearly demonstrates the drawbacks of entropy to discriminate structurally different 2D patterns. In other words, entropy is invariant to spatial rearrangement of composing elements. This is in contrast to our intuitive perception of the complexity of patterns and is problematic for the purpose of measuring the order and complexity of multi-state 2D CA configurations for aesthetic evaluation.

In the research presented in this paper, two modalities of the same algorithms are used to detect absolute symmetry in an input image if present, as well partial (sub-) symmetries.

The next section explains the swarm intelligence algorithm which will be used in detecting symmetrical patterns.

5 Stochastic Diffusion Search

The swarm intelligence algorithm used in this work is Stochastic Diffusion Search (SDS) [3, 11] which is a probabilistic approach for solving best-fit pattern recognition and matching problems. SDS, as a multi-agent population-based global search and optimisation algorithm, is a distributed mode of computation utilising interaction between simple agents. Its computational roots stem from Geoff Hinton's interest 3D object classification and mapping. See [19] for Hinton's work and [12] for the connection between Hinton mapping and SDS. SDS algorithm has been used in various fields including optimisation, generative arts and medical imaging (e.g. [1, 2, 4, 5]). SDS, contrary to many swarm intelligence algorithms, has a strong mathematical framework describing its behaviour and convergence. The full mathematical model and proof of SDS convergence are elaborated in [3].

5.1 SDS Architecture

Similar to other swarm intelligence algorithms, SDS commences a search or optimisation by initialising its population. In any SDS search, each agent maintains a hypothesis, *h*, defining a possible problem solution. After initialisation, the two phases of SDS—Test and Diffusion phases—are followed (see Algorithm 1 for a high-level description of SDS).

In the test phase, SDS checks whether the agent hypothesis is successful or not by performing a partial hypothesis evaluation and returning a domain independent boolean value. Later in the iteration, contingent on the strategy employed, successful hypotheses diffuse across the population and in this way information on potentially good solutions spreads throughout the entire population of agents.

In other words, in the Test phase, each agent performs *partial function evaluation,pFE*, which is some function of the agent's hypothesis, $pFE = f(h)$; and in the Diffusion phase, each agent recruits another agent for interaction and potential communication of hypothesis.

Algorithm 1 SDS Algorithm

```
01: Initialising agents
02: While (stopping condition is not met)
03:     Testing hypotheses
04:         Determining agents status (active/inactive)
05:     Diffusing hypotheses
06:         Exchanging of information
07: End While
```

5.2 *Standard SDS and Passive Recruitment*

In standard SDS, *passive recruitment mode* is employed. In this mode, if the agent is inactive, a second agent is randomly selected for diffusion; if the second agent is active, its hypothesis is communicated (*diffused*) to the inactive one. Otherwise there is no flow of information between agents; instead a completely new hypothesis is generated for the first inactive agent at random (see Algorithm 2). Therefore, recruitment is not the responsibility of the active agents. Higher rate of inactivity boosts exploration, whereas a lower rate biases the performance towards exploitation. Details of the test phase and the fitness function is described later in this paper.

Algorithm 2 Passive Recruitment Mode

```
01: For each agent ag
02:     If ( !ag.isActive )
03:         r_ag = pick a random agent
04:         If ( r_ag.isActive )
05:             ag.hypothesis = r_ag.hypothesis
06:         Else
07:             ag.hypothesis = generate random hypothesis
08:     End If
09: End For
```

5.3 *Partial Function Evaluation*

One of the concerns associated with many optimisation algorithms (e.g. Genetic Algorithm, Particle Swarm Optimisation and etc.) is the repetitive evaluation of a computationally expensive fitness functions. In some applications, such as tracking a rapidly moving object or generation of CA patters, the repetitive function evaluation significantly increases the computational cost of the algorithm. Therefore, in addition to reducing the number of function evaluations, other measures can be used in an attempt to reduce the computations carried out during the evaluation of each possible solution, as part of the overall optimisation (or search) processes.

The commonly used benchmarks for evaluating the performance of swarm intelligence algorithms are typically small in terms of their objective functions computational costs [17, 43], which is often not the case in real-world applications (examples of costly evaluation functions are seismic data interpretation, selection of sites for the transmission infrastructure of wireless communication networks and radio wave propagation calculations of one site, etc.).

Costly objective function evaluations have been investigated under different conditions [22] and the following two broad approaches have been proposed to reduce the cost of function evaluations:

- The first is to estimate the fitness by taking into account the fitness of the neighbouring elements, the former generations or the fitness of the same element through statistical techniques introduced in [13].
- In the second approach, the costly fitness function is substituted with a cheaper, approximate fitness function.

When agents are about to converge, the original fitness function can be used for evaluation to check the validity of the convergence [22].

The approach that the standard SDS algorithm uses is similar to the second method. Many fitness functions are decomposable to components that can be evaluated separately. During the test phase of SDS, in partial function evaluation (*pFE*, which is some function of the agent's hypothesis, $pFE = f(h)$), the evaluation of one or more of the components may provide partial information to guide the subsequent optimisation process.

In other words, instead of evaluating the hypothesis in its entirely, part of it, which is called *micro-feature*, is selected and evaluated accordingly. Therefore, during the test phase, only the randomly selected micro-features of the hypotheses are evaluated and the status of each agent is thus determined. Thus, if the micro-feature of each hypothesis consists of, say, $\frac{1}{10}$ of the entire hypothesis, the computational expense for the evaluation process of each hypothesis would be $\frac{9}{10}$ computationally cheaper.

Next, details of the process through which SDS performs its spatial-independent symmetry detection is presented.

6 Experiments

This section explains the design of the experiments conducted along with the results of applying SDS to identify partial or full symmetries on the cellular automata generated patterns. The inputs to the system are sample patterns used as proof of principle to show the functionality of the method; afterwards, some real world cellular automata

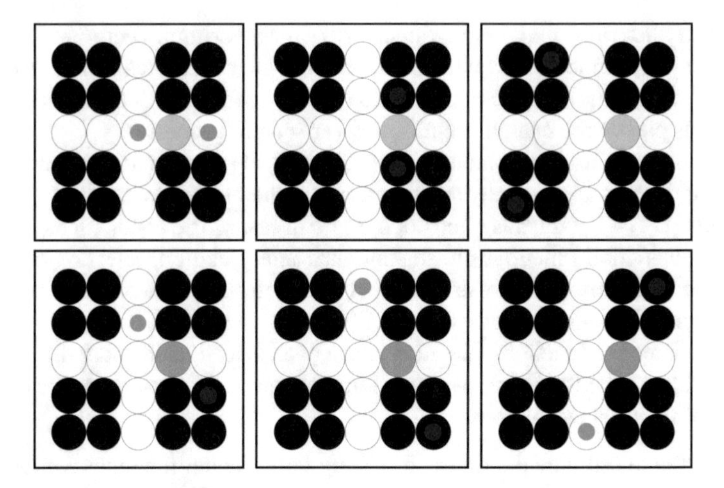

Fig. 5 Sample hypothesis set to be (3, 2); active hypotheses are shown in *green* and the inactive ones are displayed in *red*; the selected micro-features are highlighted in *blue*

generated patterns are fed in the system to evaluate the overall performance of the algorithm in detecting the aforementioned types of symmetries.

In order to adopt SDS to identify symmetries, the following important considerations are taken into account:

- the search space comprises of the entire cells on the grid
- SDS hypothesis is a cell index. For instance, in a 5×5 grid, the coordinate (2, 2) could be the hypothesis and micro-features[1] can be selected by specifying the x_d and y_d distances from the hypothesis; therefore, assuming the (x_d, y_d) distance is (2, 0), this micro-feature should be compared against its corresponding element with $(-2, 0)$ distance from the hypothesis.
- the environment in cellular automata is torus, which means if moving downwards along the search space when we reach the last raw, the next row to be visited is the top row. The same is applicable when moving between columns (see Fig. 1 shows the 2D representation of the cellular automata and its real structure as torus).

The patterns in Fig. 5 show the hypothesis (3, 2) and the various possible micro-features, some of which resulting in the hypothesis' status to be *true* while some others lead to the hypothesis' status to be *false*. The hypothesis in these figures are set to be (3, 2) and various micro-features are selected to test the symmetry of the pattern along various axes of symmetry. The torus structure of cellular automata is demonstrated in the choice of some of the corresponding micro-features; see, for example, Fig. 5 top-right corner, where the micro-feature is chosen at $(-1, -1)$ distance. Thus the corresponding cell is chosen at (1, 1) distance from the hypothesis,

[1]Micro-features are used in the test phase of SDS to determine the status of the agent (i.e. active or inactive).

which means moving out of the 2D canvas from the right border and entering again from the left.

The process through which SDS commences with the initialisation phase and then cycle through the two phases and test and diffusion is explained next.

6.1 Initialisation Phase

During the initialisation phase each one of the agents in the population is assigned a hypothesis which is a random (x, y) coordinate from the search space. Additionally, the status of all agents are initially set to *false*.

6.2 Test Phase

In the test phase, each agent, which is already allocated an (x, y) hypothesis, picks a random x_d and y_d distances from the hypothesis cell as its micro-feature; the randomly selected micro-feature is then compared against corresponding mirrored cells to checks if the mirror cell has the same value. If the values are the same, the status of the agent is set to *true*, otherwise *false*

6.3 Diffusion Phase

The process in the diffusion phase is the same as the one detailed in the algorithm description where each inactive agent picks an agent randomly from the population; if the randomly selected agent is active, the inactive agent adopts the hypothesis of the active agent (i.e. the (x, y) coordinate), otherwise the inactive agent picks a random coordinate from the search space.

After n number of iterations agents converge on the (x, y) coordinates with the most symmetrical quality.

6.4 Experiments and Discussion

One of the main features of SDS is partial function evaluation which here manifests itself in: each time comparing one cell on one side of the symmetrical point to its corresponding cell on the other side. Therefore even when an agent is active, in the next iteration it picks another micro-feature and checks the point from "a different perspective "to ensure that the symmetry still holds. In other words, using this approach, the algorithm allocate its resources "wisely" and repeatedly tests the already maintained points of interest against any asymmetrical discovery.

For the experiments reported in this work, the population size is empirically calculated using the following formula:

$$\text{pSize} = \frac{w^2}{4} \tag{13}$$

where pSize is the population size and w is the width of the search space. Using this set-up, the agents land on fourth of the search space; therefore for a 5×5 search space, $pSize = 6$.

As illustrated in the figures, some agents became active on (x, y) coordinates which do not represent the full four-fold symmetry; these agents will eventually pick different micro-features in the next iterations and become inactive; consequently, when they are inactive, they need to choose random agents; given that the number of active agents on the centre of symmetry increases over time (thanks to the diffusion phase), it is likely that an active agent is chosen. This would lead to the inactive agents picking micro-features from the centre of symmetry in their next iterations and become/stay active. Note that in these experiments, alpha is used for the transparency of the agents' colour; therefore as shown on the figures, the cell with the largest number of active agents can be distinguished from others.

There are occasions when more than one centre of symmetry exists, or there exist some partial symmetries (also called sub-symmetry) in the image along with full centre of symmetry; in this case another flavour of the recruitment strategy is deployed which is called *context-sensitive mechanism*. This strategy frees up some of the agents who are active and share the same hypothesis, and therefore allows the algorithm to constantly check for traces of symmetry in the input pattern.

Algorithm 3 Context Sensitive Mechanism

```
01: If ( ag.activity )
02:     r_ag = pick a random agent
03:     If ( r_ag.activity AND
04:            ag.getHypothsis == r_ag.getHypothsis )
05:         ag.setActivity ( false )
06:         ag.setHypotheis ( randomHypothsis )
07:     End If
08: End If
```

In other words, the use of context sensitive mechanism biases the search towards global exploration. Thus, if an active agent randomly chooses another active agent that maintains the same hypothesis, the selecting agent is set inactive and adopts a random hypothesis. This mechanism frees up some of the resources in order to have a wider exploration throughout the search space as well as preventing cluster size from overgrowing; this process goes on while ensuring the formation of large clusters in case there exists a perfect match or good sub-optimal solutions (see Algorithm 3).

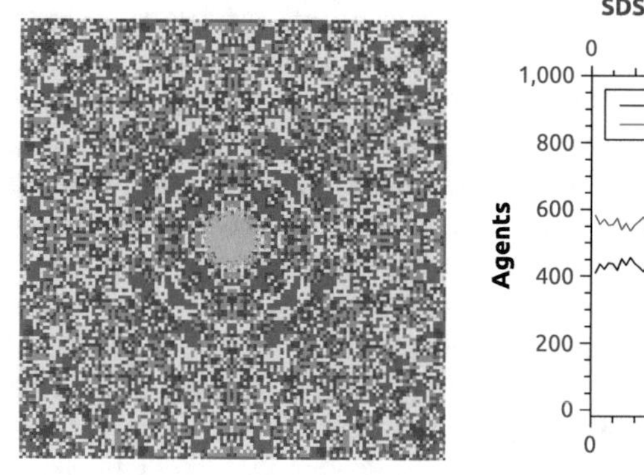
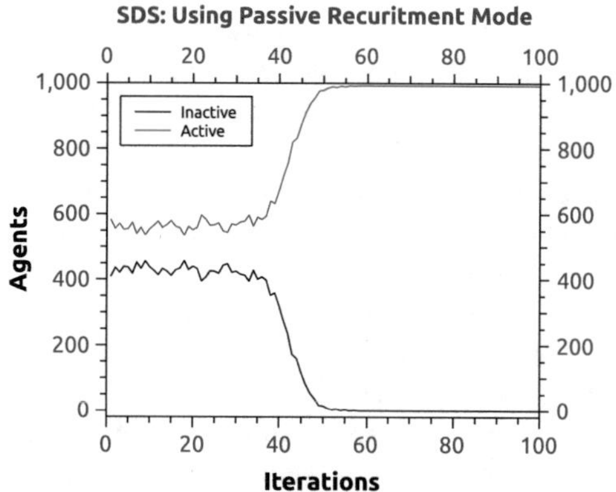

Fig. 6 Passive recruitment mode

The next set of experiments use some complex patterns, generated by cellular automata techniques. Initially an experiment is run that utilises the *passive recruitment mode* without the introduced *context-sensitive mechanism* and later, the impact of *context-sensitivity* is discussed.

The graph in Fig. 6 illustrate the behaviour of the agents' activities; this graph demonstrates that after the initialisation phase, the number of active and inactive agents are balanced; however over time, and due to the presence of a centre of symmetry in the pattern, the number of active agents increases and the number of inactive agents decreases. Therefore, ultimately, once the absolute center of symmetry (where symmetry holds irrespective of which micro-feature is chosen) is identified, the entire agent population becomes active and the number of inactive agents drops to zero.

Using context-sensitive mechanism, the graph in Fig. 7 illustrates the behaviour of SDS algorithm using this mode, where the populations are biased towards global

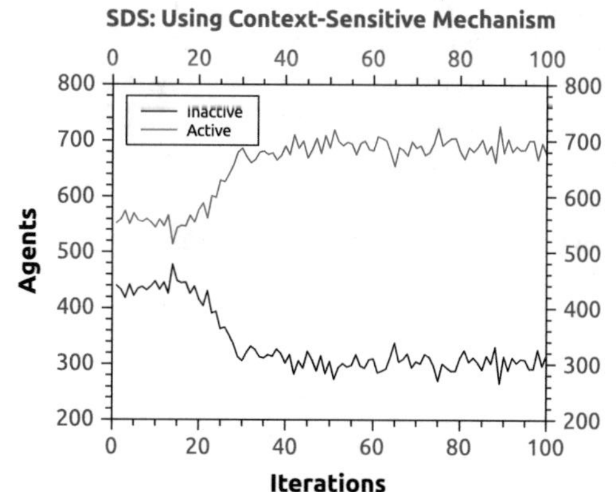

Fig. 7 Context sensitive mechanism

exploration. In this graph, while the increase of active agents and the decrease of inactive agents are visible, it is evident that there are always agents which are released back from the centre of symmetry to the search space to explore the possibility of the *presence of further (sub-)symmetrical points*. This feature is particularly useful in dynamic environment, and where there are more than one absolute points of symmetry (the next experiment uses an input image with a few points of symmetries). The figure shows many active (green) and inactive (red) agents throughout the search space and the graph illustrates that the number of inactive agents never drops to zero.

The next experiment, which uses a more symmetrically complex CA-generated pattern, demonstrates the crucial difference between using SDS with and without the context-sensitive mechanism. As stated before, context-sensitive mechanism reduces the greediness of the agents and allows the agents to explore the search space for any undetected symmetry, while the pure passive mechanism is greedy and once it finds the absolute point of symmetry (where symmetry holds no matter which micro-feature is picked), it gradually pulls all the agents towards the point and stops them from locating possible partial symmetries in the canvas.

The new input to be used in this experiment has two identically CA grown patterns one on the top-left corner and another on on the bottom-left corner. When running the SDS algorithm, it becomes clear that the passive recruitment strategy (see Fig. 8) initially locates two points of symmetry (at $n = 100$ iterations), however later (at $n = 200$ iterations) all agents are drawn towards the absolute point of symmetry (note that the search spaces in cellular automata are torus).

Using the context sensitive approach, the largest partial symmetries are also identified and highlighted (see Fig. 9). The graphs at the bottom of Figs. 8 and 9 clearly show the behaviour of the agents in both modes. As displayed in the graph of Fig. 9, while the number of active and inactive agents are distinguishably far from one another, yet it is shown that the number of active agents does not reach the maximum possible[2] and the number of inactive agents does not drop to zero. This mechanism insures the identification of other (sub-) symmetrical points in the input. Therefore, depending on the functionalities needed, either of these approaches could be used.

Another observation to be expanded in the future work is the direct proportionality of the agents' activity to the 'strength' of the symmetry. Therefore, while context-sensitive mechanism finds partial symmetries, it is able to '*rank*' the various clusters of agents which are formed over the pattern. This could lead to introducing the ratio of active/inactive agents as a measure for *order* and *complexity* along with Shannon's entropy [37] and information gain [6, 8, 42] which are among the very few measures used in cellular automata for measuring symmetry.

[2] Given the size of the side of search space is $ssSize = 129$, the population size for this pattern is $pSize = \frac{129^2}{4} = 4{,}160$.

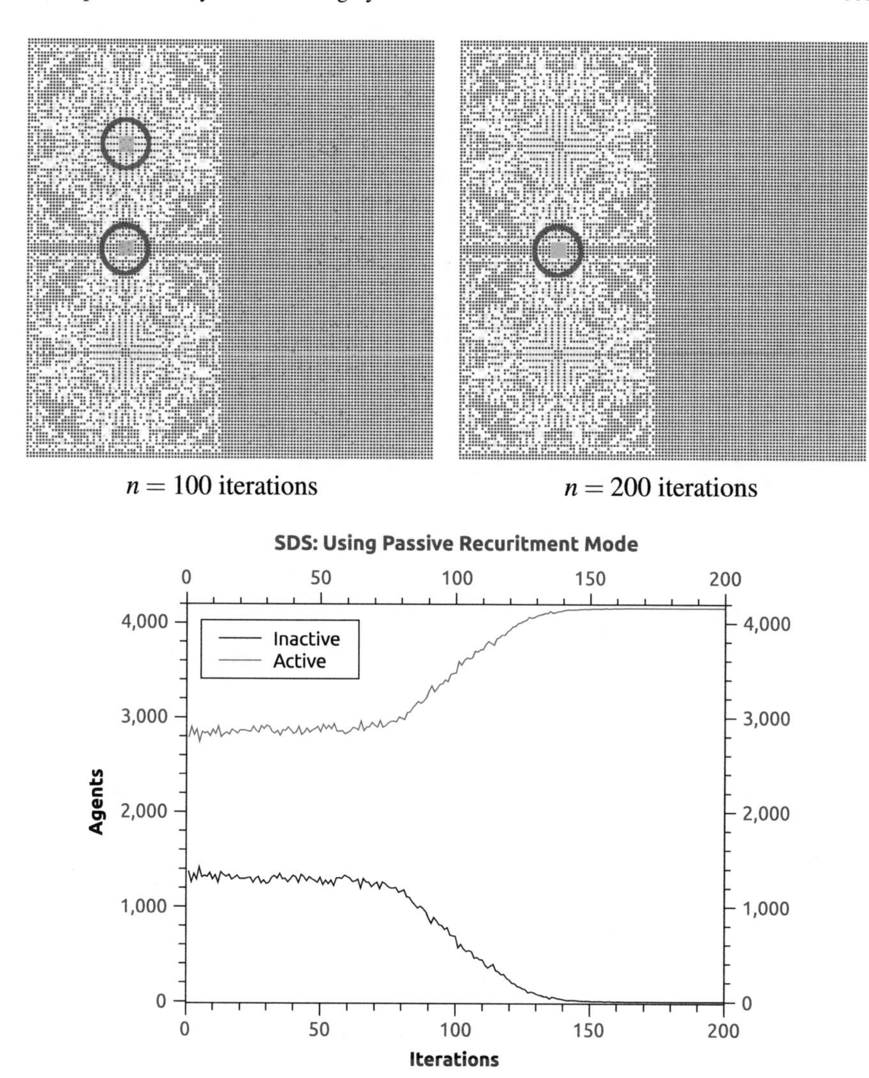

Fig. 8 Passive Recruitment mode: finding absolute symmetry

The swarm intelligence, in neither of the modalities (i.e. passive mode and passive with context-free mode) technique has classified Fig. 4b and c as unstructured. as they do not posses dominant centre(s) of symmetries.

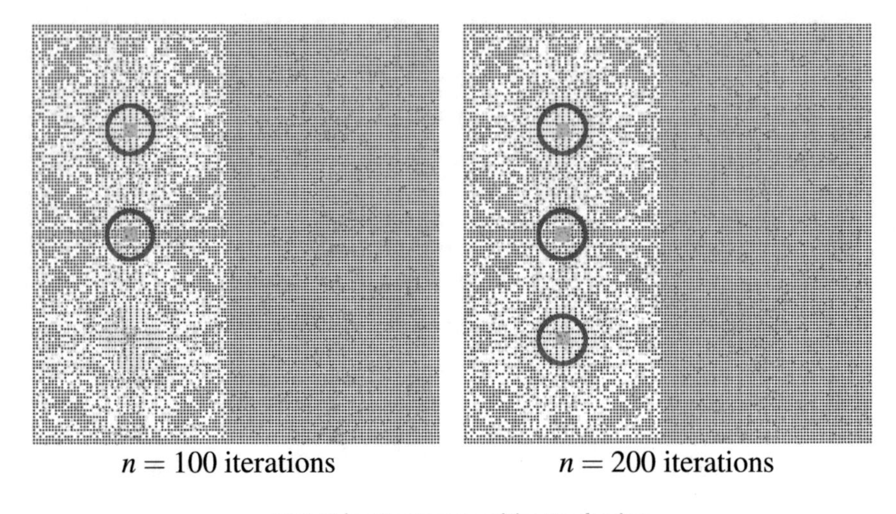

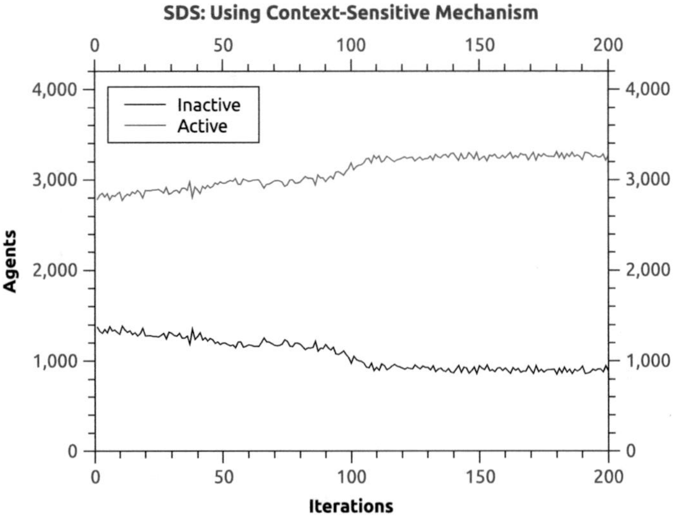

Fig. 9 Context-Sensitive mechanism: finding partial symmetry

7 Comparison Between Information Gain and Swarm Intelligence

In this work, various CA generated images were used as input to both information gain as well as the swarm intelligence algorithm in order to detect the structural characteristics (i.e. order-symmetry and randomness).

It is shown that information gain is able to discriminate structural characteristics ranging from full symmetry, partially structured and fully random patterns (see Fig. 4). One of the main feature of this measure is to quantify the existing structure of the image.

However information gain is unable identify the centre of symmetry and if there are multiple centres of symmetries (see Fig. 9), without having a complete centre of symmetry, it classifies the image as completely symmetrical.

The swarm intelligence technique used (Stochastic Diffusion Search or SDS), is capable of identifying all the centres of symmetry irrespective of the structure of the image as evaluated by information gain. On the other hand, SDS is unable to quantify the structure of the input images and shows weakness when fed with images which lack any "visible" centres of symmetries.

These results show the complementary nature of information gain and SDS to work alongside each other offering insight into both the structure of the images as well as the presence of any symmetries.

8 Conclusion

CA provide perspective and powerful tools in generating computer graphics. The multi-state CA rule space is a vast set of possible rules which can generate interesting patterns with high aesthetic qualities. The interaction of CA rules at local level generates emergent global behaviour, that can sometimes demonstrate attractive complexity. Some characteristics of CA, such as the regularity and complexity of the rules that are employed locally, suggest that they could be well suited to generating computer graphics.

This paper demonstrates the capability of a swarm intelligence algorithm—Stochastic Diffusion Search—in detecting absolute symmetries (when present) and the centre of partial symmetrical patterns within the input image. Additionally entropy and information gain were introduced and applied to several CA generated patterns with varying structural characteristics (order-symmetry and randomness) (Figs. 4 and 10).

Evaluating the symmetry of cellular automata generated patterns is often a difficult task partly due the the large size of the search space, and partly due to the constantly changing, dynamic environment in which the cellular automata patterns are generated. These factors contribute to making the detection of symmetrical patterns computationally expensive.

One of the main features of Stochastic Diffusion Search is partial function evaluation which is particularly useful when dealing with large problems with high dimensions and costly evaluation function (e.g. in this case, the expensive computational cost of detecting symmetry in cellular automata generated patters). The performance of this algorithm is explained in the paper and the results are accordingly demonstrated.

The advantages and weaknesses of the swarm intelligence technique and the information gain measure are presented and it is shown that the can be be used as complementary tools to better understand the structural complexity of the input images.

Fig. 10 Information gain measure applied on the CA generated pattern

$$H = 0.92537$$
$$\overline{G}_{i,j+1} = 0.86462$$
$$\overline{G}_{i,j-1} = 0.86462$$
$$\Delta\overline{G}_{i,j\pm1} = 0$$
$$\overline{G}_{i-1,j} = 0.86462$$
$$\overline{G}_{i+1,j} = 0.86462$$
$$\Delta\overline{G}_{i\pm1,j} = 0$$
$$\overline{G}_{i-1,j+1} = 0.89953$$
$$\overline{G}_{i+1,j-1} = 0.89953$$
$$\Delta\overline{G}_{P_d} = 0$$
$$\overline{G}_{i+1,j+1} = 0.89953$$
$$\overline{G}_{i-1,j-1} = 1.98465$$
$$\Delta\overline{G}_{S_d} = 0$$

Following the introduction of this novel technique, among the future research topics are: conducting a comparison with other evolutionary and non-evolutionary techniques, computing the correlation between the size of search space and the computational complexity of the process, ranking the quality of the symmetries detected, and applying this method to dynamically evolving cellular automata generated patterns.

References

1. al-Rifaie, A.M., al-Rifaie, M.M.: Generative music with stochastic diffusion search. In: Johnson C., Carballal A., Correia J. (eds.) Evolutionary and Biologically Inspired Music, Sound, Art and Design, Lecture Notes in Computer Science, vol. 9027, pp. 1–14. Springer, Berlin (2015). doi:10.1007/978-3-319-16498-4_1
2. al-Rifaie, F.M., al-Rifaie, M.M.: Investigating stochastic diffusion search in dna sequence assembly problem. In: Proceedings of SAI Intelligent Systems Conference. IEEE (2015)
3. al-Rifaie, M.M., Bishop, M.: Stochastic diffusion search review. In: Paladyn, Journal of Behavioral Robotics, vol. 4(3), pp. 155–173 (2013). doi:10.2478/pjbr-2013-0021
4. al-Rifaie, M.M., Bishop, M., Blackwell, T.: Information sharing impact of stochastic diffusion search on differential evolution algorithm. J. Memet. Comput. **4**, pp. 327–338 (2012). doi:10.1007/s12293-012-0094-y
5. al-Rifaie, M.M., Bishop, M., Caines, S.: Creativity and autonomy in swarm intelligence systems. J. Cogn. Comput. **4**, pp. 320–331 (2012). doi:10.1007/s12559-012-9130-y

6. Andrienko, Yu. A., Brilliantov, N.V., Kurths, J.: Complexity of two-dimensional patterns. Eur. Phys. J. B **15**(3), 539–546 (2000)
7. Atallah, M.J.: On symmetry detection. Comput. IEEE Trans. **100**(7), 663–666 (1985)
8. Bates, J.E., Shepard, H.K.: Measuring complexity using information fluctuation. Phys. Lett. A **172**(6), 416–425 (1993)
9. Bauerly, M., Liu, Y.: Computational modeling and experimental investigation of effects of compositional elements on interface and design aesthetics. Int. J. Man-Mach. Stud. **64**(8), 670–682 (2006)
10. Behrens, R.: Design in the Visual Arts. Prentice-Hall, Upper-Saddle River (1984)
11. Bishop, J.: Stochastic searching networks. In: Proceedings of the 1st IEE Conference on Artificial Neural Networks, pp. 329–331. London, UK (1989)
12. Bishop, J., Torr, P.: The stochastic search network. Neural Networks for Images. Speech and Natural Language, pp. 370–387. Chapman & Hall, New York (1992)
13. Branke, J., Schmidt, C., Schmeck, H.: Efficient fitness estimation in noisy environments. In: Spector, L. (ed.) Genetic and Evolutionary Computation Conference, Morgan Kaufmann, Burlington (2001)
14. Carroll, M. J. (eds.): HCI Models, Theories, and Frameworks: Toward a multidisciplinary Science. Morgan Kaufmann Publishers, San Francisco (2003)
15. Cover, T.M., Thomas, J.A.: Elements of Information Theory (Wiley Series in Telecommunications and Signal Processing). Wiley-Interscience, New York (2006)
16. Dawkins, R.: The blind watchmaker. Norton & Company, New York (1986)
17. Digalakis, J., Margaritis, K.: An experimental study of benchmarking functions for evolutionary algorithms. Int. J. **79**, 403–416 (2002)
18. Gangestad, S.W., Thornhill, R., Yeo, R.A.: Facial attractiveness, developmental stability, and fluctuating asymmetry. Ethol Sociobiol. **15**(2), 73–85 (1994)
19. Hinton, G.F.: A parallel computation that assigns canonical object-based frames of reference. In: Proceedings of the 7th International Joint Conference on Artificial intelligence-Vol. 2, pp. 683–685. Morgan Kaufmann Publishers, Burlington (1981)
20. Javaheri Javid, M.A., Blackwell, T., Zimmer, R., Al Rifaie, M.M.: Spatial Complexity Measure for Characterising Cellular Automata Generated 2D Patterns. In: Pereira, F., Machado, P., Costa, E., Cardoso, A. (eds.) Progress in Artificial Intelligence: 17th Portuguese Conference on Artificial Intelligence, EPIA 2015, Coimbra, Portugal, September 8–11, 2015. Lecture Notes in Artificial Intelligence, vol. 9273, pp. 201–212. Springer, Heidelberg (2015)
21. Jiang, H., Ngo, C.W., Tan, H.K.: Gestalt-based feature similarity measure in trademark database. Pattern Recognit. **39**(5), 988–1001 (2006)
22. Jin, Y.: A comprehensive survey of fitness approximation in evolutionary computation. Soft Comput. **9**, 3–12 (2005)
23. Lee, S., Liu, Y.: Curved glide-reflection symmetry detection. IEEE Trans. Pattern Anal. Mach. Intell. **34**(2), 266–278 (2012)
24. Lewis, M.: Evolutionary visual art and design. In: Romero, J., Machado P. (eds.) The Art of Artificial Evolution, Natural Computing Series, pp. 3–37. Springer, Heidelberg (2008)
25. Leyton, M.: Symmetry, Causality, Mind. MIT Press, Cambridge (1992)
26. Linz, P.: An Introduction to Formal Languages and Automata. Jones & Bartlett Publishers, Burlington (2001)
27. Liu, Y.: Computational symmetry. In: Proceedings of the CMU Robotics Institute (2000)
28. McCormack, J.: Interactive evolution of l-system grammars for computer graphics modelling. In Green, D., Bossomaier, T. (eds.) Complex Systems: From Biology to Computation pp. 118–130. ISO Press, Amsterdam (1993)
29. Mitra, N.J., Guibas, L.J., Pauly, M.: Partial and approximate symmetry detection for 3d geometry. ACM Trans. Graph. (TOG) **25**(3), 560–568 (2006)
30. Møller A.P., Cuervo, J.J.: Asymmetry, size and sexual selection : meta-analysis, publication bias and factors affecting variation in relationships, p. 1. Oxford University Press, Oxford (1999)

31. Møller, A.P.R.T.: Bilateral symmetry and sexual selection: a meta-analysis. Am. Nat **151**(2), 174–192 (1998)
32. Park, I.K., Lee, K.M., Lee, S.U.: Perceptual grouping of line features in 3-D space: a model-based framework. Pattern Recogn. **37**(1), 145–159 (2004)
33. Podolak, J., Shilane, P., Golovinskiy, A., Rusinkiewicz, S., Funkhouser, T.: A planar-reflective symmetry transform for 3d shapes. In: ACM Trans. Graph. (TOG), vol. 25, pp. 549–559. ACM, New York (2006)
34. Railton, P.: Aesthetic Value, Moral Value and the Ambitions of Naturalism In Aesthetics and Ethics, chap. 3. University of Maryland, College Park (2001)
35. Randy, T., Steven, G.: Human facial beauty. Human Nat. **4**, 237–269 (1993)
36. Shackelford, T.K.L.R.J.: Facial symmetry as an indicator of psychological emotional and physiological distress. J. Personal. Soc. Psychol. **72**2, 456–66 (1997)
37. Shannon, C.: A mathematical theory of communication. Bell Syst. Tech. J. **27**, 379–423 & 623–656 (1948)
38. Shannon, C., et al.: The synthesis of two-terminal switching circuits. Bell Syst. Tech. J. **28**(1), 59–98 (1949)
39. Sims, K.: Evolving virtual creatures. In: Proceedings of the 21st Annual conference on Computer Graphics and Interactive Techniques, pp. 15–22. ACM, New York (1994)
40. Sun, C., Sherrah, J.: 3d symmetry detection using the extended gaussian image. EEE Trans. Pattern Anal. Mach. Intell. **19**(2), 164–168 (1997)
41. Todd, S., Latham, W., Hughes, P.: Computer sculpture design and animation. J. Vis. Comput. Animat. **2**(3), 98–105 (1991)
42. Wackerbauer, R., Witt, A., Atmanspacher, H., Kurths, J., Scheingraber, H.: A comparative classification of complexity measures. Chaos, Solitons Fractals **4**(1), 133–173 (1994)
43. Whitley, D., Rana, S., Dzubera, J., Mathias, K.E.: Evaluating evolutionary algorithms. Artif. Intell. **85**(1–2), 245–276 (1996)
44. Wolter, J.D., Woo, T.C., Volz, R.A.: Optimal algorithms for symmetry detection in two and three dimensions. Vis. Comput. **1**(1), 37–48 (1985)
45. Zabrodsky, H., Peleg, S., Avnir, D.: Symmetry as a continuous feature. IEEE Trans. Pattern Anal. Mach. Intell. **17**(12), 1154–1166 (1995)
46. Zhang, J.S., Chrzanowska-Jeske, M., Mishchenko, A., Burch, J.R.: Generalized symmetries in boolean functions: Fast computation and application to boolean matching. In: Proceedings of the IWLS. Citeseer (2004)

Power Quality Enhancement in Off-Grid Hybrid Renewable Energy Systems Using Type-2 Fuzzy Control of Shunt Active Filter

Abdeldjabbar Mohamed Kouadria, Tayeb Allaoui, Mouloud Denaï and George Pissanidis

Abstract Maintaining a clean, reliable and efficient electric power system has become more challenging than ever due to the widespread use of solid-state power electronic controlled equipment in industrial, commercial and domestic applications. Non-linear loads draw non-sinusoidal current and reactive power from the source causing voltage and current distortion, increased losses in the power lines and deterioration of the overall power quality of the distribution grid. Devices such as tuned passive filters are among the oldest and most widely used techniques to remove power line harmonics. Other solutions include active power filters which operate as a controllable current source injecting a current that is equal but with opposite phase to cancel the harmonic current. This chapter deals with the design of fuzzy control strategies for a three-phase shunt active power filter to enhance the power quality in a hybrid wind-diesel power system operating in standalone mode. The proposed control scheme is based on Interval Type 2 fuzzy logic controller and is applied to the regulation of the DC bus voltage to compensate for real power unbalances during variable load conditions. A simulation study is performed under Matlab/Simulink to evaluate the performance and robustness of the system under different wind speed conditions.

A.M. Kouadria · T. Allaoui
Energetic and Computer Engineering Lab, University of Tiaret, Tiaret, Algeria
e-mail: abdeldjabbar14@yahoo.fr

T. Allaoui
e-mail: allaoui_tb@yahoo.fr

M. Denaï (✉) · G. Pissanidis
School of Engineering and Technology, University of Hertfordshire, Hatfield, UK
e-mail: m.denai@herts.ac.uk

G. Pissanidis
e-mail: g.1.pissanidis@herts.ac.uk

Y. Bi et al. (eds.), *Intelligent Systems and Applications*,
Studies in Computational Intelligence 650, DOI 10.1007/978-3-319-33386-1_17

1 Introduction

In recent years, renewable energy technologies have experienced considerable developments due to the world's expanding energy needs. Wind energy is among the fastest growing renewable energy technologies and has the potential to play a significant role in achieving the world's future energy targets [1].

In many developed and developing countries, off-grid hybrid energy systems are increasingly being utilized to connect remote areas to a source of electric power as an alternative to the conventional standalone diesel-generator system [2]. A Wind Diesel Hybrid System (WDHS) consists of Wind Turbine Generator(s) (WTG) and Diesel Generator(s) (DG) [3, 4]. The advantage of the WDHS is the association of the continuously available diesel power with locally available, pollution-free, but intermittent and fluctuating in nature, wind energy [5]. Fuel consumption savings are maximum in WDHS with high wind penetration, in which the DGs may be shut down during periods of high wind availability.

Electric power quality has become vitally important with the proliferation of nonlinear loads such as power electronic converters in electric power distribution systems. Nonlinear loads introduce harmonics into the power network which cause a number of disturbances such as distortion of the current and voltage waveforms, electromagnetic interference, overheating of power distribution components inducing losses and reducing their lifetime [6, 7]. The impacts are expected to become even more challenging with the increasing penetration of the dispersed generation, where power electronic converters are often used to connect the generation units, such as wind turbines, photovoltaics, fuel cells and micro-turbines, etc.

Passive filters, consisting of custom-tuned LC circuits, are the oldest and perhaps the most popular and widely adopted low-cost solution to the harmonic problem [8]. However, the passive filter scheme has a limitation since it provides a fixed harmonic frequency compensation and also the addition of a passive filter interferes with the system impedance and causes resonance with other networks.

Active Power Filters (APFs) also called Active Power Line Conditioners (APLCs) are a relatively new technology which offers a more flexible alternative and provides superior filtering performance characteristics and faster transient response as compared to conventional passive filters [9]. APFs are classified into three types: Shunt active power filters, series active power filters and hybrid active power filters.

The shunt active power filter (SAPF) is used to compensate current harmonics. Universal APF is used to compensate voltage harmonics as well as current harmonics. SAPF is connected in parallel at the Point of Common Coupling (PCC) between the source and the nonlinear load. It acts as a current source and inject compensating current at PCC to make the source current sinusoidal and in phase with the source voltage [10, 11]. The series active power filter was introduced at the end of 1980. It is usually connected in series with a line through a series transformer. This filter injects the compensating voltage in series with the supply voltage. Thus, it acts as a voltage source which can be controlled to compensate voltage sags. Series filters have their application mainly where the load contains voltage sensitive devices. In many cases series active power filters are used in combination with passive LC filters [12]. The hybrid active power filter or Unified Power Quality Conditioner (UPQC) is a combination of series and shunt active power filters. It consists of two converters connected back to back with same DC-link capacitor. One inverter is connected across the load and acts as shunt APF [10, 13]. It has the advantages of both series APF and shunt APF. In other words, it can compensate both voltage and current harmonics. Therefore, this filter can be applied to almost all types of power quality problems faced by a power system network [14].

APFs basically consist of a power electronic inverter and a control circuit. The performance of these filters systems depends mainly on the converter topology employed, the adopted reference current generation strategy for harmonic compensation as well as the controller for the regulation of the DC bus voltage. The regulation of the DC bus voltage consists of maintaining the voltage across the capacitor connected to the inverter at the desired level. The role of the capacitor voltage is to compensate for inverter losses and any transient fluctuations in real power between the AC and DC sides following load changes. Various DC bus voltage control strategies have been proposed in the literature ([6, 10, 12]).

Fuzzy systems have evolved for more than four decades and have proved to be a powerful technique in dealing with uncertainties, parameter variation and especially where the system model is complex or not accurately defined for the designed control action. They have been successfully implemented in many real world applications mainly focusing on quantitative modeling and control [15]. Fuzzy logic controllers (FLCs) [7, 16, 17] have also been very popular for their applications to APFs.

This chapter presents a simplified design approach of an Interval Type-2 fuzzy logic controller (IT2FLC) [18, 19] for the regulation of the DC bus voltage of the SAPF connected to WDHS. The simulation results provide the validation of the p-q theory-based SAPF system to meet the IEEE Standard 519 which is the recommended harmonic standard for different rated nonlinear loads under balanced supply conditions.

The rest of the chapter is organized as follows: Sect. 2 present a description and detailed mathematical modeling of the hybrid power system. Section 3 presents the design steps of the proposed control strategy for the DC voltage. In Sect. 4, the overall system performance is evaluated in a series of simulation scenarios and the results are presented and discussed. Finally, Sect. 5 presents the conclusions of this contribution.

2 Hybrid Wind-Diesel Power System Configuration

The WDHS has three operation modes: Diesel Only (DO), Wind Diesel (WD) and Wind Only (WO) [4, 20]. In DO mode the DGs supply the active and reactive power demanded by the consumer load (WTGs are disconnected). In this case, the DGs operate continuously and the maximum power from the WTG is always significantly less than the system load. In WD mode, in addition to DG(s), WTG(s) also supply

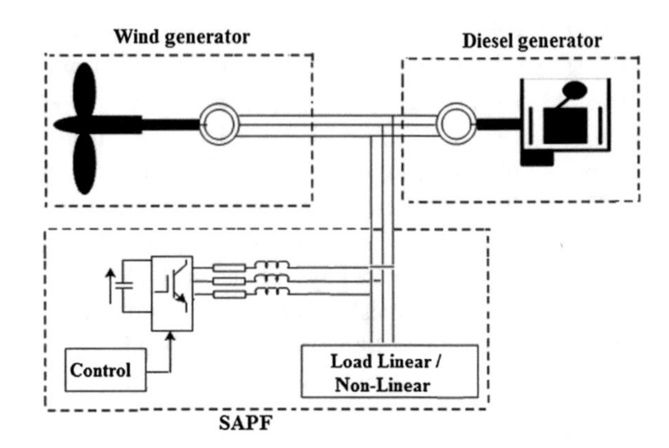

Fig. 1 WDHS and SAPF configuration

active power. In this case, the WTG power is approximately the same as the consumer load and in addition to DG(s), WTG(s) also supply active power. In WO mode, the DGs are not running, and only the WTs are supplying active power therefore no fuel is consumed in this mode.

Figure 1 shows the basic structure of a WDHS system supplying a local isolated load. The system essentially consists of a vertical axis wind turbine driving an induction generator connected to a synchronous generator driven by a diesel engine. The synchronous generator is equipped with a voltage regulator and a static exciter and the shunt active filter (SAPF).

The SAPF included in the system for compensating harmonics and reactive power consists of a three phase IGBT-based inverter connected to a DC bus capacitor. The nonlinear load is simulated with a three-phase diode bridge rectifier with R-L load.

2.1 *Wind Turbine Model*

The WT converts the kinetic energy of the wind into electric energy. The model complexity of the WT depends merely on the type of control objectives imposed. In order to study the dynamic response of multiple components and aerodynamic loading complex simulators are required. Generally, the impact of dynamic loads and interaction of large components are verified by the aero elastic simulator. However, for designing a WT controller, a simplified mathematical model is usually sufficient [3, 21]. In this work, the WT model is described by the set of nonlinear ordinary differential equation with limited degree of freedom.

The aerodynamic power captured by the rotor is given as:

$$P_a = C_p\left(\lambda, \beta\right) \frac{\rho S v^3}{2} \tag{1}$$

From Eq. (1), it is clear that the aerodynamic power (P_a) is directly proportional to the cube of the wind speed. The power coefficient C_P is a function of blade pitch angle (β) and tip speed ratio (λ). The tip speed ratio is defined as the ratio between the linear tip speed and the wind speed.

$$\lambda = \frac{\Omega R}{v} \tag{2}$$

where Ω is the angular speed of the rotor with the radius R. This function assumes a maximum for a certain value of λ. From the wind speed input v and the rotor speed Ω, the operating point of the wind turbine can be determined. When the induction machine is generating, which corresponds to its operation in the vicinity of synchronous speed, the rotor speed is relatively constant.

The models of the generators are based on the standard Park's transformation that converts all stator variables to a rotor reference frame described by a direct and quadrature (d-q) axis [2, 3, 22].

The model of GAS is composed of two electrical differential equations and a mechanical differential equation. The electrical equations, expressed in direct (d) quadrature (q) coordinate are given by [23]:

$$\begin{cases} U_{ds} = R_s i_{ds} - L_s \frac{di_{ds}}{dt} + L_m \frac{di_{dr}}{dt} \\ U_{qs} = R_s i_{qs} - L_s \frac{di_{dq}}{dt} + L_m \frac{di_{qr}}{dt} \end{cases} \tag{3}$$

$$\begin{cases} U_{dr} = R_r i_{dr} + L_r \frac{di_{dr}}{dt} + L_m \frac{di_{ds}}{dt} + P\omega_g \left(L_r i_{qr} + L_m i_{qs}\right) \\ U_{qr} = R_r i_{qr} + L_r \frac{di_{qr}}{dt} + L_m \frac{di_{qs}}{dt} + P\omega_g \left(L_r i_{dr} + L_m i_{ds}\right) \end{cases} \tag{4}$$

where R_s, R_r are stator and rotor resistances L_s, L_r are stator and rotor inductances, L_m is magnetizing inductance between the stator and rotor, P is the number of machine poles and ω_g is the angular speed of the generator.

The electromagnetic torque equation is given by:

$$T_{em} = \frac{3}{2} P \left(\lambda_{ds} i_{qs} - \lambda_{qs} i_{ds}\right) \tag{5}$$

The torque balance equation of the mechanical motion of the induction generator side is given as [4]:

$$T_g = J \frac{2}{P} \frac{d\Omega}{dt} + T_{em} \tag{6}$$

With J is the inertia, P is the number poles of the induction generator and Ω represents the rotor speed of induction generator.

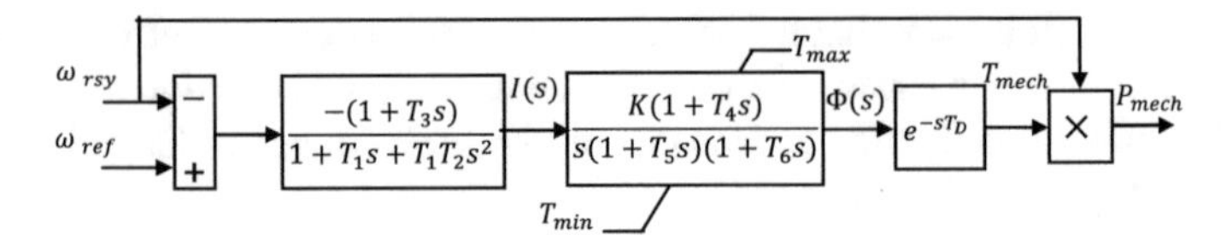

Fig. 2 Model of diesel engine

2.1.1 Modeling of Diesel Generator

The DG consists of a diesel engine (DE) and a synchronous generator (SG). The SG generates the voltage of the isolated power system and its automatic voltage regulator maintains the system voltage within prescribed levels during the three modes of operation. The DE provides mechanical power to the SG and its speed governor (speed regulator and actuator) controls the DE speed. In this work, the DE speed control is isochronous, so the diesel speed governor adjusts the fuel rate to make the DE run at a constant speed [23].

The DE is a complex device and has many nonlinear components affecting the generation unit performance. The DE model considered here [20–24] consists of a controller to monitor the steady-state speed, a first order actuator with gain K and time constant T_i to control the fuel rack position (Fig. 2).

Finally, the mechanical torque produced T_{mech} is represented by the conversion of fuel-flow to torque with a time-delay T_D.

$$T_{mech}(s) = e^{-sT_D}\Phi(s) \tag{7}$$

where s denotes the Laplace transform operator and $\Phi(s)$ represents the fuel flow.

3 Shunt APF Configuration and Control Scheme

3.1 *Shunt AFP Configuration*

The APF concept is to use an inverter to produce specific currents or voltages harmonic components to cancel the harmonic components generated by the load. The most commonly used APF configuration is the Shunt APF (SAPF) which injects current harmonics into the point of common coupling (PCC). Figure 3 illustrates the basic principle of SAPF.

3.2 *Control Scheme of the SAPF*

The SAPF control strategy is implemented in three basic stages: The first is the harmonic detection method to identify harmonic level in the system. The second part

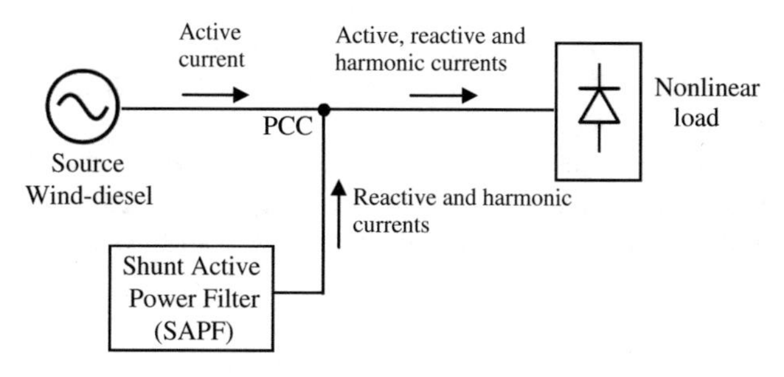

Fig. 3 Basic principle of a SAPF

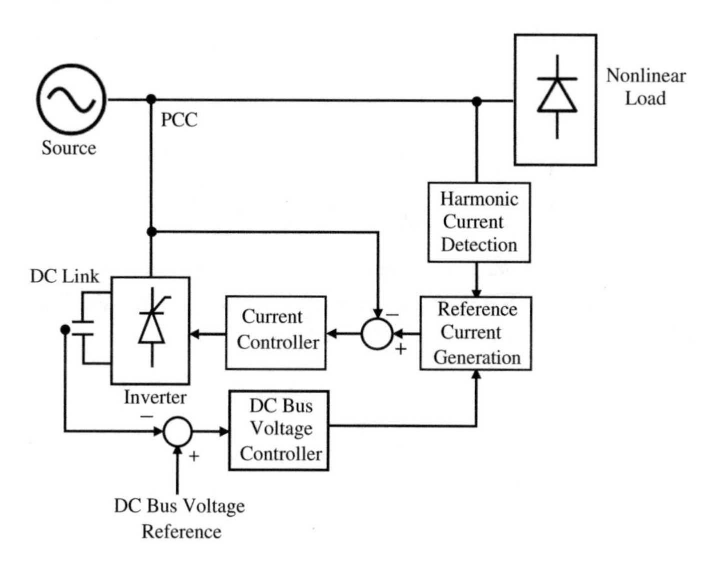

Fig. 4 SAFP control system implementation

is to derive the compensating currents and the third one is the control technique of the inverter for injecting these currents into the power system. The overall control system of the SAFP is depicted in Fig. 4.

There are different methods for generating the current reference for the shunt APF in he frequency domain. In this paper, the current reference for the active power filter is generated using the p-q theory.

3.2.1 p-q Control Strategy

This theory, derived in time-domain, also known as "instantaneous power theory" was proposed in 1983 by Akagi et al. [10, 15] to control APFs. It is based on instantaneous values in three-phase power systems with or without neutral wire, and is valid for steady-state or transitory operation, as well as for generic voltage and current power system waveforms allowing to control the APFs in real-time. Another important characteristic of this theory is the simplicity of the calculations, which involves only algebraic calculation.

This concept is based on the theory of instantaneous reactive power in the α-β reference frame. In this method, a set of voltages (v_a, v_b, v_c) and currents (i_a, i_b, i_c) from phase coordinates are first transferred to the α-β coordinates using Clark's transformation [12].

$$\begin{bmatrix} V_0 \\ V_\alpha \\ V_\beta \end{bmatrix} = \sqrt{\frac{2}{3}} \begin{bmatrix} \frac{1}{\sqrt{2}} & \frac{1}{\sqrt{2}} & \frac{1}{\sqrt{2}} \\ 1 & -\frac{1}{2} & -\frac{1}{2} \\ 0 & \frac{\sqrt{3}}{2} & -\frac{\sqrt{3}}{2} \end{bmatrix} \begin{bmatrix} V_1 \\ V_2 \\ V_3 \end{bmatrix} \tag{8}$$

$$\begin{bmatrix} I_0 \\ I_\alpha \\ I_\beta \end{bmatrix} = \sqrt{\frac{2}{3}} \begin{bmatrix} \frac{1}{\sqrt{2}} & \frac{1}{\sqrt{2}} & \frac{1}{\sqrt{2}} \\ 1 & -\frac{1}{2} & -\frac{1}{2} \\ 0 & \frac{\sqrt{3}}{2} & -\frac{\sqrt{3}}{2} \end{bmatrix} \begin{bmatrix} I_{s1} \\ I_{s2} \\ I_{s3} \end{bmatrix} \tag{9}$$

For simplicity, the zero-phase sequence component voltage and current signals are eliminated in (8) and (9). The instantaneous active power (p) and the instantaneous imaginary power (q) in a three-phase circuit are defined as:

$$\begin{bmatrix} p \\ q \end{bmatrix} = \begin{bmatrix} V_\alpha & V_\beta \\ -V_\beta & V_\alpha \end{bmatrix} \begin{bmatrix} i_\alpha \\ i_\beta \end{bmatrix} \tag{10}$$

Writing p and q in the following form:

$$\begin{cases} p = \bar{p} + \tilde{p} \\ q = \bar{q} + \tilde{q} \end{cases} \tag{11}$$

where $\bar{p}$ and $\bar{q}$ denote the average portion and $\tilde{p}$ and $\tilde{q}$ indicate the oscillating portion of p and q, respectively.

Using (10) load current in α-β frame can be calculated as:

$$\begin{bmatrix} i_{c\alpha} \\ i_{c\beta} \end{bmatrix} = \frac{1}{V_\alpha^2 + V_\beta^2} \begin{bmatrix} V_\alpha & -V_{s\beta} \\ V_\beta & V_{s\alpha} \end{bmatrix} \begin{bmatrix} p \\ q \end{bmatrix} \tag{12}$$

The compensating current signals in α-β frame can be transferred to a-b-c frame using inverse Clark's transformation as:

$$\begin{bmatrix} I_{ref1} \\ I_{ref2} \\ I_{ref3} \end{bmatrix} = \sqrt{\frac{2}{3}} \begin{bmatrix} 1 & 0 \\ -1/2 & \sqrt{3/2} \\ -1/2 & -\sqrt{3/2} \end{bmatrix} \begin{bmatrix} \tilde{I}_\alpha \\ \tilde{I}_\beta \end{bmatrix} \tag{13}$$

This method does not take into account the zero sequence components, and hence, the effect of unbalanced voltages and currents. The instantaneous reactive power (p-q) theory is widely used for three-phase balanced non-linear loads, such as rectifiers [28].

3.2.2 DC Link Voltage Regulation with Fuzzy Type 2 Controller

Fuzzy Type 2 control is basically an adaptive and non-linear controller that gives robust performance for a linear or non-linear plant with parameter variations [18, 25]. Any given linear control can be achieved with a fuzzy controller for a given accuracy.

The capacitor that powers the active filter acts as voltage source and its voltage must be kept constant to ensure that the performance of the filter is maintained and the voltage fluctuations of the semi-conductors do not exceed the limits prescribed. The Interval Type 2 Fuzzy Logic Controller (IT2FLC) is implemented as shown in Fig. 5.

Type-2 fuzzy logic systems, are an extension of ordinary Type-1 fuzzy logic systems introduced by Zadeh (1975), are characterized by fuzzy membership functions represented by fuzzy sets in [0, 1]. Unlike a Type-1 fuzzy which have crisp membership functions, T2FLC consists of five components including fuzzifier, rule base, fuzzy inference mechanism, type-reducer and defuzzifier as depicted in Fig. 6.

In an IT2FLC at least some of the fuzzy sets used in the antecedent and/or consequent parts and each rule inference output are type-2 fuzzy sets. Generally speaking, in a T2FLC, the crisp inputs are first fuzzified, usually into Type-2 fuzzy sets. The fuzzified Type-2 fuzzy sets then activate the inference engine and rule base to yield output Type-2 fuzzy sets by performing the union and intersection operations of Type-2 fuzzy set and compositions of Type-2 relations. Then a type-reduction process is applied to these output sets in order to generate Type-1 fuzzy sets (called type-reduced sets) by combining these output sets and performing a centroid calculation. Finally, the type-reduced Type-1 set is defuzzified to produce crisp output [19, 26].

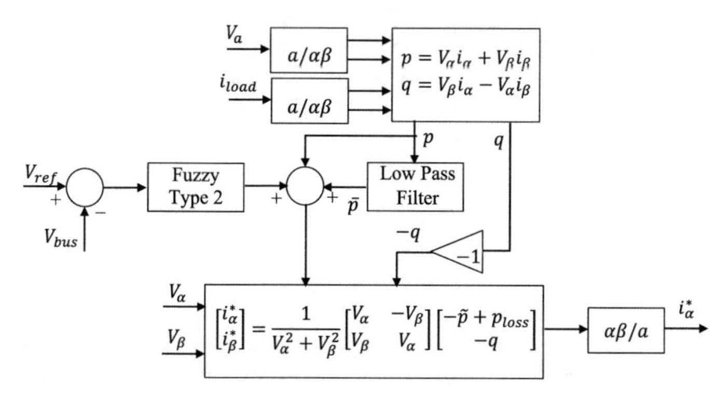

Fig. 5 Implementation of IT2FLC controller for the SAPF

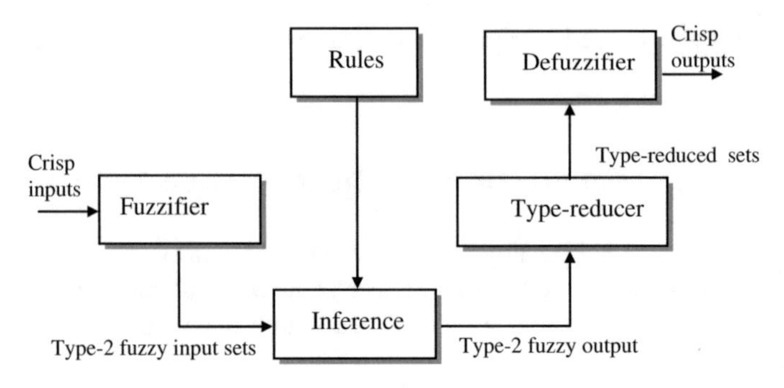

Fig. 6 Interval Type-2 fuzzy system (IT2FLS)

Table 1 Type 2 fuzzy control rules

de	e		
	N	EZ	P
N	N	N	EZ
EZ	N	EZ	P
P	EZ	P	P

A generalized rule for Type-2 fuzzy system is:

$$IF \quad x_1 \; is \, \bar{A}_{k1} \; and \; x_2 \; is \bar{A}_{k2} \; and \; \ldots \; and \; x_n \; is \, \bar{A}_{kn}$$

$$THEN \quad u_k = \sum_{i=1}^{n} p_{ik} x_i + b_k \tag{14}$$

where x_k and u_k are the input and output linguistic variables respectively; $\bar{A}_{ki}$ are Type-2 fuzzy sets for the kth rule and ith input; p_{ki} and b_k are consequent parameters of the rules which are Type-1 fuzzy sets.

The fuzzy controller inputs (DC bus voltage error (e) and change of error (de)) are implemented using Gaussian membership functions as shown in Fig. 7. The fuzzy labels are negative (N), environ zero (EZ), positive (P).

The control rules are given in Table 1.

4 Simulation Results

A series of simulation scenarios have been performed to evaluate the performance of the proposed control methods of the SAPF on the current harmonics under variable load and wind speed conditions. The parameters values used in the simulation are listed in the Appendix.

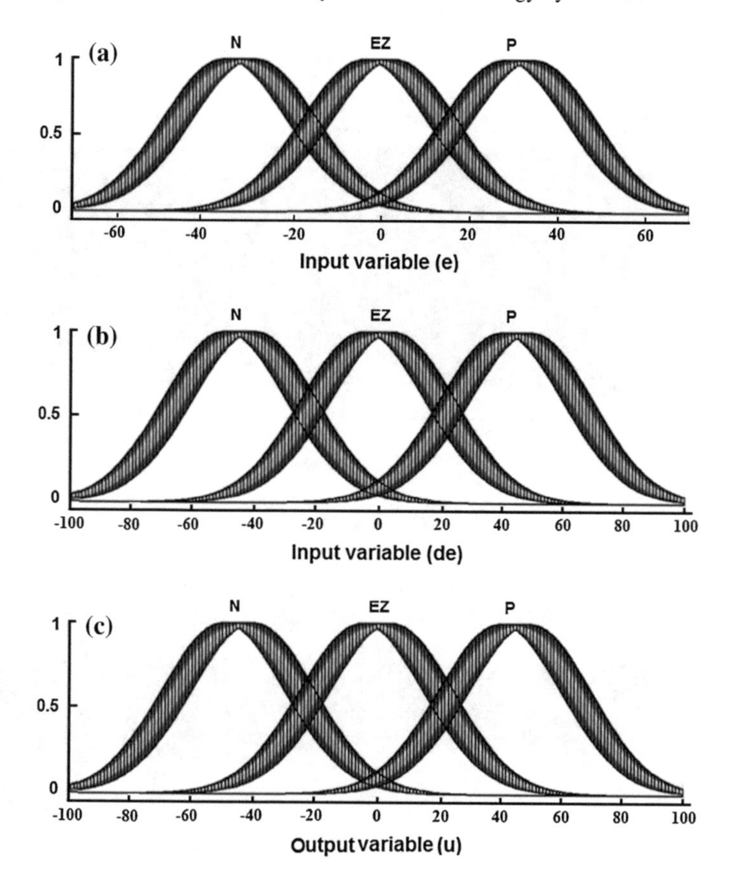

Fig. 7 The membership functions for **a** error and **b** change of error and **c** output (u)

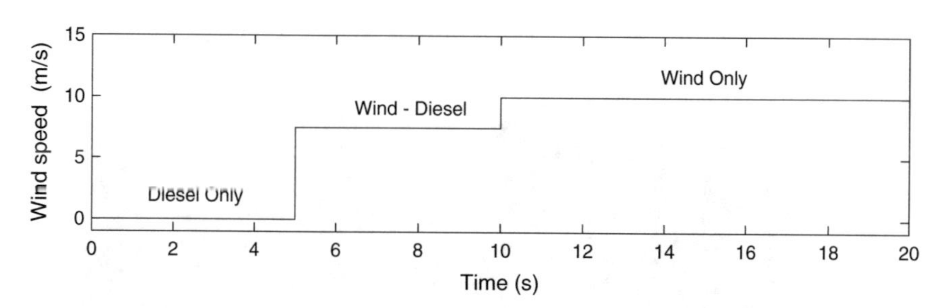

Fig. 8 Wind speed profile

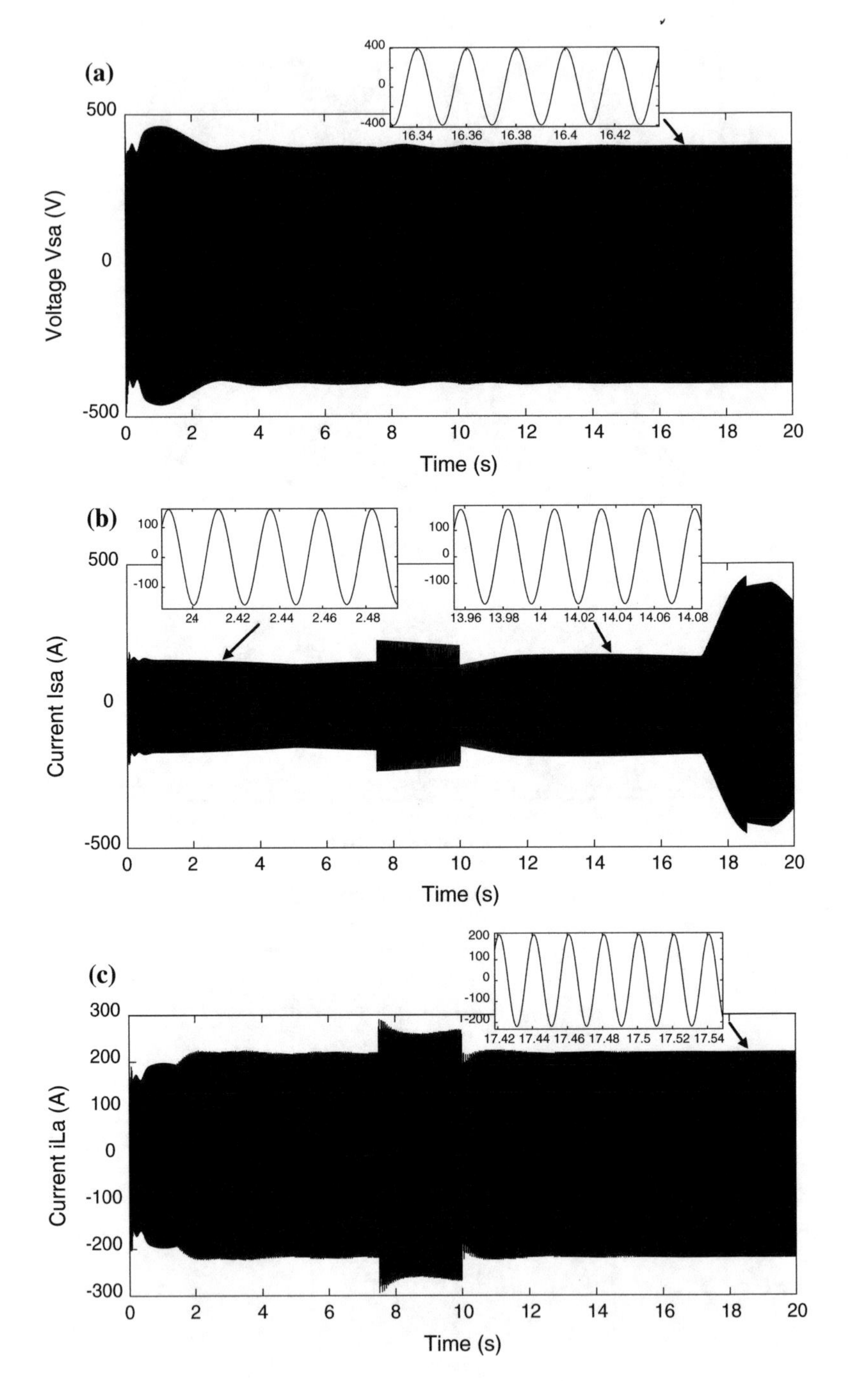

Fig. 9 Waveforms of **a** source voltage, **b** source current, **c** dump load current, **d** load current and **e** network frequency

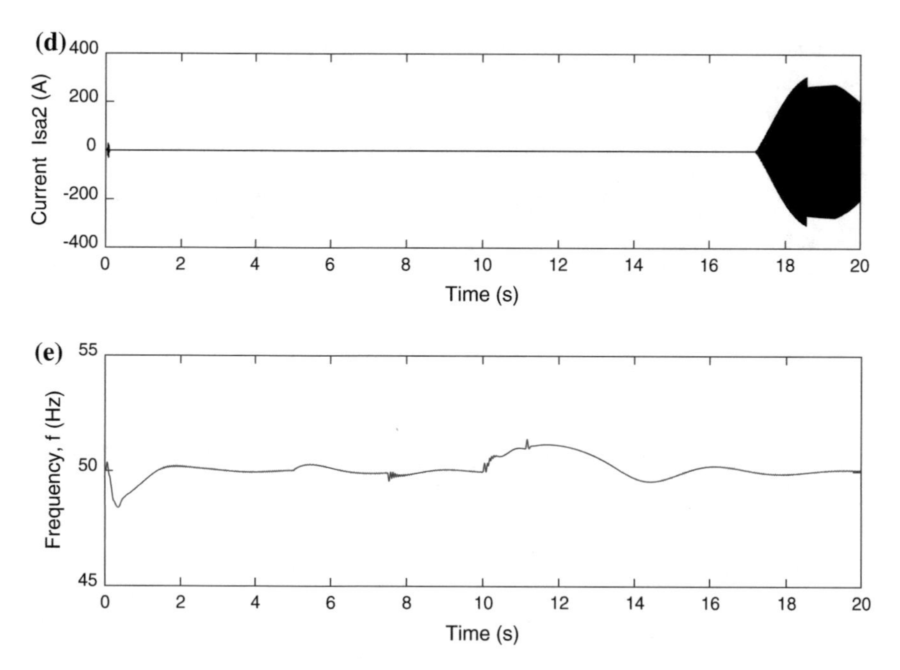

Fig. 9 (continued)

4.1 *Linear Load*

Initially, the overall wind-diesel power system is simulated with a linear load variation of 50 % applied at $t = 7$ and 10 s. The wind speed profile considered in this simulation is shown in Fig. 8.

The corresponding waveforms of the voltage source (v_{sa}), current source (i_{sa}), the main load current (i_{La}), the secondary load current (i_{sa2}) and the system frequency (f) are shown in Fig. 9.

It can be noted that in the case of linear load, the load current remains sinusoidal throughout the simulation time and the voltage and frequency are stable.

4.2 *Nonlinear Load with SAPF and Fuzzy Type 2 Controller*

Figure 10 shows the source voltage ($\boldsymbol{v_1}$), source current ($\boldsymbol{i_{sa1}}$), load current ($\boldsymbol{i_{ca}}$), filter current ($\boldsymbol{i_1}$) and the DC bus voltage ($\boldsymbol{V_{dc}}$) and Fig. 11 show the harmonic spectrum without and with the SAPF filter respectively.

The application of the APF with Fuzzy Type 2 control, has resulted in a reduction of the THD which has dropped from 28.53 to 1.9 %.

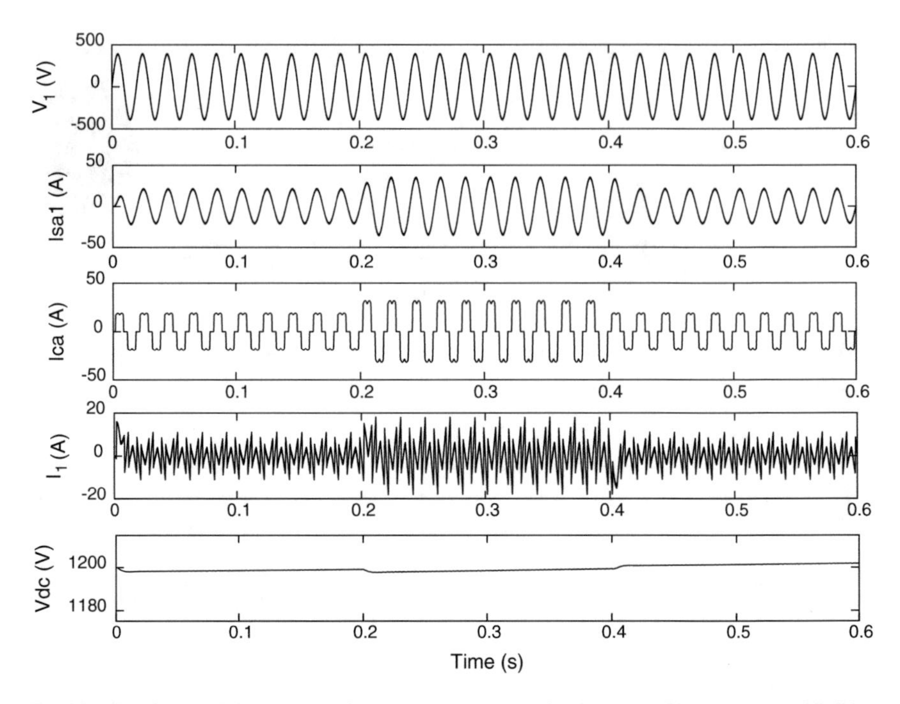

Fig. 10 Waveforms of the source voltage, source currents, load current, filter current and DC bus voltage (V_{dc})

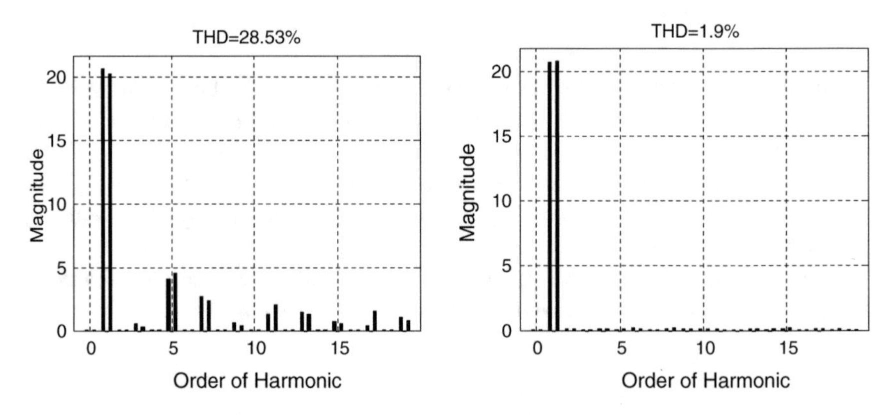

Fig. 11 Total harmonic distortion (THD) spectrum without and with APF

5 Conclusions

A control approach is presented for a three-phase shunt active power filter for the mitigation of harmonics introduced by nonlinear loads in an off-grid hybrid win-diesel power system. The control of the DC bus voltage of the shunt active power filter is based on Type-2 fuzzy systems. Improved dynamic current harmonics and reactive power compensation has been achieved.

The results show that with the proposed control scheme, the supply current waveforms were almost perfectly sinusoidal and in-phase with the supply voltage and the harmonic distortion achieved with the proposed controller (1.9 %) was well below the limit imposed by the IEEE-519-1992 recommendations of harmonic standard limits.

Appendix

Model parameters values used in the simulations.

System parameters	Values
Wind turbine (WT)	$P_n = 37\,\text{kW}, L_s = 0.087\,\Omega, V_n = 480\,\text{V}, R'_r = 0.228\, Omega,$ $f = 50\,\text{Hz}, L'_r = 0.8\,\text{mH}, L_m = 34.7\,\text{mH}, 2P = 4,$ $R_s = 0.087\,\Omega, J = 0.4\,\text{kg.m}^2$
Diesel generator (DG)	$S_n\,(kVA) = 37.5, X'_1 = 0.09, X_d = 3.23, T_{d0} = 4.4849,$ $X'_d = 0.21, T''_{d0} = 0.0681, X''_d = 0.15, T''_{q0} = 0.1,$ $X_q = 2.79, R_s(pu) = 0.017$
Consumer load	Mainload $= 100\,\text{kW}$ Nonlinearload $R_1 = 35\,\Omega, R_2 = 35\,\Omega, L = 10^{-3}H$
Active filter	$C = 2000\,\mu\text{F}, L_f = 3\,\text{mH}, V_{dc} = 1200\,\text{V}$

References

1. Stiebler, M.: Wind Energy Systems for Electric Power Generation. Springer, Berlin (2008)
2. Jurado, F., Saenz, R.: Neuro-fuzzy control for autonomous wind-diesel systems using biomass. Renew. Energy **27**, 39–56 (2002)
3. Sebastian, R.: Modelling and simulation of a high penetration wind diesel system with battery energy storage. Int. J. Electr. Power Energy Syst. **33**(3), 767–774 (2011)
4. Kassem, A.M., Yousef, A.M.: Robust control of an isolated hybrid wind diesel power system using linear quadratic Gaussian approach. Int. J. Electr. Power Energy Syst. **33**(4), 1092–1100 (2011)
5. Karaki, S.H., Chedid, R.B., Ramadan, R.: Probabilistic production costing of diesel-wind energy conversion systems. IEEE Trans. Energy Convers. **15**, 284–289 (2000)
6. Mendalek, N., Al-Haddad, K.: Modelling and nonlinear control of shunt active power filter in the synchronous reference frame, IEEE ICHQP'2000, pp. 30–35 (2000)
7. Jain, S.K., Agrawal, P., Gupta, H.O.: Fuzzy Logic controlled shunt active power filter for power quality improvement. In: IEE Proceedings in Electrical Power Applications, vol. 49, no: 5 September (2002)
8. Au, M.T., Milanovic, J.V.: Planning approaches for the strategic placement of passive harmonic filters in radial distribution networks. IEEE Trans. Power Deliv. **22**(1), 347–353 (2007)
9. Rudnick, H., Dixon, J., Moran, L.: Active power filters as a solution to power quality problems in distribution networks. IEEE Power Energy Mag. **1**(5), 32–40 (2003)

10. Singh, B., Al-Haddad, K., Chandra, A.: A review of active filters for power quality improvement. IEEE Trans. Ind. Electron. **46**(5), 960–971 (1999)
11. Kedjar, B., Al-Haddad, K.: DSP-based implementation of an LQR with integral action for a three-phase three-wire shunt active power filter. IEEE Trans. Ind. Electron. **56**(8), 2821–2828 (2009)
12. Choi, J.H., Park, G.W., Dewan, S.B.: Standby power supply with active power filter ability using digital controller. In: Proceedings of the IEEE APEC'95, pp. 783–789 (1995)
13. Kneschke, T.A.: Distortion and power factor of nonlinear loads. In: Railroad Conference, Proceedings of ASME/IEEE Joint, pp. 47–54 (1999)
14. Axente, I., Navilgone, J., Basu, M., Conlon, M.F.: A 12 kVA DSP-controlled laboratory prototype UPQC capable of mitigating unbalance in source voltage and load current. IEEE Trans. Power Electron. **25**(6), 1471–1479 (2010)
15. Mendalek, N., Al-Haddad, K., Dessaint, L.A., Fnaiech, F.: Nonlinear control technique to enhance dynamic performance of a shunt active power filter. In: IEEE Proceedings Electrical Power Applications, vol. 150, no. 4, July (2003)
16. Sharmeela, C., Mohan, M.R., Uma, G., Baskaran, J.: Fuzzy logic based controlled three phase shunt active filter for line harmonics reduction. J. Comput. Sci. **3**(2), 76–80 (2007)
17. Kouadria, M.A., Allaoui, T., Belfedal, C.: A fuzzy logic controller of three-phase shunt active filter for harmonic current compensation. Int. J. Adv. Eng. Technol. (IJAET) **7**(1), 82–89 (2014)
18. Mendel, J.M., Bob John, R.I.: Type 2 fuzzy sets made simple. IEEE Trans. Fuzzy Syst. **10**(2), 117–127 (2002)
19. Kouadria, M.A., Allaoui, T., Denaï, M.: High performance shunt active power filter design based on fuzzy interval type-2 control strategies. Int. Rev. Autom. Control (I.RE.A.CO.) **8**(5), 322 (2015)
20. Ko, H.S., Niimura, T., Lee, K.Y.: An intelligent controller for a remote wind-diesel power system design and dynamic performance analysis. In: Proceedings of IEEE Power Engineering Society General Meeting, vol. 4, pp. 2147–2151 (2003)
21. Ghedamsi, K., Aouzellag, D., Berkouk, E.M.: Control of wind generator associated to a flywheel energy storage system. Renew. Energy **33**(9), 2145–2156 (2008)
22. Ebert, P., Zimmermann, J.: Successful high wind penetration into a medium sized diesel grid without energy storage using variable speed wind turbine technology. In: Proceedings of EWEC99, Nice (1999)
23. Rezkallah, M., Chandra, A.: Control of wind-diesel isolated system with power quality improvement. In: IEEE Electrical Power and Energy Conference (EPEC) (2009)
24. Sebastian, R.: Smooth transition from wind only to wind diesel mode in anautonomous wind diesel system with a battery-based energy storage system. Renew. Energy **33**, 454–466 (2008)
25. Wu, D., Mendel, J.M.: Enhanced Karnik-Mendel algorithms for interval type-2 fuzzy sets and systems. In: Proceedings NAFIPS, San Diego, pp. 184–189 (2007)
26. Karnik, N.N., Mendel, J.M., Liang, Q.: Type-2 fuzzy logic systems. IEEE Trans. Fuzzy Syst. **7**, 643–658 (1999)

Fast Intra Mode Decision for HEVC

Kanayah Saurty, Pierre C. Catherine and Krishnaraj M. S. Soyjaudah

Abstract High Efficiency Video Coding (HEVC/H.265), the latest standard in video compression, aims to halve the bitrate while maintaining the same quality or to achieve the same bitrate with an improved quality compared to its predecessor, AVC/H.264. However, the increase in prediction modes in HEVC significantly impacts on the encoder complexity. Intra prediction methods indeed iterate among 35 modes for each Prediction Unit (PU) to select the most optimal one. This mode decision procedure which consumes around 78 % of the time spent in intra prediction consists of the Rough Mode Decision (RMD), the simplified Rate Distortion Optimisation (RDO) and the full RDO processes. In this chapter considerable time reduction is achieved by using techniques that use fewer modes in both the RMD and the simplified RDO processes. Experimental results show that the average time savings of the proposed method indeed yields a 42.1 % time savings on average with an acceptable drop of 0.075 dB in PSNR and a negligible increase of 0.27 % in bitrate.

1 Introduction

High Efficiency Video Coding (HEVC) or H.265, developed by ISO-IEC/MPEG and ITU-T/VCEG which formed the Joint Collaborative Team on Video Coding (JCT-VC) [11] became the new standard in video compression technologies in 2013. It is an evolution of the block-based hybrid motion-compensation and transform coding inherent in the previous standard, AVC/H.264. Coding performance by almost twofold is achieved by the new standard through the implementation of additional

K. Saurty
Université des Mascareignes, Pamplemousses, Mauritius
e-mail: ksaurty@udm.ac.mu

P.C. Catherine (✉)
University of Technology, La Tour Koenig, Mauritius
e-mail: ccatherine@umail.utm.ac.mu

K.M.S. Soyjaudah (✉)
University of Mauritius, Réduit, Mauritius
e-mail: ssoyjaudah@uom.ac.mu

Y. Bi et al. (eds.), *Intelligent Systems and Applications*,
Studies in Computational Intelligence 650, DOI 10.1007/978-3-319-33386-1_18

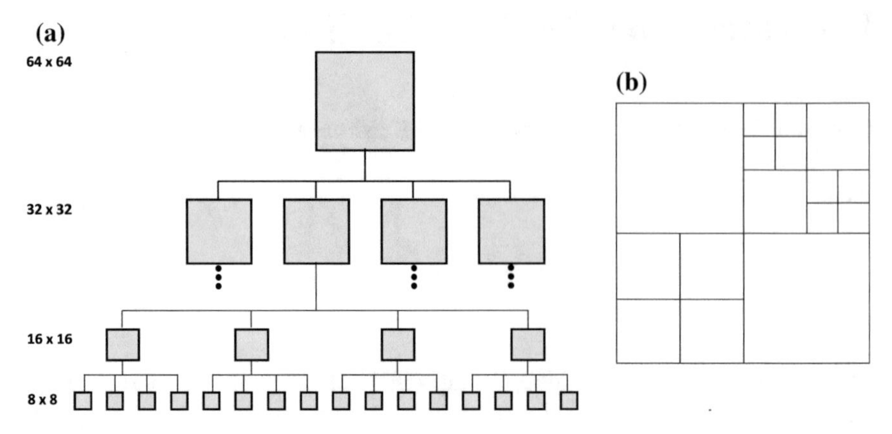

Fig. 1 **a** HEVC quad-tree structure. **b** Example of a final LCU structure

coding tools such as the newly introduced recursive tree structure, larger block transforms, adaptable loop filtering operations and advanced motion prediction [21]. In fact, video compression under HEVC aims at only half of bitrate required by the previous standard for the same quality of pictures.

The support of Coding Tree Block (CTB) of larger size compared to the conventional 16×16 macroblocks in H.264 is one of the major improvements in HEVC. The Largest Coding Unit (LCU), into which a frame is partitioned in HEVC, can adopt a size of either 16×16, 32×32 or 64×64 depending on the encoder configuration. This large sized LCU can be recursively split forming a quad-tree structure allowing sub-CUs with size of 64×64, 32×32, 16×16 and 8×8 as shown in Fig. 1a. In addition, CUs are further split into Prediction Units (PUs). In intra prediction, each CU can have a single $2N \times 2N$ PU or it can be partitioned into four $N \times N$ PUs, where N is half of the CU size. The leaf CUs in Fig. 1a may therefore be again split into 4×4 PUs. Depending on the texture of the picture, these LCUs will have many possible configurations. A large homogeneous region may be captured by a single CU of a large size which can be encoded using a smaller number of symbols compared to several smaller blocks. An example of a final LCU structure is shown in Fig. 1b illustrating different CU size combinations within a single LCU.

Support for the arbitrary Largest Coding Tree Block (LCTB) of different sizes enables the codec to be readily optimized for various contents, applications and devices. By choosing the appropriate LCTB size and the maximum hierarchical depth, the block structure can be optimized in a better way for the targeted application [17]. Another improvement in the new standard is the large number of intra prediction modes at the level of PUs. Compared to the 9 modes in AVC [23], HEVC makes use of 35 modes consisting of 33 directional modes along with the Intra Planar and the DC modes. This is illustrated in Fig. 2. The large number of prediction modes results in a more accurate intra prediction which consequently leads to smaller residuals that need to be processed. The increase in prediction modes therefore contributes to the high quality of the compression along with a lower bitrate.

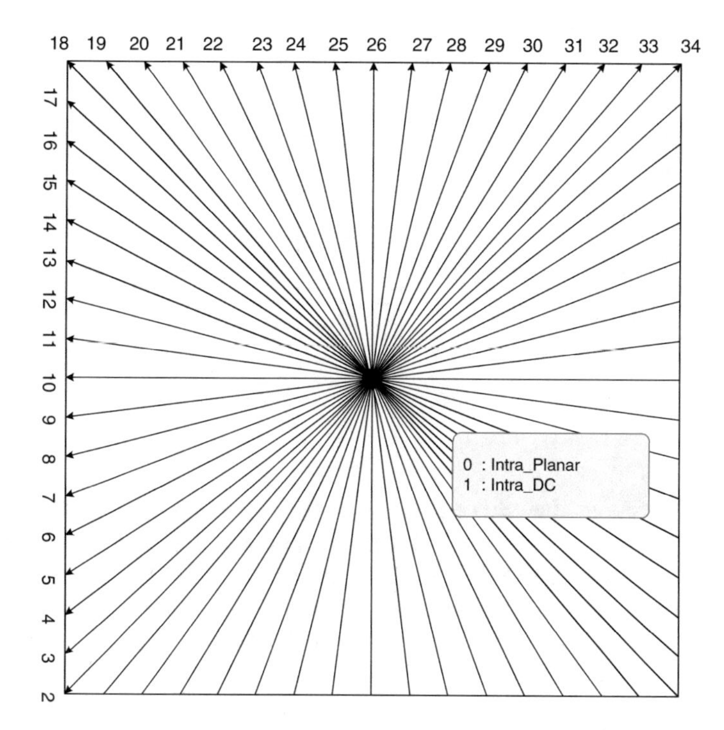

Fig. 2 HEVC intra prediction modes [10]

However, an increase in prediction modes also translates to more comparisons to be performed in order to come up with the best mode. Therefore, the benefit of higher quality derived from the increased in number of modes comes along with an increase in the encoder complexity. It therefore requires an appropriate mode selection heuristics in order to maintain a reasonable search complexity [2]. Consequently, for practical applications such as high-resolution video services and real-time processing, HEVC still requires an acceptable reduction in complexity while maintaining the high coding performance [22].

To reduce the intra coding complexity, many techniques have been proposed in other works. For example, [6] proposes an intra mode decision based on dominant edge evaluation and tree structure dependencies which yields a 20 % time savings along with an increase in bitrate of 0.9 % and a decrease of 0.02 dB in PSNR. A gradient based fast mode decision algorithm proposed in [12] results in a 20 % decrease in encoding time and a drop of 0.04 dB BD-PSNR together with a BD-Rate of 0.74 %. A decrease of 19 % in complexity is proposed by [20] at the expense of a 0.4 % increase in BD-Rate. In [14], the number of candidates for the given block are reduced by checking the boundary pixels resulting in a 13 % time savings accompanied by a difference of −0.05 dB in terms of PSNR and an increase of 1.78 % for the bitrate. A 28 % reduction is reached according to [4] together with a BD-Rate of 0.53 % and a BD-PSNR of −0.038 dB.

Table 1 Performances of other works using reduced mode

No	Algorithm	Time (%)	Δ PSNR (dB)	Δ Bit Rate (%)	Δ BD-PSNR (dB)	Δ BD- Rate (%)
1	Kumar V. [14]	−13	−0.05	1.78	-	-
2	Yongfang Shi [20]	−19	-	-	-	0.40
3	Wei Jiang [12]	−20	-	-	−0.04	0.74
4	da Silva [6]	−20	−0.02	0.9	-	1.3
5	Gaoxing Chen [4]	−28	-	-	−0.038	0.53
6	da Silva [5]	−37	−0.07	-	-	1.2
7	Guang Chen [3]	−37.61	-	-	−0.078	1.65
8	Saurty K. [19]	−38.0	−0.059	0.30	−0.063	1.15

A complexity reduction of 37.61 % is achieved in [3] with a BD-PSNR of −0.0781 dB and a BD-Rate of 1.65 %. The 37 % time savings proposed by [5] is accompanied by a −0.07 dB drop in PSNR and a BD-Rate of 1.2 %. In [19], a reduced mode approach based on the analysis of the graph obtained by plotting the HAD cost of the modes against their mode number results in 38.0 % time savings on average along with a decrease in PSNR of 0.059 dB and a 0.30 % increase in bitrate. A summary of the performances these works is also provided in Table 1. A hyphen (-) denotes data not available from the source.

In order to improve further the complexity reduction in HEVC intra prediction, a number of approaches have been proposed. They focus on the reduction in modes processed by the simplified RDO in addition to using fewer modes in the RMD process. The rest of this chapter is thus organized as follows: Sect. 2 provides an overview of intra mode decision in HEVC. Section 3 illustrates the different time-consuming components of HEVC intra prediction along with the mode selection frequency. Section 4 looks into the methods to reduce the modes taking part in the RMD process. The different approaches to reduce the number of modes entering the simplified RDO process are explained in Sect. 5. Finally, Sect. 6 concludes this chapter.

2 Intra Mode Decision in HEVC

Intra prediction in HEVC makes use of 33 angular modes along with the DC and the Planar modes for each PU [10] as shown in Fig. 2. For each PU, intra prediction is performed by extrapolating the reference pixels, obtained from the already decoded neighboring PUs, in the prediction directions for the angular modes. An example of the extrapolation for the directional mode 29 [21] is shown in Fig. 3.

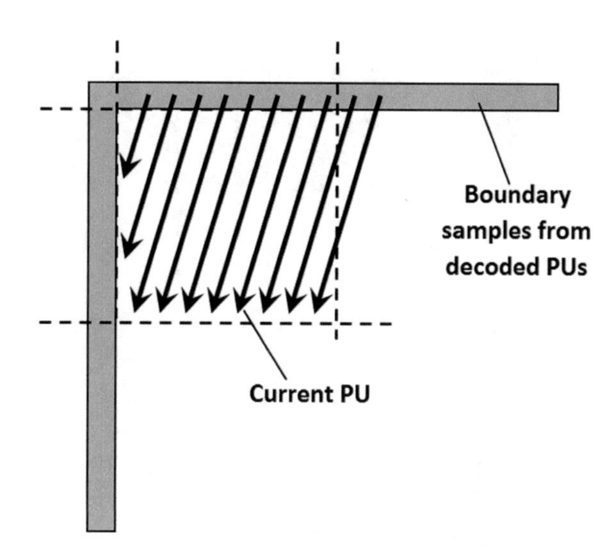

Fig. 3 Directional mode 29 for intra prediction

Table 2 Intra prediction modes combinations for a 64 × 64 CU

Size	Number of PUs	Number of modes for each PU	Total number of modes for each level
64 × 64	1	35	35
32 × 32	4	35	140
16 × 16	16	35	560
8 × 8	64	35	2,240
4 × 4	256	35	8,960
Total			11,935

The number of arithmetic operations to calculate the prediction for even one mode is quite complex and time consuming. Computing the full RDO for all of these modes in order to choose the best one would therefore require extensive and unacceptable computing resources. This is illustrated in Table 2 where the total number of possible mode combinations for a 64 × 64 CU is 11,935.

Since this process is very time consuming, HEVC already adopts a simplified three steps approach [18] which is illustrated in Fig. 4.

1. A reduced set of N candidate modes is first identified through a RMD process based on the Hadamard transform Absolute Difference (HAD) and the estimated bitrate for the various modes (35 in all). The Most Probable Modes (MPMs) obtained from the decoded neighboring PUs are added to the N candidates if they are not already included in the list.
2. A simplified Rate Distortion (RD) cost is then computed for each of the selected candidates in the previous step and the best mode identified.
3. Finally the full RD cost is computed for the best candidate obtained using a recursive transform for the selected mode.

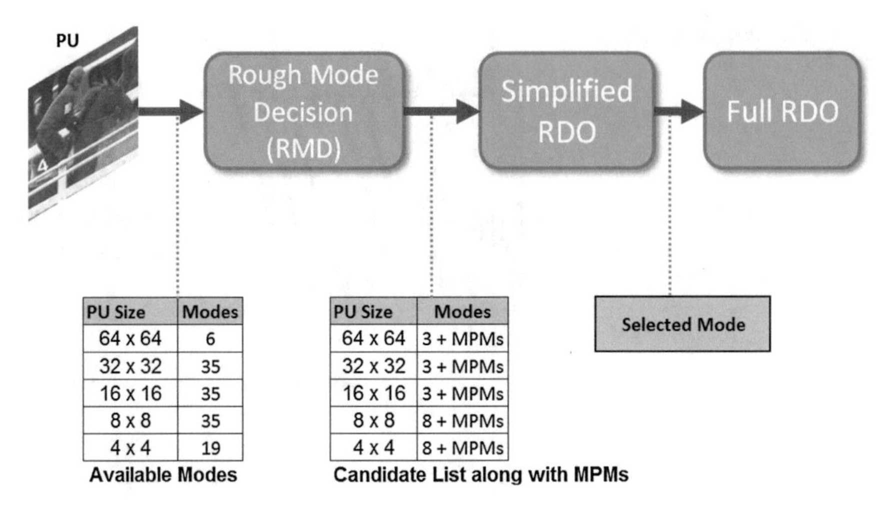

PU Size	Modes
64 x 64	6
32 x 32	35
16 x 16	35
8 x 8	35
4 x 4	19

PU Size	Modes
64 x 64	3 + MPMs
32 x 32	3 + MPMs
16 x 16	3 + MPMs
8 x 8	8 + MPMs
4 x 4	8 + MPMs

Fig. 4 HEVC mode selection process

The HAD cost for each available mode is computed based on the following formula [9]:

$$J_{HAD} = SATD + \lambda * B \tag{1}$$

where *SATD* represents the Hadamard Sum of Absolute Difference, λ is the Lagrangian constant and *B* specifies the bit cost to be considered in the decision making.

The number of modes available for each PU varies with its size. After comparing the HAD costs of these modes through the RMD process, a list of candidate modes, ordered by increasing cost, is produced.

For the 4×4 and 8×8 PUs, the RDO process selects the 8 cheapest HAD cost modes. This list is reduced to 3 candidates for larger size PUs [15, 22]. The simplified RDO process compares all the candidates in this list in order to determine the best one for this PU. However, due to this iterative mode search, the complexity is still quite high.

3 Complexity Reduction for Intra Prediction

The high complexity of HEVC for intra prediction can mainly be attributed to two main changes brought in this standard. Firstly, the high number of modes (35 in all) used for each PU compared to only 9 in the previous AVC/H.264 standard requires a larger number of iterations to to be performed prior to identifying the best mode. Secondly, the large LCU of 64×64 pixels in the new standard provides numerous combinations of CUs within one LCU, which further adds to the complexity. In the

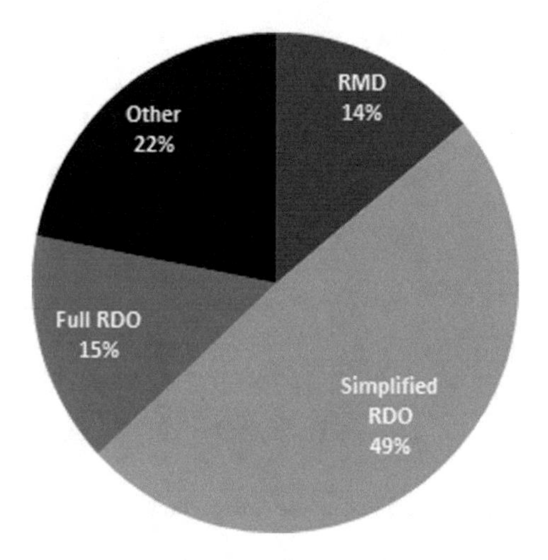

Fig. 5 Time spent during the different stages of HEVC intra prediction

latter case, a considerable time savings could be achieved by identifying the non-split CUs at an early stage of the quad tree traversal to avoid the computation associated with the smaller PU sizes [7, 13, 16, 24].

In this chapter, the focus is on time savings produced through the reduction of the number of modes taking part in the mode-decision process. Figure 5 illustrates the time spent in the different stages during intra prediction. It is observed that 14 % of the encoding time is spent in the RMD process and 49 % in the simplified RDO. In both processes, the selection is made by iterating among a list of modes. The different proposed approaches to reduce the complexity in intra prediction therefore look into techniques to minimize these iterations and yet produce an equivalent mode selection. For each PU, the full RDO process is performed on the selected mode only and this process will not therefore be affected by the mode reduction techniques.

3.1 Commonly Selected Modes

Out of the 35 possible modes from which each PU iterates to find the optimal one, some modes are more commonly selected than others. Figure 6 illustrates the average percentage selection of each of these modes for the list of sequences considered using the 4 Quantization Parameter (QP) values.

Modes 0, 1, 10 and 26 clearly comes out as the most commonly selected modes ranging from 9 % for mode 10 to 26 % for mode 0. These modes have therefore been included by default in the reduced mode approaches proposed for the RMD process. It is to be noted that each sequence may have a different percentage selection for each mode. Figure 7, for example, indicates that more than 55 % of the PUs for the *KristenAndSara* sequence select mode 10. Nevertheless, as illustrated in the figure, the four modes mentioned earlier (0, 1, 10 and 26) remain the four most commonly selected ones.

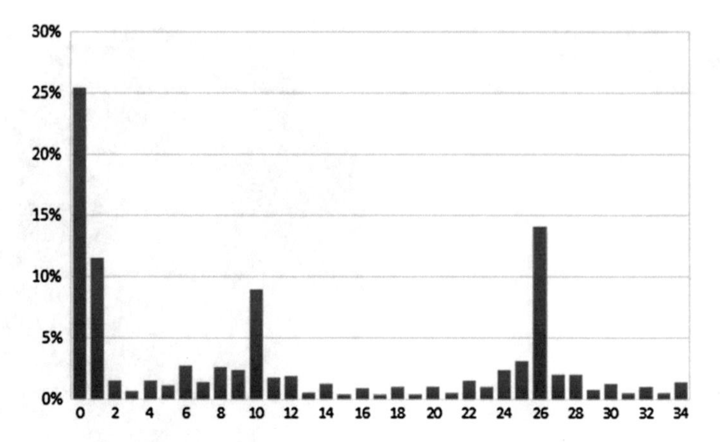

Fig. 6 Percentage occurrence of modes

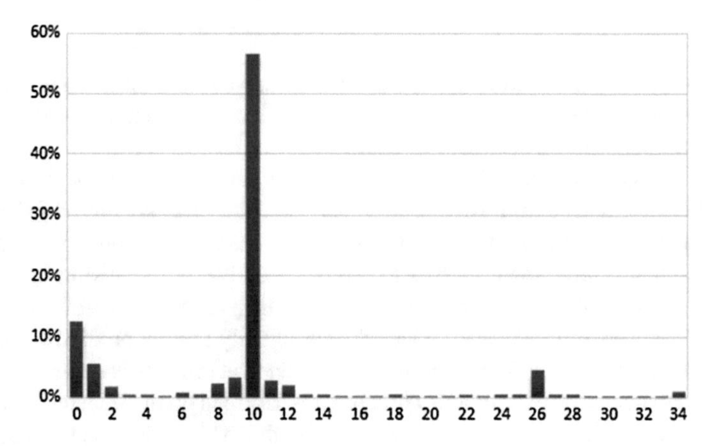

Fig. 7 Percentage occurrence of modes for the sequence *KristenAndSara*

3.2 Experimental Conditions

All experiments in this chapter were conducted using the HEVC Reference Model (HM) version 10 encoder. A list of 21 test sequences with resolution ranging from 416 × 240 (Class E) to 2560 × 1600 (Class A) have been used throughout using the *all intra high-efficiency* encoding configuration with QP values of 22, 27, 32 and 37. The performance of the proposed methods are thus reported in terms of the change in average bitrate, Peak Signal-to-Noise Ratio (PSNR) and total encoding time based on the following formulas:

$$\Delta BitRate(\%) = \frac{BitRate(proposed) - BitRate(HM)}{BitRate(HM)} * 100 \tag{2}$$

$$\Delta PSNR(dB) = PSNR(proposed) - PSNR(HM) \quad (3)$$

$$\Delta Time(\%) = \frac{Time(proposed) - Time(HM)}{Time(HM)} * 100 \quad (4)$$

The Bjøntegaard metrics [1] rate distortion performance is used in the computation of BD-Rate and BD-PSNR.

The use of fewer modes for the RMD process brings significant reduction in complexity. This is further studied in the following section.

4 Reduced Mode Approach for the RMD Process

In this approach, the aim is to provide the RDO process with the same or similar list of candidates, using a faster method. The PLANAR mode (0) and the DC mode (1) are inserted by default in the initial list as they correspond to the highly selected modes. In addition these two modes are non-angular and will not be considered in the reduced mode approach. The reduction in modes therefore applies only to the 33 angular modes. Figure 8 provides an example of the HAD cost for the different modes of an 8×8 PU. The graph shows that the HAD cost varies in a relatively continuous waveform, i.e. there is a gradual decrease or increase in HAD cost from one mode to the next and the graph may have a number of minimum values. The graph spans from mode 2 to mode 34. Modes 0 and 1 are also included in the figure. The modes falling below the dotted line are the 8 cheapest cost modes forming part of the candidate list for that PU. The graph clearly illustrates that the initial modes on the graph and those on the right hand side are well above the first 8 cheapest cost modes. The proposed approaches therefore identify these high cost regions and aim towards reducing these unnecessary HAD cost computations which in turn will

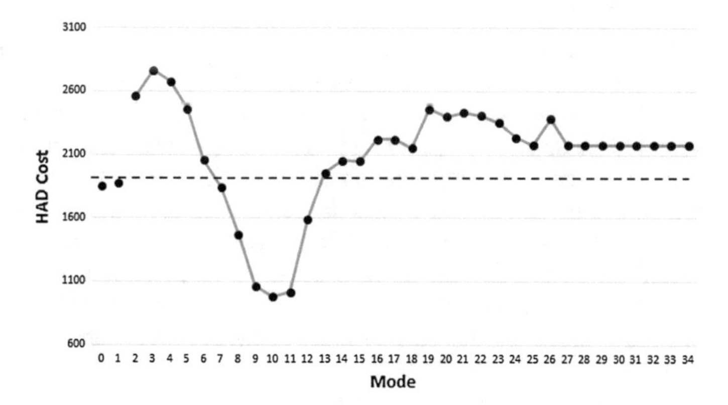

Fig. 8 Plot of HAD cost against mode number for an 8×8 PU

lead to a substantial time savings. In the following sub-sections, the approach to reduce the 35 modes and the 19 modes commonly used for the different PUs will be elaborated. The RMD for the 64 × 64 PU has not been optimized as it is already making use of a very reduced number of modes, i.e. 6.

4.1 Reducing the 35 Modes

PU sizes 8 × 8, 16 × 16 and 32 × 32 iterates among the 35 modes during the RMD process in the conventional HEVC. The proposed method to reduce the number of modes is as follows:

1. Initially, a list of modes is constructed starting with mode 2 at intervals of 4. The modes identified are 2, 6, 10, 14, 18, 22, 26, 30 and 34. The HAD cost of these 9 modes are computed to build the list of candidate modes (3 or 8 depending on PU size).
2. At this stage, additional modes which are close to the cheapest modes obtained in the previous step will also be included. The number of cheapest modes that will be considered is denoted by *lval*. The minimum value for *lval* is 1 and the maximum is the number of candidates in the list (3 or 8). The lower the value of *lval*, the fewer will be the number of modes to be considered. However, the improved time savings with a low *lval* may also lead to a performance deterioration.
3. For each of the number of lowest values as set by *lval*, the HAD cost of the 2 left adjacent neighbors and the 2 right adjacent neighbors are identified. For mode 2 and mode 34, which are located at the two extreme ends of the list of angular modes, only the 2 right and 2 left modes respectively are included. The HAD cost of these additional modes, provided they were not processed earlier, are computed and the candidate list is updated accordingly.
4. Finally the HAD cost for mode 0 and for mode 1 are computed and the candidate list is again updated.

The resulting candidate list will therefore contain the same number of modes as in the conventional HEVC although the processing of some of the 35 modes have been skipped. Thus, if only the cheapest cost mode is chosen in step 2 above (*lval* = 1), a maximum of 15 modes will go through the HAD cost computation. Since the number of modes cannot be exactly determined when choosing the value of *lval*, the N_{lval}^{modes} notation will be used, where *modes* denotes the number of available modes for that PU and *lval* indicates the number of low values used in the scheme.

The detailed results of using the N_3^{35} scheme is shown in Table 3. An encoding time gain of 4.1 % is observed with a trivial decrease in PSNR (−0.0009 dB) along with a 0.07 % increase in bitrate. This decrease in encoding time prior to entering the RDO process is quite significant at this stage.

The results obtained with different values of *lval* are indicated in Table 4 for values of *lval* between 1 and 4. For the 16 × 16 and 32 × 32 PUs, only the 3 cheapest modes will be included in the candidate list while for the 8 × 8 PU this number is increased

Table 3 Performance of HM using the N_3^{35} scheme

Resolution	Sequence	Δ PSNR (dB)	Δ Time (%)	Δ BitRate (%)
2560 × 1600	PeopleOnStreet	−0.0007	−3.8	0.03
	Traffic	−0.0021	−3.8	0.06
1920 × 1080	BQTerrace	−0.0009	−3.5	0.06
	Cactus	−0.0018	−3.7	0.01
	Tennis	−0.0009	−4.5	0.11
1280 × 720	FourPeople	0.0000	−3.9	0.01
	KristenAndSara	−0.0031	−3.5	0.03
	vidyo1	0.0024	−4.3	0.08
	vidyo3	−0.0025	−4.0	0.09
	vidyo4	−0.0004	−4.3	−0.02
832 × 480	BasketballDrill	0.0006	−3.5	0.21
	BQMall	−0.0008	−3.7	0.05
	Flowervase	−0.0032	−4.8	0.12
	Keiba	0.0019	−3.7	−0.01
	PartyScene	−0.0019	−2.8	0.00
	RaceHorses	−0.0015	−3.5	0.03
416 × 240	BasketballPass	−0.0045	−3.3	0.14
	BlowingBubbles	0.0044	−5.2	0.21
	Flowervase	−0.0006	−5.8	0.07
	Keiba	0.0091	−4.7	0.18
	Mobisode2	−0.0134	−5.3	0.04
Average		−0.0009	−4.1	0.07

Table 4 Results of using different schemes for PUs considering 35 modes

Scheme	Δ PSNR (dB)	Δ Time (%)	Δ BitRate (%)
N_1^{35}	−0.0035	−5.4	0.13
N_2^{35}	−0.0014	−4.3	0.10
N_3^{35}	−0.0009	−4.1	0.07
N_4^{35}	0.0003	−3.8	0.08

to 8 modes. It is observed that an encoding time gain of 5.4 % is obtained by using the N_1^{35} scheme (maximum of 15 modes) along with a slight degradation in quality. Considering that the whole RMD process consumes 14 % of the total encoding time, this 5.4 % overall gain is significant. In the subsequent proposed approaches, the N_1^{35} scheme will be used when time savings is required and the N_3^{35} scheme will be used when quality is of concern rather than the conventional 35 modes PUs.

4.2 Reducing the 19 Modes

PUs of size 4 × 4 pixels make use of only 19 modes in the RMD process according to conventional HEVC due to the high number of such PUs for a single LCU. This smallest PU size is responsible for almost 25 % of the total RMD time. A similar approach to the 35 modes, in Sect. 4.1, is therefore adopted to reduce the number of modes considered. However, in this case, only 1 adjacent neighbor to the left and 1 adjacent neighbor to the right are included. Similarly, for modes 2 and 34, only one neighbor is considered. The same notation, N_{lval}^{modes} will be used. The result of using the N_2^{19} scheme is provided in Table 5 and the summary of results for the different values of *lval* are shown in Table 6.

For the 19 modes PUs, the time savings obtained is relatively small. The reason being that for the N_1^{19} scheme a maximum of 13 modes are already chosen out of 19. It is found, however, that the N_2^{19} scheme provides a good performance of 0.90 % reduction in encoding time together with a small *decrease* in bitrate and a negligible drop in PSNR of 0.0005dB. This scheme will therefore be adopted throughout the

Table 5 Performance of HM using the N_2^{19} scheme

Resolution	Sequence	Δ PSNR (dB)	Δ Time (%)	Δ BitRate (%)
2560 × 1600	PeopleOnStreet	0.0011	−1.03	0.010
	Traffic	0.0008	−1.03	−0.006
1920 × 1080	BQTerrace	0.0002	−0.72	0.013
	Cactus	−0.0012	−0.72	−0.027
	Tennis	−0.0004	−1.01	0.002
1280 × 720	FourPeople	−0.0012	−0.75	−0.022
	KristenAndSara	−0.0004	−0.45	−0.011
	vidyo1	0.0001	−0.86	0.076
	vidyo3	−0.0021	−0.92	−0.076
	vidyo4	−0.0019	−0.85	−0.086
832 × 480	BasketballDrill	−0.0043	−0.96	−0.060
	BQMall	−0.0019	−0.93	−0.007
	Flowervase	−0.0010	−1.59	0.035
	Keiba	0.0023	−0.93	0.025
	PartyScene	−0.0017	−0.72	−0.022
	RaceHorses	−0.0031	−0.92	−0.059
416 × 240	BasketballPass	−0.0005	−0.62	0.131
	BlowingBubbles	0.0003	−0.29	0.092
	Flowervase	−0.0003	−1.17	0.019
	Keiba	0.0009	−1.18	−0.011
	Mobisode2	0.0029	−1.17	−0.044
Average		−0.0005	−0.90	0.000

Table 6 Results of using different schemes for PU considering 19 modes

Scheme	Δ PSNR (dB)	Δ Time (%)	Δ BitRate (%)
N_1^{19}	−0.0020	−0.98	0.0099
N_2^{19}	−0.0005	−0.90	−0.0003
N_3^{19}	0.0002	−0.82	0.0247

remaining experiments as it represents almost a 1 % guaranteed gain in encoding time with practically no deterioration in quality at the very start, prior to entering the RDO process.

5 Reducing Modes in the RDO Process

As shown in Fig. 4, HEVC makes use of a candidate list consisting of 3 modes for PU of size 16 × 16 and higher while 8 modes are selected for smaller size PUs. The proportion of time spent in the simplified RDO process for the different PU sizes is illustrated in Fig. 9. This process, which represents 49 % of the time spent for the whole intra prediction, produces the final mode decision for a PU.

A reduction in the number of modes used in this process will therefore lead to a significant time savings. Three approaches are considered:

1. Using fewer modes from the candidate list
2. Using a fraction of the original candidate list based on a threshold
3. Using a fraction of the original candidate list based on a threshold along with a reduced MPM

It can also be noted that the reduced mode approach for the RMD process in Sect. 3 has been integrated in the subsequent experiments. More specifically, the reduced RMD schemes N_1^{35} and N_2^{19} are used as they together provide more than 6 % time savings prior to entering the RDO process.

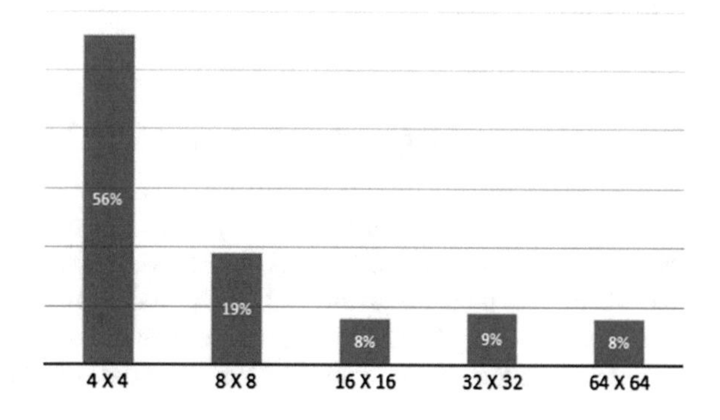

Fig. 9 Percentage time spent in the simplified RDO for each PU size

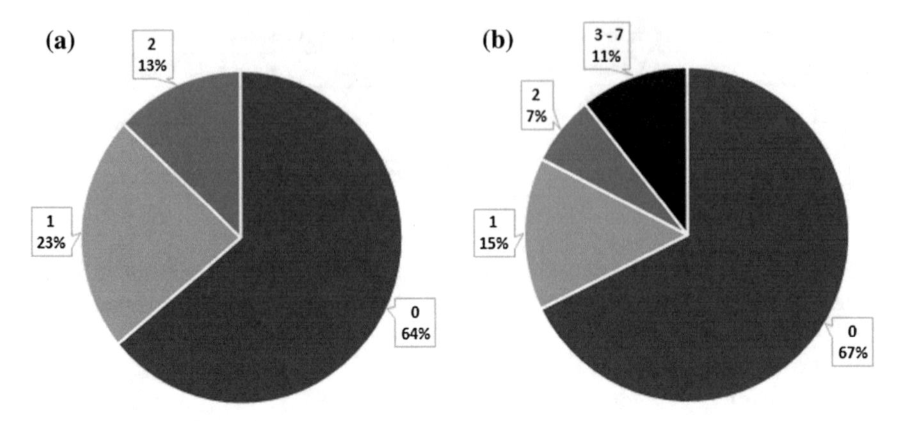

Fig. 10 Distribution of selected RDO modes **a** among the 3 RMD modes for 16×16 , 32×32 and 64×64 PUs and **b** among the 8 RMD modes for 4×4 and 8×8 PUs

5.1 Relevance of Mode Positioning in the Candidate List

The simplified RDO iterates among the modes from the candidate list along with the MPMs to identify the best mode for a given PU. The candidate list produced by the RMD process is ordered in ascending order of HAD cost starting from the cheapest. Four sequences have been used to collect the position of the selected mode in the original candidate list and the results are shown in Fig. 10a for the 4×4 and 8×8 (8 modes) and in Fig. 10b for larger size PUs (3 modes).

In this experiment, only the selected modes coming from the RMD process have been taken into account without considering the MPMs. The figures indicate that more than 75 % of all selections come from the first and second positions of the candidate list. In fact, for the small PU sizes, 67 % of them select position 0 (the cheapest cost) from the candidate list while 64 % of the larger PU sizes adopt this selection. Position 0 and position 1 (the first two modes) of the candidate list share together almost 80 % of the selections for smaller PUs and 87 % for the larger ones. It implies that these two positions should imperatively be considered in mode reduction approaches.

5.2 Using Fewer Candidates in the RDO Process

As discussed in Sect. 5.1, the starting modes in the candidate list play a major role in the mode decision process. Three distinct sets of experiments are therefore conducted to evaluate the impact of using 1, 2 and 3 modes only from the candidate list for all PUs in a sequence. The MPMs are then included accordingly. The results of using only the first mode along with the MPMs are thus displayed in Table 7.

Table 7 Performance of HM using only the first RMD mode position along with MPMs

Resolution	Sequence	Δ PSNR (dB)	Δ Time (%)	Δ BitRate (%)	BD-PSNR	BD-Rate
2560 × 1600	PeopleOnStreet	−0.074	−37.8	0.03	−0.081	1.42
	Traffic	−0.062	−37.9	−0.03	−0.058	1.20
1920 × 1080	BQTerrace	−0.073	−39.8	−0.21	−0.058	1.04
	Cactus	−0.056	−37.9	−0.31	−0.040	1.25
	Tennis	−0.029	−38.9	0.19	−0.038	1.28
1280 × 720	FourPeople	−0.072	−38.3	0.03	−0.089	1.37
	KristenAndSara	−0.030	−40.4	0.01	−0.025	0.42
	vidyo1	−0.060	−38.2	0.15	−0.079	1.47
	vidyo3	−0.045	−38.8	0.39	−0.086	1.37
	vidyo4	−0.049	−38.4	0.03	−0.049	1.18
832 × 480	BasketballDrill	−0.058	−37.2	−0.01	−0.060	1.18
	BQMall	−0.089	−37.8	−0.16	−0.081	1.36
	Flowervase	−0.096	−39.6	0.04	−0.096	1.62
	Keiba	−0.056	−38.6	−0.44	−0.035	0.71
	PartyScene	−0.128	−37.2	−0.40	−0.093	1.34
	RaceHorses	−0.077	−37.1	−0.32	−0.056	0.97
416 × 240	BasketballPass	−0.085	−37.0	0.09	−0.102	1.70
	BlowingBubbles	−0.099	−39.0	−0.32	−0.077	1.38
	Flowervase	−0.086	−40.0	−0.09	−0.077	1.37
	Keiba	−0.047	−40.6	0.02	−0.046	0.74
	Mobisode2	−0.022	−37.9	0.78	−0.058	1.64
Average		−0.066	−38.5	−0.03	−0.066	1.24

An average complexity reduction of 38.5 % is achieved using the first candidate mode only along with the MPMs for the RDO process. This reduction is associated with a 0.03 % *reduction* in bitrate and a drop of 0.066 dB in terms of PSNR. For the list of sequence tested, the average BD-PSNR is only −0.066 dB while the average BD-Rate is 1.24 %. The time savings for each of the sequences show little deviation from the average value indicating the small change in the number of modes coming from the additional MPMs. In the conventional approach, the candidate list may already contain most of the MPMs leaving only the few remaining ones to be added. When only one mode is chosen from the RMD process, most of the MPMs will have to be added to the candidate list and the final list may contain up to 4 modes, including the MPMs. Experiments have thus been conducted by using different number of modes from the candidate list and a summary of their results are depicted in Table 8.

Increasing the number of candidate modes produces lower complexity reduction, as will be expected, and the minor change in bitrate (*decrease*) implies that the compression factor has not been affected compared to conventional HEVC. The BD-PSNR and the BD-Rate also show lower values as additional modes are included.

Table 8 Summary of performance using fewer modes from the candidate list along with MPMs

RMD modes used	Δ PSNR (dB)	Δ Time (%)	Δ BitRate (%)	BD-PSNR (dB)	BD-Rate (%)
0	−0.066	−38.5	−0.03	−0.066	1.24
1	−0.037	−32.5	−0.08	−0.034	0.65
2	−0.024	−26.9	−0.01	−0.024	0.46

Table 9 Summary of performance using fewer modes in the candidate list with the N_2^{35} and N_2^{19} scheme

RMD modes used	Δ PSNR (dB)	Δ Time (%)	Δ BitRate (%)	BD-PSNR (dB)	BD-Rate (%)
0	−0.066	−37.8	−0.09	−0.052	1.15
1	−0.036	−32.1	−0.15	−0.028	0.54
2	−0.021	−26.2	−0.07	−0.017	0.33

The results show that considerable complexity reduction can therefore be achieved without increasing the bitrate at varying degrees of PSNR drop by using fewer modes in the candidate list.

The same experiment is conducted, but this time with the N_2^{35} and N_2^{19} reduced RMD schemes. Under these conditions, the number of modes in the RMD process is slightly increased for the 35 modes PUs to achieve a better quality in the output and this is illustrated by the results in Table 9. Although the complexity reduction is only slightly lower (less that 1 % when using N_2^{35}) compared to the N_1^{35} scheme used earlier, the bitrate is further reduced along with marginal change in PSNR. The BD-PSNR and the BD-Rate values are much lower compared to Table 8, confirming a substantial gain in quality although the complexity reduction is lower by less than 1 %.

5.3 Using Variable Number of Candidates in the RDO Process

In Sect. 5.2, a fixed number of modes is utilized from the candidate list produced by the RMD process. Figure 10 pointed out that modes situated at the beginning of the RMD list have a higher probability of being selected and modes in the higher positions are not so frequently used. However, when these higher position modes are selected they contribute significantly to the quality in the compression algorithm. This is readily observed from Tables 8 and 9 where the drop in PSNR becomes less significant compared to conventional HEVC when additional modes are used in the RDO process. A variable number of modes is therefore proposed based on a threshold value. The threshold value is set as a percentage of the difference in HAD

Table 10 Distribution of selected RMD modes with respect to %R for different PU sizes

%R	4 (%)	8 (%)	16 (%)	32 (%)	64 (%)
0–10	70	64	71	62	54
10–20	6	7	3	1	11
20–30	5	5	3	4	6
30–40	4	4	2	4	3
40–50	3	3	1	3	3
50–60	3	4	3	4	0
60–70	3	3	1	1	3
70–80	2	2	2	2	3
80–90	2	2	3	2	3
90–100	3	5	12	17	14

cost between the first mode and the last mode in the candidate list produced by the RMD process. The difference in cost which is termed as the Range, R, is given by:

$$R = HAD_{MaxMode} - HAD_0 \tag{5}$$

where HAD_i represents the HAD cost and i denotes the mode number starting from 0. *MaxMode* will take the value of 7 for the smaller size PUs and 2 for the larger ones. When the threshold value is used, only modes with HAD cost difference of less than %R from the candidate list will be retained for the RDO process along with the MPMs. The distribution of the selected RMD modes with respect to %R for the different PU sizes of the *BasketBallPass* sequence with QP = 22 is displayed in Table 10. The other sequences with different QP values show a similar trend.

Table 10 reveals that most of selected modes have a difference of less than 10 %R. Consequently experiments with threshold values of 5 %R and 10 %R are conducted. The detail results for a threshold value of 5 %R is given in Table 11. A 38.3 % gain in encoding time is achieved with a 0.09 % *decrease* in bitrate along with a drop of only 0.060 dB in PSNR. The bitrate has decreased for most of the sequences and a higher drop in PSNR is observed for the lower resolution ones.

The performances using different values of %R is provided in Table 12. The results show a slight decrease in time savings when the threshold is increased, indicating that more modes are being processed by the RDO. With a threshold of 20 %R, a 36 % complexity reduction is achieved with a drop of only 0.043 dB in PSNR along with a *decrease* of 0.08 % in bitrate. However, although the bitrate remains relatively static (but still with negative values), the quality in terms of PSNR shows much improvement as %R is increased.

The performances using 5 %R as threshold with different values of *lval* for the N^{35}_{lval} scheme are shown in Table 13. The results for the different schemes show only marginal improvement in bitrate with practically no change in quality compared to the N^{35}_1 scheme.

Table 11 Performance of HM using 5 %R along with MPMs

Resolution	Sequence	Δ PSNR (dB)	Δ Time (%)	Δ BitRate (%)
2560 × 1600	PeopleOnStreet	−0.064	−37.9	−0.06
	Traffic	−0.055	−37.8	−0.08
1920 × 1080	BQTerrace	−0.067	−39.6	−0.25
	Cactus	−0.049	−37.7	−0.25
	Tennis	−0.026	−38.8	−0.02
1280 × 720	FourPeople	−0.066	−38.0	−0.02
	KristenAndSara	−0.029	−40.4	−0.03
	vidyo1	−0.053	−38.2	0.08
	vidyo3	−0.041	−38.9	0.30
	vidyo4	−0.045	−38.4	−0.07
832 × 480	BasketballDrill	−0.050	−37.1	−0.12
	BQMall	−0.079	−37.6	−0.21
	Flowervase	−0.086	−39.9	−0.06
	Keiba	−0.045	−38.4	−0.32
	PartyScene	−0.119	−36.8	−0.44
	RaceHorses	−0.071	−36.9	−0.39
416 × 240	BasketballPass	−0.066	−37.5	0.11
	BlowingBubbles	−0.091	−36.8	−0.29
	Flowervase	−0.081	−39.4	−0.13
	Keiba	−0.041	−39.9	−0.12
	Mobisode2	0.030	−37.8	0.24
Average		−0.060	−38.3	−0.09

Table 12 Summary of performances using different threshold values

%R	Δ PSNR (dB)	Δ Time (%)	Δ BitRate (%)
5	−0.060	−38.3	−0.09
10	−0.055	−37.6	−0.12
20	−0.043	−36.0	−0.08

Table 13 Summary of performance using 5 %R as threshold for different values of the N_{lval}^{35} scheme

Reduced RMD scheme	Δ PSNR (dB)	Δ Time (%)	Δ BitRate (%)
N_1^{35}	−0.060	−38.3	−0.09
N_2^{35}	−0.062	−37.6	−0.18
N_3^{35}	−0.060	−36.8	−0.18

5.4 Selecting Candidates Based on %R Cost Threshold Along with a Reduced MPM

In Sects. 5.2 and 5.3, complexity reduction methods have been implemented and performances above 38 % have been achieved with acceptable loss in quality. In [8], it is proposed to skip the MPM coding for the 64 × 64 PUs. In order to increase the time savings further, only the MPMs for PU size of 4 × 4 and 8 × 8 are considered. By skipping the MPMs for large size PUs, the number of RDO iterations will be automatically reduced to increase the time savings. However, PUs of small sizes contribute considerably to the quality of the output and therefore these MPMs are maintained. The detailed results of using a 5 %R along with the proposed reduced MPMs is thus provided in Table 14.

An average reduction of 42.1 % encoding time is thus achieved along with a negligible increase of 0.27 % bitrate and a loss in quality of 0.075 dB. The reduction of complexity shows little variation throughout. It is noted, however, that the sequences *Mobisode2*, *Tennis* and *Video1* have the highest increase in bitrate whereas the two

Table 14 Performance of HM using 5 % R along with reduced MPMs

Resolution	Sequence	Δ PSNR (dB)	Δ Time (%)	Δ BitRate (%)
2560 × 1600	PeopleOnStreet	−0.081	−42.0	0.20
	Traffic	−0.071	−41.7	0.30
1920 × 1080	BQTerrace	−0.072	−43.6	−0.12
	Cactus	−0.059	−41.6	0.07
	Tennis	−0.048	−42.5	0.92
1280 × 720	FourPeople	−0.082	−42.3	0.31
	KristenAndSara	−0.036	−43.4	0.11
	vidyo1	−0.076	−42.2	0.79
	vidyo3	−0.059	−42.7	0.71
	vidyo4	−0.068	−42.2	0.63
832 × 480	BasketballDrill	−0.067	−41.4	−0.02
	BQMall	−0.086	−42.2	0.01
	Flowervase	0.116	−43.4	0.33
	Keiba	−0.058	−42.3	0.00
	PartyScene	−0.118	−41.3	−0.37
	RaceHorses	−0.084	−41.0	−0.23
416 × 240	BasketballPass	−0.082	−41.5	0.23
	BlowingBubbles	−0.092	−41.0	−0.03
	Flowervase	−0.128	−42.7	0.43
	Keiba	−0.057	−43.1	0.10
	Mobisode2	−0.029	−40.8	1.36
Average		−0.075	−42.1	0.27

Table 15 Summary of performances using 5 %R and 10 %R as threshold for different values of the N_{lval}^{35} scheme

%R	Reduced RMD scheme	Δ PSNR (dB)	Δ Time (%)	Δ BitRate (%)
5	N_1^{35}, N_2^{19}	−0.075	−42.1	0.27
	N_2^{35}, N_2^{19}	−0.073	−41.5	0.17
	N_3^{35}, N_2^{19}	−0.073	−40.8	0.15
10	N_1^{35}, N_2^{19}	−0.067	−41.6	0.26
	N_2^{35}, N_2^{19}	−0.066	−40.8	0.15
	N_3^{35}, N_2^{19}	−0.067	−40.0	0.11

Flowervase sequences and the *PartyScene* sequence are found to have impacted negatively on the PSNR.

A summary of the performance of HM using different values of %R with different reduced RMD schemes is provided in Table 15. The table shows that a performance of at least 40 % is achieved for all the different combinations of schemes and the different %R used. The drop in quality ranges from −0.066 dB to −0.075 dB while increase in bitrate ranges from 0.11 to 0.27 %.

The performance comparisons of the proposed method for the *Traffic* sequence (2560 × 1600) is shown in Fig. 11. It confirms that the PSNR loss of the proposed algorithm over the full range of QP values is marginal for a complexity reduction of 42.1 %. The experimental results also indicate that the proposed method can efficiently reduce the encoding complexity with little performance loss.

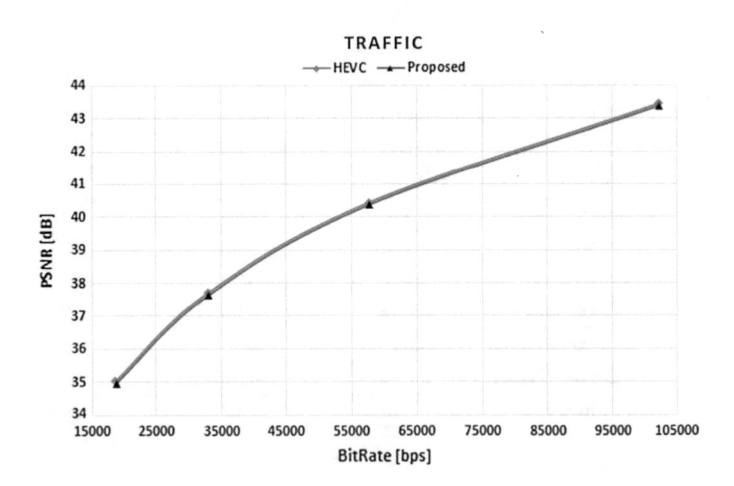

Fig. 11 Coding performance of proposed method based on a 5 %R cost threshold along with a reduced MPM for the *Traffic* sequence

Fig. 12 Output from conventional HM for the sequence Traffic for QP = 22

Fig. 13 Output from proposed method for the sequence Traffic for QP = 22

Figures 12 and 13, relating to the *Traffic* sequence using a QP value of 22, show extracts from the outputs of the conventional HM and the proposed method respectively. The visual quality of the outputs shows no perceptible difference.

The corresponding LCU in each figure is highlighted. It shows that the LCUs in both the conventional HM and the proposed method exhibit similar structure. It is also observed that CUs are of larger sizes in the proposed approach which partly explains the decrease in PSNR observed.

6 Conclusion

In this chapter, various fast mode decision methods for HEVC intra prediction have been proposed. The proposals made were directed towards reducing the number of modes taking part in the RMD and the RDO processes.

The approach used led to a maximum of 5.4 % overall reduction in complexity by reducing modes taking part in the RMD process for the 35 modes PUs and 0.98 %

for the 19 modes PUs. The approaches used for the RDO process include reducing the number of candidates in the RMD list, using a HAD cost threshold based on the difference between the first and last mode in the candidate list and skipping the MPMs for the large size PUs. On the basis of the initial encoding time reduction during the RMD and of techniques to reduce the number of iterations in the RDO process, various approaches have been proposed where complexity reduction of more than 38 % are achieved with an acceptable loss in bitrate and PSNR.

By using only the first RMD mode, a 38.5 % reduction in encoding time is achieved with a PSNR drop of 0.066 dB along with a 0.03 % *decrease* in bitrate. Using a 5 %R HAD cost threshold, a 38.3 % time savings is achieved with a drop in PSNR of only 0.060 dB and a 0.09 % *decrease* in bitrate. The maximum complexity reduction is reached when a 5 %R HAD cost threshold is used together with a reduced MPM. The performance indicates a 42.1 % time savings along with a drop of 0.075 dB in terms of PSNR and a negligible increase of 0.27 % in bitrate.

References

1. Bjøntegaard, G.: Calculation of Average PSNR Differences between RDCurves. Document VCEG-M33, VCEG (2001)
2. Bossen, F., Bross, B., Suhring, K., Flynn, D.: HEVC complexity and implementation analysis. IEEE Trans. Circuits Syst. Video Technol. **22**(12), 1685–1696 (2012)
3. Chen, G., Liu, Z., Ikenaga, T., Wang, D.: Fast HEVC intra mode decision using matching edge detector and kernel density estimation alike histogram generation. In: 2013 IEEE International Symposium on Circuits and Systems (ISCAS), pp. 53–56 (2013)
4. Chen, G., Pei, Z., Sun, L., Liu, Z., Ikenaga, T.: Fast intra prediction for HEVC based on pixel gradient statistics and mode refinement. In: 2013 IEEE China Summit International Conference on Signal and Information Processing (ChinaSIP), pp. 514–517 (2013)
5. da Silva, T.L., Agostini, L.V., da Silva Cruz, L.A.: Speeding up HEVC intra coding based on tree depth inter-levels correlation structure. In: 2013 Proceedings of the 21st European Signal Processing Conference (EUSIPCO), pp. 1–5 (2013)
6. da Silva, T.L., da Silva Cruz, L.A., Agostini, L.V.: Fast HEVC intra mode decision based on dominant edge evaluation and tree structure dependencies. In: 2012 19th IEEE International Conference on Electronics, Circuits and Systems (ICECS), pp. 568–571 (2012)
7. Goswami, K., Kim, B.-G., Jun, D., Jung, S.-H., Choi, J.S.: Early coding unit-splitting termination algorithm for high efficiency video coding (HEVC). ETRI J. **36**(3), 407–417 (2014)
8. ISO/IEC JTC 1 SC29 WG11. Prediction modes and mode coding for large size Intra block. Document JCTVC-IO227_v9, JCTVC (2012)
9. ISO/IEC JTC 1 SC29 WG11. High Efficiency Video Coding (HEVC) Test Model 10 (HM10) Encoder Description. Document JCTVC-L1002_v3, JCTVC (2013)
10. ISO/IEC JTC 1 SC29 WG11. High Efficiency Video Coding (HEVC) Text Specification Draft 10. Document JCTVC-L1003_v9, JCTVC (2013)
11. ITU-T. High efficiency video coding. Recommendation ITU-T H.265, ITU (2013)
12. Jiang, W., Ma, H., Chen, Y.:. Gradient based fast mode decision algorithm for intra prediction in HEVC. In: 2012 2nd International Conference on Consumer Electronics, Communications and Networks (CECNet), pp. 1836–1840 (2012)
13. Kim, J., Choe, Y., Kim, Y.G.: Fast Coding Unit size decision algorithm for intra coding in HEVC. In: 2013 IEEE International Conference on Consumer Electronics (ICCE), pp. 637–638 (2013)

14. Kumar, V., Govindaraju, H., Quaid, M., Eapen, J.: Fast intra mode decision based on block orientation in high efficiency video codec (HEVC). In: 2014 International Symposium on Computer, Consumer and Control (IS3C), pp. 506–511 (2014)
15. Lainema, J., Han, W.-J.: Intra-picture prediction in HEVC. In: Sze, V., Budagavi, M., Sullivan, G.J., (eds.) High Efficiency Video Coding (HEVC), Integrated Circuits and Systems, pp. 91–112. Springer International Publishing, Berlin (2014)
16. Lin, Y.C., Lai, J.C.: Edge density early termination algorithm for HEVC coding tree block. In: 2014 International Symposium on Computer, Consumer and Control (IS3C), pp. 39–42 (2014)
17. McCann, K., Han, W.-J., Kim, I.-K.: Samsungs Response to the Call for Proposals on Video Compression Technology. Document JCTVC-A124, JCTVC (2010)
18. Piao, Y., Min, J., Chen, J.: Encoder Improvement of Unified Intra Prediction. Document JCTVC-C207, JCTVC (2010)
19. Saurty, K., Catherine, P.C., Soyjaudah, K.M.S.: A modified rough mode decision process for fast high efficiency video coding (HEVC) intra prediction. In: 2015 Fifth International Conference on Digital Information and Communication Technology and its Applications (DICTAP), pp. 143–148 (2015)
20. Shi, Y., Au, O.C., Zhang, H., Zhang, X., Jia, L., Dai, W., Zhu, W.: Local saliency detection based fast mode decision for HEVC intra coding. In: 2013 IEEE 15th International Workshop on Multimedia Signal Processing (MMSP), pp. 429–433 (2013)
21. Sullivan, G.J., Ohm, J., Han, W.-J., Wiegand, T.: Overview of the high efficiency video coding (HEVC) standard. IEEE Trans. Circuits Syst. Video Technol. **22**(12), 1649–1668 (2012)
22. Sun, H., Zhou, D., Goto, S.: A low-complexity HEVC intra prediction algorithm based on level and mode filtering. In: 2012 IEEE International Conference on Multimedia and Expo (ICME), pp. 1085–1090 (2012)
23. Wiegand, T., Sullivan, G.J., Bjontegaard, G., Luthra, A.: Overview of the H.264/AVC video coding standard. IEEE Trans. Circuits Syst. Video Technol. **13**(7), 560–576 (2003)
24. Zhang, H., Ma, Z.: Early termination schemes for fast intra mode decision in high efficiency video coding. In: 2013 IEEE International Symposium on Circuits and Systems (ISCAS), pp. 45–48 (2013)

WSN Efficient Data Management and Intelligent Communication for Load Balancing Based on Khalimsky Topology and Mobile Agents

Mahmoud Mezghani and Mahmoud Abdellaoui

Abstract In this paper, we have used the multi-agent system (MAS) and the Khalimsky theory to optimize the routing paths and to improve the data management in wireless sensor network (WSN). We have used hierarchical architecture composed by three layers. In the first layer, the sensors are randomly deployed and grouped into clusters for data gathering. The second layer contains the leaders having the highest remaining energy in the clusters. These sensors ensure the fusion and the aggregation of collected data. The third layer contains a set of sensors deployed according to the Khalimsky topology to transmit data packets to the Sink using an optimal path. Each sensor node incorporates an agent to manage the communications and the collected data. In the third layer, mobile agents are used to balance the selection of nodes forming paths between the Sink and the gateways. Our approach showed a reduction of the energy consumption up to 40 % compared with LEACH protocol. Mobile agents were more effective than stationary agents. They allowed to save approximately 5 % of energy. Also, they allowed a balance of tasks distribution and reduced energy gap between nodes which consequently permit the prolongation of the network lifetime.

1 Introduction

Technological advances in wireless networking and integration of microprocessors have created a new generation of platforms for distributed embedded computing and large-scale wireless sensor networks supporting varied applications. This technology has revolutionized the way we live, work and interact with the physical environment around us. Today, the tiny and inexpensive wireless sensors can be literally scattered on roads, structures, walls or machines, creating a second digital networks able to

M. Mezghani
National Engineering School of Sfax (ENIS), Sfax, Tunisia
e-mail: mahmoud.mezghani@gmail.com

M. Abdellaoui (✉)
National School of Electronics and Telecommunication of Sfax (ENET'COM), Sfax, Tunisia
e-mail: mahmoudabdellaoui4@gmail.com

Y. Bi et al. (eds.), *Intelligent Systems and Applications*,
Studies in Computational Intelligence 650, DOI 10.1007/978-3-319-33386-1_19

detect a variety of physical phenomenas. Numerous applications are then invented such as the detection and monitoring of disasters, the environmental monitoring, intelligent building, monitoring and preventive maintenance of machines, health care, intelligent transport, etc. [1, 2].

Wireless sensors are characterized by limited processing capacity, low storage memory, and especially limited energy resource. In the most cases, the recharge of the batteries in wireless sensor networks is practically impossible due to the location of sensors in hostile environment. In fact, the sensor power consumption plays an important role in the life of the network which has become the predominant performance criteria in this area [3].

Several research studies have appeared with a purpose to optimize the energy consumption of nodes through the use of innovative conservation techniques to improve network performance, including the maximizing of its life. Generally, energy economization shall be only through effective solution for managing the different activities that consume the most energy. The radio component is the most consumer of energy in a sensor. It is therefore imperative that the data transmission is more consumer than the processing operations. Consequently, to minimize the energy consumption in the WSN, it is necessary to optimize any communication between the sensor nodes [4, 5].

In the literature, several energy conservation strategies are proposed at all levels from the physical layer to the application layer. Especially, several studies focus on the network topology, data routing and efficient management of collected informations from the source to the base station. Actually, the hierarchical clustered topologies are the most adopted allowing efficient data routing protocols. Data aggregation and fusion are widely used for data management efficiency minimizing the number of transmitted packets in the network [4]. The Multi-agent System (MAS) has been widely utilized in many applications and become very popular in the WSN area. The MAS technology have been identified as one of the most suitable technologies for WSN modeling, since each node may be modeled as an autonomous and collaborative agent [6–8]. However, researchers use two kinds of agents to execute tasks in WSN: stationary agents and mobile agents. In WSN using stationary agents, data is collected by sensors and transferred to the cluster head for processing as data fusion and aggregation. So, in this type of network a large amount of data travels between the nodes. By using mobile agent only on demand data is sent to any node. Mobile agents are used to perform the following functions: eliminating data redundancy among sensors by local processing at node level; eliminating spatial redundancy by data aggregation; reducing communication overhead by concatenating data.

We devote this paper to energy efficiency of communications and data management in WSN. We propose a multi-layers, clustered and hierarchical topology as deployment architecture for WSN based on Khalimsky theory. The clusterization is ensured by the Voting-Based Algorithm (VCA) [9]. We use a MAS composed by three types of agents incorporated in all sensors. We affect different roles to ensure an optimized data management with data fusion and aggregation. A mobile agents based on Khalimsky paths determination are used to accomplish the routing tasks of informations and to balance the energy conservation among all nodes.

The remainder of this paper is organized as follows: related work and problem statement are introduced in Sect. 2. Then, we describe our proposed WSN architecture based on the Khalimsky theory and MAS for efficient data management and intelligent data routing balancing tasks between sensor nodes in Sect. 3. Simulation analysis is discussed in Sect. 4. Finally, Sect. 5 gives some conclusions and prospects.

2 Related Work

The design of WSN is influenced by many factors such as fault tolerance, production costs, energy consumption and network topology. These factors represent the basis for the design of wireless sensor network protocols. Depending on the adopted architecture, the routing protocols in WSN can be classified into three main classes: flat, hierarchical and geographical protocols supporting four forms of information routing process as shown in the Fig. 1 [4, 5].

In flat architecture, the sensors may communicate directly with the base station using a high power (see Fig. 1a) or via a multi-hop model with very low power (see Fig. 1b). SPIN, Directed Diffusion, CADR, COUGAR, Flooding, etc. are standard flat protocols. This type of protocols is also called Data-Centric protocols. They don't use the identification of nodes. The data is typically transmitted from each sensor node in the deployment region with significant redundancy [10, 11].

In hierarchical architecture, the node representing the cluster called cluster-head, directly transmits data to the base station (see Fig. 1c) or via a multi-hop model between cluster-heads (see Fig. 1d). LEACH, PEGASIS, H-PEGASIS, TEEN, APTEEN, etc. are the most popular hierarchical routing protocols. The main idea is to form cluster and then use local cluster-heads as a gateway to reach the destination. This architecture type saves energy because the transmissions are performed only by the cluster-head rather than by all sensor nodes [10, 11].

Geographic routing protocols use the location information for guiding the discovery data transmission paths. They use the directional data transmission to avoid flooding across the network. Therefore, with network topology location-based information, routing is optimized and network management becomes easy. The disadvan-

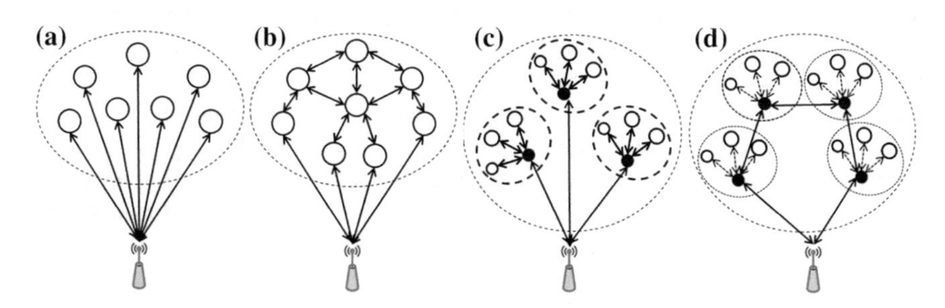

Fig. 1 Data routing forms for flat, hierarchical and geographical protocols in WSNs

tage of these routing protocols is that each node must know the locations of other nodes. MECN, GAF and GEAR are typical examples of geographical protocols. The location of the nodes could be provided using a GPS or by exchange of enormous amounts of messages [10, 11].

In hierarchical protocols the information redundancy rate reduction enables efficient data routing optimization and minimization of energy consumption in WSNs. To provide greater efficiency for hierarchical protocols, researchers have combined the data fusion and aggregation techniques to reduce more the transmitted data packets number by efficient informations management method [12–14].

Actually, several researches have worked on implementing MAS on WSN [4, 6–8]. The MAS technology promises a considerable potential for modeling WSN, since each node may be modeled as an autonomous agent, which is collaborative and even self organizing. Agents can be used for the implementation of basic services in WSNs such as data aggregation, fusion, and transmission following a requested query or triggered event. MAS have been identified as one of the most suitable technologies to contribute to this domain due to their appropriateness for modeling autonomous self-aware sensors in a flexible way.

In this context, we focus our work on the development of a hierarchical architecture type of WSN based on khalimsky theory. Our main objectives are minimizing the energy consumption and optimizing the informations management in the network to prolong its lifetime. For that, we used the Khalimsky topology and the VCA clustering algorithm to develop a multi-layers clustered hierarchical topology for WSN. For an optimal informations management and for minimizing redundancy rate, we used the techniques of data fusion and aggregation and the multi-agents system at different steps: informations collection, intra-cluster management, and packets delivery to the base station. Using both strategies directed flooding and shortest path routing algorithm based on MAS technique and Khalimsky theory, our work allowed a reduction of flooding rates and of unnecessary informations redundancy in the network, and a considerable minimization of the energy consumption [4, 5].

In the literature, despite results that minimize energy consumption and optimize information management in WSN, we find that the routing path is not always optimal and there's an imbalance of operation and power between nodes. The cluster-heads and edge nodes near the Sink are often the most used and exhausted causing the stop of the entire network. To solve this problem of imbalance between nodes in WSN, we dedicate this paper to present our work using cooperative agents embedded in each node to manage informations and energy consumption. We also use mobile agents to ensure, on the one hand, an optimal routing of data, and secondly, a balance of nodes operation and energy consumption to extend the network lifetime.

In the following, we present a summary of our previous work [4, 5]. We describe the Khalimsky topology which reduces the flooding rates and the information redundancy in the routing phase. Then, we present the impact of the combined use of MAS and khalimsky theory for the development of an efficient hierarchical WSN in terms of energy consumption and data management. Finally, we present our solution using mobile agents to balance the nodes functioning and the selection of the routing paths to prolong the network lifetime.

2.1 WSN Flooding Algorithm Based on Khalimsky Topology

In our previous article [5], we present a Khalimsky-based model minimizing flooding rates in WSN. The objective of that research was the using of Khalimsky topology for developing a hierarchical architecture restricting the set of nodes used during the information flooding. It is known that the flooding is often used in WSN for the construction of the network architecture, for information routing in data-centric protocols or for updating the routing table. This technique overloads the network by huge redundant information. It is therefore necessary to reduce the redundancy rate by optimizing the information management to minimize the power consumption.

In this section, we present a summary of Khalimsky topology used on the deployment of a hierarchical WSN architecture. We describe our data centric routing algorithm minimizing flood rates by restricting the set of nodes forming the paths between a source and a destination. We also describe the shortest path routing algorithm (SPRA) using the coordinates of Khalimsky nodes to calculate optimal routes between two nodes.

Let $\mathfrak{B} = \{\{2n+1\}; n \in \mathbb{Z}\} \cup \{\{2n-1, 2n, 2n+1\}; n \in \mathbb{Z}\}$ be a family of subsets of the set of integers $\mathbb{Z}$. Then $\mathfrak{B}$ is a base of a topology on $\mathbb{Z}$ called the *Khalimsky* topology. The *Khalimsky* plane is the Cartesian product of two *Khalimsky* topology $(\mathbb{Z}, \kappa) \times (\mathbb{Z}, \kappa)$. This topology on $\mathbb{Z}^2$ is characterized as follows. A point at its two coordinates of the form $(2k; 2k')$ is closed point (they are shown in the Fig. 2 by the black squares), if both coordinates are of the form $(2k+1; 2k'+1)$ point is an open point (shown in the Fig. 2 by the white squares). These types of points are called "*pure points*." Other points are called "*mixed points*" (presented by two lines). Figure 2 is a connectivity graph of the Khalimsky topology.

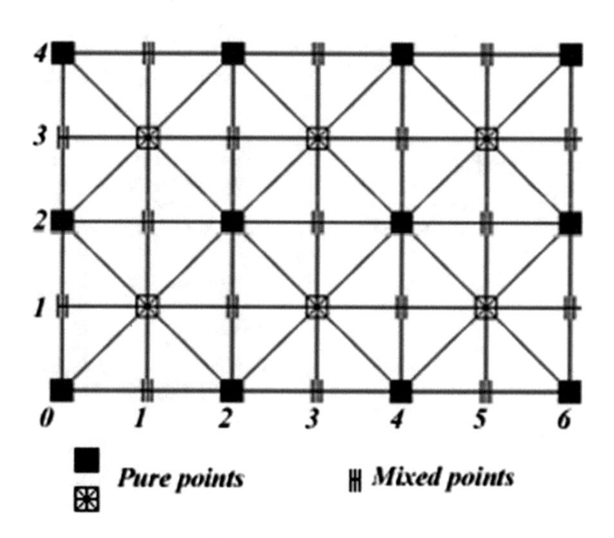

Fig. 2 Khalimsky topology graph

According to this definition, a WSN based on Khalimsky topology is formed by the set of sensor nodes scattered as in the graph of Fig. 2. Each point in the Khalimsky graph represents a sensor node. We distinguish that the "Mixed" nodes have up to 4 neighbors and can only communicate with "Pure" nodes. The "Pure" nodes have up to 8 neighbors. The differentiation of the types of the nodes in the Khalimsky topology reduces the links number between them and therefore it minimizes the communication rates and the informations redundancy.

In the following description, we define: $S_{(X_S,Y_S)}$ and $Q_{(X_Q,Y_Q)}$ the source and the destination sensor nodes, A_S^Q the link between two sensors formed by radio links of intermediate adjacent nodes, M_S^Q the minimal link between two sensors, $d_\infty(S, Q)$ is the hope number between $S_{(X_S,Y_S)}$ and $Q_{(X_Q,Y_Q)}$, and $BL(S, Q) = \bigcup M_S^Q$ the whole of intermediate sensors joining two nodes by minimal paths. Figure 3 show an example of a $BL(S, Q)$ calculated between the Sink $S_{(0,0)}$ and an node $Q_{(X_Q,Y_Q)}$.

Khalimsky topology uses five cases to calculate the optimized paths separating two nodes. The nodes forming the paths are in the zone I; II; III or IV; as well as, on the lines $D_{(y=x)}$ or $D_{(y=-x)}$ as shown in Fig. 4.

In the following theory, P represents an intermediate sensor node in the optimal whole of sensors joining a source node S and a destination node Q.

- **case 1** If S is a pure node and $\mid X_Q - X_S \mid = \mid Y_Q - Y_S \mid$, then Q is a pure node and

$$P \in BL\,(S,\,Q) \Leftrightarrow \begin{cases} |X_P - X_S| + |X_Q - X_P| = |X_Q - X_S| \\ |Y_P - Y_S| + |Y_Q - Y_P| = |Y_Q - Y_S| \end{cases}$$

The set of sensors joining S and Q are on the lines $D_{(y=x)}$ or $D_{(y=-x)}$.

- **case 2** If S is a mixed node and $\mid X_Q - X_S \mid = \mid Y_Q - Y_S \mid$, then Q is a mixed node and $P \in BL(S, Q) \Leftrightarrow BL(S, Q) = BL(S_1, Q_1) \cup BL(S_2, Q_2) \cup \{S, Q\}$

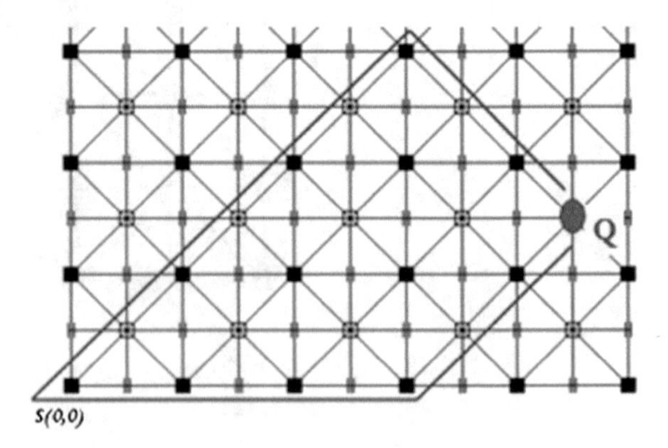

Fig. 3 Grid of $BL(S, Q) = \bigcup M_S^Q$ in Khalimsky topology graph

with $S_1\left(X_S, Y_S - \frac{Y_S - Y_Q}{|Y_S - Y_Q|}\right)$, $S_2\left(X_S - \frac{X_S - X_Q}{|X_S - X_Q|}, Y_S\right)$, $Q_1\left(X_Q - \frac{X_Q - X_S}{|X_Q - X_S|}, Y_Q\right)$, $Q_2\left(X_Q, Y_Q - \frac{Y_Q - Y_S}{|Y_Q - Y_S|}\right)$.

When S and Q are a mixed node, it's necessary to join two pure node from (S_1, Q_1) and $(S_2, Q_2)$to calculate $BL(S, Q)$.

- **case 3** If S, Q are pure nodes and $|X_Q - X_S| \neq |Y_Q - Y_S|$, then there is two cases:

1. If $|X_Q - X_S| > |Y_Q - Y_S|$ ($d_\infty(Q, S) = |X_Q - X_S|$), then

$$P \in BL(S, Q) \Leftrightarrow \begin{cases} |X_P - X_S| + |X_Q - X_P| = |X_Q - X_S| \\ |Y_P - Y_S| \leq |X_P - X_S| \\ |Y_Q - Y_P| \leq |X_Q - X_P| \end{cases}$$

In this case, the nodes joining S and Q are in the zone I or III as shown in Fig. 4.

2. If $|X_Q - X_S| < |Y_Q - Y_S|$ ($d_\infty(Q, S) = |Y_Q - Y_S|$), then

$$P \in BL(S, Q) \Leftrightarrow \begin{cases} |Y_P - Y_S| + |Y_Q - Y_P| = |Y_Q - Y_S| \\ |X_P - X_S| \leq |Y_P - Y_S| \\ |X_Q - X_P| \leq |Y_Q - Y_P| \end{cases}$$

In this case, the nodes joining S and Q are in the zone II or IV as shown in Fig. 4.

- **case 4** If Q and S are not both pure nodes or mixed nodes and $|X_Q - X_S| \neq |Y_Q - Y_S|$, then there is two cases:

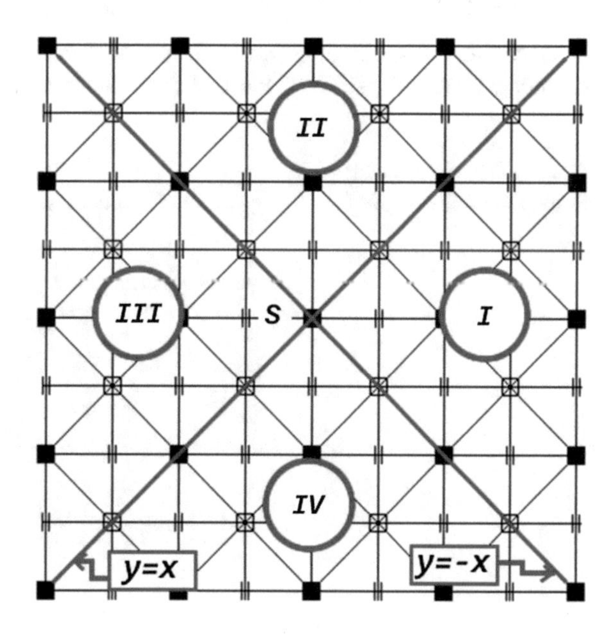

Fig. 4 Optimized paths zones in Khalimsky topology

1. P is a pure node, Q is a mixed node and $|X_Q - X_S| > |Y_Q - Y_S|$, then

$$P \in BL\,(S,Q) \Leftrightarrow P \in BL(S,Q') \cup \{Q\}$$

with $Q'\left(X_Q - \frac{X_Q - X_S}{|X_Q - X_S|}, Y_Q\right)$ a pure node.

2. $P$is a pure node, Q is a mixed node and $|X_Q - X_S| < |Y_Q - Y_S|$, then

$$P \in BL\,(S,Q) \Leftrightarrow P \in BL(S,Q') \cup \{Q\}$$

with $Q'\left(X_Q, Y_Q - \frac{Y_Q - Y_S}{|Y_Q - Y_S|}\right)$ a pure node.

•**case 5** If Q and S are both mixed nodes and $|X_Q - X_S| \neq |Y_Q - Y_S|$, then there is two cases:

1. If $|X_Q - X_S| > |Y_Q - Y_S|$, then

$$P \in BL\,(S,Q) \Leftrightarrow P \in BL(S',Q') \cup \{S,Q\}$$

with $S'\left(X_S - \frac{X_S - X_Q}{|X_S - X_Q|}, Y_S\right)$ and $Q'\left(X_Q - \frac{X_Q - X_S}{|X_Q - X_S|}, Y_Q\right)$ are pure nodes.

2. If $|X_Q - X_S| < |Y_Q - Y_S|$, then

$$P \in BL\,(S,Q) \Leftrightarrow P \in BL(S',Q') \cup \{S,Q\}$$

with $S'\left(X_S, Y_S - \frac{Y_S - Y_Q}{|Y_S - Y_Q|}\right)$ and $Q'\left(X_Q, Y_Q - \frac{Y_Q - Y_S}{|Y_Q - Y_S|}\right)$ are pure points.

According to these five cases, all informations sent between two nodes by flooding, can not get out of the $BL(S,Q) = \bigcup M_S^Q$ as shown in Fig. 5.

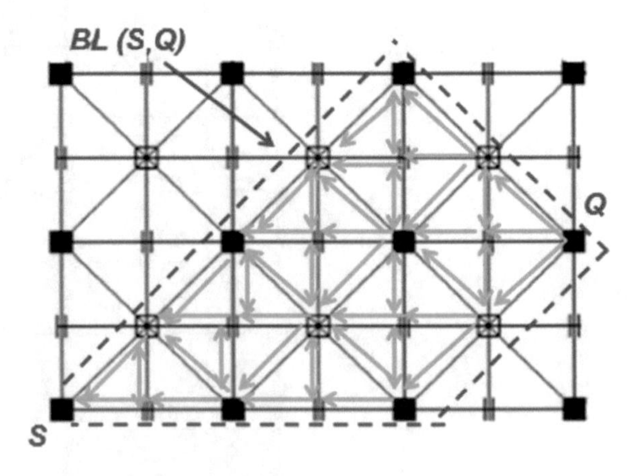

Fig. 5 Khalimsky flooding paths in $BL(S,Q) = \bigcup M_S^Q$

The next pseudo-code represents the Khalimsky based flooding algorithm. It shows the five different cases described above. For each case, the information is transmitted hope by hope with multicast manner to one, two or three nodes until reaching the destination. The $BL(S, Q)$ calculation is carried out implicitly during the transmission of data.

Algorithm 1 Khalimsky flooding algorithm ()

1: **if** $(X_Q + Y_Q\ MOD\ 2 \neq 0)$ **then** /* *Q is a mixed node:* 2^{sd},4th *and* 5th *case* */
2: $\forall\delta = \{-1, 0, 1\}$

$$Q_1(X_1, Y_1) = \begin{cases} X_1 = X_Q - E(\frac{X_0 - X_S}{d_\infty(S,Q)}) + \delta \\ Y_1 = Y_Q - E(\frac{Y_0 - Y_S}{d_\infty(S,Q)}) + \delta \end{cases}$$

3: **if** $(\exists Q_1(X_1, Y_1))$ **then**
4: $Send(Packet, Q_1)$
5: **end if**
6: **else**
7: **if** $(|X_Q - X_S| = |Y_Q - Y_S|)$ **then** /*1st *case* */
8: **if** $(d_\infty(S, Q)! = 0)$ **then**

$$Q_1(X_1, Y_1) = \begin{cases} X_1 = X_Q - \frac{X_0 - X_S}{d_\infty(S,Q)} \\ Y_1 = Y_Q - \frac{Y_0 - Y_S}{d_\infty(S,Q)} \end{cases}$$

9: **if** $(\exists Q_1(X_1, Y_1))$ **then**
10: $Send(Packet, Q_1)$
11: **end if**
12: **end if**
13: **else if** $(|X_Q - X_S| \neq |Y_Q - Y_S|)$ **then** /*3rd *case* */
14: **if** $(|X_Q - X_S| < |Y_Q - Y_S|)$ **then**
15: $\forall\delta = \{-1, 0, 1\}$

$$Q_1(X_1, Y_1) = \begin{cases} X_1 = X_Q + \delta \\ Y_1 = Y_Q - \frac{Y_0 - Y_S}{d_\infty(S,Q)} \end{cases}$$

16: **if** $(\exists Q_1(X_1, Y_1))$ **then**
17: $Send(Packet, Q_1)$
18: **end if**
19: **else if** $(|X_Q - X_S| > |Y_Q - Y_S|)$ **then**
20: $\forall\delta = \{-1, 0, 1\}$

$$Q_1(X_1, Y_1) = \begin{cases} X_1 = X_Q - \frac{X_0 - X_S}{d_\infty(S,Q)} \\ Y_1 = Y_Q + \delta \end{cases}$$

21: **if** $(\exists Q_1(X_1, Y_1))$ **then**
22: $Send(Packet, Q_1)$
23: **end if**
24: **end if**
25: **end if**
26: **end if**

Simulation tests of the Khalimsky topology are achieved using the TOSSIM simulator [15, 16]. The modeling of the energy consumption is performed by the POWERTOSSIM-Z tool [15, 16]. The obtained results showed a reduction of flooding rates by up to 31 %. Due to this reduction the energy consumption of the entire network is minimized approximately by 70 %.

2.2 WSN Intelligent Communication Algorithm Based on Khalimsky Topology and Multi-agents System

In our previous article [4], we presented a hierarchical topology based on Khalimsky theory and MAS model. Our objective is to reduce the energy consumption in WSN by minimizing the redundant information and by optimizing the routing paths. The proposed architecture is composed by three layers. The first layer contains a set of randomly deployed sensors grouped into clusters by using the Voting-based Clustering Algorithm (VCA) [9]. The second layer contains the nodes having the highest power called cluster-heads of the clusters in the first layer. The last layer contains a set of nodes deployed according to Khalimsky topology as described in the previous section. Figure 6 shows the proposed architecture.

In each sensor, an agent is incorporated to manage the energy, the collected or received informations and its role in the network. In the lower layer, the member agents (MA) act as simple members to collect data from the environment. In the middle layer, the cluster-head agents (CHA) manage the informations of the clusters using the fusion and the aggregation techniques. The CHA forward its data to the nearest gateway agents (GA) incorporated in the deployed nodes in the upper layer

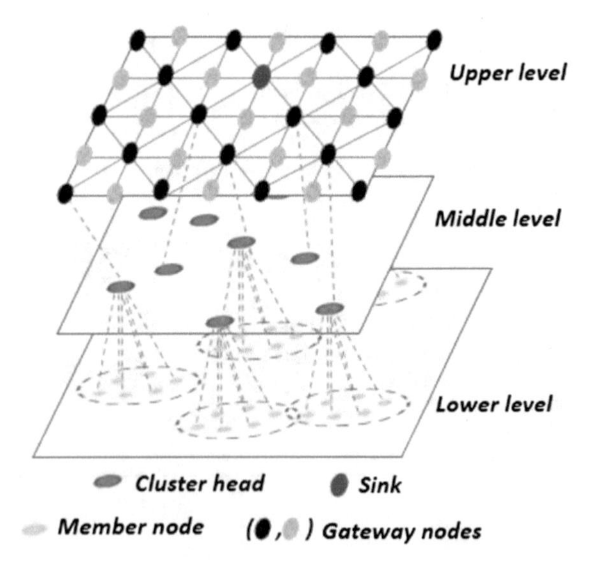

Fig. 6 Hierarchical multi-layers WSN architecture

according to Khalimsky topology. These GA route received data to the Sink using optimal paths by flooding or by selecting of one optimal path in the $BL(S, Q)$.

Clustering nodes of the lower layer is provided by the voting-based clustering algorithm (VCA) proposed in [9]. VCA is independent of network size and topology architecture. It is fully distributed algorithm. The clustering is an iterative process based on the residual energy of each node in the lower level. To determine the cluster heads, each node i broadcasts its residual energy value e_i and collects those of its neighbors V_i. After taht, each node must assigns the note n_{ij} proportional to their residual energy by the following formula:

$$n_{ij} = \frac{e_j}{\sum_{k \in V_i} e_{k,i}} \tag{1}$$

Then, each sensor diffuse notes that it has assigned to the sensors in its neighborhood, those will take delivery of the notes as well as those assigned by other sensors. Therefore, each sensor i determines its score s_i by the sum of the scores assigned by its neighbors V_i using the following formula:

$$s_i = \sum_{k \in V_i} n_{k,i} \tag{2}$$

The node having the highest score is auto-designated as Cluster-Head. All nodes within the reach of one or more cluster-head, must join one having the minimum of members in its group. Note that the energy consumption rate of a cluster head depends on the number of neighbors that subscribe to it. To balance the energy distribution, more sensors should subscribe to a high-energy cluster head. We define a fitness function to represent such properties of a cluster head. The fitness of a sensor V_i is defined as:

$$fitness(v_i) = \frac{e_i}{degree_i} \tag{3}$$

In the following paragraph, we present some terminology and definitions necessary for understanding the VCA algorithm as shown in Algorithm 2:

- N: Total number of nodes
- V_i: The ith sensor ($i \in [1..N]$)
- e_i: Residual energy of sensor v_i
- $degree_i$: Node degree (number of neighbors) of sensor v_i
- S: An $[0, L]^2$ square area in which N sensors are dispersed
- S_{CH}: The set of nodes that are cluster heads
- S_{WD}: The set of nodes that withdrew from voting
- S_{nbr}: The set of neighboring nodes

Algorithm 2 Pseudo code of VCA algorithm ()

1: **procedure** INIT
2: $S_{nbr} \leftarrow$ {v | v is within my communication range}
3: $S_{nbr} \leftarrow \{\}$; $S_{WD} \leftarrow \{\}$; $S_{uncovered} \leftarrow S_{nbr}$; $currentvote \leftarrow \sum_{v_j \in S_{nbr}} v(v_j, nodeID)$
4: broadcast currentvote and fitness to S_{nbr}
5: **end procedure**
6: **procedure** VCA
7: **while** $(clusterhead = \{\})$ **do**
8: **if** $(S_{CH} = \{\})$ **then**
9: $newvote \leftarrow \sum_{v_j \in S_{nbr}} v(v_j, nodeID)$
10: **if** $(newvote! = currentvote)$ **then**
11: currentvote=newvote
12: broadcast voteupdate(nodeID, newvote)
13: **end if**
14: collect v_updates from neighbors in t_1 sec
15: **if** $currentvote = max\{vote(v) \mid v \in S_{uncovered}\})$ **then**
16: CH(nodeID) $\leftarrow$ true, Cluster_head $\leftarrow$ nodeID
17: send a CH advertisement and return
18: **end if**
19: **end if**
20: collect incoming CH advertisements in t_2 sec
21: $S_{CH} \leftarrow S_{CH}$ U $\{v \mid CHoverheardfromv\}$
22: **if** $(S_{CH}! = \{\})$ **then**
23: broadcast a WITHDRAW packet
24: **end if**
25: collect incoming WITHDRAW in t_3 sec
26: $S_{WD} \leftarrow S_{WD}$ U $\{v \mid WITHDRAWheardfromv\}$
27: $S_{uncovered} \leftarrow \{v \mid v \in S_{nbr}\ \&\ v \notin S_{CH}\ \&\ v \notin S_{WD}\}$
28: **if** $(S_{CH}! = \{\})$ **then**
29: head $\leftarrow$ highest_fitness(S_{CH})
30: **if** $(fitness(head) >= max\{fitness(u) \mid u \in S_{uncovered}\})$ **then**
31: cluster_head $\leftarrow$ head
32: join_cluster(cluster_head);
33: return;
34: **end if**
35: **end if**
36: **end while**
37: **end procedure**

Simulation results have shown an improved network efficiency in terms of energy consumption and informations management. The Khalimsky topology combined with multi-agents system allowed to reduce the flooding rates by 34 %. The communication rates are reduced by more than 25 % and the energy consumption is minimized by 22 % compared with LEACH protocol.

However, a load imbalance between nodes is denoted. The overly used nodes, are rapidly exhausted causing the loss of paths despite the presence of nodes having sufficiently remaining energy. The load imbalance is due to the random selection of an optimal routing path in the $BL(Source, Destination)$. In fact, the load balancing nodes is a significant factor to prolong the network lifetime.

3 MAS Distributed Architecture

In order to prolong the lifetime of the WSN and ensure optimal management of information minimizing redundancy rates, we used the multi agent systems technique adapted to our sensor nodes deployment architecture according to the Khalimsky topology. In our approach, we used mobile agents and fixed agents. The fixed agents are the member agents (MA) and cluster-head agents (CHA) used for data gathering, collection, fusion and aggregation. The mobile agents (MBA) used for balanced acheminement data packets between the sink and the other agents. Compared to our proposed approach in [4], we changed the information gathering way and communication between MA and CHA agents to reduce the number of transmitted data packets. We have added the communication possibility of CHA with all GA in their range. Using the remaining energy of neighbors, the CHA are able to balance the use of GA agents to transmit their data. The MBA agents are used to route data between GA agents and the Sink according to the Khalimsky topology. The MBA agents are used to balance the remaining energy of GA agents forming the optimal paths.

3.1 *Member Agent Role*

In this section, we present the roles of the member agents (MA) incorporated in the clustered nodes. In each node, the agent is responsible to optimize the use of the resources to optimize the information management and to minimize the energy consumption. The information gathering is performed periodically or following an elevation of a request from the CHA. The MA gathers information and store it in its memory. If the difference between old stored value and the new gathered value is greater than a threshold, the information will be updated, otherwise the old value is maintained. The MA transmits the information stored in the memory, when an application request received from the CHA or in the case of gathered information value greater than emergency threshold. This operating mode reduces the power consumption of the information gathering and processing. Figure 7 shows the operating mode of MA agent.

3.2 *Cluster Head Agent Role*

In this section, we present the CHA agents roles incorporated in the leader nodes of clusters. In each cluster, the members and the cluster-head share a data base containing the gathered informations. The shared data base is represented by the memories of all MA in the cluster. The MA agents and the CHA agent cooperate among themselves for informations fusion and aggregation. Figure 8 shows the intra-cluster MAS architecture.

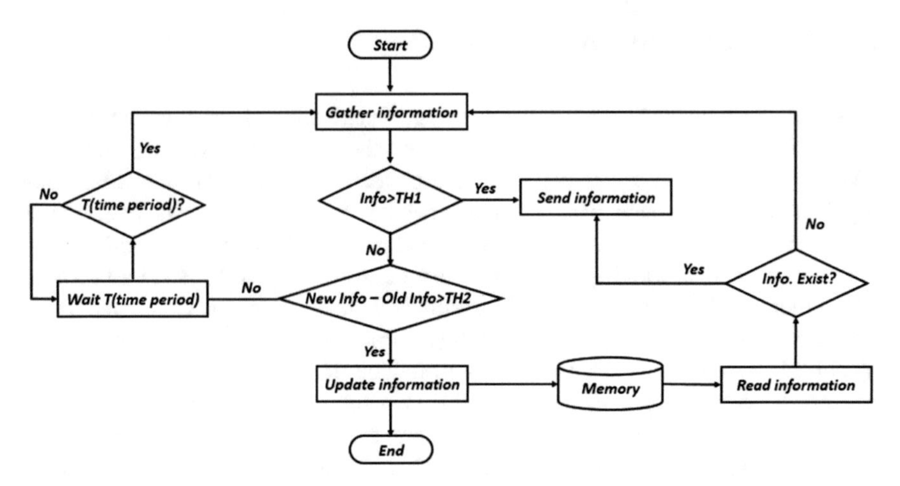

Fig. 7 Member agent process algorithm

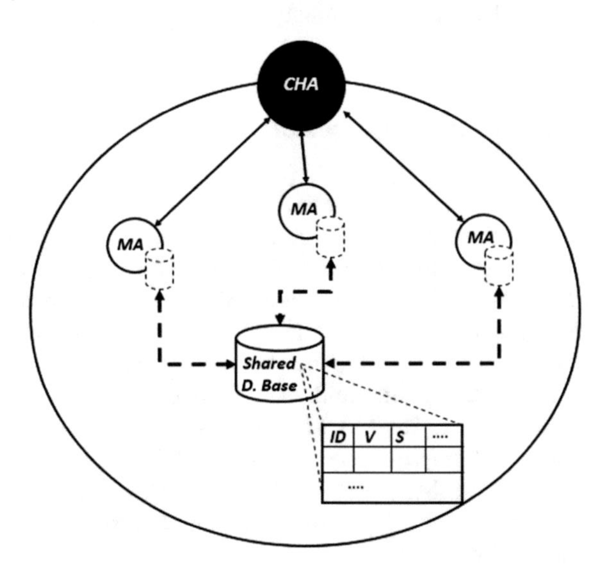

Fig. 8 Intra-cluster MAS architecture

Periodically, the CHA agent accesses to the shared database in the cluster to collect gathered informations by sending request queries to the different MA agents. When a data packet is ready to be sent to the sink, the CHA agent selects a GA among those in its range. The CHA agent uses the remaining energy as prominent factor to organize the selection of GA agents. This GA selecting way allows to balance the GA agents load and then balances the remaining energy of the sensors. When the remaining energy resource of the CHA become critical, the MA having the highest power is auto-designated as a new CHA of the cluster. The new CHA broadcasts her ID to its neighbors mentioning that he becames the new leader.

3.3 Mobile Agent Role

Currently, several approaches attempt to provide the benefits of the mobile agents in the management of WSNs [6, 7]. Particularly, the mobile agents can provide the improved efficiency in the data routing process between source-destination pair. In our approach, we used the mobile agents to load balance gateway nodes in the upper layer of our proposed WSN architecture (see Fig. 6). On each hop, the mobile agent (MBA) is interested just by the remaining energy of neighboring nodes. The MBA continue his way through the node having the highest remaining energy in the $BL(source, destination)$ of Khalimsky topology. The MBA agents are born when the CHA agents need to send data, and dies when arriving at the sink. A mobile agent is able to transmit founded data in a visited hop with him. The founded data is concatenated in the packet payload when there's free space.

4 Experiments and Results

To evaluate our proposed MAS combined with Khalimsky topology, we have used the TOSSIM simulator and POWERTOSSIM-Z to model the energy consumption of WSN composed by MICAZ sensors [15, 16]. The power consumption represents the value of consumed energy by the network sensor nodes during the simulation to transmit/receive and to process the data. According to the work [17, 18], the consumed energy by the radio component for the transmission and reception, is defined as below:

$$E_{TX} = L * E_{elec} + \varepsilon * L * d^2 \tag{4}$$

$$E_{RX} = L * E_{elec} \tag{5}$$

$$E_{TOT} = E_{TX} + E_{RX} \tag{6}$$

where, L is the length of packet in bits, $E_{elec} = 1.16\,\mu J/bit$ is the required electronic energy to run the transmitter/receiver circuitry, the parameter $\varepsilon = 5.46\,pJ/bit/m^2$, and d is the crossover distance.

The MICAz is based on the low-power 8-bit microcontroller ATmega128L with a clock frequency of 7.37 MHz. It run TinyOS and embed a IEEE 802.15.4 compliant CC2420 transceiver with a claimed data rate of 250 kbps. Table 1 presents the energy consumption costs of MICAZ platform.

Our objectives are the minimizing of the energy consumption and the balance load of sensor nodes. We are interested in this simulation study on three metrics: the remaining energy of sensors, the number of alive nodes and the energy consumption of the network over time. Based on these metrics, we have compared the performance of our novel approach using the mobile agents (KMBASPA) with our approach (KSASPA) proposed into [4] and the LEACH protocol [10, 11]. We used

Table 1 Energy consumption costs of the MICAZ platform operations

Energy costs	MICAZ
Compute for 1 T_{clk}	3.5 nJ
Transmit 1 bit	0.60 μJ
Receive 1 bit	0.67 μJ
Listen for 1 T_{clk}	9.2 nJ
Sleep for 1 T_{clk}	3 pJ (10^{-3})

a WSN network composed by 100 sensor nodes. In the lower layer there are 50 randomly deployed nodes. The VCA algorithm has determined 13 cluster heads which constitute the middle layer. The Sink is placed at the center of the upper layer which contains 50 nodes deployed according to the Khalimsky topology as shown in Fig. 6. The obtained results are based on the number of packets transmitted by each node over time. Figure 9 shows the energy consumption of our approach using the mobile agents (KMBASPA) compared with our previous approach (KSASPA) using stationary agents and LEACH protocol. According to the curves allure, the KSASPA and the KMBASPA consumed less energy than the LEACH protocol. Combined with Khalimsky topology the MAS allowed to reduce the energy consumption to about 40 % after 1500 s of simulation compared with LEACH protocol. We denote that the data routing algorithm with mobile agents is more effective. The mobile agents allowed to retain approximately 5 % of the dissipated energy better than our approach using fixed agents.

In our approach based on fixed agents, the simulation results showed imbalanced energy consumption between sensors. It is natural that border nodes near the Sink consume more energy in comparison to the rest of nodes due to receiving the data of other nodes and forwarding them to the Sink. We denote that the average gap of power consumption between nodes in our approach using mobile agents (KMBASPA) is lower compared with LEACH protocol and our solution using only stationary agents

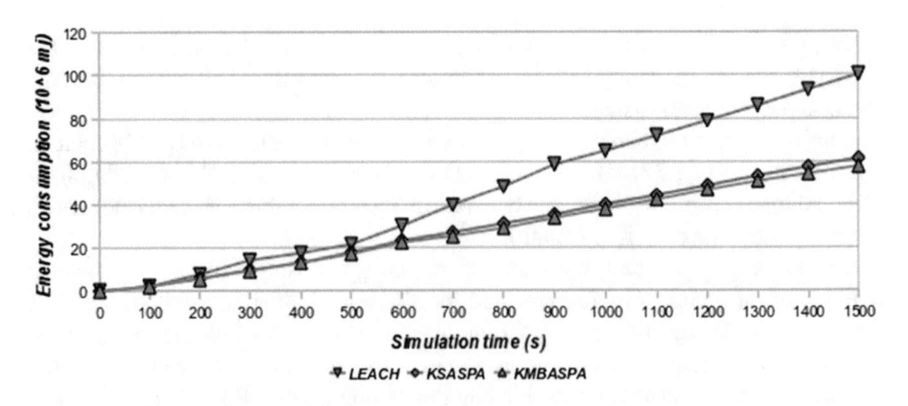

Fig. 9 Energy consumption of KMBASPA, KSASPA and LEACH protocols

(KSASPA). This gap becomes increasingly important over time for the LEACH protocol and our KSASPA approach. The average gap of power consumption is defined as below:

$$AVG = \sqrt{\frac{\sum E_{consumption} - E_{Moy}}{N}} \tag{7}$$

$$E_{Moy} = \frac{\sum E_{consumption}}{N} \tag{8}$$

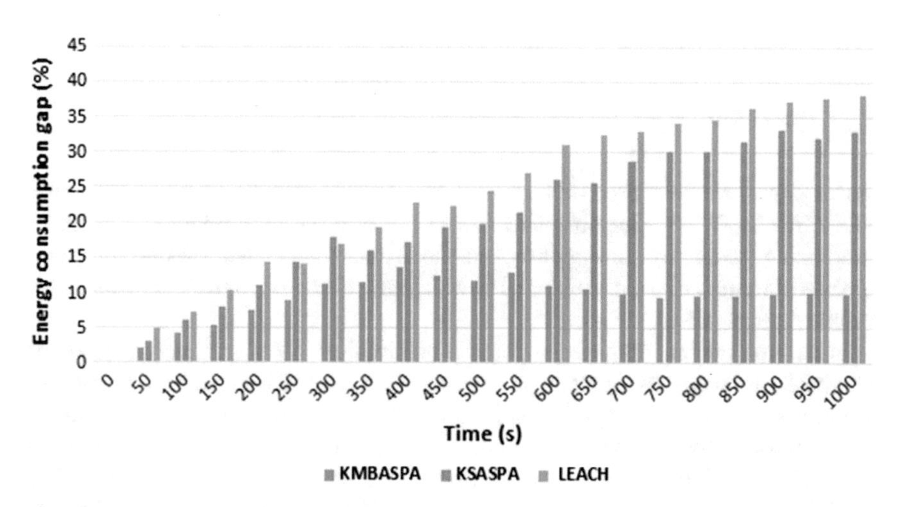

Fig. 10 Average gap of power consumption

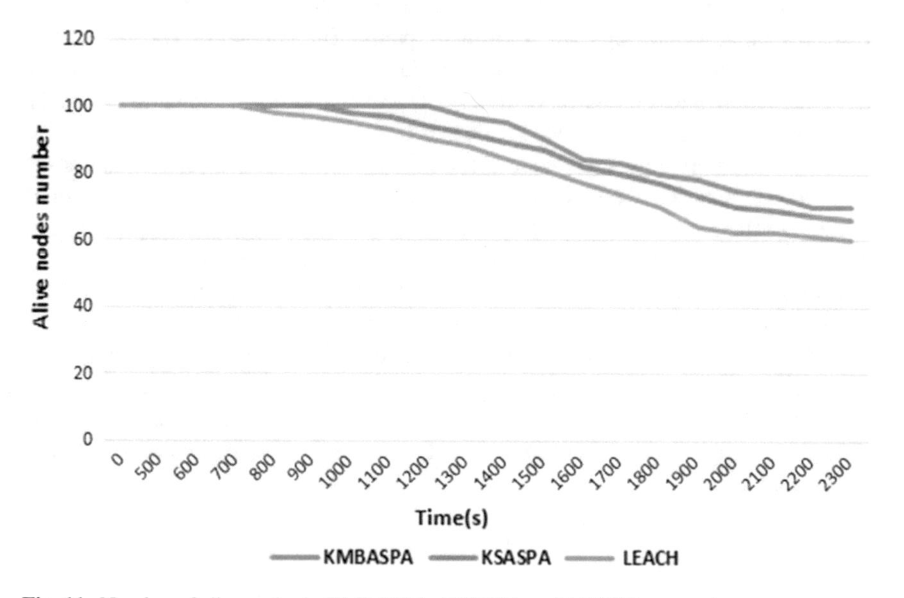

Fig. 11 Number of alive nodes in KMBASPA, KSASPA and LEACH approaches

Figure 10 shows the improvement of the mobile agents on energy consumption balance in the network.

Figure 11 shows the improvement of KMBASPA by the system lifetime graph. The curves show the number of nodes that remain alive over time. It is noticed that the nodes of the KMBASPA approach stay alive for a longer period than in LEACH and KSASPA protocols. The prolonged lifetime is due to the load balanced of nodes in the KMBASPA approach. Therefore, in the KSASPA and LEACH approaches, it appeared several nodes having a high amount of energy and several other nodes quickly exhausted.

5 Conclusion

In this paper, we have used an approach based on Khalimsky topology combined with MAS in WSNs. Our objective was to increase the performance of WSN by minimizing the energy consumption and a load balancing nodes. We have proposed two models called Khalimsky-based Fixed Agents (KSASPA) and Khalimsky-based Mobile Agents (KMBASPA).

In the first model, we have proposed a hierarchical WSN architecture based on Khalimsky topology using stationary agents incorporated in sensor nodes for data management, fusion, aggregation and routing. The member agents (MA) in clusters gathers informations from the environment and send them to the cluster-head agents (CHA). The CHA ensures data aggregation and forward it to a gateway agent (GA). The GA encapsulates collected data from different CHA and forward fusioned data to the Sink using a Khalimsky optimal path. The used path for data routing is randomly selected among several optimal Khalimsky paths. The simulation results showed encouraging effectiveness of this strategy reducing the power consumption and the data redundancy rates compared with LEACH and conventional Flooding protocols. However, a load imbalance between nodes is observed.

The second model uses the same hierarchical WSN architecture with stationary and mobile agents. This model is designed to provide load balancing between nodes to improve the performance and to prolong the lifetime of the WSN by the use of mobile agents. Simulation results have shown a reduction of remaining energy gap between sensor nodes. The balanced load allowed the extension of the nodes lifetime for more longer period than in LEACH and KSASPA approaches.

As future work, we are interested to provide more load balancing by using more base stations. The use of several base stations can reduces the access to the border nodes. Also, other multi-agent architectures can improve the performance of WSN in duty cycling including sleep moments for sensor nodes, particularly in dense deployment areas.

References

1. Akyildiz, I.F., Su, W., Sankarasubramaniam, Y., Cayirci, E.: Wireless sensor networks: a survey. Comput. Netw. **38**, 393–422 (2002)
2. Mezghani, M., Abdellaoui, M.: Multitasks-Generic platform via WSN. Int. J. Distrib. Parallel Syst. (IJDPS) **2**(4), 54–67 (2011)
3. Anastasi, G., Conti, M., Di Francesco, M., Passarella, A.: Energy conservation in wireless sensor networks: a survey. Ad Hoc Netw. **7**, 537–568 (2009)
4. Mezghani, M., Abdellaoui, M.: WSN Intelligent Communication based on Khalimsky Theory using Multi-agent Systems. In: IEEE SAI Intelligent Systems Conference, November 10–11, 2015, pp. 871–876. London, UK (2015)
5. Mezghani, M., Gargouri, R., Abdellaoui, M.: Optimization of WSNs flooding rates by khalimsky topology. Trans. Netw. Commun. **2**(6), 27–41 (2014)
6. Bendjima, M., Fehman, M.: Optimal itinerary planning for mobile multiple agents in WSN. Int. J. Adv. Comput. Sci. Appl. (IJACSA) **3**(11), 13–19 (2012)
7. Lal, A.S., Eapen, R.A.: A review of mobile agents in wireless sensor network. Int. J. Adv. Res. Electr. Electron. Instr. Eng. (IJAREEIE) **3**(7), 1842–1845 (2014)
8. Rachid, B., Hafid, H., Pham, C.: Multi-agent topology approach for distributed monitoring in wireless sensor networks. In: The Tenth International Conference on Wireless and Mobile Communications (ICWMC), pp. 1–9. IARIA (2014)
9. Qin, M., Zimmermann, R.: VCA: an energy-efficient voting-based clustering algorithm for sensor networks. J. Univ. Comput. Sci. **12**(1), 87–109 (2007)
10. Kacimi, R.: Techniques de conservation d'énergie pour les réseaux de capteurs sans fil, pp. 8–32. Université de Toulouse-France, Thèse de doctorat (2009)
11. Beydou, K.: Conception d'un protocole de routage hierarchique pour les réseaux de capteurs, pp. 38–89. Université de FRANCHE-COMTE, Thèse de doctorat (2009)
12. Chen, H., Mineno, H., Mizuno, T.: Adaptive data aggregation scheme in clustered wireless sensor networks. Computer Communications, vol. 31, pp. 3579–3585. Elsevier (2008)
13. Kalpakis, K. Tang, S.: A combinatorial algorithm for the maximum lifetime data gathering with aggregation problem in sensor networks. Computer Communications, vol. 32, pp. 1655–1665. Elsevier (2009)
14. Poorter, E.D., Bouckaert, S., Moerman, I., Demeester, P.: Non-intrusive aggregation in wireless sensor networks. Ad Hoc Networks. Elsevier (2009). doi:10.1016/j.adhoc.2010.07.017
15. Mezghani, M., Omnia, M., Abdellaoui, M.: TinyOS 2.1 with nesC: TOSSIM and PowerTOSSIM-Z Emulation and Simulation Environments to Networked Domotic Embedded Systems. In: 12th International conference on Sciences and Techniques of Automatic control and computer engineering (STA2011), 18–20 December 2011. Sousse, Tunisia (2011)
16. Perla, E., Ó Catháin, A., Carbajo, R.S.: PowerTOSSIM-z: realistic energy modelling for wireless sensor network environments. In: Proceedings of the 3nd ACM workshop on Performance monitoring and measurement of heterogeneous wireless and wired networks, 31 October 2008, pp. 35–42. Vancouver, BC, Canada (2008)
17. de Meulenaer, G., Gosset, F., Standaert, F-X., Pereira, O.: On the energy cost of communication and cryptography in wireless sensor networks. Networking and Communications, 2008. In: WIMOB '08. IEEE International Conference on Wireless and Mobile Computing, 12–14 October 2008, pp. 580–585 (2008)
18. Rabindra, B., Jae-Woo C.: Energy-efficient data aggregation for wireless sensor networks. Sustainable Wireless Sensor Networks, 14 December 2010, pp. 453–484. InTech. doi:10.5772/13708

Author Biographies

Mahmoud Mezghani received his master's degree in Computer Systems and Multimedia, from the Higher Institute of Computer Sciences and Multimedia of Sfax-Tunisia (ISIMS), and his research master's degree in Safety and Security of Industrial Systems—Real Time Computing and Electronics speciality, from the Higher Institute of Applied Sciences and Technology of Sousse-Tunisia (ISSAT-Sousse) in 2005 and 2007 respectively. Since 2011, he was PH.D. student in Electrical Engineering in National Engineering School of Sfax-Tunisia. Her research interests focus on the intelligent cooperative wireless sensor networks for optimal management of informations and data routing.

Mahmoud Abdellaoui received the B.S. degree in Electronic Engineering from High National Engineering School of Electronics and their Applications (ENSEA), Cergy, france, in 1988, and the M.S. degree from Lille University, france, in 1989. He received his Ph.D. degree in Electronic Engineering from Ecole Centrale de Paris, in 1991. He received his Habilitation degree in Electrical Engineering from National Engineering School of Sfax, Tunisia, in 2007. From October 1991 to September 1995, he was a Post doctor with Electronics and Telecommunications Research group in ENSEA-Cergy University, france. From September 1996, he has been an Assistant Professor at INSAT and since September 2002 at ISECS Sfax-University, Tunisia. In October 2007, he was promoted to the rank of Associate Professor at Higher National Institute of Electronics and Communications, Sfax, Tunisia. Since 2011, he becames a full Professor at National Engineering School of Electronics and Telecommunications of Sfax, Tunisia (ENET'COM).

His current research interests include wireless communications like Link and System capacity, Multistandards MIMO-OFDM systems; Mobile Networking (mobile IP networks, VoIP, networking wireless networks based on ISPDPLL); Wireless Sensor Network (WSN). He's the author of many published articles in: International Journal on Wireless Personal Communication (Springer); International Journal of Computer Sciences; American Journal of Engineering and Applied Sciences; International Journal of Electronics and Communications (AEU)-ELSEVIER; IEE Electronics letters; IEEE-MTT, etc. Dr. Abdellaoui has been invited as a reviewer of IEEE communications letters, International Journal of Electronics and Communications (ELSEVIER), Wireless Personal Communication (Springer), etc.

Indirect Method of Learning Weights of Intuitionistic Statement Networks

Tomasz Rogala

Abstract The paper presents an indirect method learning weights for intuitionistic statement networks. The intuitionistic statement network is a graphical model which represents some associations between statements in the form of the graph. An important element of this model apart from structure represented by the graph are also its parameters that determine the impact of individual statements to other statements. The values of weights are computed for different values of primary statements associated with external sources of observed data. Primary statement values are represented by intuitionistic values and can represent accurate values, which correspond to logical true or false or they represent approximate values. The presented method allows to identification of weights' values for every instance of evidence of primary statements regardless whether it is an accurate or an approximate value. The estimation of these weights is performed with the application of substitutional generative model represented by Bayesian network. The presented method allows the identification of weights also based on incomplete data.

1 Introduction

Technical diagnostics deals with the process of technical state recognition on the base of all available information on considered object including among others observed data about processes, knowledge about its operation and diagnostic relations between symptoms and any faults or malfunctions. Because technical objects or processes are becoming more and more complex it is necessary to acquire, store and process still growing data sets. Also, due to the fact the knowledge is often in the possession of many independent experts and can be also given in an implicit form of learning data sets from passive or active diagnostic experiments, the knowledge should be integrated from many different sources. The analysis of large data sets and also the

T. Rogala (✉)
Institute of Fundamentals of Machinery Design, Silesian University of Technology,
Konarskiego 18A, 44-100 Gliwice, Poland
e-mail: tomasz.rogala@polsl.pl

Y. Bi et al. (eds.), *Intelligent Systems and Applications*,
Studies in Computational Intelligence 650, DOI 10.1007/978-3-319-33386-1_20

process of integration of knowledge for the purpose of diagnostics is nowadays often supported by information systems among others with the use of expert systems.

Diagnostic expert systems can support maintenance engineers and users of complex technical installations and objects in recognizing of their technical state. Because the efficiency of this diagnosis process strongly depends on the quality of incorporated knowledge an essential issue is the choice of adequate knowledge representation is an essential issue. Adequate representation of knowledge contributes to, among others, the effective way of modeling of diagnostic knowledge. If it is easily to interpret then the process of constructing a knowledge base for users is becoming simpler and can be done with limited help of knowledge engineers.

Graphical models are a convenient means of representation of domain knowledge. They are commonly used in many areas of science, they combine the advantages of effective knowledge representation, tools for knowledge acquisition and inference engines. An example of such a model are intuitionistic statement networks [3] that allow modeling of the approximate diagnostic knowledge, have easy interpretation and allow building knowledge bases by many independent experts [3].

These kind of models can be constructed by users and knowledge engineers, although some aspects of network definition can be identified on the base of available learning data set. An important element of this model, apart from the structure represented by the graph, are also parameters that determine the impact of individual statements with known values to other statements. Values of these parameters can be determined with the application of described method indirectly using a substitutional model. The presented method also allows identification of weights based on incomplete data and for each instance of value from external data sources.

The paper is organized as follows. The Sect. 2 is about theoretical background of intuitionistic statement networks. The state-of-the-art is about the possibility of calculating weights on the base of available external data discussed in Sect. 3. The Sect. 4 describes the proposed method of parameters learning on the base of substitutional model for the case when primary statements accept accurate and approximate values. This section is also about the general conditions which should be met in order to apply the method. And finally it is also about the way of calculation of these parameters and their application during the operation of intuitionistic statement networks. An example of proposed method is presented in Sect. 5.

2 Intuitionistic Statement Networks

Intuitionistic statements networks [3–5] is a special graphical representation of knowledge in the form of a graph, and includes a built-in mechanism of inference. The individual vertices of the graph represent individual variables associated with data from external sources, the data that are seeking values of diagnostic inference about the considered domain, as well as other intermediate variables, which can be as additional vertices for the construction of the network model. A special feature of these networks is the ability to construct knowledge base of diagnostic system by many independent experts who construct separate parts of the network by building a partial models of this network [3, 9]. An essential condition for the development of a knowledge base constructed by partial models of intuitionistic statement networks is to use the same set of the variables which are interpreted by individual experts, involved in the development of the knowledge base of the expert diagnostic system, in the same semantically consistent manner. In order to introduce a clear semantic description each of variables are associated with a statements.

2.1 Statements

A convenient way of communication between the users and expert systems are statements. The statement is an assertion about given facts or opinions [2]. Two types of statements can be distinguished: simple which consist of only one content and a value associated with this content, or complex represented by many mutually exclusive content variants and adequate set of values associated with these variants. Simple statements are applied in intuitionistic statement networks models. Simple statements are represented as a pair of $s = < c(s), b(s) >$, where $c(s)$ is the content of the statement and $b(s)$ is the value of the content of the statement. During the operation of intuitionistic statement networks the content of the statement is invariant, but the value of statement is variable.

Because some of statement values are defined on the base of observed data and the others statement values are calculated as result of the inference process, two types of statements may be distinguished: primary and secondary. Primary statements are established directly by external measurement signals or are setting up arbitrary by object operator or knowledge engineer. The values of secondary statements are in turn established during the inference process on the basis of values of primary statements. The values of these statements are often the conclusions of the process of diagnosis.

In intuitionistic statement networks, the values of statements are represented by intuitionistic values. It allows to represent knowledge also in approximate way. The approximate value of such statement s is described as intuitionistic values $b(s) = < p(s), n(s) >$ which does not complies the law of excluded middle [4]. The $p(s)$ value represents the degree of validity of statement and $n(s)$ represents degree of non-validity. Both refer adequately to positive information and negative information as follows [1, 4]:

$$b(s) = < p(s), n(s) > \quad for \quad p(s), n(s) \in [0, 1] \tag{1}$$

An example of intuitionistic statement s is presented below:

- $c(s)$—Velocity vibration severity is acceptable,
- $b(s) = \{0.82, 0.1\}$.

Because some inconsistency in statement values or contradictions in the modeled knowledge domain may occur, the disagreement parameter level $d(s)$ for values $p(s)$ and $n(s)$ has been introduced [4]:

$$d(s) = \begin{cases} 0, & \text{if } p(s) + n(s) \leq 1; \\ p(s) + n(s) - 1, & \text{if } p(s) + n(s) > 1. \end{cases} \tag{2}$$

The parameter $d(s)$ does not equal 0 when the values of statements are contradicted. It means that the parameter $d(s)$ should be treated as special indicator of contradictions which may occur in the intuitionistic network [3]. Contradictions may results from the contradicted external data associated with primary statements or can results from contradictions in the network structure. More information about this issue can be find in [3–5].

The characteristics of intuitionistic values shows that statements shall have no influence on values of other statements during the inference process once the degree of validity or invalidity are to less to recognize the statement as valid or nonvalid. For example when $n(s) = p(s) = 0$ or the degree of validity or nonvalidity does not exchange some minimum value the statement can be treated as unrecognized [4] because of an irrelevant value of this statement e.g. $p(s) + n(s) \leq 0.2$.

In Fig. 1 a space of intuitionistic value is shown. This space can be divided into values which:

- correspond to the conviction of the truthfulness of the content of statement
- on the basis of which we do not know whether the content of statements corresponds to the conviction of its truthfulness or lack in the truthfulness,

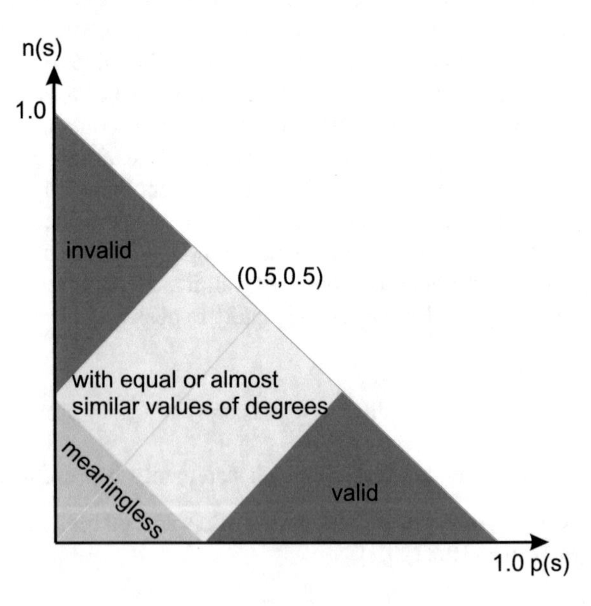

Fig. 1 A spacedivision into some subspaces with adequate interpretation

- correspond to the conviction of the lack of the truthfulness of the content of statement
- for which it can be said that these values do not represent any of the above conditions and do not carry any information about the content of the statement.

Also, one can distinguish values of intuitionistic statements that correspond to the logical truth and false. These values can be described as follows:

$$b(s) = \{1, 0\} - true, \tag{3}$$

$$b(s) = \{0, 1\} - false. \tag{4}$$

2.2 Statement Network

This subsection shows the definition of intuitionistic statement networks presented in [4]. An intuitionistic statement networks is represent by a graph G_{in} represented by the following set [4]:

$$G_{in} = < C, V, E, A > \tag{5}$$

where:

- C—is a set of vertices which represents statement contents [4],

$$C = \{c(s) : s \in S\}, \tag{6}$$

- V—is a set of vertices which consist of two subsets $V = V_p \cup V_n$. These subsets represent values of degrees of validity

$$V_p = \{p(s) : s \in S\} \tag{7}$$

  and nonvalidity

$$V_n = \{n(s) : s \in S\} \tag{8}$$

  of different statements.
  These vertices could appear as: conjunctive AND, disjunctive OR and FIXED where:

  – v_{fixed} is a vertex whose values are arbitrary defined by external sources (processed of observed signals defined by expert). These vertices accept the accurate values represented by intuitionistic values if they correspond to logical true or false which can be written adequately as:

$$\{1.0, 0.0\} \quad and \quad \{0.0, 1.0\}, \tag{9}$$

and they will be named as hard evidences or part of hard evidences when the evidence comprises of many primary statements [6]. In the case where v_{fixed} accepts the values which corresponds to the approximate values they will be named soft evidences [6].
  - v_{and} and v_{or} are vertices whose values are determined on the base of applying adequate implications in intuitionistic statement networks presented in the further part of this section,

- E—is a set of edges connecting vertices $c \in C$ with vertices $v \in V$ where the set E is composed of two subsets E_p and E_n where [4]:
  - E_p—is a subset of p-edges $e_p = < c(s), p(s) >$ connecting contents $c(s)$ of statements s with values of degrees of their validity $p(s)$,
  - E_n—is a subset of n-edges $e_n = < c(s), n(s) >$ connecting contents $c(s)$ of statements s with values of degrees of their nonvalidity $n(s)$,
- A—is a set of edges between the set of vertices V. These edges plays fundamental role in description of the relationship between different statement values.

The part of intuitionistic statement networks which define the constant structure of this network is depicted by H_{in} [4]

$$H_{in} = < C, V, E > \tag{10}$$

where the D_{in}

$$D_{in} = < V, A > \tag{11}$$

is an element which refers to this part of the graph which defines the mutual relations with the use of necessary and sufficient conditions [4]. An edge $a \in A$

$$(a = < s_i, s_j > \in A) \Rightarrow (s_i \rightarrow s_j) \ for \ (i \neq j) \tag{12}$$

defines the implication relations between statements s_1 and s_2. This relation shows that, the statement s_1 is a sufficient condition for statement s_2 and that the recognition of the validity of statement s_1 implies the validity of statement s_2. The necessary condition means that the statement s_2 is a necessary condition to statement s_1 and that the recognition of nonvalidity of statement s_2 implies the nonvalidity of statement s_1.

In the intuitionistic statement networks the necessary and sufficient conditions are modeled with the use of a set of inequalities [3]. If two statements s_1 and s_2 are connected by an edge $a = < s_1, s_2 >$ where s_1 is an ascendent and s_2 is subsequent, and the value of statement s_1 is known $b(s_1)$, then by using the sufficient condition, it can be written the following implication [4]:

$$(s_1 \rightarrow s_2, b(s_1 = < p(s_1), n(s_1) >) \Rightarrow p(s_2) \geq p(s_1) \tag{13}$$

while for known value of subsequent statement $b(s_2)$ the necessary conditions can be written [4]:

$$(s_1 \rightarrow s_2, b(s_2 = < p(s_2), n(s_2) >) \Rightarrow n(s_1) \geq n(s_2). \tag{14}$$

In the case once subsequent statement s is associated with vertex v_{or} and many antecedents $s_1, s_2, \ldots$ the implication is defined as [4]:

$$((\vee_i s_i) \rightarrow s) \Rightarrow p(s) \geq max_i\{p(s_i)\}. \tag{15}$$

Once one subsequent statement s is associated with vertex v_{and} and many antecedents $s_1, s_2, \ldots$ the implication is defined as [4]:

$$((\wedge_i s_i) \rightarrow s) \Rightarrow p(s) \geq min_i\{p(s_i)\}. \tag{16}$$

The influence on the change of value of individual statement on the change of value of another statement may be modified by adequate weights w assigned do the set of edges a and play important role when the value of secondary statement is established on the base of many primary or others intermediate statements. The weight parameter is defined as:

$$w_k = \{w(a_k)\ \ a_k \in A,\ \ w_k = [0, 1]\} \tag{17}$$

Taking into account the considered weights assigned to individual edges the implications shown in Eqs. (13–16) accept the following form:

$$(s_1 \rightarrow s_2, b(s_1 = < p(s_1), n(s_1) >) \Rightarrow p(s_2) \geq w_{1,2} p(s_1) \tag{18}$$

$$(s_1 \rightarrow s_2, b(s_2 = < p(s_2), n(s_2) >) \Rightarrow n(s_1) \geq w_{2,1} n(s_2) \tag{19}$$

$$((\vee_i s_i) \rightarrow s) \Rightarrow p(s) \geq max_i\{w_k p(s_i)\}, \tag{20}$$

$$((\wedge_i s_i) \rightarrow s) \Rightarrow p(s) \geq min_i\{w_k p(s_i)\}, \tag{21}$$

Weights appearing in the above inequalities depend on the different division of a set of statements into the sets of primary and secondary statements and depending of values of primary statements. The weights assigned to edges in this paper are assumed as one and only free parameters of intuitionistic statement networks. The values of these weights can be determined on the base of:

- opinions of domain engineers, which define the structure of relations between the vertices V (defining the relations between statement values)
- external data which contains information about the historical change of values of individual statements including also the information about statements which contents are related to technical state of considered object.

The second possibility is a subject of further considerations in this paper.

3 State-of-the-Art

Because the efficiency of expert systems based on the use of the intuitionistic statements networks depends on the quality of knowledge, an important factor, is the ability to acquiring this knowledge from different sources. For the predefined model of intuitionistic network, the unknown values of weights parameters are the one and only free parameters that can be established by domain expert or/and on the base learning data from active or passive experiments.

So far, the evaluation of weights on the base of learning data for intuitionistic statement network was not the subject of intense study for these type of networks. Analyzing the field of possible solutions following solutions have been distinguished:

- identification of weights in direct way on the base of available implications in the form of inequality equations and available learning data set,
- evaluation of dependencies between values of statements on the base of different data mining methods without taking into consideration the predefined structure of intuitionistic statement network,
- estimation of weights in indirect way by constructing substitutional model which takes into account previously prepared structure of intuitionistic statement network and thus enabling to calculate weights even for incomplete learning data set.

Determination of weights in a direct way is a method that refers to the definition of intuitionistic statement networks and requires solving the set of inequality equations expressed by a set of implications presented in Eq. (18) and available learning data in the form of intuitionistic values. The main disadvantage of applying this approach is inability to take into account incomplete data due to the fact that intuitionistic statement network is not generative model. Furthermore, the use of this method leads obtaining weights in the form of interval values.

Similar disadvantage is characterized by other different data mining methods which make it possible to evaluate the impact between the considered statements but, which do not have possibilities to include predefined structures of dependency between the statements expressed in the form of an intuitionistic statement network.

The third approach, which is previously defined main objective of this paper consist on the use of the equivalent substitutional model in the sense of the same dependencies or independencies between considered statements, as it is, in the original graph. If the substitutional model has the generative properties, then it makes it possible to take into account incomplete data while learning. One of the main disadvantage of this approach is the necessity to transform the intuitionistic statement network in the form of substitutional graph representation.

4 Method of Learning Weights

4.1 General Conditions

Particular attention while searching for a candidate for substitutional model was paid to the family of graphical network models, which also allow to represent the relationship between statements in the form of a graph. For this purpose belief network model was chosen. The structure of the relationship between the statements considered in belief networks is examined through the prism of conditional independencies and dependencies between statements.

The following general conditions for the transformation of intuitionistic statements networks into the form of equivalent substitutional network are formulated as follows:

- (a) substitutional model Z_{in} have to contain the same statements as in original model G_{in} although the constant part of original model, e.g. set E, which relates to edges connecting contents with values of statements, may be omitted. Every statement in the substitutional model should be represented by one vertex with only two mutually exclusive possible states of values e.g. true and false.
- (b) dependencies and/or the conditional dependencies in the original model have to be equivalent to the dependencies and/or conditional dependencies in substitutional model Z_{in}.

To satisfy the condition (a) defining the set of the content variants of statements **c** is sufficient. If there are no semantically opposite contents of statements then there is no reason to represents the edges E of original model by substitutional model.

The condition (b) requires that the conditional dependencies and/or dependencies between the statements in the intuitionistic statement network were also met in a substitutional network without going into the nature of this relationship. This means that it is not important whether it is a causal relationship, implicational based on adequate and sufficient conditions, but it is important that the same independencies between the statements will also be preserved in a substitutional model.

4.2 Conditional Dependencies in Intuitionistic Statement Networks

Dependencies between the statements of intuitionistic networks (specifically the values of these statements) are conditional. For disjoint sets A, B, C, whose elements are pairs of values $b(s_i) = \{p(s_i), n(s_i)\}$ of intuitionistic statements s_i and whose sets A and B contain pairs of values of secondary statements and set C contains pairs of values of primary statements the following condition can be formulated.

If the pairs of statements belonging to the set A and B are independent for all instances of values occurring in pairs of statements in set C, and are dependent once the elements of set C belong to the secondary statements, then a pair of statements belonging to a set of A and B are conditionally independent with respect to C and is denoted as $ind = A \perp B|C$. Described conditional independence in intuitionistic statements networks can also be interpreted as follows:

- conditional independence occurs when there is no path connecting any pair of statements between the set of A and B, where a path from s_x to s_z designated $s_x \rightarrow s_z$ of a graph of intuitionistic statement network is named any connection between pairs of $\{p(s_x), n(s_x)\}$ and $\{p(s_z), n(s_z)\}$ and any intermediate pairs of values that occur on the path according to the direction of the edge, and that none of these indirect pairs do not belong to the set of primary statements.

The structure of the proposed substitutional model is builded on a complete set of the independence set $IND_{in} = \{ind_1, ind_2, \ldots\}$ coming from intuitionistic model when the primary statements receiving only the accurate external data. For these parts of the substitutional network, whose primary statements accept approximate values, the construction of the substitutional graph is based on the applying of the same interactions between primary and selected secondary statements. In order to achieve some set of independencies or the same interactions between primary and secondary statements in belief networks, they require to take into account the conditions of conditional independencies. There are known generally two criteria designed for acyclic directed graph including belief networks which relate to this issue: the global criterion Markov criterion d-separation [7, 8].

According to the d-separation criteria the dependencies between vertices associated with statements S_x and S_z in the substitutional statement networks occur once the following conditions are satisfied [6]:

- S_x and S_z occur in serial $S_x \rightarrow S_y \rightarrow S_z$ or divergence $S_x \leftarrow S_y \rightarrow S_z$ structure of the substitutional model graph and there is no other intermediate vertex between S_x and S_z which has the established accurate value,
- S_x and S_z occur in the convergence structure $S_x \rightarrow S_y \leftarrow S_z$ of the substitutional model and intermediate vertices or its descendant have known values in the form of accurate values (hard evidences).

Due to the necessity of determining the weight values for cases in which the original primary statements accept approximate values which correspond to the soft evidence the above mentioned conditions do not provide a conditional independence between the statements, as described above. The primary statements with approximate values do not in fact block the passing information in belief networks when the vertex is an intermediate vertex regardless of the structure in which it occurs: a convergent, divergent or serial structure [6].

It also means that values of primary statements with approximate values can be affected by values of other primary statements. In order to avoid this and preserve the condition of independencies between primary statements which accept external independent sources of data,

$$S_{fix,i} \perp S_{fix,j} \quad for \quad i \neq j, \tag{22}$$

only interactions between primary and secondary statements should be taken into account. For this purpose the following rule for constructing substitutional network should be applied:

- (a) for every pair of values of primary statement in intuitionistic statement network s_i, $\{p(s_i), n(s_i)\}$ with approximate value (with soft evidence) apply a vertex as a root in the divergence structure of the substitutional graph,

Additionally for remaining pair of values of secondary statements and pair of values of primary statements with accurate values (hard evidences) the following rules should be applied with the following order:

- (b) for every pair of values of primary statement s_i, $\{p(s_i), n(s_i)\}$ with known accurate values (hard evidence), which in the original graph are placed as intermediate vertices (or they are leafs or roots of this graph) apply a divergence or serial structure in the substitutional model,
- (c) for every pair of vertices of secondary statements which in the original graph are placed as intermediate vertices (or they are leafs or roots of this graph) apply convergence structure in substitutional model.

This approach does not guarantee the independence between the primary statements when they are neighbors. In order to achieve independency between them additional vertex not associated with any statement acting as additional separator is needed.

Application of the above rules is limited only to the case in which between the connected pairs of statements in intuitionistic statement network both a sufficient and necessary condition occur. In the case where it is satisfied only one condition, the interaction between statements cannot be reflected in belief networks in which the relationship between the vertices is bidirectional regardless of the direction of an edge.

Figure 2 shows an example of transformation of the network structure of intuitionistic statement network into the structure of substitutional belief networks. In the presented example the different rules were applied depending on the accurate or approximate values of primary statements. For those parts of the network that are located in the area of impact of primary statement with accurate value rule b and c were applied. For other parts of the network which are under the impact of primary statements with approximate values rule a and c were applied.

4.3 Mapping Intuitionistic Value to Belief Value

In order to calculate the weight values based on substitutional model, in which the primary statements values has the form of point value it is necessary to determine the function to map intuitionistic values into point values representing the degrees of belief in substitutional network. Comparing the accurate values expressed by intuitionistic values and belief degree it can be noted that:

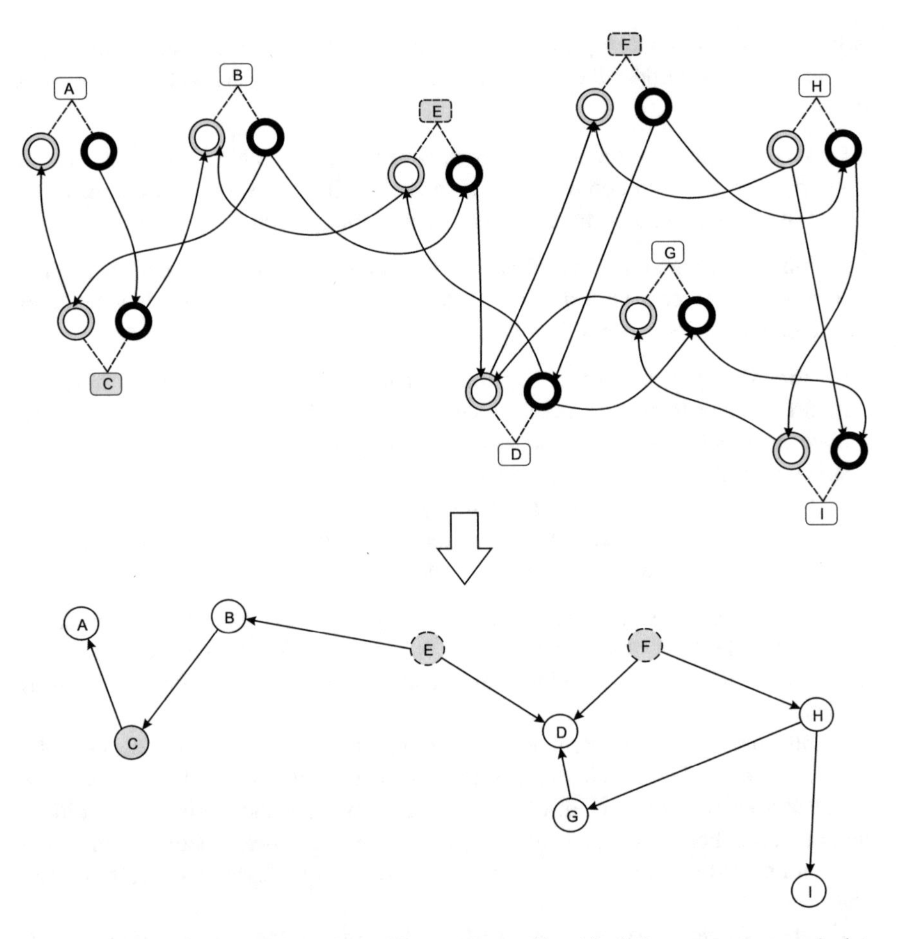

Fig. 2 An example of transformation of the structure of intuitionistic statement networks into the structure of substitutional network represented by belief network. The vertices colored in *gray* corresponds to the primary statements. Primary statements which accept accurate values are depicted by *solid line* around the vertex and the primary statements which accept approximate results are surrounded by *dashed lines*

$$
\begin{aligned}
P(s_{i,BN} = true) = 1.0 \quad corresponds \;\; to \quad b(s_{i,ISN}) = < 1.0, 0.0 >, \\
P(s_{i,BN} = false) = 1.0 \quad corresponds \;\; to \quad b(s_{i,ISN}) = < 0.0, 0.1 > .
\end{aligned}
\tag{23}
$$

Additionally looking for equivalent description of intuitionistic value and belief value for state which corresponds to unknown information about the true or false of considered statement it can be noted that:

$$
P(s_{i,BN} = true) = 0.5 \;\; corresponds \;\; to \;\; b(s_{i,ISN}) = < 0.5, 0.5 >, \tag{24}
$$

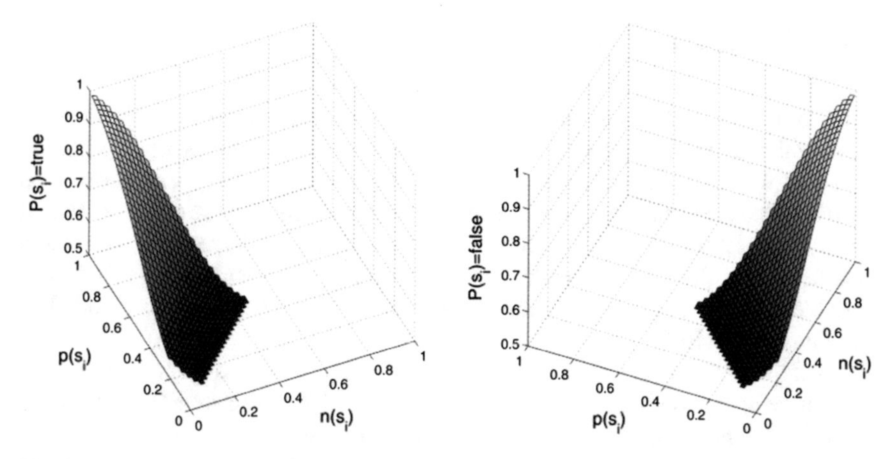

Fig. 3 Example mapping functions of intuitionistic value to belief value. On *left* for $n(s) - p(s) > 0$. On *right* for $n(s) - p(s) < 0$

and which can be extended to the range of intuitionistic values which satisfy the following equality:

$$p(s) - n(s) = 0 \quad for \ \ 0.2 \leq p(s) + n(s) \leq 1.0. \tag{25}$$

It is assumed that the intuitionistic values in the remaining space can be mapped to the point value in the following way (Fig. 3):

$$P(s_i = true) = \begin{cases} -\frac{1}{4}cos(n(s_i) - p(s_i)) + \frac{3}{4}, & (n(s_i) - p(s_i)) > 0; \\ 1 - P(s_i = false), & (n(s_i) - p(s_i)) < 0. \end{cases} \tag{26}$$

$$P(s_i = false) = \begin{cases} -\frac{1}{4}cos(n(s_i) - p(s_i)) + \frac{3}{4}, & (n(s_i) - p(s_i)) < 0; \\ 1 - P(s_i = true), & (n(s_i) - p(s_i)) > 0. \end{cases} \tag{27}$$

$$P(s_i = 0.5) = \left\{ -\frac{1}{4}cos(n(s_i) - p(s_i)) + \frac{3}{4}, \ (n(s_i) - p(s_i)) = 0. \right. \tag{28}$$

The presented mapping functions satisfies the conditions presented in Eqs. (23–25) affect only the intuitionistic space constrained by the following inequalities:

$$0.2 \leq p(s_i) + n(s_i) \leq 1.0. \tag{29}$$

4.4 Calculation of Weights

For a known set of learning data of historical value of statements, it is possible to identify the network parameters of substitutional model. Procedure for determining values of the network parameters represented by a Bayesian networks is based on the use of the counting procedures or other learning methods of learning conditional probabilities tables assigned to individual vertices. Details of the methods of learning parameters for Bayesian networks can be found among other in [6, 7]. It is also possible to learn network parameters for the case of complete as well as incomplete learning data set [6].

For the predefined network structure of substitutional model it is possible to run the procedure for calculating the weights for individual edges. Procedure for calculating the weights which comprises the cases where the original statements takes accurate or approximate values is as follows:

- transform continuous space of intuitionistic values of primary statements into discrete counterparts,
$$(p(s), n(s)) \rightarrow (p(s), n(s))_d, \tag{30}$$
in the area of $0.2 > (p(s) + n(s)) > 1$,
- for every discrete subarea of this space apply mapping function,
$$vb(s) = \{f((p(s), n(s))_d) : (p(s), n(s))_d \rightarrow [0, 1]\} \tag{31}$$
which meet the requirements presented in Eqs. 23–25 e.g. function shown in Sect. 4.3
- for each instance of discrete value $vb(s)_i$ where $vb(s)_i = \{vb(s_i) : s_i \in S_{primary}\}$ means the value of statement represented by the belief degree in Bayesian network,
  - calculate posterior probabilities for jth secondary statements,
  $$P(vb_j)|vb_i) \tag{32}$$
  where $\{vb(s_j) : s_j \in S/\, S_{primary}\}$
  - for each kth primary statement vb_k where $vb(s)_k = \{vb(s_k) : s_k \in S_{primary}\}$ and $k \neq i$.
  Calculate
  $$d_{j,k} = P(vb_j|vb_i/vb_k) \quad where \quad P(vb_k) = 0.5, \tag{33}$$
- for every secondary statement, which value affects the greater number of primary statements, the weight assigned to edge between the statement of jth and kth statement is determined on the basis of:
$$w_{j,k} = \frac{d(j,k)}{\sum_{n=1}^{M} d(j,n)} = \frac{P(vb_j|vb_i/vb_k)}{\sum_{n=1}^{M} P(vb_j|vb_i/vb_n)} \tag{34}$$

where n is the number of edges directed towards the statement jth from other primary statements or other secondary statements which appears as intermediate vertex between kth primary statements and considered jth secondary statement.

4.5 The Operation of Intuitionistic Statement Networks with Edge Weights

Because the values of weights assigned to each edge are functions of configuration of division of statements to primary or secondary subsets, and depends on the value of the primary statements they need to by dynamically changed during the use of intuitionistic statement networks. As a rule, for diagnostic purposes the set of primary statements is usually known primary. Due to this fact more attention will be paid to the appropriate assignment of edge weights depending on the value of the primary statements.

In Fig. 1 a space of intuitionistic values is shown. Depending on primary statement kth and considered secondary statement jth the following edge weighs are assigned to individual edges:

$$w(j,k)_t = \begin{cases} 0, & 0.2 < (p(s_k) + n(s_k)); \\ \in [0, 1], & 0.2 \leq (p(s_k) + n(s_k)) \leq 1.0; \\ w(j,k)_{t-1}, & (p(s_k) + n(s_k)) > 1.0. \end{cases} \quad (35)$$

where the values [0; 1] are established on the base of procedure which is shown in Sect. 4.4. The weight $w(j,k)_t = w(j,k)_{t-1}$ depicts, that the value of weight at the time t does not change and equals to the weight from the previously calculated weight because once some contradictions according to Eq. (2) in input data occurred.

5 Example of Application

This section provides an example of application of using method presented. The method is shown on an example of application of intuitionistic statement network applied for the purpose of simple diagnostics of car battery. The network model of intuitionistic statement network is shown in Fig. 4. The network comprises of the following set of statements:

- S_{00} Car battery works properly.
- S_{01} Excessive voltage drop during engine start.
- S_{02} Charging voltage during engine operation is low.
- S_{03} Car battery is discharged.
- S_{04} Car battery is worn.
- S_{05} Internal impedance of car battery is too high.

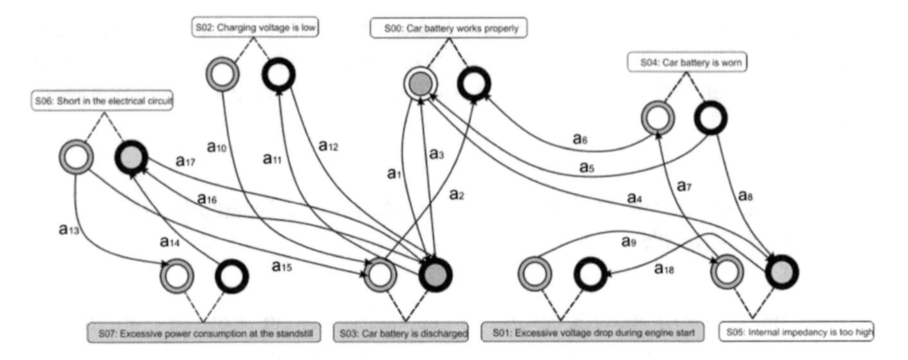

Fig. 4 Intuitionistic statement network for car battery diagnostics

- S_{06} Short in the electrical internal circuit.
- S_{07} Excessive power consumption at the standstill.

The division into the set of primary and secondary statements is as follows:

$$S_{07}, S_{03}, S_{01} \in S_{primary}$$
$$S_{secondary} \in S/S_{primary}$$

In the considered example, the edges defined between vertices of values of every pair of associated statements represents both sufficient implication and necessary implication. Analysis of conditional independencies in intuitionistic network model indicates the presence of following independencies:

$$b(S_{07}), b(S_{06}) \perp b(S_{02}), (b(S_{00}), b(S_{04}), b(S_{05}), b(S_{01})|b(S_{03})$$
$$b(S_{02}) \perp b(S_{07}), b(S_{06}), (b(S_{00}), b(S_{04}), b(S_{05}), b(S_{01})|b(S_{03})$$
$$b(S_{04}), b(S_{05}), b(S_{01}), b(S_{00}) \perp b(S_{07}), b(S_{06}), b(S_{02})|b(S_{03})$$

which means that primary statement S_{07} in this case has an influence only on the statement S_{06}, the primary statement S_{01} has influence on statements S_{04}, S_{05}, S_{00}, while the primary statement S_{03} has an influence on statements S_{00}, S_{02}, S_{06}, S_{04}, S_{05}.

Having a set of point values which comes from the maintenance events of the considered class of car batteries, these data can be used for the purpose of weights parameter identification. Available learning set is presented in Table 1.

Using the rules about building a substitutional model based on the information on conditional independencies and dependencies in intuitionistic statements networks a new structure of substitutional model represented by Bayesian network can be obtained. The structure of the graph of this model is shown in Fig. 5. Due to the fact that in the considered example the values of primary statements could have approximate values, primary statements are placed as root vertices in the divergent

Table 1 Set of learning examples

Case	S_{00}	S_{01}	S_{02}	S_{03}	S_{04}	S_{05}	S_{06}	S_{07}	No.
1	Yes	No	No	No	No	No	No	No	9203
2	Yes	Yes	No	No	No	No	No	No	190
3	Yes	No	Yes	No	No	No	No	No	12
4	No	No	Yes	Yes	No	No	No	Yes	92
5	No	Yes	No	No	Yes	Yes	No	No	12
6	No	Yes	No	Yes	Yes	Yes	No	No	50
7	No	No	No	Yes	No	No	Yes	Yes	21
8	No	Yes	No	Yes	Yes	Yes	Yes	Yes	3

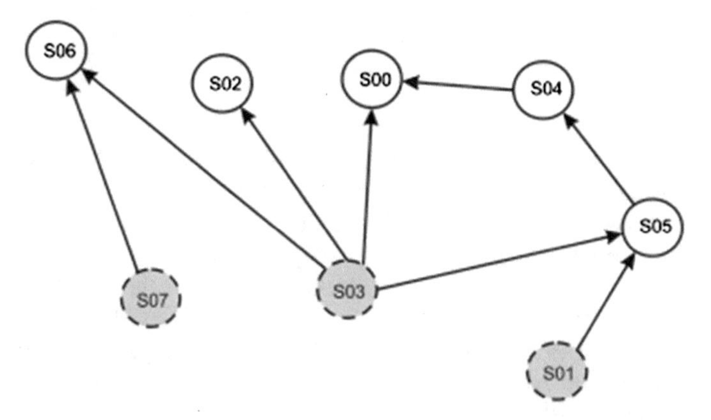

Fig. 5 Structure of substitutional graph intended to car battery diagnostics

structure. For all other statements that are not neighbors of vertices connected with primary statements a serial structure is applied. The collection of independence conditions between statements in substitutional network is as follows:

$$\begin{aligned} &b(S_{07})_{bn} \perp b(S_{03})_{bn}, b(S_{00})_{bn}, b(S_{04})_{bn}, b(S_{05})_{bn}, b(S_{01})_{bn}, b(S_{02})_{bn}) \\ &b(S_{03})_{bn} \perp b(S_{07})_{bn}, b(S_{01})_{bn} \\ &b(S_{01})_{bn} \perp b(S_{07})_{bn}, b(S_{06})_{bn}, b(S_{03})_{bn}, b(S_{02})_{bn} \end{aligned} \tag{36}$$

The above set of independencies shows, which values of the secondary statements are not determined by specific primary statements. Comparative analysis of this set with the set of conditional independencies of intuitionistic statement networks shows that the set of interactions between individual primary statements and selected subsets of secondary statements are the same.

After the elaboration of the structure of substitutional model, the discretization of intuitionistic space of value of primary statements can be performed. Discretization of such spaces has been done according to division presented in Fig. 6. Next, apply-

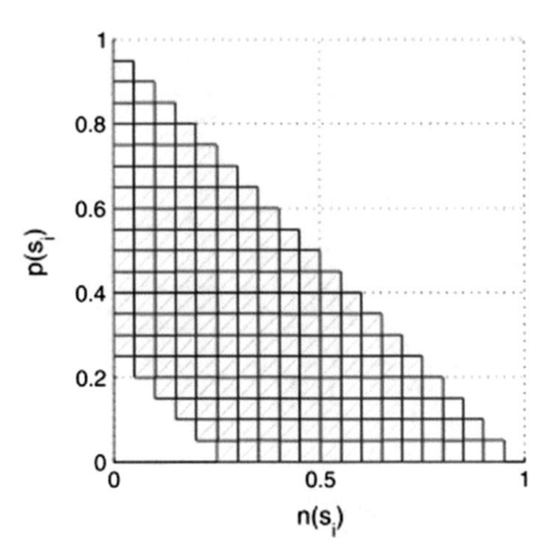

Fig. 6 Discrete division of the space of intuitionistic primary statement

ing the mapping function presented in Eqs. (26–28), new point values of primary statements in substitutional model are obtained.

Figure 7 shows the results obtained for weights for edges a_{14} and a_{16} which connect primary statements $S07$ and $S03$ with secondary statement $S06$. It may be noted that the impact on the values of statement $S06$ is the highest when the primary statements have both values (for AND vertex) or individual values (for OR vertex)—close to the higher degree of validity of statement $S07$ and/or high degree of nonvalidity for statement $S03$. On the base of further analysis of weights variability a sensitivity of secondary statement $S06$ on the values of primary statements $S07$ i $S03$ can be carried out. It may be also noted that the value of primary statement $S06$ is high sensitive to

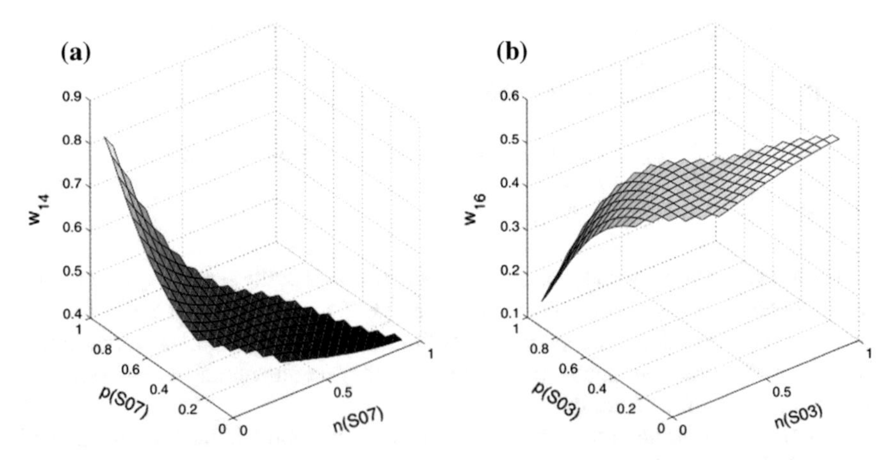

Fig. 7 Results of weights assigned to edges a_{14} and a_{16} which results from the soft evidences of primary statements $S07$ (on *left*) and $S03$ (on *right*) on the secondary statement $S06$

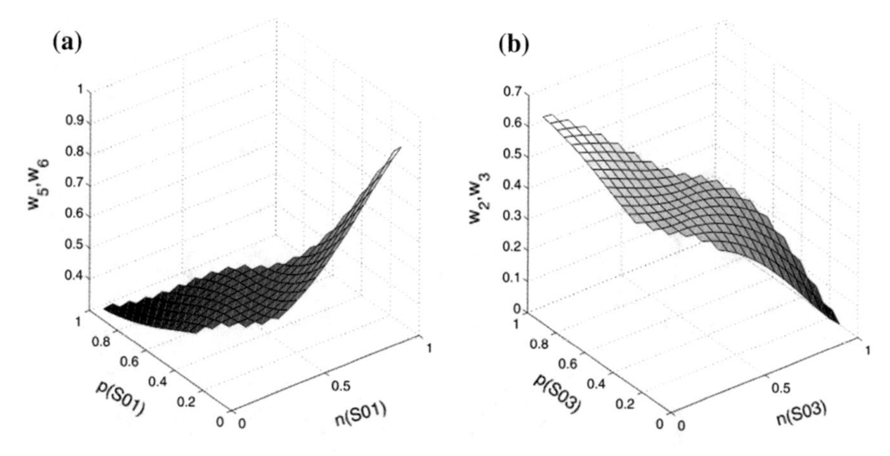

Fig. 8 Results of weights assigned to edges a_2, a_3 and a_5, a_6 which results from the soft evidences of primary statements $S01$ (in **a**) and $S03$ (in **b**) on the secondary statement $S00$

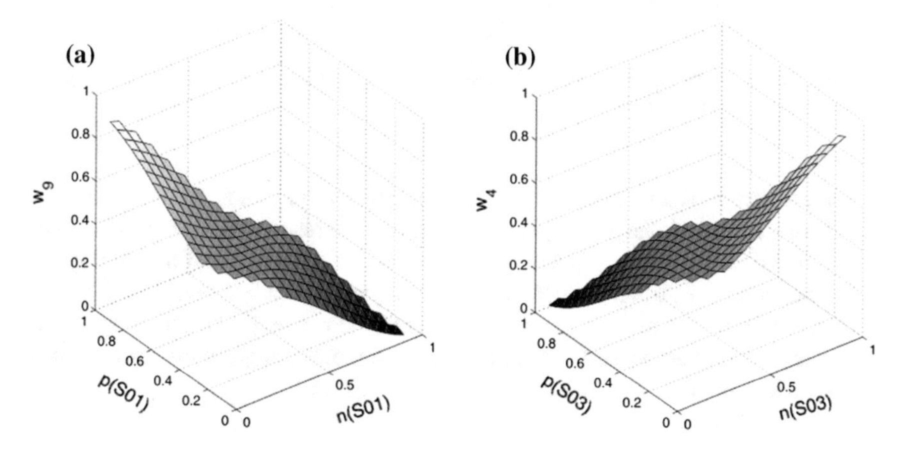

Fig. 9 Results of weights assigned to edges a_9 and a_4 which results from the soft evidences of primary statements $S01$ (in **a**) and $S03$ (in **b**) on the secondary statement $S05$

change of primary statements in the area which corresponds to high values of degree of validity of statements $S03$ and $S07$.

Figure 8 presents the results obtained for weights of edges a_5, a_6 and a_2, a_3. The analysis of values of these weights indicate that the main impact on value of the secondary statement $S00$ is the highest when the primary statements have both values (for AND vertex of $S00$) or individual values (for OR vertex of $S00$)—close to the higher degree of nonvalidity of statement $S01$ and/or high degree of validity for statement $S03$. Another results of impact of statements $S01$ and $S03$ on statement $S05$ is presented in Fig. 9.

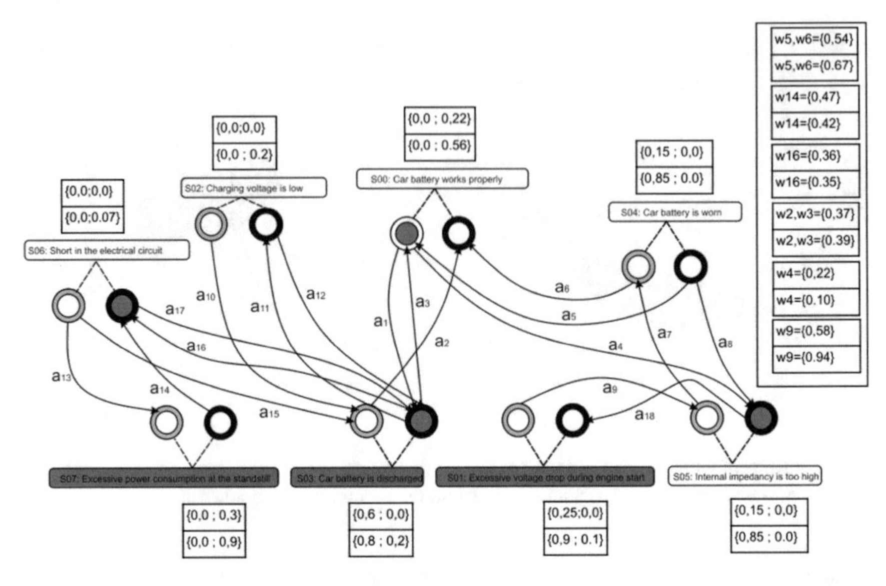

Fig. 10 Results of operation of intuitionistic statement network for selected outcomes of primary statements and calculated weights

Others weight values assigned to the other edges are equal to 1 as they concern only to the case when one primary statement impact the result of secondary statement.

Figure 10 shows selected results for secondary statements for different approximate outcomes of primary statements. The values of calculated weights for these outcomes are presented in the upper right side of the figure. In the subsequent rows of tables of values associated with individual statements are given the result values. The values of not presented weights values are equal to 1.

6 Conclusion

The paper presents the method of calculating weights assigned to the edges of intuitionistic statement networks on the base of learning data. The values of weights are computed for different values of primary statements associated with external sources of observed data. The results of these weights depend also on the different division of statement set into subsets of primary and secondary statements which shows seeking values of diagnostic inference. Because the division into subsets of primary ad secondary statements do not often change during the operation of the intuitionistic statement networks particular attention was paid to the description of determining weights for different values of the primary statements. Values of primary statements can appear as accurate values which correspond to logical true or false or they can appear as approximate results. The presented method allow to identification of weights values for every instance of evidence regardless whether it is accurate or approximate value. When the values of weights are unknown the impact of individual primary statements on secondary statements is equal and depends only on the values of observed values. By calculating the values of weights of intuitionistic statement networks for every instance of approximate or accurate value of evidence the impact of individual primary statements on the secondary statements can be determined.

The described method of calculating weights of values for intuitionistic statement networks is based on the basis of substitutional network model. The substitutional model is equivalent to the intuitionistic networks in the sense of the same subsets of interactions between individual primary statements with approximate observed values and secondary statements as well as in the sense of conditional dependencies in the case when the primary statements have accurate observed values. The substitutional model is represented by belief network model. Thanks to the known method of parameter learning for Bayesian networks, the conditional probability tables can be easily determined with the use of available learning data set. Next, the application of the described method makes it possible to calculate the values of weights for intuitionistic statement networks for every instance of value of primary statements. The presented method is simple but computationally demanded due to the necessity of multiple substitutional model recalculation. One of the main advantage of presented method is the possibility of learning weights parameters on the base of incomplete learning data set.

Acknowledgments Described herein are selected results of study, supported partly from the budget of Research Task No. 4 entitled *Development of integrated technologies to manufacture fuels and energy from biomass, agricultural waste and others* implemented under The National Centre for Research and Development (NCBiR) in Poland and ENERGA SA strategic program of scientific research and development entitled *Advanced technologies of generating energy.*

References

1. Atanassov, K.T.: On Intuitionistic Fuzzy Sets Theory. Springer, Berlin (2012)
2. Cholewa, W.: Expert systems in technical diagnostics. In: Korbicz, J., et al. (eds.) Fault Diagnosis. Models, Artificial Inteligence, Applications, pp. 591–631. Springer, Berlin (2004)
3. Cholewa, W.: Multimodal statement networks for diagnostic applications. In: Sas, P., Bergen, B. (eds.) Proceedings of the International Conference on Noise and Vibration Engineering ISMA 2010, 20–22 September, pp. 817–830. Katholique Universiteit Lueven (2010)
4. Cholewa, W.: Gradual forgetting operator in intuitionistic statement networks. In: Angelov, P., Attassov, K.T., Kacprzyk, J., et al. (eds.) Proceedings of the 7th IEEE International Conference Intelligent Systems IS'2014, 24–26 September. Intelligent Systems: Mathematical Foundations, Theory, Analyses, Warsaw, vol. 1, pp. 613–620 (2014)
5. Cholewa, W.: Dynamical statement network. In: Awrejcewicz, J. (ed.) Applied Non-Linear Dynamical Systems. Springer Proceedings in Mathematics & Statistics, vol. 93, pp. 351–361. Springer, Berlin (2014)
6. Jensen, V.J.: Bayesian Networks and Decision Graphs. Springer, New York (2002)
7. Lauritzen, S.L., Spiegelhalter, D.J.: Local computations with probabilities on graphical structures and their application to expert systems. J. R. Stat. Soc. **50**(2), 157–224 (1998)

8. Pearl, J.: Probabilistic Reasoning in Intelligent Systems: Networks of Plausible Inference. Morgan Kaufmann, San Francisco (1988)
9. Rogala, T., Amarowicz, M.: Multimodal statement networks for organic rankine cycle diagnostics - use case of diagnostic modeling. In: Angelov, P. and Attassov, K.T., Kacprzyk, J., et al. (eds.) Proceedings of the 7th IEEE International Conference Intelligent Systems IS'2014, 24–26 September. Intelligent Systems: Mathematical Foundations, Theory, Analyses, Warsaw, vol. 1, pp. 605–612 (2014)

Combined Data and Execution Flow Host Intrusion Detection Using Machine Learning

Tajjeeddine Rachidi, Oualid Koucham and Nasser Assem

Abstract We present in this chapter a novel method for detecting intrusion into host systems that combines both data and execution flow of programs. To do this, we use sequences of system call traces produced by the host's kernel, together with their arguments. The latter are further augmented with contextual information and domain-level knowledge in the form of signatures, and used to generate clusters for each individual system call, and for each application type. The argument-driven cluster models are then used to rewrite process sequences of system calls, and the rewritten sequences are fed to a naïve Bayes classifier that builds class conditional probabilities from Markov modeling of system call sequences, thus capturing execution flow. The domain level knowledge augments our machine learning-based detection technique with capabilities of deep packet inspection capabilities usually found, until now, in network intrusion detection systems. We provide the results for the clustering phase, together with their validation using the Silhouette width, the cross-validation technique, and a manual analysis of the produced clusters on the 1999 DARPA data-set from the MIT Lincoln Lab.

Keywords Host intrusion detection systems · System calls · System call arguments · Clustering · Classification

1 Introduction

Intrusion occurs when a legitimate program or application is compromised through one of its vulnerabilities. This occurs (i) when a virus attaches itself to the program at rest on disk, (ii) at loading time when the program dynamically links to

T. Rachidi (✉) · O. Koucham · N. Assem
Al Akhawayn University in Ifrane, 53000 Ifrane, Morocco
e-mail: T.Rachidi@aui.ma

O. Koucham
e-mail: O.Koucham@aui.ma

N. Assem
e-mail: N.Assem@aui.ma

Y. Bi et al. (eds.), *Intelligent Systems and Applications*,
Studies in Computational Intelligence 650, DOI 10.1007/978-3-319-33386-1_21

unsafe libraries [1], or (iii) when program flaws such as buffer overflows [2] are exploited, using stack smashing [3] like techniques, by a remote malicious adversary to pass on malware/payload to the application at run-time. The passed/injected payload becomes then part of the application underlying run-time process. Since the life-cycle of a typical intruding malware consists in activities such as discovery, replication (pro-creation), disguise, privilege escalation, and/or assault, this translates into basic operations such as repeated scan of the network, repeated access to the file system and/or boot sector, spawning of offspring etc. that are not part of the original intended activity of the application program. *Host* intrusion detection systems (HIDS) must therefore be able to detect the unusual activity within a host, flag it as intrusive, and trigger a response agent.

Contrary to *Network* Intrusion Detection Systems (NIDS), which consist in placing sensors in selected places on the network to inspect network packets for anomalous traffic that can be part of an intrusion [4], HIDS are deployed as software agents on a host machine; and rely on process/program execution activity logs, system logs, and resource usage within the host to perform detection, and to trigger a response from a response agent according to user defined policies.

While NIDS can detect a wide range of attacks such as port scans, distributed denial of service (DDOS) attacks, and vulnerability exploit attempts [4], which HIDS cannot do, they are limited when dealing with encrypted network communications, which effectively blind the NIDS [5]. This is a crucial limitation in future IPv6 networks, where all traffic is encrypted and eventually encapsulated/tunneled [5]. HIDS, on the other hand, circumvent the encrypted and tunneled network communication problems, since they operate on clear data, eventually coming through the network, after they have been processed by the TCP/IP stack of the target host.

Through activity logs produced by the host's kernel, HIDS are therefore able to trace actions and monitor the behavior of each individual user/process/program. They are also able to operate naturally on semantically richer and higher quality data about the system they are monitoring, while NIDS have to parse, reassemble, and interpret the traffic in order to analyze application-level information [5, 6] without any kind of knowledge about the target system. This makes HIDS less vulnerable to evasion or stealthy techniques because of the difficulty to desynchronize the view that the intrusion detection system has of the monitored application [6]. Last but not least, HIDS can perform a more precise response [6] since the compomized process is easily identified within its execution environment.

Traditionally, HIDS have been classified into *anomaly-based* or *misuse-based* systems depending on the approach used to detecting the intrusive behavior. Misuse-based intrusion detection systems rely on a set of signatures manually generated by domain experts and stored in a knowledge base, in order to directly recognize suspicious behaviors [7]. Anomaly-based intrusion detection systems, on the other hand, create a model for the normal behavior of the monitored systems or of its users called a profile [8]. The assumption is that any behavior that deviates significantly from the normal profile is flagged as suspicious/intrusive.

Each of these approaches has its own advantages and disadvantages: Misuse-based intrusion detection systems depend strongly on the quality of the knowledge base they rely upon. Signatures that are too specific may not capture variations of attacks, while signatures that are too general can trigger a high rate of false alerts [8]. More importantly, such systems can miss novel or polymorphic attacks [8], as it becomes quickly intractable to keep up with the ever-increasing number of vulnerabilities [9]. Yet, misuse-based intrusion detection systems have significant characterization power, and can generally pinpoint the exact nature of an attack, and when it is triggered. Anomaly-based intrusion detection systems, on the other hand, do not require an up-to-date knowledge base in order to keep up with new vulnerabilities and therefore can cope with zero-day type of attacks [8]. Their advantage lies in their ability to learn and classify behaviors. Having said this, such systems require a long phase of training in order to learn/build the model for the normal profile. An attacker who slowly changes its behavior can make the anomaly-based HIDS learn the anomalous behavior and fail to detect attacks. Moreover, such systems do not react well to sudden changes in the environment leading to a significant rise in the number of false alerts. Despite these shortcomings, anomaly-based HIDS are considered by many as the way forward, for they are able to cope with the exponential number of threats, while circumventing the encryption challenge.

Clearly, the modeling of user application and/or process behavior [10] from the set of audit trails and system logs is at the heart of anomaly-based HIDS. HIDS operating on system call traces have proven to be particularly effective, as requests made by processes to the operating system are considered a good basis for capturing the behavior of a process/application/user [11]. Moreover, system call analysis allows for direct access to application clear data coming from the network or the standard input device through *read()* and *scanf()* like call arguments.

In this chapter, we investigate a host-based anomaly intrusion detection system that takes advantage of system calls, their arguments, host contextual information and domain-level knowledge in order to build better profiles of the monitored processes. Relying on supervised machine learning techniques, we aim at better capturing the behavior of processes by constructing both normal and intrusive process profiles from sequences of system calls and their arguments combining both data and flow techniques. Specifically, we use system call arguments such as names of files accessed, the length and the nature of the passed data themselves, together with host contextual information such as file access rights, process user ids; and domain-level knowledge in the form of signatures, expressed as regular expressions, to build clusters for each system call through adequate metrics that we developed for this purpose. System call sequences are then rewritten using the obtained clustering models. The new sequences are then fed to a naïve Bayes supervised classifier (SC2.2) [12] that builds class conditional probabilities from Markov modeling of very long system call sequences.

The key assumption behind our two-phase approach is that clustering of system calls based on arguments, contextual information, and domain knowledge will distinguish between instances of the same system call (activities) within a process based on the data the system call operates on, allowing for learning a more precise profile of intrusive processes and therefore minimizing the ability of intruders to evade intrusion detection and decreases the rate of false alerts. In a sense, using arguments of system calls together with the domain-level knowledge (such as known sequences of shell code bytes) give extra information that only some signature-based deep packet inspection techniques in NIDS can perform on clear traffic.

The rest of the chapter is organized as follows: Sect. 2 reviews related work. Section 3 presents our proposed system architecture. Sections 4 and 5 present the details of the data-driven clustering and flow-driven classification stages respectively. The experimental results on the 1999 DARPA intrusion detection data-set are presented in Sect. 7. Finally, Sect. 8 concludes the chapter.

2 Related Work

One of the earliest attempts to use process system calls for intrusion detection is described in [11]. Therein, system calls were shown to be a reliable mean to characterize processes, and thus to build profiles of normal behavior. Specifically, short-range correlations between system calls in a process were used to capture its normal behavior. Deviations from this profile were considered to be signs of potential attacks/intrusions. The proposed method scans traces of normal behavior and builds up a database using a sliding window that records local system calls patterns. As the authors outlined, this definition of normal behavior does not take into account several aspects of the process behavior such as parameter values passed to the system calls. Further works capitalized on this first approach by introducing new methods based on relative frequency comparison [13], rule induction techniques [13] or Hidden Markov Models (HMMs) [13, 14]. While results show that HMMs can generate accurate results on average, they remain demanding computationally because of the convergence process used to reach the optimal profile (HMM). Suggestions to improve the performance of HMM-based intrusion detection such as merging multiple sub-HMMs have been discussed [14]. It is not sure, however, how combinations of models obtained from disjoint sub-sequences yield a final model that correctly accounts for long sequences. Furthermore, the learning has been made out of normal sequences exclusively. The underlying assumption is that all that is not normal is intrusive. This approach does not capture reality where various levels of threat are needed to guide the response agents to appropriate action.

A novel method that introduces system call arguments analysis to improve intrusion detection was exposed in [15]. The authors build several models to characterize system call arguments, and identify anomalous occurrences. An anomaly score is then computed, which combines the probability values returned by each model, and is compared against user adjustable thresholds. In this approach the order of occurrence of system calls is not taken into account, meaning that the detection is solely based on the arguments of system calls. Consequently, attacks that do not significantly affect system call arguments yet introduce variations in the usual sequence of system calls invocations are not detected. Building on [15], [16] improves the solution by removing inefficient models and introducing the notion of sequence of system calls through Markov chains. In there, the authors introduce also clustering in order to construct system call clusters and perform classification on the clusters. The clusters are learned exclusively from the positive cases (normal executions) and classification is thus done based on a threshold rather than by comparing two classes. Only system call arguments are considered and no contextual information or domain-level knowledge is used. While using unsupervised learning exclusively from the positive cases is more practical as negative cases are not always easy to find, we believe that performing clustering on positive and negative cases followed by supervised learning theoretically produces better results in terms of detection granularity.

More recently, [17] used system manipulation call patterns and semantics to perform detection and to ensure portability of the technique across different operating systems. The approach has the merit of being tested on a newly created data-set *The ADFA Linux data-set* for modern operating systems, containing modern attacks. A three-phase detection approach based on classification of system calls has also been proposed in [18]. Finally, HIDS that leverage system calls have been proposed for specific environments, such as distributed applications running on collaborative nodes [19], embedded systems [20], and the could [21].

In our approach, we focus on traditional host environment, and use ordered sequences of system calls and their arguments, together with contextual information and domain-level knowledge to build a system for classifying sequences of system calls made by processes into two distinct classes "intrusive" or "normal". Our approach therefore combines both data and flow models for detection with capabilities of deep packet inspection techniques available in NIDS through the use of domain-level knowledge. We believe this last feature constitutes a step towards the unification of not only NIDS and HIDS, but also signature-based and anomaly-based approaches.

3 System Overview

Our approach consists in two-phases: A data-driven clustering phase and a flow-driven classification phase. The clustering computes clusters per system call and per process. These clusters reflect the differences in system call instances throughout a single process: Depending on the argument passed to it: the *read()* system call, for instance, can be part of an intrusion or a normal user/application behavior. We also use key arguments that have never been used before, and build the necessary distances on carefully categorized attributes resulting in meaningful clusters, computed over both the positive and the negative cases. Additionally, we incorporate new attributes based on contextual information such as identifiers and file modes to better characterize the profile of processes. We also introduce domain-level knowledge through distances that mirror common indicators of suspicious behaviors. For instance, it is conventional for most UNIX systems to allow users to create files of any size in the */tmp* directory [22]. This functionality is crucial for programs such as *gcc*, *vi* or *mail* as they need to generate temporary files. Attackers can use this directory to store and execute malicious scripts. To detect such attacks, we rely on metrics based on pattern matching in order to identify clusters of system calls (for example *execve()* calls) that correspond to execution of files in the */tmp* directory. Consequently, system calls that use the */tmp* directory are assigned different clusters and are thus characterized as suspicious by our system. The domain-level knowledge essentially captures intrusion signatures and therefore extends the HIDS with features commonly found in deep inspection NIDS. Using novel attributes for clustering enables us to derive semantically better models for the clusters. Not only do these models reflect differences in system call instances, but also they describe meaningful characterizations of suspicious behaviors based on domain-level knowledge.

The classification phase leverages the obtained clusters and performs classification on the rewritten sequences using SC2.2, which has proven its performance against standard classifiers on sequences of system calls taken from University of New Mexico (UNM) and the Massachusetts Institute of Technology (MIT) Artificial Intelligence Laboratory data sets.

Figure 1 depicts the system architecture of our approach. We shall at first describe in Sect. 4 the clustering process, the elaboration of attributes from call arguments, the choice of metrics for each attribute and the matching process. In Sect. 5, we describe the Markov model we build for sequences of system calls together with classification process.

4 Data-Driven Clustering of System Calls

4.1 Formalism

Every execution of a process p consists of a sequence $s_p = (s_{1p}, s_{2p}, \ldots, s_{np})$ made of system calls $s_{1p}, s_{2p}, \ldots, s_{np} \in S$ where S is the set of all system calls used by the host. For each system call instance s_{ip}, we extract attributes from its arguments and contextual information and construct the tuple $< a_1^{s_{ip}}, a_2^{s_{ip}}, \ldots, a_m^{s_{ip}} >$, where $a_i^{s_{jp}}$ is the (ith) attribute of the (jth) system call s_{jp} belonging to an execution of the process p. In general, attributes can be of several kinds:

Arguments: These are path or file names (e.g., */usr/lib/libadm.so.1*), the set of possible values (e.g., O_RDONLY, O_WRONLY, or O_RDWR file access mode for *open()* system call), discrete numeric values (e.g., start of virtual address mapping in *mmamp()*), or user and group identifiers *uid* and *pid*, etc.

Contextual information: These are call return values, file/device access rights in octal representation, file/device owner and group identifiers, real and effective process owner, or group identifiers.

Domain-level knowledge: These are the set of signatures that can be applied on attributes (e.g., pattern matching on path attributes, shellcode on buffer attributes).

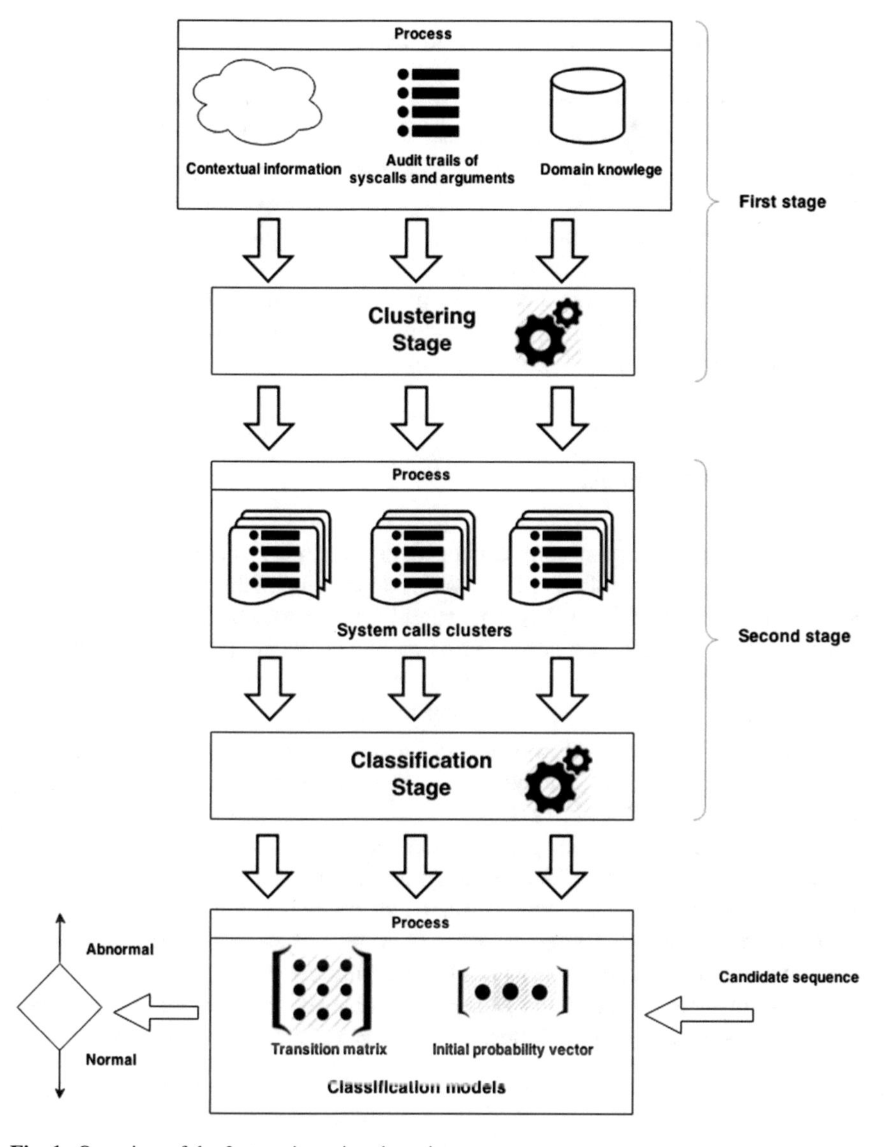

Fig. 1 Overview of the 2-stage intrusion detection system

During a first phase, clustering is performed per system call and per process. We define for a process p the set of samples $T^{s_{ip}}$ containing samples of the form $< a_1^{s_{ip}}, a_2^{s_{ip}}, \ldots, a_m^{s_{ip}} >$ belonging to s_{ip}. It follows then that for a process p calling l unique system calls we obtain l sets of samples $T_p = \{T^{s_{1p}}, T^{s_{2p}}, \ldots, T^{s_{lp}}\}$. Clustering is performed on each set $T^{s_{ip}}$ resulting in a set of clusters $C^{s_{ip}} = \{C_1^{s_{ip}}, C_2^{s_{ip}}, \ldots, C_t^{s_{ip}}\}$, where t is the number of clusters obtained after clustering. For each cluster $C_k^{s_{ip}}$

we define a model $M_k^{s_{ip}}$ which captures the information relative to the cluster, and describes a *new* system call s_{ip}^k.

During a second phase, each system call s_{ip} of p is matched against the clusters in $C^{s_{ip}}$ and is labeled as s_{ip}^k if $C_k^{s_{ip}}$ is the most suitable cluster. The match is done using the models $M^{s_{ip}} = \{M_1^{s_{ip}}, M_2^{s_{ip}}, \ldots, M_t^{s_{ip}}\}$. Thus for a candidate execution of a process p which consists of a sequence $s_p = (s_{1p}, s_{2p}, \ldots, s_{np})$ we obtain using the clusters derived in the first phase a new sequence $s'_p = (F_p(s_{1p}), F_p(s_{2p}), \ldots, F_p(s_{np}))$ where F_p is the transformation that matches s_{ip} to its closest cluster for process p. We are interested in how to derive $C^{s_{ip}}$ and $M^{s_{ip}}$ from the corresponding $T^{s_{ip}}$ (first phase) and how to match a sample of s_i to s_{ip}^k corresponding to the closest cluster $C_k^{s_{ip}} \in C^{s_{ip}}$ relying on the models in $M^{s_{ip}}$ (second phase).

4.2 Attributes Metrics

As explained before, each system call is assigned a number of attributes. Of course, the number and nature of the attributes varies depending on the amount of information given by the audit facility used to monitor the system calls. Generally speaking, we define for each attribute a set of categories wich partitions the range of values that it can take. The partitioning is done relying on domain-level knowledge. Thus, for an attribute $a_j^{s_{ip}}$ belonging to s_{ip} we define (1) a set of categories $A_j^{s_{ip}} = \{A_{j1}^{s_{ip}}, A_{j2}^{s_{ip}}, \ldots, A_{jl}^{s_{ip}}\}$ (2) a metric $d_j^{s_{ip}}$ that expresses distances between categories with parameter $D_{between}$ and within categories with parameter D_{within}. Thus we assign for each attribute $a_j^{s_{ip}}$ a tuple $(A_j^{s_{ip}}, d_j^{s_{ip}}, D_{between}, D_{within})$ with $D_{between}$ and D_{within} user adjustable. In reality, many system calls share the same type of attributes and so we are not required to define categories and metrics for every attribute. The categories can be either described explicitly or implicitly. The first case involves expressing ranges or sets of possible values. In the second case, we express constraints on attributes and categories arise naturally depending on whether the attributes obey the constraints or not. An example of implicitly defined categories will be described later when discussing path attributes. Thus, the generic distance between attributes that is used follows:

$$d_j^{s_{ip}}(a_j^{s_{ip}}, b_j^{s_{ip}}) = \begin{cases} 0 & \text{if } a_j^{s_{ip}} = b_j^{s_{ip}} \\ D_{within} & \text{if same category} \\ D_{between} & \text{if different category} \end{cases} \quad (1)$$

where, $a_j^{s_{ip}}$ and $b_j^{s_{ip}}$ are values of the jth attribute of s_{ip} belonging to two different samples of the same process p.

Combining the distances between attributes, we are able to derive a general distance between system call instances. This distance is parametrized by user adjustable weights for tuning. The distance between two instances s_{ip}^1, s_{ip}^2 of system call s_{ip} in process p is:

$$d(s_{ip}^1, s_{ip}^2) = \frac{\sum_{j=1}^{m} weight_j \times d_j(a_j^{s_{ip}}, b_j^{s_{ip}})}{\sum_{j=1}^{m} weight_j} \quad (2)$$

where $< a_1^{s_{ip}}, a_2^{s_{ip}}, \ldots, a_m^{s_{ip}} >$ are attributes of s_{ip}^1, $< b_1^{s_{ip}}, b_2^{s_{ip}}, \ldots, b_m^{s_{ip}} >$ are attributes of s_{ip}^2, $d_j(a_j^{s_{ip}}, b_j^{s_{ip}})$ is the (jth) component of the distance based on the (jth) attributes of the system calls instances s_{ip}^1 ($a_j^{s_{ip}}$) and s_{ip}^2 ($b_j^{s_{ip}}$), and $weight_j$ is a user-defined weight for the jth component.

We will now describe categories and metrics associated with commonly found attributes. The categories were chosen so that they reflect domain-level knowledge, and be relevant with regards to detecting malicious activities. For all the following definitions, i and j refer to different instances of the same attribute.

4.2.1 Path Attributes

Paths are particularly interesting attributes since they are affected by a number of attacks and represent a good indicator of normal and abnormal behavior. There are a number of characteristics that can be used to build the model e.g., the path depth, the path length, the characters distribution, the file extensions, etc. While previous efforts have developed a variety of models to represent path attributes, some are still riddled with inefficiencies. The authors in [15] use a combination of three models: string length, character distribution and structural inference. The string length model computes the average μ and variance σ^2 of the string lengths in the training data, the character distribution model computes a discrete probability distribution of the characters as histograms, and the structural inference aims at capturing the format of the strings by deriving a probabilistic grammar using HMMs. As discussed in [16], the structural inference model is the weakest as it introduces excessive simplifications and causes a significant number of false alarms. The string length model, on the other hand, performs well, and is a good indicator of suspicious behavior.

Based on these remarks, we divided path attributes into four components. We use path lengths and path depths during clustering. Paths that are too dissimilar in terms of length and depth are assigned bigger distances. Moreover, we introduce comparison between paths based on Levenshtein distance which represents the minimum numbers of single-character modifications, that is insertion, deletion or substitution that are required to change a path into the other. The idea is to cluster together paths that are similar in terms of tree structure and file names. Finally, path attributes are also equipped with a distance called *path exclusion* that defines implicit categories. A set of regular expressions are used to describe constraints, and paths that match the same regular expressions are clustered together. There can be as many categories as groups of defined regular expressions. Regular expressions constitute a degree of freedom given to the user in order to introduce domain-level knowledge and influence on the resulting clusters. The sub-components of the path attribute are directly used in Eq. (2) as attributes. We now define the four components that are used with path attributes:

Path length distance: $d(i,j)_{length} = \frac{|\Delta length|}{(length_{max} - length_{min})}$, where $length_{max}$ and $length_{min}$ are the maximum and minimum lengths taken from all samples belonging to the process and system call under clustering respectively.

Path depth distance: $d(i,j)_{depth} = \frac{|\Delta depth|}{(depth_{max} - depth_{min})}$, where $depth_{max}$ and $depth_{min}$ are the maximum and minimum depths taken from all samples belonging to the process and system call under clustering respectively.

Path distance: $d(i,j)_{path} = \frac{levenshtein(path_i, path_j)}{length_{max}(path_i, path_j)}$ where $length_{max}(path_i, path_j)$ is the maximum length of $path_i$ and $path_j$ and $levenshtein(path_i, path_j)$ is the Levenshtein distance between $path_i$ and $path_j$.

Path exclusion distance: $d(i,j)_{path\_excl} = \begin{cases} D_{within} & if\ path_i \& path_j\ match \\ D_{between} & otherwise \end{cases}$

4.2.2 Identifiers Attribute

Identifiers concern attributes such as real/effective user/group identifiers, file user/group owners, etc. Starting from the observation that these values are generally associated with specific groups: Super user ($id = 0$), special system users ($id < 1000$) and real users ($id > 1000$) we defined three categories. The distance is computed using Eq. (1) and the following categories $A_{super\_user} = \{0\}$, $A_{system\_users} = \{1, \ldots 999\}$, and $A_{real\_users} = \{1000, \ldots\}$.

4.2.3 File Mode Attribute

File modes are usually associated with path attributes. We can classify file modes depending on the permission levels and on the presence of special bits, i.e. setuid, setgid, or sticky bits. We thus define two main categories that divide the file modes into those with special bits and those without, and each category is further divided into three categories depending on the permissions level. We then combine both distances into one final distance where each distance is computed using Eq. (1) and the categories $A_{suid/sgid/sticky} = \{1000, \ldots\}$ and $A_{normal} = \{0, \ldots, 1000\}$ corresponding to special bit access rights associated with files/devices in octal form on the one hand, and the categories $A_{first\_level} = \{0, \ldots, 7\}$, $A_{second\_level} = \{8, \ldots, 77\}$, and $A_{third\_level} = \{78, \ldots, 777\}$ for permissions on the other hand.

4.2.4 Return Value Attribute

Return values reveal information about the status of the system call after stopping. Null return values are usually associated with a normal termination, while non-null values indicate errors. Categories for this attribute are thus straightforward and the distance is computed using Eq. (1) and the return value categories $A_{abnormal} = Z^*$ and $A_{normal} = \{0\}$.

4.2.5 Nominal Attributes

Nominal attributes are attributes which can take a limited number of values. An example would be the *flags* argument used by the *open()* system call. In this case and when there is no logical division of the possible values, we define as many categories as number of possible values and use Eq. (1) to compute the distance.

4.2.6 Process Creation Arguments Attributes

Execution arguments are handled differently than the rest of the attributes. We aim at clustering together *execve()* calls that are similar in terms of arguments. Thus, we take into account the number of arguments of each call and the cumulative similarity between each of the arguments based on the edit distance (Levenshtein). When the two system calls have a different number of arguments, the difference in number of arguments is assumed to have maximum edit distance (distance = 1). The distance is thus expressed as:

Arguments distance: $d(i,j)_{arguments} = \frac{1}{l_{max}}\left(\sum_{i=1}^{l_{min}} \frac{levenshtein(arg_i^1, arg_i^2)}{length_{max}(arg_i^1, arg_i^2)} + (l_{max} - l_{min})\right)$

where arg_i^1 is the ith argument of the first *execve()* call, arg_i^2 is the ith argument of the second *execve()* call, l_{min} is the minimum number of arguments of the two *execve()* calls l_{max} is the maximum number of arguments of the two *execve()* calls, and $length_{max}(arg_i^1, arg_i^2)$ is the maximum length of arg_i^1 and arg_i^2.

4.3 *Clustering and Modeling*

Unlike classification, clustering is an unsupervised method which aims at partitioning a set of data with no pre-defined classes into meaningful sub-classes called clusters [23]. As outlined before, our motivation behind using clustering comes from the observation that system calls do not behave uniformly throughout the execution of a whole process. The idea is then to find meaningful sub-groups of system call samples that exhibit similarities and which will help us build more accurate models. Unlike unsupervised classification methods mentioned in [15, 16], building clusters on both positive and negative samples allows not only to distinguish different behaviors of a system call but also to isolate general characteristics of attacks using the carefully designed metrics defined in the previous section which take into account domain-level knowledge.

In order to build clusters, we use agglomerative hierarchical clustering, which follows a bottom-up approach [23] starting from single-instance clusters and merging at each round the two nearest clusters until reaching some stopping criterion. Distance between clusters was tested using different approaches, namely single-linkage, complete-linkage and average-linkage clustering. For the single-linkage clustering, the distance between two different clusters is defined as the minimum of the distance between any points in the two clusters:

$$d_{min}(C_i, C_j) = min_{p \in C_i, p' \in C_j} d(p, p') \quad (3)$$

For the complete-linkage clustering, the distance between two different clusters is defined as the maximum of the distance between any points in the two clusters.

$$d_{max}(C_i, C_j) = max_{p \in C_i, p' \in C_j} d(p, p') \quad (4)$$

For the average-linkage clustering, the distance between tow different clusters is the average of the distances between any points in the two clusters.

$$d_{avg}(C_i, C_j) = \frac{1}{|C_i|\,|C_j|} \sum_{p \in C_i} \sum_{p' \in C_j} d(p, p') \quad (5)$$

We experimented with two different stopping criteria based on a user-defined maximum number of clusters and on a maximum inter-cluster distance. At the end of the clustering process, we are able to generate for each set of samples $T^{s_{ip}}$ the corresponding set of clusters $C^{s_{ip}} = \{C_1^{s_{ip}}, C_2^{s_{ip}}, \ldots, C_t^{s_{ip}}\}$.

Starting from the clusters in $C_p^{s_i}$ we generate models $M_p^{s_i}$ that capture the characteristics of the clusters and provide a convenient way to match candidate sequences to clusters. We have chosen to retain the following components in the models (1) Path length mean ($\mu_{cluster\_length}$), (2) Path length variance ($\sigma^2_{cluster\_length}$), (3) Path depth mean ($\mu_{cluster\_depth}$), (4) Path depth variance ($\sigma^2_{cluster\_depth}$), (5) List of all paths and associated probability distribution, (6) List of all file modes and associated probability distribution, (7) List of all file owners and associated probability distribution, (8) List of all file groups and associated probability distribution. (9) List of execution arguments and associated probability distribution, (10) List of all return values and associated probability distribution, (11) List of all identifiers and associated probability distribution, and (12) List of all other nominal attributes (flags, etc.) and associated probability distribution.

Depending on the system call that is clustered, some or all of the elements will be used in the derived models. Finally, the models are saved and used in the next stages (sample matching and classification). Notice that the clustering/modeling process needs to be done only once in order to generate the models.

4.4 Samples Matching

Samples matching aims at transforming an execution of a process p which consists of a sequence $s_p = (s_{1p}, s_{2p}, \ldots, s_{np})$ into a new sequence $s'_p = (s'_{1p}, s'_{2p}, \ldots, s'_{np})$ where $s'_{ip} = s^k_{ip}$ corresponding to s_{ip}'s closest cluster $C_k^{s_{ip}} \in C^{s_{ip}}$. As in clustering, we are interested in defining metrics that determine the distance between a sample and a cluster. We rely on the models $M^{s_{ip}}$ derived during clustering and define metrics based on the system calls' attributes and the description of the cluster found in the model. The distance between a system call s_i^p with attributes $< a_1^{s_{ip}}, a_2^{s_{ip}}, \ldots, a_m^{s_{ip}} >$ and a cluster $C_k^{s_{ip}}$ with a model $M_k^{s_{ip}}$ composed of l components is defined as:

$$d(s_{ip}, C_k^{s_{ip}}) = \frac{\sum_{j=1}^{l} weight_j \times d_j(s_{ip}, M_k^{s_{ip}})}{\sum_{j=1}^{l} weight_j} \tag{6}$$

where $d_j(s_{ip}, M_k^{s_{ip}})$ is the (jth) component of the distance based on the (jth) component of the model, and $weight_j$ is a user-defined weight for the (jth) component of the model. We are looking for a cluster C_k such that:

$$\begin{cases} F_p(s_i) = s_{ip}^k \\ k = arg\ \min_x d(s_{ip}, C_x^{s_{ip}}) \end{cases} \tag{7}$$

The distances associated with each component, that is $d_j(s_{ip}, M_k^{s_{ip}})$ are defined as follows:

Path length: We use Chebyshev's inequality [15] which takes into account the variance in order to get a smoother distance.

$$d_{path\_length}(s_{ip}, M_k^{s_{ip}}) = \begin{cases} |1 - \frac{\sigma_{length}^2}{(length_i - \mu_{length})^2}|, & \text{if } length_i \neq \mu_{length} \\ 0, & \text{if } length_i = \mu_{length} \end{cases}$$

where μ_{length} is the mean of path lengths associated with $M_k^{s_{ip}}$ and σ_{length}^2 is the variance of path lengths associated with $M_k^{s_{ip}}$.

Path depth: Similarly, we use Chebyshev's inequality as follows:

$$d_{path\_depth}(s_{ip}, M_k^{s_{ip}}) = \begin{cases} |1 - \frac{\sigma_{depth}^2}{(depth_i - \mu_{depth})^2}|, & \text{if } depth_i \neq \mu_{depth} \\ 0, & \text{if } depth_i = \mu_{depth} \end{cases}$$

whereμ_{depth} is the mean of path depths associated with $M_k^{s_{ip}}$ and σ_{depth}^2 is the variance of path depths associated with $M_k^{s_{ip}}$.

Path: We use the smallest Levenshtein distance as follows:

$$d_{path}(s_{ip}, M_k^{s_{ip}}) = \min_{f \in cluster_{filenames}} levenshtein\ (filename_i, f)$$

where $cluster_{filenames}$ is the list of file names associated with $M_k^{s_{ip}}$.

Execution arguments: For each argument, we find the closest matching argument in the cluster's list of arguments and we take into account the associated probability.

$$d_{exec\_args}(s_{ip}, M_k^{s_{ip}}) = \begin{cases} \hat{arg_i} = \min_{arg \in cluster_{args}} levenshtein(arg_i, arg) \\ \frac{1}{l} \sum_{i=1}^{l} \frac{levenshtein(\hat{arg_i}, arg)}{length_{max}(\hat{arg_i}, arg)} \times (1 - P(\hat{arg_i})) \end{cases}$$

where P is the probability distribution associated with the component, $length_{max}$ $(\hat{arg_i}, arg)$ is the maximum length of $\hat{arg_i}$ and arg, and l is the number of execution arguments in the candidate sample.

Other components: For all the other components, the distance is computed as follows:

$$d_{other}(s_{ip}, M_k^{s_{ip}}) = \begin{cases} 1 - P(value_i) & \text{if } value_i \in valuesList_{cluster} \\ 0 & \text{otherwise} \end{cases}$$

where $cluster_{filenames}$ is the list of the component's values associated with $M_k^{s_{ip}}$ and P is the probability distribution associated with the component.

Thus, each system call s_{ip} gets attributed its closest cluster $C_k^{s_{ip}}$ and is replaced by the system call s_{ip}^k. A system call for which no clusters have been derived is left unchanged. The new sequence of system calls is then used as input to the classification stage.

5 Execution Flow Modeling and Learning

5.1 *Naïve Bayes Classifier*

In [12] we proposed a classifier, called SC2.2, which classifies arbitrarily long sequences of system calls made by a process using a naïve Bayes classification approach, with class conditional probabilities derived from Markov chain modeling of sequences. The problem of classification consists in finding, for a given test sequence $s = (s_1, s_2, \ldots, s_n)$ the corresponding class c. A naïve Bayesian classifier chooses the class c with the highest posterior probability given the sequence s. Bayes theorem defines the posterior probability of class c given sequence s as:

$$P(C = c|S = s) = \frac{P(S = s|C = c) * P(C = c)}{P(S = s)} \tag{8}$$

where S and C are random variables denoting sequences and classes. $P(S = s|C = c)$ is the class conditional probability of sequence s given the class c, $P(C = c)$ is the prior probability of class c and $P(S = s)$ is the prior probability of sequence s. We are thus trying to find the class c among all candidate classes that maximizes $P(C = c|S = s)$. As $P(S = s)$ is independent of the class c, the classification problem is thus reduced to finding $C(s)$ such that

$$C(s) = \arg\max_{c} P(S = s|C = c) \times P(C = c) \tag{9}$$

A naïve Bayesian classifier postulates independence of features given the class so that the class conditional probability of sequence s given the class c can be expressed as

$$P(S = s|C = c) = \prod_{i=1}^{n} P(S = s_i|C = c) \tag{10}$$

Our method replaces the class conditional independence hypothesis by using a Markov model to estimate the class conditional probability of a sequence s. The prior probability of class c is estimated from the training data as in a traditional naïve Bayes classifier.

5.2 *Observable Markov Chain Model*

An *order n* observable Markov process is a system where the dynamics can be modeled as a process that moves from state to state depending (only) on the previous n states. In a first order process with M states there are M^2 possible transitions between states. Each transition is associated with a probability called the state transition probability that refers to the probability of moving from one state to the other. Each state is also assigned an initial probability. A first order observable Markov model with m states can thus be defined by an $M \times M$ transition probabilities matrix *TP* and an *M-dimensional* initial probability vector *IP*. Both *TP* and *IP* are built from the training data during the training phase. By modeling a sequence $S = (s_1, s_2, \ldots, s_n)$ as a Markov chain we are able to express its occurrence probability in class c by:

$$P_c(s_1 s_2 \ldots s_n) = P_c(s_1) \times P_c(s_2|s_1) \times \cdots \times P_c(s_n|s_{n-1})) \tag{11}$$

where $P_c(s_1)$ is the initial probability of system call s_1 and $P_c(s_j|s_i)$ is the probability of system call s_j being invoked after system call s_i. The class conditional probability s given c can then be estimated from the training data as:

$$P(S = s|C = c) = P_c(s_1) \prod_{i=2}^{n} P_c(s_i|s_{i-1}) \tag{12}$$

5.3 *Classification*

Classification of a sequence s consists in finding the class c that satisfies Eq. (9). In the case where both classes have equal posterior probabilities, the majority class (normal) is used to resolve the issue. However, this could be set differently by a system administrator depending on the policy used. Using a naïve Bayes-like classifier that relies on Markov modeling comes with some issues regarding zero conditional probabilities and products of a very large number of probabilities. These issues impact on the quality (accuracy, detection rate, etc.) such a classifier. The m-estimate is used in order to deal with zero transitional probabilities while a test, backtrack, scale and re-multiply algorithm (TBSR) is used to deal with vanishing numbers due to the computation of class conditional probabilities of long sequences. Details of the solutions used can be found in [12]

It is worth noting that when benchmarked SC2.2 yields a detection rate that is better than most standard classifiers in a majority of the data sets while maintaining a comparable or slightly better accuracy score. SC2.2 was compared against naïve Bayes multinomial (NBm), the decision tree C4.5, RIPPER (Repeated Incremental Pruning to Produce Error Reduction), support vector machine (SVM) and logistic regression (LR). The difference between posterior probabilities while making the classification was also significant enough to give an accepted level of confidence in the results. Moreover, performance results indicate that SC2.2 does well in terms of speed. Classification performance on a set of 184 sequences (20602 system calls) averages 1,7 seconds per run (with 10 runs for the 10-fold cross validation method) using a top of the shelf computer (Intel i5 @ 2.40 Ghz, 64 bits, 6 Go of RAM). For all these reasons, SC2.2 was a natural choice for the classification step.

6 Experimentation

6.1 Data Sets

In order to test our solution we used the 1999 DARPA intrusion evaluation data set from the Massachusetts Institute of Technology (MIT) Lincoln Lab [24]. The data-set contains three weeks of training data and two weeks of testing data. While the data comes in different formats, we chose to work with the Basic Security Modules (BSM) audit files. BSM is a software utility running on Solaris operating systems which helps in collecting security-related information on the machine [25]. The software handles more than 200 kernel and user events which are executed on the system including system calls. For each event, BSM associates various parameters including execution time, arguments, paths, return values and general process information. Information associated with events is organized in records and stored in a binary file which can be processed using operating system utilities such as *auditreduce* and *praudit*. More details about BSM records' structure can be found in the Solaris system manual [26].

Only the second week of training data contains attacks. These are organized in four categories [27], namely:

- Probes or scan attacks (PSA): Attempts at gaining information about the information system before an attack. These include attacks such as port sweeps, pings on multiple host addresses, etc.
- Denial of Service (DoS): Attacks in which the attacker disrupts the handling of legitimate requests or denies legitimate users access to a machine.
- Remote to Local (R2L) attacks: Attacks which allow an attacker to gain local access as a user of a target machine through sending specially crafted packets. These attacks include for instance exploiting buffer overflows in network server softwares (*imap, sendmail*, etc.)
- User to Root (U2R) attacks: Attacks in which an attacker who has access to a machine with a normal user account gains root access to the system through exploiting some vulnerability. Examples include exploiting buffer overflows, race conditions, etc.

DoS and PSA attacks do not generate process activity in the target host. These are only visible in network dumps, and are not suitable for detection using system calls or their arguments. We thus focus on attacks that target a remote or a local service and that can be detected with our solution. Specifically, we chose to focus on three programs, namely *fdformat*, *eject* and *ps*, a short description of which is as follows:

Eject: Eject is used by removable media devices that lack an eject button or which are managed by Volume Management in Solaris 2.5. A vulnerability in the volume management library, *libvolmgt.so.1*, due to insufficient bounds checking on arguments allows an attacker to overwrite the internal stack space of the program. An attacker is thus able to exploit this vulnerability in order to gain root access on the targeted machine. Since buffer overflow attacks usually use a lengthy payload which includes the shell code, we expect some system call arguments to be of significant length. Such system calls will thus get clearly separated from the rest in the clustering phase due to the unusual path length mean, as well as the edit distance.

Fdformat: Fdformat is used to format floppy disks and PCMCIA memory cards. As it makes use of the *libvolmgt.so.1* library, this program is subject to the same vulnerability as in the *eject* program. We thus expect the attack to be similarly detected.

Ps: The ps program that comes with Solaris 2.5 suffers from a race condition that allows an attacker to execute code with root privilege and gain root privileges. This vulnerability is exploitable if the attacker has access to temporary files which is the case when the permissions on */tmp* or */var/tmp* are set incorrectly. In addition to the significant deviation introduced by the shell code in the path length mean and the edit distance, we also expect to see differences from normal behavior with regards to the file mode leading to a clear separation of normal and abnormal system calls in the clustering phase.

Table 1 Execution statistics of processes under study in the training data-set

Process	Normal executions	Attacks
fdformat	5	3
eject	7	3
ps	106	3

6.2 Methodology

Training data-sets from weeks 1–3 were used during the training process, while testing data-sets from week 4 and week 5 were used for testing. Using the *praudit* utility we transformed the BSM audit files into raw XML files then used *Python* scripts in order to extract execution traces for each of the three chosen programs while identifying those corresponding to the attacks (see Table 1). We also extracted for each process in the training data-set the list of all unique system calls used along with their instances. We then applied our two-stage HIDS to each process by first generating clusters of each system call then transforming the associated execution traces in the training and testing data-sets to generate new execution traces based on clusters. The resulting labeled execution traces were then fed to our classifier (SC2.2) which uses a 10-fold cross validation method in order to make sure that no training data are used to testing [12].

7 Results

7.1 Clustering Validation

Before matching the samples in the testing data-sets and performing classification, we need to check whether the clusters generated in the first stage are suitable. In order to validate the clustering, we rely on three indicators (1) We ensure that we get an optimal number of clusters by computing the Silhouette width [28] which provides a measure of how well an instance lies within its cluster and with regards to the other clusters. (2) We assess the performance of our models by cross-validating the instances. This is done by verifying that the inputs are assigned to the clusters they contributed to create using the generated models. (3) We inspect the clusters to determine whether there is a semantically clear segregation of instances indicating abnormal and normal behavior. The results of each of these validation indicators are discussed below.

Silhouette Information

The Silhouette width is a measure of how much an instance belongs to the cluster where it has been assigned. It can be computed using the following formula [28]:

$$sil(i) = \frac{b(i) - a(i)}{max(a(i), b(i))} \tag{13}$$

where $a(i)$ is the average distance to other elements in the cluster assigned to i, and $b(i)$ is the smallest average distance to other clusters with respect to i.

If $sil(i) \simeq 1$, the corresponding sample is considered to be well clustered. If $sil(i) < 0$, the corresponding sample is in the wrong cluster whereas if $sil(i) \simeq 0$, the sample is considered to be on the border between two clusters. To assess the clustering, we compute the average Silhouette width over all the clustered samples. We then choose the number of clusters that maximizes the average Silhouette width. Table 2 shows for each system call and for each process the corresponding maximum average Silhouette width along with the associated optimal number of clusters. As the results indicate, the average Silhouette for all the system calls is well above 0 which indicates a satisfactory clustering.

Cross-Validation of the Clusters

As discussed before, we build models in order to represent the clusters using a number of attribute. It is important for the samples matching and classification stages that these models constitutes a good description of the clusters. In order to assess the adequacy of the models, we perform cross-validation on the generated clusters. We use the models in order to check whether the inputs get assigned to the same clusters they helped build. A good assignments success should give us the confidence that new samples will be correctly assigned to the suitable cluster. Table 2 provides the cross-validation results for each system call and each process. For all the system calls, the assignment success is 100 % which indicates that the models are good representatives of the clusters.

Table 2 Cluster validation

System call	Silhouette width	Cross-validation success (%)	Number of clusters	Weights
Eject				
close()	0,457	100	14	all weights to 1
execve()	0,691	100	9	all weights to 1
mmap()	0,467	100	4	all weights to 1
open()	0,527	100	4	all weights to 1
Fdformat				
access()	0,671	100	3	all weights to 1
execve()	0,646	100	2	all weights to 1
mmap()	0,462	100	5	all weights to 1
Ps				
execve()	0,743	100	9	all weights to 1
mmap()	0,538	100	3	all weights to 1
open()	0,548	100	2	all weights to 1
stat()	0,490	100	13	all weights to 1

Table 3 Sample clusters generated for the *stat()* system call in the (*fdformat* process)

Clusters	Paths	File modes
Cluster 1	/devices/pseudo/vol@0:volct /var/spool/cron/atjobs	2066
Cluster 2	/devices/sbus@1f	1066
Cluster 3	/usr/bin/chmod /usr/bin/cp etc.	1005
Cluster 4	vol/dev/aliases /vol/dev/aliases/floppy0 etc.	1006
Cluster 5	usr/lib/libadm.so.1 /usr/lib/libc.so.1 etc.	1007
Cluster 6	/vol/dev/aliases/\246\034\300...	1006
Cluster 7	/vol/rdsk/aliases/\246\034\300...	1006

Table 4 Sample clusters generated for the *execve()* system callin the (*eject* process)

Clusters	Paths	Arguments
Cluster 1	/export/home/geoffp/eject	–
Cluster 2	/usr/bin/eject	\034\300\023...
Cluster 3	/usr/bin/ksh	–
Cluster 4	/export/home/violetp/ejectexp	–

Clusters Semantics

The overall goal of the clustering phase is to separate between normal and abnormal samples. Since the clustering is driven by domain-level knowledge, we should be able to recognize suspicious behaviors by manual inspection of the clusters. As an example, Table 3 shows seven (7) clusters for the *stat()* system call based on the file path and the file access control information (File modes). Table 4 shows four (4) clusters for the *execve()* system call based on the file path of the executable, as well as the execution arguments. The clusters clearly show that instances carrying attacks, in this case shellcode payloads that cause long parameters, are separated into specific clusters (clusters 6 and 7 in Table 3 and cluster 2 in Table 4) from the rest of the normal instances.

7.2 Classification Results

In order to test the performance of our classifier, we count the numbers of test sequences in the testing data which were correctly and incorrectly classified by the system. These are used to compute the following measures, where normal sequences are considered as *negative* and the intrusion sequences as *positive*:

- True positives (TP): The number of intrusive sequences correctly classified.
- True negatives (TN): The number of normal sequences correctly classified.
- False negatives (FN): The number of intrusive sequences incorrectly classified.
- False positives (FP): The number of normal sequences incorrectly classified.

We then measure the performance of our classifier by computing its *accuracy*, which is the fraction of input sequences correctly predicted, its *detection rate*, which is the fraction of intrusive sequences correctly predicted, and its *false positive rate* (FPR), which is the fraction of normal sequences predicted as intrusive as follows:

$$Accuracy = \frac{TP + TN}{\#\, of\ input\ sequences} \tag{14}$$

$$Detection\ rate = \frac{TP}{TP + FN} \tag{15}$$

$$FPR = \frac{FP}{TN + FP} \tag{16}$$

Table 5 compares our results to those of *SysCallAnomaly* which are presented in [15] along with results in [16], which uses original *SysCallAnomaly* against the same data-set and proposes an improved version stripped of the structural inference model. We also compared the performance of our two-stage classifier with bare classification using SC2.2.

A first remark concerns the difference of results between [15] and [16] when using *SysCallAnomaly* against the same data-set with [16] showing slightly worse results. While the discrepancy can be explained by differences in tuning, analysis done in [16] shows that many models used in the original *SysCallAnomaly* implementation, especially the structural inference model, yield a high false positive rate. Our two-stage classifier shows better accuracy across all the tested processes compared with results in [16]. The classifier performs also better in terms of false positive rate whereas the performance of SysCallAnomaly in [16] improves as expected. Comparison between

Table 5 Classification results of our two-stage HIDS compared with other classifiers

Program	SysCall Anomaly [15]	SysCall Anomaly 2 [16]	SysCallAnomaly without HMM [16]	SC2.2	Two-stage HIDS
Eject					
Accuracy	100	90.00	90.00	100	100
Detection rate	100	100	100	100	100
FPR	0	25	25	0	0
Fdformat					
Accuracy	100	80	80	100	100
Detection rate	100	100	100	100	100
FPR	0	25	25	0	0
Ps					
Accuracy	100	98.1	99	84.46	100
Detection rate	100	100	100	100	100
FPR	0	12.5	16.66	16.08	0

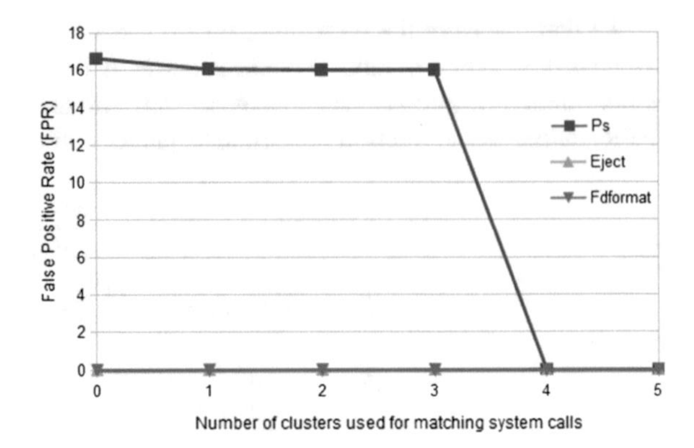

Fig. 2 False positive rate progress with respect to the number of clusters used for matching system calls

the two-stage classifier and bare SC2.2 shows also improvements in the ps process which is the most diverse out of all the tested processes. This indicates that the first stage of clustering helps discriminate even more the positive and negative classes.

Figure 2 shows the progress of the false positive rate (FPR) for each process with regards to the number of clusters used in the clustering and matching phases (up to five clusters). We chose the system calls to be clustered in the order of their frequency in the corresponding process. As the graph shows, all processes show a 0 % false positive rate when at least the four most frequent system calls are used in the clustering and matching stages. Choosing the system calls based on their frequency order does not make any a priori judgment on the attacks but begs the question of the existence of a smaller subset of system calls that can produce a 0 % false positive rate. This is particularly interesting as using smaller sets of clusters improves performances in terms of speed in the matching stage as well as reduces the number of different system calls used in the classification stage models leading to better memory and time efficiency.

8 Conclusion

In this chapter, we presented a novel two-stage host intrusion detection system that uses both data and execution flow of programs. We combined data driven clustering of system calls and supervised learning for program flow to perform detection of both normal and intrusive programs. We used contextual information and introduced domain-level knowledge at the clustering stage to allow for tuning the detection to different environments and capitalize signature like detection respectively. We are currently investigating the impact of key parameters of the clustering phase on the performance of the overall detection, and the aggregation of the detection models constructed for each program into a system-wide detection model.

References

1. Kwon, T., Su, Z.: Automatic detection of unsafe dynamic component loadings. IEEE Trans. Softw. Eng. **38**(2), 293–313 (2012). doi:10.1109/TSE.2011.108
2. Cowan, C., Wagle, P., Pu, C., Beattie, S., Walpole, J.: Buffer overflows: attacks and defenses for the vulnerability of the decade. DARPA Inf. Surviv. Conf. Expo (DISCEX) **2**, 119–129 (2000)
3. Pincus, J., Baker, B.: Beyond stack smashing: recent advances in exploiting buffer overruns. IEEE Secur. Priv. **2**(4), 20–27 (2004). doi:10.1109/MSP.2004.36
4. Peyman, K., Ghorbani, A.A.: Research on intrusion detection and response: a survey. Int. J. Netw. Secur. **1**, 84–102 (2005)
5. Lazarevic, A., Kumar, V., Jaideep, S.: Intrusion detection: a survey (2005)
6. Vigna, G., Kruegel, C.: Host-based intrusion detection systems. In: Bigdoli, H. (ed.) Handbook of Information Security. Wiley (2005)
7. Patcha, A., Park, J.M.: An overview of anomaly detection techniques: existing solutions and latest technological trends. Comput. Netw. **51**(12), 3448–3470 (2007)
8. Garcia-Teodoro, P., Díaz-Verdejo, E.J., Maciá-Fernández, G., Vázquez, E.: Anomaly-based network intrusion detection: techniques, systems and challenges. Comput. Secur. **28**(1–2), 18–28 (2009)
9. Symantec: Vulnerability trends @ONLINE (2013). http://www.symantec.com/threatreport/topic.jspid=vulnerability_trends&aid=total_number_of_vulnerabilities
10. Di Pietro, R., Mancini, L.V.: Intrusion Detection Systems, 1st edn. Springer Publishing Company, Berlin (2008)
11. Forrest, S., Longstaff, T.A.: A sense of self for unix processes. In: IEEE Symposium on Research in Security and Privacy, pp. 120–128 (1996)
12. Assem, N., Rachidi, T., El Graini, M.T.: Intrusion detection using bayesian classifier for arbitrarily long system call sequences. Int. J. Comput. Sci. Inf. Syst. **9**(1), 71–81 (2013)
13. Warrender, C., Forrest, S., Pearlmutter, B.: Detecting intrusion using system calls: alternative data models. In: In Proceedings of the IEEE Symposium on Security and Privacy (1999)
14. Hoang, X., Hu, J.: An efficient hidden markov model training scheme for anomaly intrusion detection of server applications based on system calls. IEEE Int. Conf. Net. **2**, 470–474 (2004)
15. Kruegel, C., Mutz, D., Valeur, F., Vigna, G.: On the detection of anomalous system call arguments. In: Snekkenes, E., Gollmann, D. (eds.) Computer Security - ESORICS 2003. Lecture Notes in Computer Science, vol. 2808, pp. 326–343. Springer, Berlin, Heidelberg (2003)
16. Zanero, S.: Unsupervised learning algorithms for intrusion detection. Ph.D. thesis, Politecnico di Milano (2008)
17. Creech, G., Hu, J.: A semantic approach to host-based intrusion detection systems using contiguousand discontiguous system call patterns. IEEE Trans. Comput. **63**(4), 807–819 (2014)
18. Lin, Y.D., Lai, Y.C., Lu, C.N., Hsu, P.K., Lee, C.Y.: Three-phase behavior-based detection and classification of known and unknown malware. Secur. Commun. Netw. **8**, 2004–2015 (2015)
19. Ma, C., Shen, L., Wang, T.: An anomalous behavior detection method using system call sequences for distributed applications. KSII Trans. Internet Inf. Syst. (TIIS) **9**(2), 659–679 (2015)
20. Yoon, M.K., Mohan, S., Choi, J., Christodorescu, M., Sha, L.: Intrusion detection using execution contexts learned from system call distributions of real-time embedded systems. arXiv preprint arXiv:1501.05963 (2015)
21. Kührer, M., Hoffmann, J., Holz, T.: Cloudsylla: Detecting suspicious system calls in the cloud. In: Stabilization, Safety, and Security of Distributed Systems, pp. 63–77. Springer (2014)
22. Garfinkel, S., Spafford, G., Schwartz, A.: Practical Unix & Internet Security, 3rd edn. O'Reilly Media, Inc., Sebastopol (2003)
23. Tan, P.N., Steinbach, M., Kumar, V.: Introduction to Data Mining, 1st edn. Addison-Wesley Longman Publishing Co. Inc, Boston (2005)
24. Laboratory, M.L.: Darpa intrusion detection evaluation data set @ONLINE (2013). http://www.ll.mit.edu/mission/communications/cyber/CSTcorpora/ideval/data/1999data.html (1999)

25. Maheshkumar, S., Gürsel, S.: Formulation of a heuristic rule for misuse and anomaly detection for u2r attacks in solaris operating system environment. In: Security and Management, pp. 390–396. CSREA Press (2003)
26. Oracle: Sunshield basic security module guide @ONLINE. http://docs.oracle.com/cd/E19455-01/806-1789/6jb25l4bv/index.html. (2013)
27. Kendall, K.: A database of computer attacks for the evaluation of intrusion detection systems. Master's thesis, MIT (1999)
28. Rousseeuw, P.: Silhouettes: a graphical aid to the interpretation and validation of cluster analysis. J. Comput. Appl. Math. **20**(1), 53–65 (1987)

Erratum to: An Unsupervised Text-Mining Approach and a Hybrid Methodology to Improve Early Warnings in Construction Project Management

Mohammed Alsubaey, Ahmad Asadi and Charalampos Makatsoris

Erratum to:
"An Unsupervised Text-Mining Approach and a Hybrid Methodology to Improve Early Warnings in Construction Project Management" in: Y. Bi et al. (eds.), *Intelligent Systems and Applications*, Studies in Computational Intelligence 650, DOI 10.1007/978-3-319-33386-1_4

The chapter was inadvertently published without including the updates in the affiliations, a change in the corresponding author's name from Harris Makatsoris to Prof. Charalampos Makatsoris and contact, also some more additional corrections within the chapter. The original chapter has been updated with the changes.

The updated original online version for this chapter can be found at
DOI 10.1007/978-3-319-33386-1_4

M. Alsubaey · A. Asadi
Department of Mechanical, Aerospace and Civil Engineering,
Brunel University, London, UK
e-mail: alsubaey.mohammed@gmail.com

A. Asadi
e-mail: AhmadAsadi@outlook.com

C. Makatsoris (✉)
School of Aerospace, Transportation and Manufacturing,
Sustainable Manufacturing Systems Centre, Cranfield University,
Bedfordshire MK43 0AL, UK
e-mail: h.makatsoris@cranfield.ac.uk

Y. Bi et al. (eds.), *Intelligent Systems and Applications*,
Studies in Computational Intelligence 650, DOI 10.1007/978-3-319-33386-1_22

Printed in the United States
By Bookmasters